VIRGIN OLIVE OIL

PRODUCTION, COMPOSITION, USES AND BENEFITS FOR MAN

FOOD AND BEVERAGE CONSUMPTION AND HEALTH

Additional books in this series can be found on Nova's website
under the Series tab.

Additional e-books in this series can be found on Nova's website
under the e-book tab.

VIRGIN OLIVE OIL

PRODUCTION, COMPOSITION, USES AND BENEFITS FOR MAN

ANTONELLA DE LEONARDIS
EDITOR

New York

Library of Congress Cataloging-in-Publication Data

ISBN: 978-1-63117-656-2

Published by Nova Science Publishers, Inc. † New York

CONTENTS

PREFACE

The population living in the countries bordering the Mediterranean Sea produce and consume olive oil from millennia. Recently, copious scientific literature has provided evidence that a regular consumption of olive oil is associated with longevity, healthier aging, cardiovascular health, prevention and protection against cancer.

Thus, while some decades ago olive oil was considered to be merely an ethnic food, today its worldwide recognition is rising given the fact that it is widely considered a functional food able to provide health and well-being.

Olive oil is considered to be a crucial component of the so-called 'Mediterranean Diet' that since 2010 it has been inscribed in the UNESCO's list of Intangible Cultural Heritage of Humanity.

Consequently, everything revolving around the olive tree attracts great interest from scientists, consumers, and producers alike. Moreover, in recent years, consumption and production of olive tree products, such as olive oil as well as olive fruits, food containing olive oils, leaves, bioactive extracts and single molecules, have become increasingly more popular in countries far from the Mediterranean area.

At the same time, scientific research on olive tree products are increasing exponentially involving academics of several disciplines, especially agronomy, arboriculture, engineering, economy, food technology, medicine and pharmacology.

This book is a collection of overviews and original research on various aspects relative to virgin olive oil as well as its well-known and innovative related products, with a special focus on the effects of such products on human health. Chapters are presented by contributors of international standing and leaders in the field.

The main topic of the book is, of course, virgin olive oil of which the latest findings on composition, extraction processes, varieties, growing, geographical characterizations, sensory qualities, culinary performance and medical activity are described. Table olives and other by-products are also addressed. However, in all the chapters, the benefits of olive tree products to human health are always emphasized and expanded.

Exceptional benefits of virgin olive oil derives mainly from both its typical composition and specific physical process of extraction (Chapters 1-3).

In Chapter 1 an accurate description of the chemical composition of virgin olive oil is given. Specifically, acidity, peroxide index, phenolic compounds, fatty acid and triglycerides composition, sterols, and other minor unsaponifiable compounds are discussed.

It emerges that the key elements that distinguish olive oil from other edible oils/fats are, above all, the high content of monounsaturated fatty acids (oleic acid) and the significant presence of natural antioxidants, such as tocopherols, carotenoids, and mainly the phenolic compounds.

Virgin olive oil is a fruit juice, produced exclusively by mechanical extraction processes.

In Chapter 2, after a brief historical presentation of the traditional process, an update on some recent processing techniques is reported. The effects of a few technical variants on the concentration of total phenolic compounds and anti radical activity is also discussed, focusing on how the new techniques could affect public health in the context of cardiovascular diseases, metabolic syndrome and cancers.

Extraction technology of virgin olive oil has remained substantially unchanged for centuries. In Chapter 3 the historical evolution of olive oil production with mechanical procedures is analyzed by using a novel industrial archaeological methodology. Specifically, the evolution of extraction mechanical procedures from the second half of the nineteenth century until the mid-twentieth century is reconstructed by studying the invention and patents implemented from the Historical Archive of the Spanish Office of Patents and Trademarks.

Another distinguishing feature of virgin olive oil is the fact that each oil can express its own identity. Indeed, olive oil composition is strongly affected by olive cultivars, the geographical origin and agronomical factors (Chapters 4-9).

In Chapter 4 the composition of fatty acids, sterols and polyphenols in most monovarietal oils of eastern Spain are presented. A detailed historical reconstruction of the olive tree is given in the introduction in order to show the origin of some Spanish varieties. Fatty acid composition has proven to be a good varietal indicator, especially the content of the major fatty acids (oleic, linoleic and palmitic acids) as well as the oleic/palmitic ratio. However, sterols, α-tocopherol, and other minor components are also able to group varietal virgin olive oils allowing, in a few cases, a clear and unequivocal differentiation.

Chapter 5 discusses a comparative study among a few Tunisian olive cultivars that is carried out by determining the volatile compounds polyphenols, *ortho*-diphenols and flavonoides in the leaves, fruits and stems. Significant differences among the cultivars studied have been found in function to the environmental conditions of growth.

Thus, variety and geographical origin may be used as the criteria for discriminating different commercial virgin olive oil. In order to guarantee the authenticity of the commercial virgin olive oil, the European Union has set up the labels of origin 'Protected Designation of Origin' and "Protected Geographical Indication".

Statistical procedures may be useful to prove the close relationship existing between quality and geographical origin of a virgin olive oil. An example is given in Chapter 6 in which Italian and Western Greek olive oils, analyzed by using both conventional analytical parameters and innovative instrumental techniques (^{1}H and ^{13}C-NMR), are compared by the application of multivariate statistical analysis. The application of chemometric techniques is essential, due to the huge number of chemical-physical variables that had to be analyzed simultaneously.

To respond to the increasing demand of olive oil and its related products, olive orchards are implanted in new areas and intensive olive cultivations are introduced with the aim of increasing production and limiting costs.

In Chapter 7 olive variety suitability and the training procedure for modern olive growing is reported. Research is carried out in Tunisia and specifically, a comparative trial has been set to test different table olives cultivars cultivated with different tree-trainings and pruning systems (central leader form, open-vase form and free form). Both the free and open-vase forms have shown to be highly productive, easy to develop, received little or no pruning and the most economical training system for intensive tree conditions.

Olive oil quality and composition is mainly influenced by olive fruit characteristics, and, therefore, all aspects that influence their development have a crucial effect on olive products.

In Chapter 8 the influence of diverse agronomic factors on olive fruits and olive oil production, composition and quality is reviewed and discussed giving special importance to olive farming systems, fertilization and irrigation, as well as the incidence of olive pests and diseases. Specifically, different agricultural practices, production regimes, fertilization and irrigation systems are focalized. Moreover, the occurrence of pests (olive fruit fly, olive moth, olive black scale, among others), and diseases (olive anthracnose, and verticillium) are discussed in terms of chemical degradation caused in olive oils, loss of productivity, olive oils declassification, changes in chemical composition, and loss of bioactivity and stability, which altogether leads to unprecedented economic damage.

Chapter 9 describes the influence of agronomical and technological factors of production on both the quantitative and qualitative composition of phenolic compounds of virgin olive oil. Specifically, the influence of cultivar, fruit ripeness, climatic conditions and agronomical techniques are more widely considered. Moreover, regarding the technological aspects of production, phenolic compounds in virgin olive oil are related to the activity of endogenous enzymes present in the olive fruit.

In the fruits phenolic compounds are an important part of the fruit defense chemical system, where they exhibit antimicrobial activity and protection against the oxidative damages.

Although the olive's phenols are found in the virgin olive oil only in minimal part (at level of 100-400 mg/kg of oil), they are responsible for the oil oxidative stability and sensory properties.

The phenolic composition of olive oil is complex and includes hydroxytyrosol, tyrosol, the dialdehydic form of elenolic acid linked to either hydroxytyrosol or tyrosol, oleuropein aglycon, lignans and many others.

The current interest in olive oil polyphenols is based on their important nutritional and biological properties (Chapter 10-12).

In Chapter 10 the recent findings on the *in vitro* chemo-preventive activities of hydroxytyrosol are discussed. This compound has demonstrated to be able to counteract the main hallmarks of cancerogenesis acting as an antioxidant, anti-proliferative and an anti-inflammatory agent.

Chapter 11 summarizes the epidemiological studies and investigation trials focusing on the effects of virgin olive oil in the inflammatory process and/or inflammatory-related diseases. Also in this case, the cellular and molecular anti-inflammatory mechanisms of virgin olive oil have been particularly associated with its high amounts of phenolic compounds, as well as, to its composition in mono- and poly-unsaturated fatty acids.

Chapter 12 highlights the fact that olive-oil consumption is inversely correlated with the incidence of stroke. Specifically, diverse effects of olive oil and its chemical constituents on various molecular mechanisms that affect vascular cell function are discussed. Also in this context, oleic acid and the polyphenolic compounds appear to be the crucial factors that reduce the risk factors of cardiovascular disease.

Virgin olive oil also differentiates itself from all the other oil/fats for both its typical organoleptic characteristics and its flexibility to be used in culinary preparation (Chapter 13, 14).

Highly specialized tasting panels usually conduct the measurement of organoleptic properties of virgin olive oil. Traditional methods may confound sensory precision with other aspects related to the bias of the tasters, such as their beliefs, motivations, and preferences.

In Chapter 13 a new method in olive oil tasting based on 'Signal Detection Theory' is presented. The proposed procedure, based on a dissociative model, allows for the obtaining of independent measures of sensory and cognitive factors. From a practical point of view, this dissociation between sensory and decision process may contribute to an optimization of the evaluation of the quality of the olive oil, facilitating the comparison between evaluations of experts and regular consumers.

In several countries of the Mediterranean area, both in domestic and commercial appliances, virgin olive oil is preferred not only as a salad dressing, but also as one of the ingredients in the preparation of typical food.

In Chapter 14 the performance of olive oil as a cooking fat, as an ingredient in baked products, as a canning agent of vegetables and fish preserves and finally, as a component of various sauces is reported. In general, though sometimes contrasting results have been found, virgin olive oil seems to be a valid ingredient in the preparation of food, especially when the cooking time is not extensive and the temperature is not too high.

Monounsaturated oleic acid and the antioxidant substances contained in virgin olive oil are also present in the olive fruits that can also be eaten processed as table olives. Table olives contain further beneficial components as emerged from Chapter 15 in which the focus is on the dietary fibre content in olive fruit. The main components of the soluble fibre are pectic polysaccharides, such as homogalacturonans, rhamnogalacturonans and arabinans. These show many health benefits, such as hypocholesterolemic, hypoglycemic, prebiotic and anticancer activity.

Finally, as result of olive oil production a large amount of residue, such as mill wastewater, olive pomace, leaves and sediments are originated. These residual products are rich in bioactive compounds and so are considered very attractive as potential (source) of natural antioxidants (Chapter 16-18).

Chapter 16 focuses on chemical composition and health properties of olive tree by-products. Moreover, this chapter gives a description of the potential technologies that allow to recovery and purify the antioxidant polyphenols from olive by-products.

While, an innovative technology for the selective recovery of phenolic compounds from the olive oil mill wastewaters, called 'molecular imprinted technology', is presented in Chapter 17. This technique allows the recovery of target bioactive compounds with high purity and efficiency and at the same time, it also presents other interesting properties, such as low cost, physical resistance and reusing.

Although several techniques have been developed for olive-oil mill wastes, few industrial plants currently invest in purification and utilization of olive oil mill by-products that would prevent environmental pollution.

In Chapter 18 alternative use for the recycling olive oil mill wastewaters and solid wastes is presented. Solid wastes are low-cost lignocellulosic materials and can be used as an efficient fuel with low amounts of N and S. Instead, the wastewaters have favorable chemical properties (organic carbon, potassium and phosphorus content) and disposal on soils may be considered as an appropriate option to solve management problems, restore soil fertility and promote productivity.

In conclusion, topics presented in this book show the versatility and usefulness of several olive tree products. Virgin olive oil is certainly the most important product and the most valuable. However, even other related products, including olive fruits and even the waste by-products have proven to be a source of beneficial substances for Man and for the environment.

The editor thanks all contributors for their careful and professional participation to the realization of the book.

Dr. Antonella De Leonardis
Editor
Università degli Studi del Molise
Via De Sanctis
86100 Campobasso, Italia
Tel: 0874-404641
antomac@unimol.it

In: Virgin Olive Oil
Editor: Antonella De Leonardis

ISBN: 978-1-63117-656-2
© 2014 Nova Science Publishers, Inc.

Chapter 1

CHEMISTRY AND BIOACTIVE COMPONENTS OF OLIVE OIL

Raquel de Pinho Ferreira Guiné[*]
CI&DETS Research Centre and Department of Food Industry,
Polytechnic Institute of Viseu, Portugal

ABSTRACT

Olive oil has been consumed since immemorial times and its production assumes a very important role in some countries, particularly in the Mediterranean surroundings. In fact this food product is one of the key components of the Mediterranean Diet, which, for its recognized properties and benefits, as well as cultural importance, has been inscribed by UNESCO in 2010 on the Representative List of the Intangible Cultural Heritage of Humanity.

Olive oil has proved to have many nutritional and medicinal effects. The health benefits of olive oil are extensive and new positive attributes are being discovered very frequently. In addition to bolstering the immune system and helping to protect against viruses, olive oil has also been found to be effective in fighting against diseases such as: cancer, heart disease, oxidative stress, blood pressure, diabetes, obesity, rheumatoid arthritis or osteoporosis.

These positive effects are due to the particular composition of olive oil, unique among the vegetables oils, and to the presence of compounds of nutritional importance and bioactive molecules. Therefore, the aim of this chapter is to focus on the chemical composition of the olive oil, the compounds present and their relation to effects produced.

Keywords: Chemical composition, fatty acids, phenolic compounds, bioactive components, triglycerides

[*] Department of Food Industry, ESAV, Quinta da Alagoa, Estrada de Nelas, Ranhados, 3500-606 Viseu, Portugal. Email: raquelguine@esav.ipv.pt.

1. INTRODUCTION

Olive oil is an integral ingredient of the Mediterranean diet. Its consumption dates back to biblical times, and the cultivation of the olive tree as well as the production of olive oil from the mature olives constitute an essential component of farming practices in the Mediterranean basin [1]. Olive cultivation is widespread throughout the Mediterranean region being important for the rural economy, local heritage as well as for the environment [2,3].

Virgin olive oil, obtained from the fruit *Olea europaea* L., is the only edible oil produced in large scale world-wide by mechanical or physical methods. Its consumption has been increasing in the past years due to its unique sensory characteristics, besides the nutritional and therapeutic properties reported [3]. In fact, olive oil is a very versatile product, long known to many generations in the Mediterranean areas as essential to the population's health and diet. However, at present its use is no longer limited to the Mediterranean countries, being widely appreciated around the world for its nutritional, health and sensory properties [2]. Virgin olive oil possesses singular sensory attributes, being characterized by a unique flavour, which represents one of its most important qualitative aspects, paying a major role in consumer approval [2].

The association of food with health is universal and well patent in the saying "you are what you eat". In the later decades, the importance of nutrition in the public's mind has grown and this tendency is still maintained. Surveys show that the protection/prevention against illness, tackling obesity, and the nutritional quality of foods is of even higher priority than the improvement to the taste, colour, and texture of food [4].

Over the years, the so-called Mediterranean diet has become widely associated with improved health and well-being as well as protection against cardiovascular diseases and colon, breast and skin cancers [5]. The Mediterranean diet includes an important consumption of fruits and vegetables complemented with a high intake of olive oil and other olive products.

Olive oil is a product of great importance due to its nutritional value, which has been acknowledged internationally [6]. Therefore, olive oil is an essential part of people's diet, because of its flavour and culinary value as well as its nutritional properties and biological effects on human health [7].

Virgin olive oil is one of the edible fats most highly valued by people in the Mediterranean area because it can be consumed without any refining process, since it is obtained from olives exclusively by mechanical processes, and thus retains its natural flavor and aroma. It also has highly appreciated nutritional characteristics [8]. Furthermore, olive oil appears to be a functional food with various biocomponents. Among these are monounsaturated fatty acids that may have special health benefits and also phytochemicals [9]. Some of these effects are associated with extra virgin olive oil content in phenolic compounds, high amounts of oleic acid, tocopherols and phytosterols [10]. Evidence showed that olive oil is a source of at least 30 phenolic compounds, which are strong antioxidants and radical scavengers. Recent findings confirm that olive oil phenols are powerful antioxidants, both *in vitro* and *in vivo*, besides possessing other potent biological activities that could in part account for the health benefits of the Mediterranean diet [9]. In fact, they have been suggested to play a preventive role in the development of cancer and heart disease [11].

Moreover, apart from the antioxidant activity, phenolic compounds from olive oil have other roles, namely relating to nutritional properties and sensory quality [11].

The beneficial effects are attributed to a favorable fatty acid profile and to the presence of some minor components that are also responsible for the unique flavor and taste of olive oil [5]. Accumulating evidence suggests that olive oil may have several health benefits that include a reduced risk of coronary heart disease, the prevention of several varieties of cancers and the modification of immune and inflammatory responses [9]. Furthermore, olive oil contains compounds with potent antimicrobial activities against bacteria, fungi, and mycoplasma [5]. In addition, olive oil has anti-inflammatory activities, having been demonstrated that newly pressed extra-virgin olive oil contains oleocanthal, with similar pharmacological activity as the drug ibuprofen [5].

The basic aspect that distinguishes olive oil from other vegetable oils is its high proportion of monounsaturated fatty acids, such as oleic acid which represents about two thirds of the total fatty acids content, as well as the modest presence of polyunsaturated fatty acids [7]. A healthy diet must contain a limited amount of saturated fatty acids so as to reduce the total cholesterol content and a high amount of monounsaturated fatty acids which prevent the risk of cardiovascular diseases, reduce the insulin body-requirement and decrease the plasma concentration of glucose. Moreover, there seems to be a relationship between the intake of olive oil, the richest food in the monounsaturated fatty acid oleic acid, and breast cancer risk and progression [12].

Olive oil contains natural antioxidants such as tocopherols, carotenoids, sterols and phenolic compounds that represent 27% of the unsaponifiable fraction. The main phenols identified in olive oil are gallic, caffeic, vanillic, p-coumaric, syringic, ferulic, homovanillic, p-hydroxybenzoic and protocateuric acids, tyrosol and hydroxytyrosol. Phenolic compounds of olive oil have multiple effects, including the stability to oxidation of extra virgin olive oil during storage. It has been claimed that hydroxytyrosol is the most active antioxidant compound in virgin olive oil. Furthermore, phenolic compounds have the capacity to inhibit or delay the growth rate of several bacteria and microfungi [7].

The unsaponifiable matter in olive oil constitutes about 1 to 2%, being the major portion represented by phytosterols, which are recognized by their biological effects, such as cytostatic activity, blood cholesterol control or cancer prevention [10].

The U.S. Food and Drug Administration permitted a claim on olive oil labels stating: "Limited and not conclusive scientific evidence suggests that eating about two tablespoons (23g) of olive oil daily may reduce the risk of coronary heart disease" [13].

2. CHEMICAL COMPOSITION OF OLIVE OIL

The chemical composition of virgin olive oil is influenced by genetic factors associated to the cultivar as well as environmental factors such as edaphological characteristics and climatic conditions. In this way, the characteristics of olive oil are greatly influence by the region of production [14,15]. Apart from these factors, the quality of olive oil is also strongly related to the physiological conditions of the fruit from which it was extracted, so that the stage of ripening may directly or indirectly affect the quality. Furthermore, there is an indirect effect provided by the action of external agents of deterioration which increase during fruit ripening [16]. This is because as ripening advances certain metabolic processes take place which involve changes in the profile of certain compounds such as triglycerides, fatty acids, polyphenols, tocopherols, chlorophylls and carotenoids. These changes, besides influencing the oxidative stability and the nutritional value of the final product, also affect the sensory characteristics, with particular emphasis on aroma.

Gomez-Rico et al. [17] studied the influences of agronomic practices and found that irrigation positively affected both fruit and olive oil quality.

Mendoza et al. [15] studied 88 virgin olive oil samples original from Spain, from three successive crop seasons, produced from the mixture of two varieties of olives: Morisca and Carrasqueña. The results showed that 88.5% of selected original grouped cases are correctly classified according to the ripening stage (85.7% green, 80% spotted and 78.9% ripe) having this been based on the most discriminating variables: avenasterol, linolenic acid, beta-sitosterol and gadoleico.

Baccouri et al. [18] studied the influence of the olive ripening stage on the quality indices, the major and the minor components and the oxidative stability of the two main monovarietal Tunisian cultivars (Chétoui and Chemlali) virgin olive oils. Their results indicated a very good correlation between the oxidative stability and the concentrations of total phenols, practically secoiridoids and α-tocopherol.

Virgin olive oil is mainly composed of triacylglycerols (between 97 and 98%), minor variable amounts of free fatty acids and minor glyceridic compounds (partial glycerides, phospholipids and oxidized triacylglycerols) and finally around 1% of unsaponifiable constituents varying in structure and polarity [18,19].

The oxidative stability, sensory quality and health properties of virgin olive oil are possible due to a well-balanced chemical composition [20].

2.1. Acidity

One of the parameters that allow the classification of olive oils is the acidity. Extra-virgin olive oil comes from virgin oil production only and contains no more than 0.8% acidity, expressed per g oleic acid/100g olive oil [21] having a superior taste. Virgin olive oil comes from virgin oil production only too but the acidity is slightly higher, less than 2% and is judged to have a good taste. When the acidity of the olive oil exceeds 2%, and/or the sensory evaluation is lower the sample is graded as ''lampante virgin olive oil'' and is recommended to be refined prior to human consumption. Oils labeled as Olive oil are usually a blend of refined and virgin olive oils, and contain no more than 1% acidity, and commonly lack a strong flavor. *Olive pomace oil* is refined pomace olive oil often blended with some virgin oil. It has a more neutral flavor than virgin olive oil maintaining the same fat composition, thus giving it the same health benefits. Refined olive oil is the olive oil obtained from virgin olive oils by refining methods that do not lead to alterations in the initial glyceridic structure. It has a free acidity, expressed as oleic acid, of not more than 0.3 grams per 100 grams (0.3%) and its other characteristics correspond to those fixed for this category in this standard.

Table 1 shows some values of acidity obtained in some studies about olive oil composition.

Table 1. Acidity of olive oils

Cultivar	Acidity (g oleic acid/100 g)	Reference
Morisca and Carrasqueña	0.29 – 0.53	[15]
Chétoui and Chemlali	0.23 – 0.42	[18]
Cornicabra	0.08 – 0.55	[8]
Arbequina, Manzanilla, Nevadillo and Ascolana	0.12 – 1.03	[22]
Hor Kesra, Sredki, Chladmi, Betsijina and Aloui	0.19 – 0.31	[10]
Halhali, Egriburun, Hasebi, Karamani and Saurani	0.52 – 0.85	[23]

2.2. Peroxide Index

The established upper limit for the peroxide value in extra virgin olive oil and virgin oil is 20.0 miliequivalent O_2 per kg [21].

Table 2 shows some values of the peroxide index for different samples of olive oil, produced from different cultivars of olives.

Table 2. Peroxide index of olive oil samples

Cultivar	Peroxide index (meqO$_2$/kg)	Reference
Morisca and Carrasqueña	5.71 – 6.47	[15]
Chétoui and Chemlali	2.93 – 16.67	[18]
Cornicabra	1.9 – 19.1	[8]
Arbequina, Manzanilla, Nevadillo and Ascolana	7.84 – 11.0	[22]
Hor Kesra, Sredki, Chladmi, Betsijina and Aloui	4.19 – 5.83	[10]
Halhali, Egriburun, Hasebi, Karamani and Saurani	2.01 – 7.08	[23]

2.3. Phenolic Compounds

The amount of phenolic compounds present in extra virgin olive oil is a major factor influencing its quality because the natural phenols improve the resistance to oxidation and also because, to certain extent, they are also responsible for the characteristic taste [20].

Olive oil presents considerable amounts of natural antioxidants which have proven to be important in the prevention of many diseases [24].

Baccouri et al. [18] observed that the concentration of phenolic compounds progressively increased to a maximum at the "reddish" and "black" pigmentation stage of the olives used to make the oil, decreasing thereafter. This trend was also observed by Salvador et al. [25]. One other factor that proved to affect the amount of phenolic compound was irrigation of the fruits [17,18, 26].

Oleuropein belongs to a specific group of coumarin-like compounds, the secoiridoids, which are abundant in olives. The secoiridoids are present exclusively in plants belonging to the family of *Olearaceae* and are characterised by the presence of elenolic acid, either in the glucosidic or aglyconic forms. The most abundant secoiridoids of virgin olive oil are the dialdehydic form of elenolic acid linked to hydroxytyrosol or tyrosol and an isomer of the oleuropein aglycon [20].

Some phenolic acids that have been identified and quantified in virgin olive oil include gallic, protocatechuic, p-hydroxybenzoic, vanillic, caffeic, syringic, p- and o-coumaric, ferulic and cinnamic acids [27,28].

(+)-Pinoresinol is a common component of the lignan fraction of various plants, but in the olive oil tree the forms found are (+)-1-acetoxypinoresinol and (+)-1-hydroxy-pinoresinol and their corresponding glucosides. The quantity of lignans in virgin olive oil may be up to 100 mg/kg [20].

Flavonoids are subdivided into different classes: flavones, flavonols, flavanones, and flavanols. Carrasco-Pancorbo et al. [29] and Morelló et al. [30] have reported the presence in olive oil of flavonoids such as luteolin and apigenin.

Table 3 shows the amount of total phenolic compounds found in olive oil samples. Table 4 reports some values regarding the concentration of some specific phenolic compounds in olive oil.

Manai-Djebali et al. [10] reported values of *ortho*-diphenols between 21 and 369 mg/kg olive oil, for five cultivars studied.

Table 3. Total phenolic compounds in olive oil samples

Cultivar	Total phenols (mg cafeic acid/kg)	Reference
Morisca and Carrasqueña	270 – 336	[15]
Chétoui and Chemlali	46 – 568	[18]
Cornicabra	180 – 614	[8]
Arbequina, Manzanilla, Nevadillo and Ascolana	99 – 373	[22]
Hor Kesra, Sredki, Chladmi, Betsijina and Aloui	253 – 1400	[10]
Halhali, Egriburun, Hasebi, Karamani and Saurani	64 – 321	[23]

Table 4. Concentration of different phenolic compound in olive oil

Phenolic compounds	Amount present (mg/kg)	
	Baccouri et al. [18]	Manai-Djebali et al. [10]
Hydroxytyrosol	1.22 – 75.62	3.29 – 19.0
Tyrosol	2.54 – 30.88	4.98 – 27.3
Decarboxymethyl oleuropein aglycon + (+)-1-acetoxypinoresinol	1.78 – 126.33	-
Decarboxymethyl ligstroside aglycon	15.08 – 80.30	-
Oleuropein aglycon	11.11 – 397.93	17.6 – 667
Ligstroside aglycon	3.56 – 24.96	4.08 – 150
Vanillic acid	0.16 – 1.84	0.14 – 0.51
Coumaric acid	1.01 – 6.58	0.14 – 0.46
4-(acetoxyethyl)-1,2-dihydroxybenzene	-	0.17 – 0.38
Pinoresinol	-	3.34 – 15.1
Secoiridoids	16.47 – 510.11	38.0 – 383
Simple phenols	7.72 – 109.42	-

2.4. Fatty Acids

The resistance of virgin olive oils to oxidation is higher than that of other edible oils because of their contents of natural antioxidants and lower unsaturation levels. The higher the number of double bonds in fatty acids, the shorter is the induction period for oil autoxidation [22].

The nutritional benefits of olive oil are primarily related to the fatty acid composition, mainly due to the high content of oleic acid and also to the balanced ratio of saturated and polyunsaturated fatty acids [24]. The high content of oleic acid in olive oil helps slowing down the penetration of fatty acids into arterial walls. Oils richer in monounsaturated fatty acids (MUFAs) and poorer in saturated fatty acids (SFAs) are preferred because of the proven beneficial effect of MUFAs on serum cholesterol levels [18]. Baccouri et al. [18] found ratios between monounsaturated and polyunsaturated fatty acids (MUFA/PUFA) ranging between 3.65 and 7.83, while Mendoza et al. [15] found values for the ratio monounsaturated/saturated fatty acids in the range 4.14 – 4.40. Table 5 shows the fatty acids composition of some studied olive oils.

Torres and Maestri [22] reported values of SFAs in the range 14.9 – 19.2% of total fatty acids, MUFAs varying from 64.3 to 75.0% and PUFAs from 8.37 to 16.5%, for samples of olive oil from 4 different cultivars of olives: Arbequina, Manzanilla, Nevadillo and Ascolana.

Salvador et al. [8] studied the chemical composition of 181 Cornicabra virgin olive oils from five successive crop seasons and found values of free fatty acids ranging between 0.08 and 1.83%, as percentage of oleic acid.

Table 5. Fatty acids in olive oil

Fatty acids	Limits for extra virgin olive oil [21]	Amount present (%)		
		Salvador et al. [8]	Baccouri et al. [18]	Mendoza et al. [15]
Palmitic, C16:0	7.5 – 20.2	6.99 – 11.05	8.86 – 19.33	12.60 – 13.00
Palmitoleic, C16:1	0.3 – 3.5	0.49 – 1.11	0.32 – 2.90	1.12 – 1.28
Margaric, C17:0	-	0.04 – 0.07	0.04 – 0.05	0.09 – 0.11
Margaroleic, C17:1	-	0.08 – 0.11	0.01 – 0.06	0.09 – 0.11
Stearic, C18:0	0.5 – 5.0	2.61 – 4.43	1.99 – 3.95	3.29 – 3.40
Oleic, C18:1	55.0 – 83.0	76.5 – 82.5	59.25 – 72.71	69.40 – 71.60
Linoleic, C18:2	3.5 – 21.0	3.07 – 6.62	8.80 – 18.41	9.50 – 11.50
Linelenic, C18:3	≤ 1.0	0.48 – 0.95	0.33 – 0.71	0.65 – 0.72
Arachidic, C20:0	≤ 0.6	0.28 – 0.62	0.44 – 0.56	0.35 – 0.38
Gadoleic, C20:1	-	0.29 – 0.39	0.01 – 0.04	0.21
Behenic, C22:0	≤ 0.2	0.12 – 0.21	0.05 – 0.19	0.13 – 0.14
Lignoceric, C24:0	≤ 0.2	-	0.01 – 0.08	0.03 – 0.04

2.5. Triglycerides

Because of the specificity of the triacylglycerol composition in different kinds of fats and oils, it is being increasingly used in the food industry to confirm authenticity. According to

Manai-Djebali et al. [10] the oils from five different cultivars are characterized by three primary triacylglycerols: OOO, POO and OLO, seven secondary triacylglycerols: OLL, OLnO, PLO+SLL, PPL, PPO, SOO and SLS+POS, and they still contain small amounts (≤ 1%) of OLLn, PLLn, PLL LLL, LLnLn and PPP.

Table 6 shows the concentrations of the relevant triglycerides found in olive oil, according to the findings of different researchers.

2.6. Sterols

Sterols are important constituents of olive oils because they relate to the quality of the oil and are widely used to check authenticity [8].

The amount of unsaponifiable matter in olive oil is about 1–2%. The major portion is represented by phytosterols, which are recognized by their biological effects, such as cytostatic activity blood cholesterol control or cancer prevention [10].

Salvador et al. [8] studied different samples of Cornicabra virgin olive oil and concluded that they all showed high campesterol content, ranging from 4.1 to 4.3, exceeding the threshold of 4% established by EU Regulations. They also reported a high Δ 5-avenasterol content, with a mean value of 6.2 ± 1.2%.

Table 7 shoes the composition reported by different authors in relation to the contents in sterols for different olive oil samples.

Table 6. Triglycerides in olive oil

Triglycerides*		Amount present (%)		
		Baccouri et al. [18]	Manai-Djebali et al. [10]	Mendoza et al. [15]
ECN40	LLLn	-	-	0.00
ECN42	LLL	-	0.10 – 0.75	0.15 – 0.19
	OLLn	-	0.20 – 0.41	0.26 – 0.31
	PLLn	-	0.06 – 0.13	0.09 – 0.11
ECN44	LLO	0.07 – 3.04	1.81 – 6.40	2.28 – 2.79
	OLnO	-	1.43 – 3.09	1.82 – 2.12
	PLL	0.00 – 0.45	0.36 – 0.57	0.53 – 0.60
ECN46	OLO	12.23 – 36.59	14.0 – 18.7	12.60 – 13.90
	PLO+SLL	-	5.14 – 10.1	6.97 – 8.15
	PPL	-	0.41 – 1.12	0.64 – 0.82
ECN48	OOO	30.77 – 56.32	30.0 – 46.0	34.00 – 36.40
	POO	5.08 – 37.34	19.7 – 26.3	24.70 – 25.30
	PPO	0.00 – 2.54	2.23 – 4.63	3.90 – 3.97
	PPP	-	0.41 – 0.67	0.44 – 0.45
ECN50	SOO	0.27 – 2.86	4.04 – 4.69	5.56 – 5.87
	SLS+POS	-	0.79 – 1.33	1.60 – 1.76

*ECN: equivament carbon number calculated from triacylgyicerols and fatty acid composition; P: palmitic; S: stearic; O: oleic; L: linoleic; Ln: linolenic acids.

Table 7. Sterols in olive oil

Sterols	Amount present (%)		
	Salvador et al. [25]	Mendoza et al. [15]	
		State of ripeness	
		Green	Ripe
Cholesterol	0.06 – 0.60	0.13	0.14
Brassicasterol	0.00 – 0.57	-	-
24-Methylencholesterol	0.00 – 0.31	0.22	0.21
Campesterol	3.82 – 4.64	2.38	2.50
Campestanol	0.00 – 0.43	0.12	0.12
Stigmasterol	0.38 – 1.75	0.79	1.05
Δ 7-Campesterol	0.00 – 2.27	0.58	0.62
Δ 5.23-Stigmastadienol	0.00 – 0.38	-	-
Clerosterol	0.53 – 1.05	1.05	1.00
Beta-Sitosterol	81.2 – 88.2	82.10	83.30
Sitostanol	0.00 – 1.27	0.52	0.50
Δ 5-Avenasterol	3.34 – 10.97	10.50	9.10
Δ 5,24-Stigmastadienol	0.26 – 1.09	0.71	0.61
Δ 7-Stigmastenol	0.09 – 1.30	0.24	0.25
Δ 7-Avenasterol	0.018 – 0.44	0.61	0.60
Total sterols (mg/kg)	1014 – 2055	1666	1665

2.7. Minor Unsaponifiable Compounds

The biological properties of olive oil are related, among other, to the presence of minor components, such as squalene and phytosterols, antioxidant compounds, such as tocopherols and particularly phenols [18,31]. In particular, among the natural antioxidants, phenolic compounds, α-tocopherol and β-carotene, are reported to play a key role in preventing oxidation and have been correlated to the storage stability of virgin olive oils [32]. Psomiadou and Tsimidou [33] have shown that the most representative component of the olive oil unsaponifiable fraction, which is squalene, does not significantly affect the oil stability (regarding ramification), despite its positive effects on human health [31].

Squalene is the major olive oil hydrocarbon. In fact, this terpenoid hydrocarbon makes up more than 90% of the hydrocarbon fraction [31]. Baccouri et al. [18] observed that olive oil produced from Chemlali cultivar of olives presented very high amounts of squalene, reaching 10.48 g/kg of oil. However, along the maturity process, this level deceased remarkably down to 2 g/kg. A similar trend was also observed for Chétoui oils (obtained under irrigation regime), in which the squalene content fell progressively from 5.99 down to 3.58 g/kg as ripening progressed. Conversely, in Chétoui oils obtained in rainfed conditions, the trend was different, first increasing until it reached a maximum of 8.27 g/kg and then decreasing.

Tocopherols usually present in extra virgin olive oil are: α-, β- and γ-tocopherols [16]. Both tocopherols and polar phenolic compounds are responsible for the oxidative stability of olive oil and, therefore, for its prolonged shelf life. Regarding this aspect, no doubt α-tocopherol plays a major role [34]. Baccouri et al. [18] and Beltran et al. [16] concluded from their works that for extra virgin olive oil α-tocopherol is by far the most abundant isoform of vitamin E. In the work of Baccouri et al. [18] the contents in α-tocopherol ranged from 120.55 to 478.13 mg/kg, while β-tocopherol and γ-tocopherol varied in the ranges 2.20 – 12.83 mg/kg and 4.83 – 18.15 mg/kg, respectively. Salvador et al. [8] found values of α-tocopherol ranging between 19 and 380 mg/kg in 151 samples of commercial virgin olive oil, while Manai-Djebali et al. [10] reported values in the range 139 – 368 mg/kg for five different cultivars. Arslan and Schreiner [23] analysed olive oils from five different cultivars and reported the following ranges of values for α-, β- and γ-tocopherols: 133.2 – 343.0, 0.1 – 3.4, 0.8 – 14.8 mg/kg, respectively.

Baccouri et al. [18] observed a drastic decrease in the chlorophyll content along drying, evidenced by the negative correlations between the chlorophyll concentrations and the ripening index. The values observed for chlorophyll concentration varied from 21.40 to 1.88 mg/kg of oil. In general, during fruit ripening, the chlorophylls, which are present in all unripe fruit, break down following the transformation of chloroplasts into chromoplasts [35]. Salvador et al. [8] reported values for chlorophyll concentration in the range 1.7 – 27.1 mg/kg for commercial virgin olive oils. Torres and Maestri [22] found values of chlorophyll between 2.46 and 6.38 mg/kg, Manai-Djebali et al. [10] between 1.15 and 6.22 mg/kg, and Arslan and Schreiner [23] in the range 7.0 – 17.7 mg/kg.

The concentrations of carotenoid pigments decrease very markedly during ripening, as reported by Baccouri et al. [18], with values coming down from 11.33 to 1.23 mg/kg of oil. The concentrations for carotenoids observed by Salvador et al. [8] for commercial virgin olive oils varied between 2.3 and 14.0 mg/kg. Torres and Maestri [22] reported a range of values for the carotenoids concentration of 1.55 – 3.61 mg/kg while Manai-Djebali et al. [10] fund values varying between 1.07 and 3.82 mg/kg and Arslan and Schreiner [23] reported values in the interval 5.3 – 12.6 mg/kg.

Olive fruits are green at first but as ripening process they darken to purple-black at the same time as the oil content increases. As ripening progresses, photosynthetic activity decreases and the concentrations of both chlorophylls and carotenoids decrease progressively. At full ripening stage, the violet or purple color of the olive fruit is due to the formation of anthocyanins [36].

The total amount of pigments in olive oil is an important quality parameter because it correlates with color, which is a basic attribute for evaluating olive oil quality. Besides, pigments are also involved in autoxidation and photooxidation mechanisms. Studies have demonstrated that chlorophylls act as prooxidants under light storage, whereas β-carotene minimizes lipid oxidation due to its light-filtering effect [22].

CONCLUSION

Accumulating evidence suggests that olive oil may have health benefits, such as reduction in the risk of coronary heart disease, prevention of several varieties of cancers and the modification of immune and inflammatory responses. These health benefits of olive oil result from its unique composition, since it appears to be a functional food with various bioactive components, like for example monounsaturated fatty acids. Besides, it is also a good source of phytochemicals, including polyphenolic compounds, being possible to find in it at least 30 phenolic compounds, which are strong antioxidants and radical scavengers. Recent studies reveal that olive oil phenols are powerful antioxidants, both *in vitro* and *in vivo*, apart from other potent biological activities that could partially account for the observed health effects of the Mediterranean diet.

ACKNOWLEDGMENT

The author would like to thank the valuable contribution of the reviewers of the present chapter: Prof. Maria João Barroca (PhD), Department of Chemical and Biological Engineering, Polytechnic Institute of Coimbra, Portugal; and Prof. Ana Cristina Correia (Msc), Department of Food Industry, Agrarian School of Viseu, Portugal.

REFERENCES

[1]   Alonso-Salces RM, Héberger K, Holland MV, Moreno-Rojas JM, Mariani C, Bellan G, Reniero F, & Guillo C (2010) Multivariate analysis of NMR fingerprint of the unsaponifiable fraction of virgin olive oils for authentication purposes. *Food Chemistry*, 118, 956-965.

[2]   Haddada FM, Manai H, Daoud D, Fernandez X, Lizzani-Cuvelier L, & Zarrouk M (2007) Profiles of volatile compounds from some monovarietal Tunisian virgin olive oils. Comparison with French PDO. *Food Chemistry*, 103, 467-476.

[3]   Kandylis P, Vekiari AS, Kanellaki M, Kamoun NG, Msallem M, & Kourkoutas Y (2011) Comparative study of extra virgin olive oil flavor profile of Koroneiki variety (Olea europaea var. Microcarpa alba) cultivated in Greece and Tunisia during one period of harvesting. *LWT - Food Science and Technology*, 44, 1333-1341.

[4]   Rodgers S, & Young NWG (2008) The Potential Role of Latest Technological Developments Including Industrial Gastronomy in Functional Meal Design. *Journal of Culinary Science & Technology*, 6, 170-187.

[5]   Yamada P, Zarrouk M, Kawasaki K, & Isoda H (2008) Inhibitory effect of various Tunisian olive oils on chemical mediator release and cytokine production by basophilic cells. *Journal of Ethnopharmacology*, 116, 279-287.

[6]   Christopoulou E, Lazaraki M, Komaitis M, & Kaselimis K (2004) Effectiveness of determinations of fatty acids and triglycerides for the detection of adulteration of olive oils with vegetable oils. *Food Chemistry*, 84, 463-474.

[7]   Kachouri F, & Hamdi M (2006) Use Lactobacillus plantarum in olive oil process and improvement of phenolic compounds content. *Journal of Food Engineering*, 77, 746-752.

[8]   Salvador MD, Aranda F, Gómez-Alonso S, & Fregapane G (2001b) Cornicabra virgin olive oil: a study of five crop seasons. Composition, quality and oxidative stability. *Food Chemistry*, 74, 267-274.

[9]   Oi-Kano Y, Kawada T, Watanabe T, Koyama F, Watanabe K, Senbongi R. & Iwai K (2007) Extra virgin olive oil increases uncoupling protein content in brown adipose tissue and enhances noradrenaline and adrenaline secretions in rats. The *Journal of Nutritional Biochemistry*, 18, 685-692.

[10] Manai-Djebali H, Krichène D, Ouni Y, Gallardo L, Sánchez J, Osorio E, Daoud D, Guido F, & Zarrouk M (2012) Chemical profiles of five minor olive oil varieties grown in central Tunisia. *Journal of Food Composition and Analysis*, 27, 109-119.

[11] Kachouri F, & Hamdi M (2004) Enhancement of simple polyphenols in olive oil derived from a OMW fermentation by. Lactobacillus plantarum. *Process Biochemistry*, 39, 841-845.

[12] D'Imperio M, Dugo G, Alfa M, Mannina L, & Segre AL (2007) Statistical analysis on Sicilian olive oils. *Food Chemistry*, 102, 956-965.

[13] Drummond L (2010) Sunday Telegraph (Australia), October 17, Features; p. 10.

[14] Criado MN, Morello JR, Motilva MJ, & Romero MP (2004) Effect of growing area on pigment and phenolic fractions of virgin olive oils of the Arbequina variety in Spain. *Journal of American Oil Chemists' Society*, 81, 633-640.

[15] Mendoza MF, Gordillo CM, Expóxito JM, Casas JS, Cano MM, Vertedor DM, & Baltasar MNF (2013) Chemical composition of virgin olive oils according to the ripening in olives. *Food Chemistry*, 141, 2575–2581.

[16] Beltran G, Aguilera MP, Del Rio C, Sanchez S, & Martinez L (2005) Influence of ripening process on the natural antioxidant content of Hojiblanca virgin olive oils. *Food Chemistry*, 89, 207-215.

[17] Gomez-Rico A, Salvador MD, Moriana A, Perez D, Olmedilla N, Ribas F, & Fregapane G (2007) Influence of different irrigation strategies in a traditional Cornicabra cv. olive orchard on virgin olive oil composition and quality. *Food Chemistry*, 100, 568-578.

[18] Baccouri O, Guerfel M, Baccouri B, Cerretani L, Bendini A, Lercker G, Zarrouk M, & Miled DDB (2008) Chemical composition and oxidative stability of Tunisian monovarietal virgin olive oils with regard to fruit ripening. *Food Chemistry*, 109, 743-754.

[19] Boskou D (1996) *Olive oil: Chemistry and technology.* Champaign, IL: AOCS Press.

[20] Bendini A, Cerretani L, Carrasco-Pancorbo A, Gómez-Caravaca AM, Segura-Carretero A, Fernández-Gutiérrez A, & Lercker G (2007) Phenolic molecules in virgin olive oils: A survey of their sensory properties, health effects, antioxidant activity and analytical methods. An overview of the last decade. *Molecules*, 12, 1679-1719.

[21] EEC (2003) Characteristics of olive and olive pomace oils and their analytical methods EEC Regulation 1989/2003. *Official Journal of the European Communities*, 295, 57-66.

[22] Torres MM, & Maestri DM (2006) The effects of genotype and extraction methods on chemical composition of virgin olive oils from Traslasierra Valley (Córdoba, Argentina). *Food Chemistry*, 96, 507-511.

[23] Arslan D, & Schreiner M (2012) Chemical characteristics and antioxidant activity of olive oils from Turkish varieties grown in Hatay province. *Scientia Horticulturae*, 144, 141-152.

[24] Ayadi MA, Grati-Kamoun N, & Attia H (2009) Physico-chemical change and heat stability of extra virgin olive oils flavoured by selected Tunisian aromatic plants. *Food and Chemical Toxicology*, 47, 2613-2619.

[25] Salvador MD, Aranda F, & Fregapane G (2001a) Influence of fruit ripening on Cornicabra virgin olive oil quality: a study of four crop seasons. *Food Chemistry*, 73, 45-53.

[26] Tovar MJ, Romero MP, Alegre S, Girona J, & Motilva MJ (2003) Composition and organoleptic characteristics of oil from Arbequina olive (*Olea europaea* L.) trees under deficit irrigation. *Journal of the Science of Food and Agriculture*, 82, 1755-1763.

[27] Buiarelli F, Di Berardino S, Coccioli F, Jasionowska R, & Russo MV (2004) Determination of phenolic acids in olive oil by capillary electrophoresis. *Annali di Chimica*, 94, 699-705.

[28] Carrasco-Pancorbo A, Segura-Carretero A, & Fernández-Gutiérrez A (2005) Co-electroosmotic capillary electrophoresis determination of phenolic acids in comercial olive oil. *Journal of Separation Science*, 28, 925-934.

[29] Carrasco-Pancorbo A, Gómez-Caravaca AM, Cerretani L, Bendini A, Segura-Carretero A, & Fernández-Gutiérrez A (2006) Rapid quantification of the phenolic fraction of spanish virgin olive oils by capillary electrophoresis with uv detection. *Jounal of Agricultural and Food Chem*istry, 54, 7984-7991.

[30] Morelló JR, Vuorela S, Romero MP, Motilva MJ, & Heinonen M (2005) Antioxidant activity of olive pulp and olive oil phenolic compounds of the arbequina cultivar. *Jounal of Agricultural and Food Chem*istry, 53, 2002-2008.

[31] Owen RW, Mier W, Giacosa A, Hull WE, Spiegelhalder B, & Bartsch H (2000) Phenolic compounds and squalene in olive oils: The concentration and antioxidant potential of total phenols, simple phenols, secoiridoids, lignans and squalene. *Food and Chemical Toxicology*, 38, 647-659.

[32] Rahmani M, & Saari-Csallany A (1998) Role of minor constituents in the photooxidation of virgin olive oil. *Journal of the American Oil Chemists Society*, 75, 837-843.

[33] Psomiadou E, & Tsimidou M (1999) On the role of squalene in olive oil stability. *Journal of Agricultural and Food Chemistry*, 47, 4025-4032.

[34] Mateos R, Dominguez MM, Espartero JL, & Cert A (2003) Antioxidant effect of phenolic compounds alpha-tocopherol and other minor components in virgin olive oil. *Journal of Agricultural and FoodChemistry*, 51, 7170-7175.

[35] Kozukue N, & Friedman M (2003) Tomatine, chlorophyll, b-carotene and lycopene content in tomatoes during growth and maturation. *Journal of the Science of Food and Agriculture*, 83, 195-200.

[36] Roca M, & Minguez-Mosquera MI (2001) Change in the natural ratio between chlorophylls and carotenoids in olive fruit during processing for virgin olive oil. *Journal of the American Oil Chemists' Society*, 78, 133-138.

In: Virgin Olive Oil
Editor: Antonella De Leonardis

ISBN: 978-1-63117-656-2
© 2014 Nova Science Publishers, Inc.

Chapter 2

OLIVE OIL: PRODUCTION, BIOACTIVE PROPERTIES AND PUBLIC HEALTH

*Hui Jun Chih**

School of Public Health, Faculty of Health Sciences, Curtin University, Australia

ABSTRACT

This chapter begins with a brief historical review of traditional machinery-based extraction of virgin olive oil, followed by an update on the development of some recent processing techniques around the world, including the non-traditional olive oil producing nation, Australia. The effects of production on the main bioactive properties of the extracted oil, such as concentration of total phenolic compounds and antiradical activity, will be presented. The implications of employing these techniques on public health are also mentioned, especially in the context of cardiovascular diseases, metabolic syndrome, cancers, as well as environmental impact. This chapter concludes with suggestions for future research.

Keywords: Olive oil, mechanical extraction, malaxation, processing aids, bioactive compounds, public health

INTRODUCTION

Virgin olive oil is a natural juice extracted from olive fruit (*Oleaeuropaea*) by mechanical pressing of the fruit. The oil contains a good ratio of fatty acids as well as the bioactive unsaponifiable compounds. Its unique composition delivers protective effect against cardiovascular diseases, metabolic syndrome and cancer. Increased awareness of the health benefits of olive oil have resulted inan increase in the world consumption of olive oil. Innovative and effective processing techniques are necessary to continue the delivery of

---

* Corresponding Author address: School of Public Health, Faculty of Health Sciences, Curtin University, GPO BOX U1987, Perth, Western Australia, 6845, Australia. Email: h.chih@curtin.edu.au.

virgin olive oil with bioactive compounds for improved public health outcomes. As this chapter provides a brief summary of olive oil extraction, rooms for innovative industrial practice to optimize public health impacts are recommended.

BRIEF HISTORY OF THE TRADITIONAL MACHINERY EXTRACTION OF OLIVE OIL

Extraction of virgin olive oil, a natural juice from the fruit *Oleaeuropaea*, dated back to 5000 B.C. [1]. In the early days, olives were ground in a mortar with stones and the paste was pressed by heavy rocks to release the oil. Separation of the oil and the vegetable water was done solely by gravity force. Millstone crusher was developed during the Roman Age to replace the use of mortar with additional presses to facilitate the separation of oil. The invention of screw press in 50 B.C. by the Greeks was regarded as a major milestone in the history of the development of olive oil production. Heavy labour forces were required to operate the wooden and iron screw presses [1].

In the early 17th century, hydraulic press was firstly developed to assist the production of virgin olive oil. Labour requirements were minimized through the development of electrically driven hydraulic pumps, cage presses, column presses and super presses with pressure up to 350-500 atmospheres. Invention of percolation system and centrifugation system came in place in mid 1900s after realizing the disadvantages of the pressing system. The machinery involved in the pressing system was complex and required large investment of labour forces. In addition, the discontinuous system reduced the production capacity [1]. The innovative percolation and centrifugation system promoted continuous production of the oil and thus improved the efficiency of virgin olive oil production. In particular, the development of automated centrifugal decanter in 1960s was an major achievement in the development of virgin olive oil production [1].

The three-phase olive oil decanter is an example of the automated centrifugal decanter. It consists of a crusher, malaxer and centrifuge. The high centrifuge speed (3500- 3600 rpm) applied in a three-phase decanter greatly enhanced the separation of the liquid oil and water phase from the solid pomace phase [1]. The centrifuged products include the virgin olive oil, oily by-product and wastewater. Addition of lukewarm water to the three-phase decanter in the ratio of 1:1 olive paste:water (w/w) is a common practice in the industry to further improve the extraction of virgin olive oil. However, this step reduces the concentration of phenolic compounds in the oil and its subsequent storage stability. This has a direct impact on public health as the phenolic compounds are important for the antioxidant activity of olive oil, and are responsible for the anticancerous properties [2]. In addition, the wastewater produced from the three-phase decanter poses a disposal problem for the industry. With every 100 kilograms of olives processed, approximately 100 liters of vegetation wastewater is produced [3]. The organic wastewater, consists of sugars, polyalcohols, pectins, lipids and notable amounts of aromatic compounds (tannins and polyphenols), makes it highly polluted to the environment. This is a public health hazard. Indeed, its biochemical oxygen demand (BOD) is 89-100 g/L and the chemical oxygen demand (COD) is 80-200 g/L [4]. The high BOC and COD values indicate an urgency to reduce (if not totally eliminate) the wastewater, particularly as the production of olive oil is expected to increase.

The phototoxic effect of the large amount of wastewater produced from the three-phase decanters initiated the modification of the decanter. In this regard, a two-phase decanter was invented for the production of olive oil. As no water is added to the two-phase decanter, it does not produce problematic wastewater high in organic matters.

The yield of oil extraction is similar between the two types of decanter, with the two-phase decanter producing an extraction yield of 1 % greater than that extracted by the three-phase decanter [1]. The higher extraction yield is favoured by many producers as popularity of the health benefits of olive oil saw the increase in its demand. In addition to the oil yield, concentration of phenolic compounds in the oil extracted by the two-phase decanter is also higher than that extracted by three-phase decanter. As previously reported, the concentration of total phenolic compounds in Spanish *Arbequina* olive oil extracted by two-phase decanter was reported as 80.88 mg/kg oil, compared to a lower concentration of 42.07 mg/kg oil in olive oil extracted by a three-phase decanter [5]. The significant 48 % increment in the concentration of total phenolic compounds in the two-phase decanter-extracted oil is attributable to the zero amount of water required during the olive oil extraction process. The two-phase decanter is widely used nowadays due to its greater energy- and cost-efficiency. Its minimized usage of water, lower toxic impact on the environment and the increased level of bioactive phenolic compounds [6], are some of the positive public health implications.

PROCESSING TECHNIQUES BEYOND MACHINERY AND ITS IMPACT ON PUBLIC HEALTH

Despite development in the machinery technique, mechanical extraction of virgin olive oil only confers an oil recovery of approximately 60 % [7, 8]. The continued use of such low-yield techniques is probably due to the restrictions enforced by the International Olive Council (IOC) that forbids the use of chemical solvents in the production of virgin olive oil. The act aims to ensure the quality of olive oil and preserve its health benefits. However, with increasing recognition of the health benefits of olive oil and its increasing demand around the world, there is a need to look at natural alternatives to improve the yield of oil extraction.

Through mechanical extraction, approximately 90 % of the bioactive phenolic compounds in olives end up in the by-product [8-10]. The concentration of phenolic compounds in the olive oil is directly related to the health benefits, the stability as well as the sensory property of olive oil [11-14]. Therefore, it is necessary to investigate new extraction techniques in terms of their ability to improve both the yield of olive oil and the concentration of phenolic compounds in the extracted olive oil.

Approximately a decade ago, the Australian Olive Association (2002) has pointed out in the *Research and Development Plan for the Australian Olive Industry 2003-2008* that Australia needs to "*develop best practice post-harvest technologies guidelines*" to achieve greater yield of oil extraction with high quality oil. The practices need to ensure that olive oil producers are "*maximizing the health active components of olive oil through various production and/or processing variables*" in order to compete with imports [15]. Since then, research has been conducted in Australia to understand the impact of various factors on the extraction of quality olive oil.

Yield and quality of olive oil can be affected by various factors, such as olive growing sites, cultivar, maturity level, storage of harvested olive fruits, milling equipment, malaxation length, level of oxygen gas during malaxation, malaxation temperature and processing aids such as talc, enzyme and citric acid [16]. While agronomic practices may affect the quality of olive oil, to the best of the author's knowledge, no conclusive outcome on the effectiveness of a particular processing technique in improving the extraction of oil and phenolic compounds has been drawn to date. Extending the length of malaxation period and addition of processing aids can be the potential solutions to this problem. Economically, less amount of monetary investment is involved by application of these techniques comparing to genetically modifying the cultivar, shift of growing sites and/or modification of extraction machinery.

Malaxation

Malaxation is a slow mixing process that allows coalescence of tiny free oil droplets into bigger ones. As mentioned, modifying the malaxation time during the extraction process involves less cost than modifying the equipment. For optimal oil recovery and concentration of phenolic compounds, malaxation duration of olive paste was recommended to be at least 15 minutes but should not be more than 90 minutes [17]. The effect of olive paste malaxation period on the other bioactive properties, however, was not established from the study.

Another study was later initiated in Western Australia to investigate the effect of malaxation time on the quality parameters of the extracted virgin olive oil. Extending the length of olive paste malaxation period from 30 to 60 minutes increased the oil recovery albeit insignificant [18]. It was thought that greater coalescence of the oil droplets occurred during extended malaxation. The formation of bigger oil mass could be easily separated from the solid matrix during the centrifuge phase, thus leading to increased oil recovery. While the extended malaxation period led to the greater coalescence, there was insignificant cell disruption and thus the insignificant increase in oil recovery [18]. Extending the duration of malaxation is uneconomical for the olive oil producers. Meanwhile, the other quality parameters were found to be the same after extending the malaxation period [18]. In other words, it is ineffective in increasing the concentration of bioactive compounds to deliver better health outcomes for the consumers. A search for alternative processing technique is necessary.

Processing Aids

The application of processing aids is receiving a great deal of interest from the olive oil industry with the aim of increasing the yield of oil extraction. In some occurrences, when the initial moisture content of the olive fruits is high, or the type of crusher employed exerts violent motion to the paste, emulsion is formed. Emulsion can complicate the separation of the oil droplets from the solid paste, thus reducing the yield of oil extraction. In this regard, processing aids may be a solution to this problem. Examples of some experimented processing aids include talc, enzymes and citric acid.

Talc

Micronized mineral talc has been shown effective in binding the water in olives with high moisture content. As the moisture content in the olives is bound by talc, there is less formation of emulsion. As a result, less oil droplets are trapped in the emulsion. Consequently, more oil can be extracted, leading to greater olive oil extraction efficiency. The concentration of talc commonly applied is about 1-2 % of the paste. The extracted oil is clearer and has similar qualities to the non-treated olive oil [1]. Addition of talc in a concentration of 2 and 4 % can breakdown emulsion, release the oil droplets into the system and improve the extraction yield by 2 and 5 %, respectively [1]. Application of talc is approved in Spain but not in Australia yet. To the best of the author's knowledge, the effect of long term consumption of olive oil produced with talc remains unavailable and the long-term management of its by-products may pose environmental hazard.

Enzymes

The cell wall of olive fruits consists of mainly pectin, cellulose and hemicellulose [19]. In order to effectively release the oil vacuoles kept in different layers in the olive cells, it is necessary to rupture each of the pectin, cellulose and hemicellulose layers. Physical cell rupture of the olive fruits is inefficient in breaking down these cell layers, as evident from the relatively low oil recovery from the mechanical extractions. Therefore, it is necessary to use an alternate processing technique to act on the olive fruit matrix. Natural processing aids with specific activity can target the matrix and breakdown the vacuoles of the mesocarp cells. This action will help the release of the stored oil droplets. For this purpose, enzymes such as pectinase, cellulase, hemicellulase and endopolygalacturonase may be added during the extraction of olive oil. The enzymes degrade the pectin, cellulose and hemicellulose of the olive cell wall. It is anticipated that they can promote the release of cell components (such as oil droplets and phenolic compounds) trapped in the colloidal tissues of cytoplasm [20]. Meanwhile, the enzyme endopolygalacturonase is capable of breaking down the emulsions formed during olives crushing and olive paste malaxation [16]. The quantity of "free oil" and concentration of total phenolic compounds are expected to increase in the olive oil extracted with enzymes [1].

The effect of addition of enzymes with pectolytic and cellulolytic activities to olive paste has been studied in Italy. These enzymes, particularly the ones with specific pectolytic activity effectively break down the cell structure, which can release the stored oil droplets. In addition, application of enzymes as a processing aid during the production of olive oil improves the yield of oil extracted without affecting the composition of fatty acids, sterols, aliphatic and triterpene alcohols, triterpenedialcohols and other fractions of the unsaponifiable matter of the oil [16, 20]. Another reported benefit of enzyme extraction is that they do not negatively affect the acidity level and oxidative stability of the extracted olive oil. In addition, the oil extracted with enzymes is reported to have higher concentration of volatile and phenolic compounds, with improved sensory profile [20]. The higher storage stability and greater antioxidant activity of olive oil extracted by enzyme are attributable to the higher amount of phenolic compounds extracted into the oil [21-23]. It is anticipated that enzymes, when breaking down the cell structure of olive fruits, also release the phenolic compounds into the oil. While the actual mechanism has not yet been elucidated, the outcomes have positive public health implications.

In Australia, the use of enzymes in the extraction of olive oil was first conducted by Canamasas in 2006. He applied different types of enzymes, such as pectinases and cellulases, during the extraction of olive oil from *Barnea* and *Picual* olives grown in Victoria, Australia. It is revealed that the oil yield was increased due to the enzyme treatments [24]. However, his study did not provide information on the chemical properties (particularly the concentration of bioactive phenolic compounds) and sensory profile of the extracted oil. As olive cultivars and growing sites may affect the quality of olive oil, it is possible that the observations reported by Canamasas may not be applicable to all cultivars of olive grown at different sites. It is therefore necessary to evaluate the effect of enzymes on the quality of olive oil extracted from a variety of olive cultivars and olives grown at different sites.

In response to this knowledge gap, a preliminary study was conducted in Western Australia [25]. Olives harvested from two different sites were extracted using Viscozymes, a natural mix of cell wall degrading enzymes produced from *Aspergillusniger*. The study found that there was no interaction between growing sites and concentration of enzymes used on the quality parameters [25]. It was found that after adding the enzymes and malaxing for 30 minutes, the oil recovery, concentration of total phenolic compounds and antiradical activity of enzymes extracted olive oil were all significantly greater than those of the control. Details of the findings can be found on the original article published in the International Journal of Food Science and Technology [25]. Enzymes assisted extraction is a promising natural technique that can produce great amount of bioactive olive oil. Nevertheless, its long term public health implications need to be monitored.

Citric Acid

Another relatively new natural processing aid that has yet been fully investigated is citric acid. Citric acid as a chelating agent may prevent the loss of phenolic compounds during the extraction of olive oil. It can deactivate polyphenol oxidase by chelating the copper metal in the structure of polyphenol oxidase and scavenging the available oxygen radicals [26]. As the breakdown of phenolic compounds is inhibited, the use of citric acid during olive oil processing is anticipated to maximise retention of the total phenolic compounds in the oil phase, thus improving its health profile.

This speculation was previously tested on Italian and French olive oils [27, 28]. Aliakbarian et al. reported that, according to their model produced using response surface methodology, the addition of a high concentration (30% w/v) of aqueous citric acid at 13.79 mL kg−1 to olive paste during the extraction of Italian Coratina olive oil resulted in improvement in the oil recovery, concentration of total phenolic compounds and antiradical activity [27]. In another study conducted in Western Australia, addition of 30% (w/v) aqueous food-grade citric acid at 1 : 1000 (v/w) prior to the 30 min malaxation period significantly improves the oil recovery, concentration of total phenolic compounds and antiradical activity by 46, 120 and 32% respectively, when compared to the control sample [18]. The same approach was replicated with higher concentration of citric acid and the data showed increasing trend of oil recovery, concentration of total phenolic compounds and antiradical activity (unpublished data). It is anticipated that the citric acid particles assisted the breakdown of the cell structure at the microscopic level. Further research is necessary to confirm this assumption.

The use of food-grade citric acid means that it is safe for human consumption. The widely available food-grade citric acid can be easily applied by many local olive oil producers around the world. As premium virgin olive oil is available locally, consumers can choose to buy locals thus reducing the environmental impact from overseas-imported olive oil. Another public health implication from these new processing techniques includes less wastage and environment pollution as more phenolic compounds are retained in the oil instead of being left in the by-product. In addition to meeting the International Olive Council standard for 'olive oil' [18], this promising methods delivers high bioactivity that can promote good health amongst the consumers. Further studies can be conducted to establish the optimum concentration of citric acid necessary for the extraction of premium quality olive oil. The health impact of consuming citric acid extracted olive oil can also be investigated by longitudinal epidemiological studies.

HEALTH BENEFITS OF OLIVE OIL AND ITS PUBLIC HEALTH IMPLICATIONS

The benefits of incorporating olive oil into the diet have been well recognized. Regular consumption of olive oil reduce the risk of cardiovascular diseases, metabolic diseases and cancers. The protective effects of olive oil against cardiovascular diseases are due to the uniquely high monounsaturated fatty acid (MUFA) to polyunsaturated fatty acid (PUFA) ratio as well as the presence of bioactive unsaponifiable phenolic compounds in olive oil [2, 29]. These health beneficial compounds are not found in other widely consumed vegetable oils, such as soybean, peanut and coconut oil. The phenolic compounds in olive oil are strong antioxidants which are effective in scavenging free radicals. The antioxidant effect of olive oil prevents the free radicals from initiating oxidation process. As a result, it minimizes the occurrence of metabolic diseases, including atherosclerosis, arthritis, cell mutation and cancer [2].

Cardiovascular Diseases and Metabolic Syndrome

The low incidence of cardiovascular disease and cancer in people living in the Mediterranean region has raised the involvement of their diet in such reductions. It is believed that the Mediterranean diet, which consists of 50 mL of extra virgin olive oil per day, has protective effect on human health [30]. Their study recruited 12 men with a body mass index of 22. After two hours of ingestion, there was a significant reduction in inflammation. In addition, the plasma antioxidant status was increased. The study speculated that long term consumption of olive oil with high concentration of phenolic compounds could prevent atherosclerosis.

In addition to the bioactive phenolic compounds, the high MUFA to PUFA ratio in olive oil also contributes to the health profile of olive oil. Over a three-week trial on 200 men over six research centres in five European countries, consumption of 25 mL olive oil per day was found to increase the level of high density lipoprotein and decrease the ratio of total cholesterol: high density lipoprotein as well as the concentration of triglycerides in circulation. Although the mechanism by which these benefits occur is not clear, it is thought to be due to the high MUFA to PUFA ratio in olive oil [29, 31]. When olive oil with a higher concentration of phenolic compounds was consumed, the positive health benefits was more apparent with a lower level of oxidized lipoproteins found amongst the study participants [32]. The study confirmed that daily consumption of 25 mL olive oil, in particular olive oil with high level of phenolic compounds (336 mg/kg), was associated with a reduced risk of heart diseases. The results were in agreement with the findings of Tuck and Hayball (2002), Grignaffini et al. (1994) and Salami et al. (1995). Based on their studies, it is suggested that phenolic compounds, particularly hydroxytyrosol, of olive oil inhibits oxidation of low density lipoprotein cholesterol and thus reduces the formation of atherosclerotic plaques and the risk of occurrence of coronary heart disease [2, 33, 34].

In addition to the cardiovascular protective effects of olive oil, consumption of olive oil reduces other aspects of the metabolic syndrome such as obesity and diabetes [29]. Metabolic syndrome is reported to be directly related to the risk of cardiovascular diseases [35]. In a two-year cohort study, consumption of eight grams of olive oil per day was found to significantly affect several aspects of the metabolic syndromes, such as reducing body weight and abdominal obesity, decreasing insulin resistance and improving endothelial function in 90 adults [36]. The outcome suggested that consumption of olive oil could reduce the occurrence of metabolic syndrome and therefore also the risk of cardiovascular diseases.

Olive oil also plays a role in weight management as it encourages energy expenditure after consumption [6, 37]. As suggested by Soares (2010), after consuming olive oil, more of this fat is used in the oxidative pathway and less is stored. A review of recent clinical trials also mentioned that there is no weight gain among the users [38].

Results from other studies have also suggested that olive oil has the ability to prevent age-related cognitive decline and Alzheimer's disease [39, 40]. In a study involving 8028 subjects aged 65 years old and above over a four-year period, they found that consumption of olive oil improves the visual memory of these subjects [39]. The protective effects of olive oil against cardiovascular diseases, metabolic syndrome, weight management and cognitive functions are well-documented. Therefore, dieticians encourage regular inclusion of olive oil as a source of lipid in the diet [32, 41]. Whilst consumption of olive oil can improve the health of the population, public health practitioners need to communicate this message carefully. Olive oil is a source of fat with considerable high calorie. Consumption of olive oil should be used when substituting the other types of fat. Should it be added to a diet without lowering the overall energy intake, active participation in physical activity is necessary to increase energy output.

Cancers

The antioxidant activities exerted by the phenolic compounds in olive oil can reduce the occurrence of cancer [2]. Indeed, as summarized by Escrish et al. (2006) using the epidemiological data collected from case-control and cohort studies, there was a strong association between cancer and consumption of olive oil. The phenolic compounds in olive oil were found effective in preventing damage to the DNA and initiation of cancerous cells *in vitro*. In this regard, olive oil consumption could reduce the occurrence and progression of breast, ovarian, prostate, colon and stomach cancers [29, 42-45].

In addition to the epidemiological data, *in vitro* studies were also conducted to understand the protective mechanism exerted by phenolic compounds in olive oil on cancers. Menendez et al. (2008) studied the anticarcinogenic effect of the isolated single phenols (hydroxytyrosol and tyrosol), polyphenol acid (elenolic acid), lignans ((+)-pinoresinol and 1-(+)-acetoxypinoresinol) and secoiridoids (deacetoxyoleuropeinaglycone, ligstrosideaglycone, and oleuropeinaglycone) in olive oil on cultured human breast cell lines. Of all the tested fractions, (+)-pinoresinol, 1-(+)-acetoxypinoresinol, deacetoxyoleuropeinaglycone, ligstrosideaglycone, and oleuropeinaglycone were found to induce strong tumoricidal effects. The breast cancer cell proliferation and survival were significantly prevented as the protein expression was inhibited [46]. The study highlighted the reversible effect on breast cancer by the phenolic compounds in olive oil, the oleuropeinaglycone.

Messages to the Public

As illustrated above, the quality of virgin olive oil has public health implications. The bioactive compounds, at the required concentrations and combinations, are necessary to deliver the designated health effects. Sadly, the valuable virgin olive oil is often a subject for adulteration as oil producers blend it with other types of oil, or from refined olive oil or pomace olive oil to maximize their profits [6, 47]. In addition to the fall of the unique concentration and combination of bioactive compounds found in virgin olive oil, the consequence of this fraud raises the food safety alarm. Consumption of the toxic polycyclic aromatic hydrocarbons in the refined or pomace olive oil, as Aued-Pimentel et al. (2013) stated, is a serious public health issue [47]. Authenticity detection technology is required to prevent this dishonest practice especially when the quality of virgin olive oil is paramount to the health benefits it can deliver to its consumers. While the food technologists work on developing new technologies, efforts are also necessary to educate the public about the various types of olive oil available in the market. The positive health outcomes delivered by only the virgin olive oil needs to be communicated clearly to the consumers. Without this, increased consumption of ordinary olive oil may only increase their calorie intake instead of lowering their risk of cardiovascular diseases, metabolic syndrome and cancers.

As the health benefits of olive oil are evident from many well-conducted studies, health claims about positive blood lipids profile and protection against oxidative stress are now allowed on the packaging [48]. Consumers need to understand that it is the combined actions of the various bioactive compounds, rather than an individual component isolated from olive oil, that delivers the greater health outcomes [6].

While the public enjoys olive oil, rather than popping pills of health extracts, they should note that olive oil is a type of fat. All types of fat are energy-densed. The intake of olive oil, like many other types of fat, should not exceed the daily recommended intake. Olive oil can be used to replace or substitute a certain percentage of the saturated fats but should not be used as an added fat source [48]. Excessive dietary fat intake is linked to weight gain and obesity, which are well-established risk factors to many adverse health outcomes. Public health practitioners need to employ the bottom-up approach and work with the community to ensure that the message is communicated carefully.

An article stated the influence of cooking shows in changing the dietary pattern of the general public [49]. With increasing number of food programs going on air, chefs can be some of the best candidates to educate the public about healthy cooking. The use of butter in cooking can be reduced or be replaced by olive oil. In addition, consumers who are concerned about the grassy pungent flavour of virgin olive oil maybe introduced to the late harvest fruity virgin olive oil, which are milder in flavour. While refined olive oil or pomace olive oil are more bland in flavor, consumers should be discouraged from using these oil. The refined or pomace olive oil had undergone chemical refinery processes and contains toxic and lacks sufficient concentration of bioactive compounds. It is unable to deliver the designated health benefits of virgin olive oil nor to promote public health.

CONCLUSION

This chapter revealed that natural processing aids are effective in increasing the efficiency of olive oil production and its bioactivity. The natural occurring enzymes and citric acids are food-grade. Increased consumption of these bioactive olive oil in replacement of other saturated fats, are expected to improve the physical health of the general public.

Looking into the future, the olive oil producers, food technologists, chemists and public health practitioners can work together in the following areas:

- Monitor and control adulteration or blending of premium bioactive olive oil and low grade (refined or pomace) olive oil.
- Investigate the bioactivity profile and environmental impact of the by-products extracted from processing aids-extracted oil
- Evaluate long-term health effects of consuming processing aids-extracted olive oil through well-designed longitudinal epidemiological studies
- Investigate the barriers to uptake of locally-produced olive oil
- Educate the general public about the health benefits of olive oil and reading the health claims
- Design, implement and evaluate bottom-up community-based campaigns that promote the use of virgin olive oil in the public's diet while replacing their other sources of saturated fats

With these aims in mind, local virgin olive oil production can be made sustainable to produce sufficient amount of premium bioactive olive oil to promote good health among the consumers and lower the environmental impact to the public.

REFERENCES

[1]   Di Giovacchino (2000) Technological aspects. *Handbook of olive oil: analysis and properties*, eds Harwood J & Aparicio R (Aspen Publication, Maryland).
[2]   Tuck KL & Hayball PJ (2002) Major phenolic compounds in olive oil: metabolism and health effects. *Journal of Nutritional Biochemistry* 13(11):636-644.

[3]   Alburquerque JA, Gonzálvez J, García D, & Cegarra J (2004) Agrochemical characterisation of "alperujo", a solid by-product of the two-phase centrifugation method for olive oil extraction. *Bioresource Technology* 91(2):195-200.

[4]   Alfano G, Belli C, Lustrato G, & Ranalli G (2008) Pile composting of two-phase centrifuged olive husk residues: Technical solutions and quality of cured compost. *Bioresource Technology* 99(11):4694-4701.

[5]   Gimeno E, Castellote AI, Lamuela-Raventós RM, De la Torre MC, & López-Sabater MC (2002) The effects of harvest and extraction methods on the antioxidant content (phenolics, [alpha]-tocopherol, and [beta]-carotene) in virgin olive oil. *Food Chemistry* 78(2):207-211.

[6]   García-González DL & Aparicio Rn (2010) Research in Olive Oil: Challenges for the Near Future. *Journal of Agricultural and Food Chemistry* 58(24):12569-12577.

[7]   Ranalli A, Malfatti A, Lucera L, Contento S, & Sotiriou E (2005) Effects of processing techniques on the natural colourings and the other functional constituents in virgin olive oil. *Food Research International* 38(8-9):873-878.

[8]   Artajo, Romero MP, & Motilva MJ (2006) Transfer of phenolic compounds during olive oil extraction in relation to ripening stage of the fruit. *Journal of the Science of Food and Agriculture* 86(4):518-527.

[9]   Vlyssides AG, Loizides M, & Karlis PK (2004) Integrated strategic approach for reusing olive oil extraction by-products. *Journal of Cleaner Production* 12(6):603-611.

[10]   Rodis PS, Karathanos VT, & Mantzavinou A (2002) Partitioning of olive oil antioxidants between oil and water phases. *Journal of Agricultural and Food Chemistry* 50(3):596-601.

[11]   Bendini A, Cerretani L, Carrasco-Pancorbo A, Gómez-Caravaca A, Segura-Carretero A, Fernández-Gutiérrez A, & Lercker G (2007) Phenolic Molecules in Virgin Olive Oils: a Survey of Their Sensory Properties, Health Effects, Antioxidant Activity and Analytical Methods. An Overview of the Last Decade Alessandra. *Molecules* 12(8):1679-1719.

[12]   Gutiérrez F, Arnaud T, & Garrido A (2001) Contribution of polyphenols to the oxidative stability of virgin olive oil. *Journal of the Science of Food and Agriculture* 81(15):1463-1470.

[13]   Morales MT & Tsimidou M (2000) The role of volatile compounds and polyphenols in olive oil sensory quality. *Handbook of olive oil: analysis and properties*, eds Harwood J & Aparicio R (Aspen Publisher, Maryland), pp 393-458.

[14]   Servili M, Esposto S, Fabiani R, Urbani S, Taticchi A, Mariucci F, . . . Montedoro GF (2009) Phenolic compounds in olive oil: antioxidant, health and organoleptic activities according to their chemical structure. *Inflammopharmacol* 17(2):76-84.

[15]   AOA (2002) *Research and Development Plan for the Australian Olive Industry 2003-2008* (Australian Olive Association).

[16]   Di Giovacchino, Sestili S, & Di Vincenzo D (2002) Influence of olive processing on virgin olive oil quality. *European Journal of Lipid Science and Technology* 104(9-10):587-601.

[17]   Kalua C, Bedgood D, Bishop A, & Prenzler P (2006) Changes in volatile and phenolic compounds with malaxation time and temperature during virgin olive oil production. *Journal of Agriculture Food Chemistry* 54:7641-7651.

[18] Chih H, James AP, Jayasena V, & Dhaliwal SS (2013) Effect of growing location, malaxation duration and citric acid treatment on the quality of olive oil. *Journal of the Science of Food and Agriculture* 93:1272-1277.

[19] Najafian L, Ghodsvali A, Haddad Khodaparast MH, & Diosady LL (2009) Aqueous extraction of virgin olive oil using industrial enzymes. *Food Research International* 42(1):171-175.

[20] Chiacchierini E, Mele G, Restuccia D, & Vinci G (2007) Impact evaluation of innovative and sustainable extraction technologies on olive oil quality. *Trends in Food Science & Technology* 18(6):299-305.

[21] Servili M & Montedoro G (2002) Contribution of phenolic compounds to virgin olive oil quality. *European Journal of Lipid Science and Technology* 104(9-10):602-613.

[22] Vierhuis E, Servili M, Baldioli M, Schols HA, Voragen AGJ, & Montedoro G (2001) Effect of enzyme treatment during mechanical extraction of olive oil on phenolic compounds and polysaccharides. *Journal of Agricultural and Food Chemistry* 49(3):1218-1223.

[23] Ranalli A & De Mattia G (1997) Characterization of olive oil production with a new enzyme processing aid. *Journal of American Oil Chemists Society* 74:1105-1113.

[24] Canamasas P (2006) Use of coadjuvants in olive oil extraction. in *Australian and New Zealand Olivegrower and Processor*, pp 29-32.

[25] Chih H, James AP, Jayasena V, & Dhaliwal SS (2012) Addition of enzymes complex during olive oil extraction improves oil recovery and bioactivity of Western Australian Frantoio olive oil. *International Journal of Food Science and Technology* 47:1222-1228.

[26] Choe E & Min D (2006) Mechanisms and factors for edible oil oxidation. *Comprehensive Review of Food Science and Food Safety* 5:169-186.

[27] Aliakbarian B, Dehghani F, & Perego P (2009) The effect of citric acid on the phenolic contents of olive oil. *Food Chemistry* 116:617-623.

[28] Gallina-Toschi T, Cerretani L, Bendini A, Bonoli-Carbognin M, & Lercker G (2005) Oxidative stability and phenolic content of virgin olive oil: an analytical approach by traditional and high resolution techniques. *Journal of Separation Science* 28:859-870.

[29] Medeiros DM & Hampton M (2007) Olive oil and Health Benefits *Handbook of Nutraceuticals and Functional Foods* ed Wildman REC (CRC Press Boca Raton), 2nd edn Ed, pp pp. 297-308.

[30] Bogani P, Galli C, Villa M, & Visioli F (2007) Postprandial anti-inflammatory and antioxidant effects of extra virgin olive oil. *Atherosclerosis* 190(1):181-186.

[31] Covas MI, Nyyssonen K, Poulsen HE, Kaikkonen J, Zunft HJF, Kiesewetter H, . . . Marrugat J (2006) The effect of polyphenols in olive oil on heart disease risk factors - A randomized trial. *Annals of Internal Medicine* 145(5):333-341.

[32] Covas MI, Ruiz-Gutierrez V, de la Torre R, Kafatos A, Lamuela-Raventos RM, Osada J, . . . Visioli F (2006) Minor components of olive oil: Evidence to date of health benefits in humans. *Nutrition Reviews* 64(10):S20-S30.

[33] Grignaffini P, Roma P, Galli C, & Catapano AL (1994) Protection of low-density lipoprotein from oxidation by 3,4-dihydroxyphenylethanol. *Lancet* 343(8908):1296-1297.

[34] Salami M, Galli C, De Angelis L, & Visioli F (1995) Formation of F2-isoprostanes in oxidized low density lipoprotein: Inhibitory effect of hydroxytyrosol. *Pharmacological Research* 31(5):275-279.

[35] Alessi MC & Juhan-Vague I (2006) PAI-1 and the metabolic syndrome - Links, causes, and consequences. *Arteriosclerosis Thrombosis and Vascular Biology* 26(10):2200-2207.

[36] Esposito K, Marfella R, Ciotola M, Di Palo C, Giugliano F, Giugliano G, . . . Giugliano D (2004) Effect of a Mediterranean-Style Diet on Endothelial Dysfunction and Markers of Vascular Inflammation in the Metabolic Syndrome: A Randomized Trial. *Journal of the American Medical Association* 292(12):1440-1446.

[37] Soares MJ (2010) *The Effect of Olive Oil on Postprandial Thermogenesis, Fat Oxidation and Satiety: Potential Implications for Weight Control* (Oxford: Academic Press).

[38] Bes-Rastrollo M, Soares MJ, & Martinez-Gonzalez MA (2010) Olive Oil Consumption and Weight Gain. *Olives and Olive Oil in Health and Disease Prevention*, eds Preedy VR & Watson RR (Oxford: Academic Press), pp 895-902.

[39] Berr C, Portet F, Carriere I, Akbaraly TN, Feart C, Gourlet V, . . . Ritchie K (2009) Olive oil and cognition: Results from the three-city study. *Dementia and Geriatric Cognitive Disorders* 28(4):357-364.

[40] Burgener SC, Buettner L, Buckwalter KC, Beattie E, Bossen AL, Fick DM, . . . McKenzie S (2008) Evidence supporting nutritional interventions for persons in early stage Alzheimer's disease (AD). *J Nutr Health Aging* 12(1):18-21.

[41] Perez-Jimenez F, Ruano J, Perez-Martinez P, Lopez-Segura F, & Lopez-Miranda J (2007) The influence of olive oil on human health: Not a question of fat alone. *Molecular Nutrition & Food Research* 51:1199-1208.

[42] Escrish E, Ramirez-Tortosa C, Sanchez-Rovira P, Colomer R, Solanas M, & Gaforio JJ (2006) Olive oil in cancer prevention and progression. *Nutrition Reviews* 64(10):S40-52.

[43] Sotiroudis TG & Kyrtopoulos SA (2008) Anticarcinogenic compounds of olive oil and related biomarkers. *European Journal of Nutrition* 47(S2):69-72.

[44] Menendez JA, Papadimitropoulou A, Vellon L, & Lupu R (2006) A genomic explanation connecting "Mediterranean diet", olive oil and cancer: Oleic acid, the main monounsaturated Fatty acid of olive oil, induces formation of inhibitory "PEA3 transcription factor-PEA3 DNA binding site" complexes at the Her-2/neu (erbB-2) oncogene promoter in breast, ovarian and stomach cancer cells. *European Journal of Cancer* 42(15):2425-2432.

[45] Galeone C, Talamini R, Levi F, Pelucchi C, Negri E, Giacosa A, . . . La Vecchia C (2007) Fried foods, olive oil and colorectal cancer. *Annals of Oncology* 18:36-39.

[46] Menendez JA, Vazquez-Martin A, Garcia-Villalba R, Carrasco-Pancorbo A, Oliveras-Ferraros C, Fernandez-Gutierrez A, & Segura-Carretero A (2008) tabAnti-HER2 (erbB-2) oncogene effects of phenolic compounds directly isolated from commercial Extra-Virgin Olive Oil (EVOO). *BMC Cancer* 8:377-399.

[47] Aued-Pimentel S, Silva SAD, Takemoto E, & Cano CB (2013) Stigmastadiene and specific extitntion (270 nm) to evaluate the presence of refined oils in virgin olive oil commercialized in Brazil. *Food Science and Technology (Campinas)* 33:479-484.

[48] Martín-Peláez S, Covas MI, Fitó M, Kušar A, & Pravst I (2013) Health effects of olive oil polyphenols: Recent advances and possibilities for the use of health claims. *Molecular Nutrition & Food Research* 57(5):760-771.

[49] Phillipov M (2013) Mastering obesity: MasterChef Australia and the resistance to public health nutrition. *Media, Culture & Society* 35(4):506-515.

In: Virgin Olive Oil
Editor: Antonella De Leonardis

ISBN: 978-1-63117-656-2
© 2014 Nova Science Publishers, Inc.

Chapter 3

ANALYSIS THROUGH GRAPHICAL KNOWLEDGE OF HISTORICAL OLIVE OIL PRODUCTION OBTAINED FROM MECHANICAL PROCEDURES: APPLICATION TO THE INVENTION PRIVILEGES AND PATENTS FROM THE HISTORICAL ARCHIVE OF THE SPANISH OFFICE OF PATENTS AND TRADEMARKS

José Ignacio Rojas-Sola and *Miguel Castro-García*
University of Jaén, Department of Engineering Graphics, Design and Projects,
Campus de las Lagunillas, Spain

ABSTRACT

Olive oil extraction is one of the most historical processes within the global food industry. The evolution of technology has depended on the dominant civilization and the importance that has been given to this product. An example to illustrate this is Monte Testaccio in Rome (Italy) where the remains of 26 million amphorae from olive oil transportation can be found.

Industrial archaeology as a discipline includes the study of industrial age remains and proto-industrialization. In this research, the application of this discipline is illustrated to the analysis of the historical olive oil production by mechanical procedures from the second half of the nineteenth century until the mid-twentieth century in Spain thanks to the invention privileges and patents implemented from the Historical Archive of Spanish Office of Patents and Trademarks.

Industrial archaeology studies conducted on historical olive oil production are often based on aspects of industrial operation performance. However, this chapter presents a novel methodology which is able to study the evolution from graphic information from technical documents over time. For this, it uses the documentation of implemented

---

* Corresponding author. Professor José Ignacio Rojas-Sola, Ph.D. University of Jaén, Department of Engineering Graphics, Design and Projects, Campus de las Lagunillas, s/n, Jaén 23071. Spain. Tel: +34-953-212452; Fax: +34-953-212334; Email: jirojas@ujaen.es

inventions from the Historical Archive of the Spanish Office of Patents and Trademarks in the period 1826-1966.

In general, the knowledge of any activity can be classified into two groups: explicit and tacit knowledge. Explicit knowledge is embodied in texts and, as discussed in this case, in graphics. However, the quantification of tacit knowledge is much more complex, having to resort to other types of documents and extrapolating from historical time.

Thus, this chapter will analyze the graphic documentation from the privileges and patents implemented to achieve a measurement of the knowledge contained. This analysis will provide an insight into the evolution of knowledge on historical olive oil production by mechanical procedures and, at the same time, specify which historical periods contributed to greater technological growth in this field.

In conclusion, a new methodology is obtained which is capable of quantifying the knowledge provided by graphic documentation, studying the evolution within a technological field and extrapolating future trends.

Keywords: Olive Oil, Graphical Knowledge, Industrial Archaeology, Historical Archive of the Spanish Office of Patents and Trademarks

INTRODUCTION

The family known as *oleaceae* has 29 genera and 600 species, among which is *Olea europaea L.*, commonly known as the olive. It is a species adapted to the Mediterranean climate, and has accompanied Mediterranean peoples throughout history [1]. Its cultivation arose in the Middle East, in areas such as the Iranian plateau or the coast of Syria and Palestine, over 3,000 years ago, and extended over the whole of the Mediterranean area.

Over time, the olive has gained importance as food and has marked the culture of different civilizations, reflecting its mark on the landscape through olive plantations. Now therefore there is a rich cultural heritage of this plant species, both in the form of tangible and intangible assets and the relationship between the two.

One of the most important parts of this tangible heritage is formed by olive mills, where olive oil is extracted. These facilities were designed to address the two stages of this process: milling the fruit and pressing the olive paste. The historical evolution of these stages has been constant over time, and has progressed in parallel with the most up-to-date technology of the time. However, these changes in existing designs or in the extraction technique were usually transmitted through the general knowledge of those involved, and often do not appear in historical writings or documents.

Nevertheless, in Spain, before the Industrial Revolution and from historical period of the seventeenth century, those who proposed mechanical optimizations or changes in the olive production system expressed their innovations under the name of invention privileges in a first stage, and invention patents in a second stage, in order to obtain exclusive industrial use.

These inventions are very important because at this time, Spain was a world leader in the production of olive oil. Consequently, Spain experienced an active innovation process, accompanied also by new ideas from other countries, such as Italy or France, in the form of introduction certificates. At the same time, this historical period was also that of greatest technological change when compared with modern techniques, as it was the seed of the new and current production system.

That is why from the seventeenth century onwards, the necessary conditions exist to enable the study of knowledge brought by one of the inventions of an advanced country like Spain in the production of olive oil in its most revolutionary technological phase from the perspective of production techniques. In this sense, this chapter aims to respond to the quantification of knowledge applied to the extraction of olive oil from different types of pressing within a historical period by applying an appropriate methodology for study.

METHODOLOGY

Knowledge Management

The research presented in this chapter is based on the knowledge base given by the inventions in the Historical Archive of Spanish Office of Patents and Trademarks of the Spanish Ministry of Industry, Tourism and Commerce. However, using this information creates certain problems in the study: firstly, the quantification of knowledge and secondly, the historical component of the research, as the inventors no longer exist.

Specifically, for the quantification of knowledge of the invention records we have used methodology based on Knowledge Management (KM), used in existing knowledge systems in companies today. Its purpose is to locate the knowledge capital of the company and how to harness it to so that it can be used by human resources. To this end, we designed the working methodology of this chapter in order to know the historical evolution of olive pressing.

Knowledge is divided into tacit and explicit knowledge. The concept of tacit knowledge first appears in work of the philosopher and scientist Michael Polanyi in 1966 [2]. It is knowledge that is possessed by a person and which has technical and cognitive dimensions, for example, know-how, value systems or beliefs, among others, promoting creativity and innovation. In contrast, explicit knowledge is that which can be expressed, transmitted and stored using formal systemic language [3].

Based on these two types of knowledge the SECI model was proposed in 1995, which specifies the possibility of creating any kind of knowledge from prior knowledge [4], with four knowledge conversion processes that make up this model: Socialization, Externalization, Combination and Internalization [5, 6] whose relationship is expressed schematically in Figure 1.

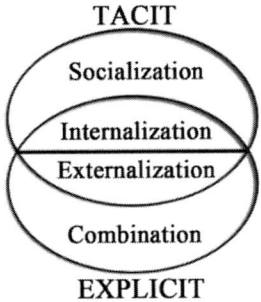

Figure 1. SECI model.

Once the classes of knowledge and their interactions have been defined, it can be quantified according to each invention patent.

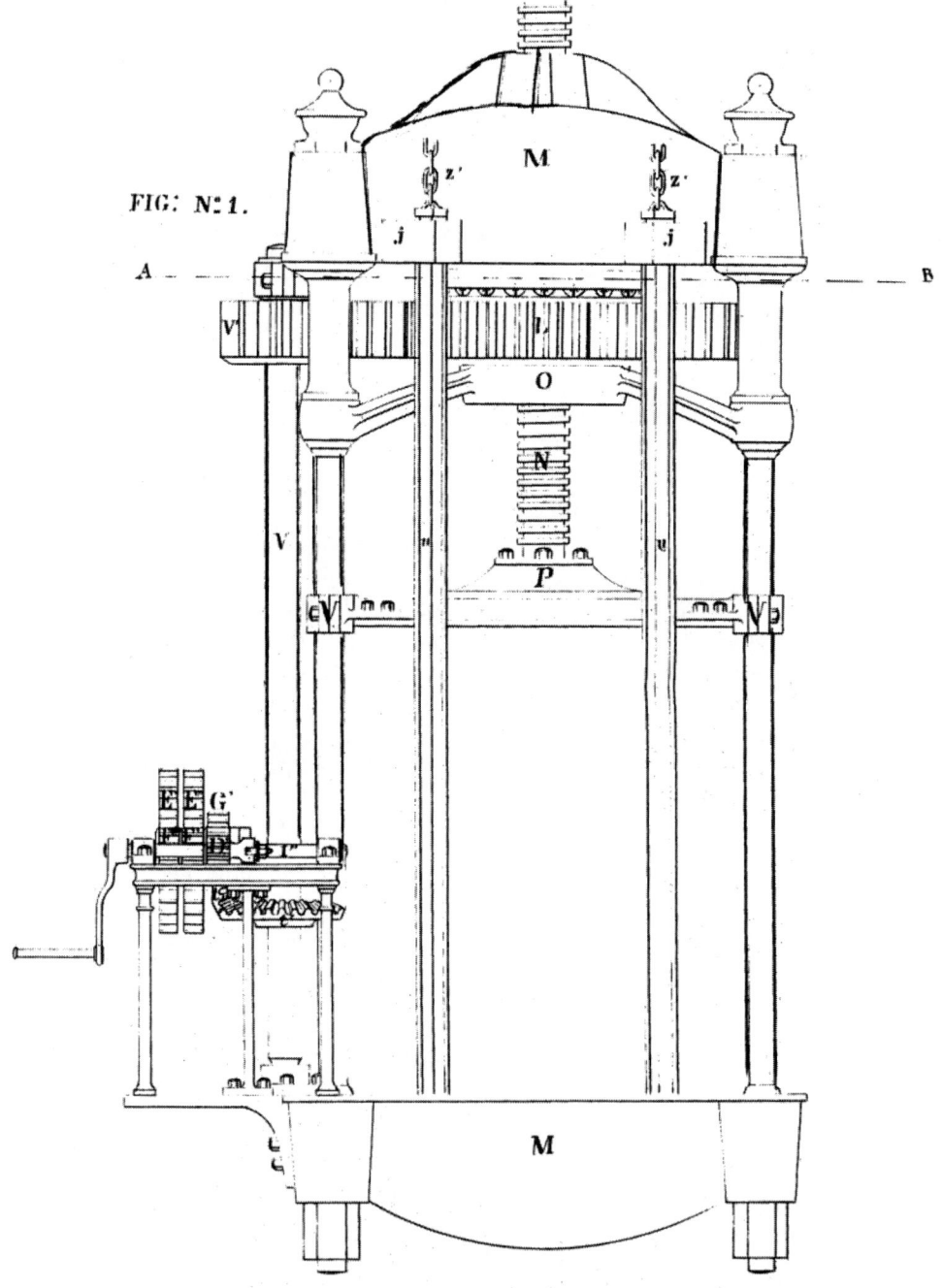

Figure 2. Historical Archive of Spanish Office of Patents and Trademarks of the Spanish Ministry of Industry, Tourism and Commerce. Record n° 1668.

Invention Privileges and Patents

The first recorded technological idea in Spain dates from 1522, when King Carlos V granted the right to operate for a fixed period or for life. This was known by the name of Royal Warrant of Invention Privilege, which was simplified to invention privilege in 1826; there were a total of 5,113 cases by the end of 1826. In 1878, a decree again changed the name of the invention privileges to invention patents, maintaining at all times the exploitation and protection of exclusivity of the inventor and introducing minor nuances with each name change.

The study of invention patents made in this research covers the period from 1826 to 1966. These patents are stored in the Historical Archive of Spanish Office of Patents and Trademarks of the Spanish Ministry of Industry, Tourism and Commerce, which contains all invention privileges and patents related to the production of olive oil. We found 209 patents, of which 40 were put into practice; this information was gathered from the specialized database made by the research team of José Patricio Sáiz González of the Universidad Autónoma de Madrid [7]. Breaking down these figures, we found that 19 were invention privileges dating from 1826 to 1878, and the other 21 were invention patents from the period between 1878 and 1940. As an example, Figure 2 shows part of the graphic representation of invention privilege number 1668, which corresponds to the plans of a press for olives, grapes and other fruit proposed by Julio Parizot en 1857.

Quantification of Knowledge

As discussed above, the measurement of the knowledge gained from each record is subject to the historical nature of the investigation. From this we infer the impossibility of evaluating certain aspects such as the individual's cognitive ability. In addition, it should be stressed that the knowledge generated at a given time can be used at that time in history, or be dilated in time. However, the big advantage of the historical component of the investigation is that technological evolution is previously known, which allows us to order developments qualitatively, by comparing the technology analyzed with that which currently exists.

Thus, we study the 40 invention records over 140 years. These records correspond to multiple purposes, for example, record number 1082 was a method for separating olive stones and extracting olive oil invented by Francisco Sanmartin in 1853, record number 4462 is an improved apparatus which accelerates olive pressing in order to make cost savings invented by Francisco Cortes Canto in 1868, or patent number 139030 consists of a series of improvements in continuous presses for olives and other fruits invented by Luis Feliu Vallespinosa in 1935.

Therefore there are various topics that deal with records relating to processing olives for olive oil extraction. To normalize this problem, it is necessary to establish a set of variables that are able to study uniformly all records and collate all the knowledge they contain. Consequently, we propose five study variables: 'p' for pressing oil paste; 'm' for the type of milling; 'e' for energy source; 'd' for standard innovations; and 's' for safety measures for both workers and the integrity of the machinery. The breakdown and study of this knowledge has previously been done carried out [8].

Specifically, the quantification of knowledge for a given total time i and a study variable x has the expression K_i^x. This total value is determined by the previous value of the knowledge gained from the last record analyzed in terms of two coefficients that express the value corresponding to explicit and tacit knowledge (Eq. 1)

$$K_i^x = K_{i-1}^x(\Phi_i^x \cdot \sigma_i^x) \tag{1}$$

where:

K_{i-1}^x is the knowledge value of x before time i.

Φ_i^x is the coefficient of explicit knowledge created by the invention record of variable x at moment i.

σ_i^x is the coefficient of tacit knowledge of the same record.

In terms of the explicit knowledge coefficient, an in-depth knowledge of the elements under analysis is necessary, in this case the extraction of olive oil using traditional methods. Its value will depend on its importance from a historical perspective, in other words, the prior knowledge of the technological evolution in the period under study.

The coefficient of tacit knowledge is related to the transformations produced by socialization and internalization, which are complex to quantify because there is no method to accurately measure the amount of knowledge a person possesses, and it would be impossible to speak to the people involved. However, thanks to the educational level of the time and existing documentation, it follows that the new tacit knowledge was achieved thanks to the knowledge generated from repetitive tasks that made up the working day. This type of tacit knowledge generation is quantified by Yang Xu and Alain Bernard [9] (Eq. 2).

$$\sigma_i^x = 1 - (1 - \alpha)^n \tag{2}$$

where α is cognitive potential and skills which a person connected with the production of olive oil generates, with value 0.2, and n represents the number of years. There is a saturation phenomenon relating to tacit knowledge in the sense that a person is unable to obtain new knowledge simply from carrying out their work. This saturation becomes evident in the 8th year and is complete from the 15th when the differential is close to zero.

Once the value of all the variables has been found it is necessary to find the final value for each of the invention records (K). Therefore, all the variables are weighted so that their value is in proportion to their importance in the classical process of olive oil extraction. To do this, we have the advantage of being able to compare the invention records of the 140-year period under study, and analyze the contributions of each record. This, along with the consensus of experts in the field, allows us to establish the weighting of each variable (Eq. 3).

$$K = 0{,}32p + 0{,}27m + 0{,}08e + 0{,}22d + 0{,}11s \tag{3}$$

We can see that the variable relating to pressing and milling technology is decisive, with a total weight of 59% of the total of K.

RESULTS AND DISCUSSION

The methodology described above allows us to study all invention records from the point of view of knowledge. However, the analysis of historical development focuses on the development of knowledge concerning presses and how this development has affected their production. To do this, it is necessary to filter which presses affect this type of olive oil extraction and to classify invention records according to the type of presses (Table 1).

Table 1. Invention privileges and patents according to type of press

Variables	Press type	Privilege or Patent number
t1	Steam-powered hydraulic	4573, 5459, 39849, 107068
t2	Animal-powered hydraulic	5459
t3	Manually-powered hydraulic	795, 2976, 4642, 5459
t4	Screw	1120, 1668, 2300, 2976, 4068, 5459, 5838, 3458, 6959, 43015, 44328, 86053, 139030
t5	Corner	5459
t6	Beam and Quintal	117
t7	Other methods	2587, 8937, 5459, 38782, 44184, 281169

This classification consists of 6 types of presses which were the most commonly used during the time of study, and a seventh category containing other methods for obtaining olive oil. In the latter category are important developments such as patents whose file numbers are 38782 and 44184, which describe the Acapulco system designed by the Marquis D. Miguel del Prado and Lisboa in 1906 and 1908 respectively. This development represented a dramatic change in the production of olive oil and was the seed of present techniques of a two-stage continuous extraction system with centrifugation.

In addition, the classification includes olive oil production in categories t1 to t6 during the period from 1857 to 1930, thanks to the studies carried out by Juan Francisco Zambrana Pineda for the Spanish Ministry of agriculture, fisheries and food in 1987 [10].

After processing the data, we obtain the evolution of knowledge of the various presses and this is linked to their production over the same time scale. Thus, Figure 3 shows the evolution of the knowledge of the 13 invention records whose scope is screw presses (t4), proving how this type of presses has received more than three times more improvement and innovation initiatives than the other presses. Furthermore, the contribution of knowledge has been positive, except for two years when the failed patents filed marked a technological lag from what already existed.

Relating to the production of olive oil, it must be indicated that it has oscillated with three high peaks in the time period under study. These peaks were preceded by the introduction of various records, which leads us to speculate about the direct relationship between the knowledge about this type of press and an improvement in production. However, this improvement is not due to an optimization of the process, but rather to an evolution in the machinery used, which prevented technological lag. As the historical evolution seen from the present shows, the technology in use at these times was out of date.

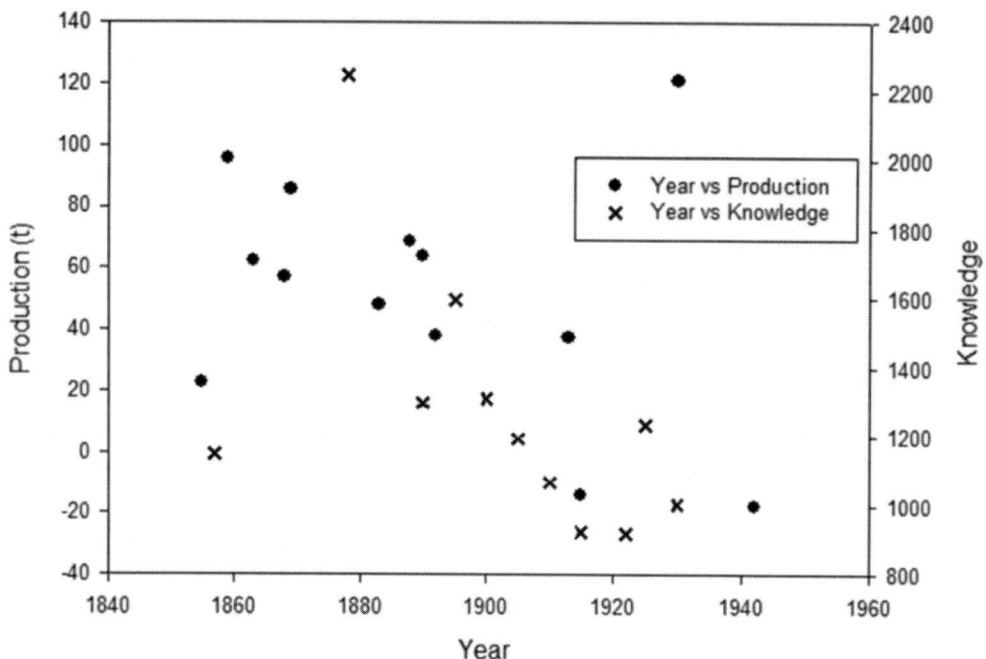

Figure 3. Evolution of knowledge about screw presses and annual production.

At the same time there are high values of the knowledge which may appear contradictory, as this is technology which has been surpassed greatly by hydraulic presses and by the Acapulco system. However, it can be verified that the new records improved olive oil extraction to such an extent that the disappearance of this technology was delayed for years.

The following results deal with steam-powered hydraulic presses (t1). These mechanisms received 4 great innovations which caused a sharp increase in the production of olive oil (Figure 4). They were well supported and had a good reputation in comparison with the continuous Acapulco system, thanks to the fact that it provided knowledge and to the political influence of the Marquis of Cabra.

In fact, the last years of the time series show a significant increase in the production of olive oil where the improvements introduced by patent number 39849 decisively affected industrial performance. Later, outside the 140-year period analyzed, the press system was surpassed by the Acapulco system when its technological advantages had been improved and proven; this was the time when the hydraulic press began its decline and subsequent disappearance.

The following press analyzed according to its historical behavior is the manually-operated hydraulic press (t3). There are 4 invention records which were implemented, a number which contrasts with other presses with a higher production, such as the animal-powered hydraulic press (t2) and the beam and quintal press (t6), which brought less to common knowledge. The reason for this is the dimension and the production potential which this manual press had, which made is suitable for domestic use or use in a small business.

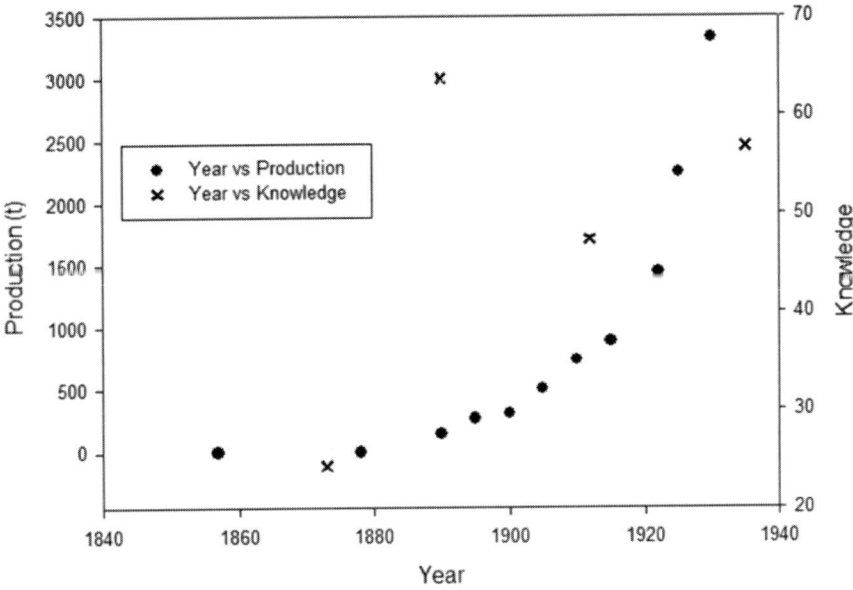

Figure 4. Evolution of steam-powered hydraulic presses and their annual production.

Production began to be regulated according to its importance 20 years after the production of steam-powered hydraulic presses. However, there is no little contribution of knowledge which produces an effect once the obsolete systems begin to be phased out (Figure 5).

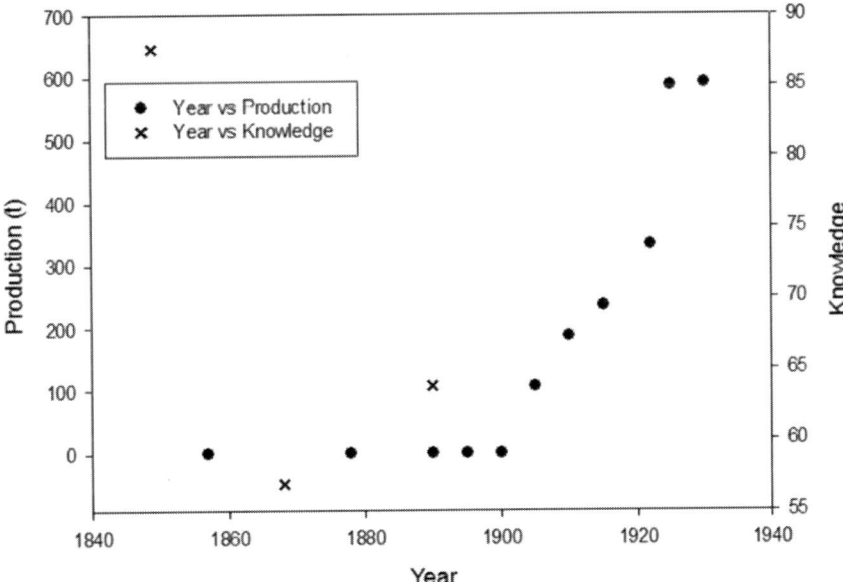

Figure 5. Evolution of the knowledge of manually-powered hydraulic presses and their annual production.

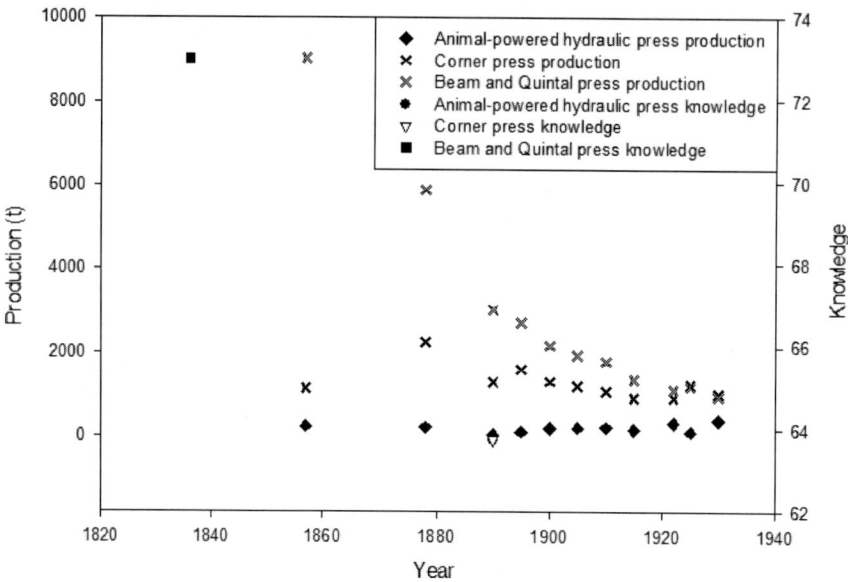

Figure 6. Evolution of the knowledge of animal-powered hydraulic presses, corner presses and beam and quintal presses and their annual production.

Lastly, we analyzed the results of the animal-powered hydraulic press (t2), the corner press (t5), and the beam and quintal press (t6). Their evolution from the perspective of knowledge and production is shown in Figure 6.

The first thing which stands out among the results obtained in the small number of invention records (3) for the three types of presses, and for this reason they were replaced as their potential was low. Types t2 and t5 share record number 5459, which affects all the others except t6, and gave a very slight improvement in their operation. Both presses show constant performance over time, suggesting that their two pressing techniques would soon become obsolete. With regard to the beam and quintal press, it shows a clear and rapid loss of production when it became obsolete, as it had been used since Roman times. It was improved by only one invention privilege in 1836, but this did not halt its decline as it did not improve its industrial performance.

CONCLUSION

Throughout history, the technical evolution of olive oil extraction processes has been progressive. However, in the period analyzed in this chapter there was a large scale change in production techniques, in the case of the beam and quintal press, and the appearance of new techniques such as the continuous Acapulco system or the steam-powered hydraulic presses. This transition generated 40 invention privileges which were put into practice, and which led to the introduction of new knowledge in olive oil extraction.

This chapter has described a methodology which is able to filter, classify and measure the knowledge of these 40 records, which is a new approach to the study of historical industrial heritage. This has allowed us to know how much knowledge was contributed by each record according to existing knowledge, and we have found that on occasions this contribution was insignificant or had negative values, which means that the reality was the lack of any innovation.

Finally, it is necessary to highlight the knowledge contributed by each of the presses in the historical production of olive oil. This comparison shows the quantitative influence of this knowledge, for example in screw presses, although on occasions the contribution of knowledge in obsolete technique did not prevent them from falling into disuse.

FUNDING

The research presented in this chapter has been carried out in the framework of the Research Project in the National Plan for Research and Development (2008-2011), entitled *"Las técnicas infográficas y la ingeniería industrial como apoyo de la historia de la tecnología convertida en herramienta museográfica para centros de interpretación y museos del aceite de oliva"* (HAR2009-06943) (Infographic techniques and industrial engineering as support for the history of technology as a museographic tool for interpretation centres and museums of olive oil), financed by the Spanish Ministry of Science and Innovation.

ACKNOWLEDGMENT

Our thanks also to the National Training Plan of University Lecturers (FPU) of the Spanish Ministry of Education, Culture and Sport.

REFERENCES

[1]   Gomez-Escalonilla Sánchez-Heredero, M., & Vidal-Hernández, J. (2006). *Variedades del olivar*. Madrid: Ministerio de Agricultura, Pesca y Alimentación.

[2]   Polanyi, M. (1983). *The Tacit Dimension*. First published Doubleday & Co, 1966. Reprinted Peter Smith, Gloucester, Mass, 1983. Chapter 1: "Tacit Knowing".

[3]   Alavi, M., & Leidner, D. E. (2001). Review: knowledge management and knowledge management systems: conceptual foundations and research issues. *MIS Quarterly, 25,* 107-136.

[4]   Nonaka, I., & Takeuchi, H. (1995). *The knowledge-creating company: how Japanese companies create the dynamics of innovation.* New York: Oxford University Press.

[5]   Jaakkola, H., Heimbürger, A., & Linna, P. (2010). Knowledge-oriented software engineering process in a multi-cultural context. *Software Quality Journal, 18,* 299-319.

[6]   Nonaka, I. (1994). A Dynamic Theory of Organizational Knowledge Creation. *Organization Science, 5,* 14-37.

[7]   Sáiz González, J. P. (2000). *Invención, patentes e innovación en España* (1759-1878). Madrid: Universidad Autónoma de Madrid.

[8]   Rojas-Sola, J. I., Castro-García, M., & Carranza-Cañadas, M. P. (2012). Contribution of historical Spanish inventions to the knowledge of olive oil industrial heritage. *Journal of Cultural Heritage, 13,* 285-292.

[9]   Xu, Y. & Benard, A. (2011). Quantifying the value of knowledge within the context of product development. *Knowledge-Based Systems, 24,* 166-175.

[10] Zambrana Pineda, J. F. (1987). *Crisis y modernización del olivar español. 1870-1930.* Madrid: Secretaría General Técnica del Ministerio de Agricultura, Pesca y Alimentación.

In: Virgin Olive Oil
Editor: Antonella De Leonardis

ISBN: 978-1-63117-656-2
© 2014 Nova Science Publishers, Inc.

Chapter 4

FATTY ACID, STEROL AND POLYPHENOL CONTENT IN MOST MONOVARIETAL OILS OF THE SPANISH EAST AREAS

Isabel López-Cortés[1], Domingo C. Salazar-García[2], Domingo M. Salazar[1] and Borja Velázquez-Martí[3]*

[1]Department of Vegetal Production, Universidad Politécnica de Valencia,
Valencia, Spain
[2]Research Group on Plant Foods in Hominin Dietary Ecology, Department of Human Evolution, Max-Planck Institute for Evolutionary Anthropology, Leipzig, Germany
[3]Department of Rural and Food Engineering, Universidad Politécnica de Valencia,
Valencia, Spain

ABSTRACT

The importance of specific sterols present in olive oil, as well as α-tocopherol and polyphenols, in the prevention of certain diseases is well known. Establishing differences in the content of these essential components on the EVOO (Extra Virgin Olive Oil) is important, and allows an evaluation of the quality and suitability of olive oil for health. In this chapter, the composition of fatty acids, sterols and polyphenols in most monovarietal oils of eastern Spain were analyzed, to quantify each of these constituents in the different varieties. This would allow suitable oils to be chosen for specific nutrition programs, or in preventive pharmacology. The specific composition of each component was obtained for each variety. The oils tested were produced in a pilot project using cold extraction, followed by centrifugation. The olives used were graded as 3-3.5 in ripeness under standard colorimetric scales, except the varieties from Blanqueta and Marfil, in which it was sought a similar ripeness time. Olives were obtained from selected trees from commerial groves, most of them several hundred years old. And some were classified as emblematic trees. Fatty acids have been studied using *n*-heptane extract to separate the fatty acid methyl esters with High Performance Liquid Chromatography (HPLC). Sterols,

---

* Corresponding author: Isabel López-Cortés, Department of Vegetal Production. Universidad Politécnica de Valencia. Camino de Vera s/n. 46022 Valencia (Spain) islocor@upv.es.

tocopherols and polyphenols were analyzed according to the procedures described by Sanchez-Casas, who modified the method produced by Slover [74]. Total phenolics were estimated according to the Folin-Ciocalteu spectrophotometric method [74]. All materials had been previously characterized by UPOV (International Union for the protectionof new Varieties of Plants) TG99/4.

Keywords: Extra Virgin Olive Oil (EVOO); Fatty acids; Sterols; Polyphenols

INTRODUCTION

Olive oil is one of the most commonly consumed foods in the Mediterranean diet. Its positive effects on health are widely recognized, both in popular culture and at a a scientific level. This is due to its complex chemical composition which comprises various antioxidant substances, that can be both nutritional and therapeutic [38]. Polyphenols, sterols and fatty acid are the most influential constituents in terms of health benefits. They are present in considerable quantities in olive oil, and play a major role in human metabolism, being simultaneously influent in factors such as stability, flavor and color of the oil.

Archaeological evidence of pollen, as well as fossilized leaves dating from the Tertiary in Mongardino (Italy), show that olive trees exist in the Mediterranean region from at least 3.2 million years ago [76]. However, it is not until the Palaeolithic period when remains from olive trees are associated to human activity. Several Middle Palaeolithic sites suggest that Neanderthals already consumed plant foods in Mediterranean Iberia [67], existing even archaeological evidence of the presence of olive tree seeds in caves from soutern Iberia such as Vanguard, Gorham and Ibex [14]. Archaeological sites from the Middle East, like Shanidar in Irak [70], and Kebara Cave in Israel [41], also show evidence of the presence of olive tree products used for consumption. Later on, discoveries from the Upper Palaeolithic clearly show that anatomical modern humans knew olive trees: remains of manipulated endocarps from Mentone, France, (ca. 30000 years BP), leaf fossils in the south of the Cycladic Islands (ca. 35000 years BP), and fossilized remains from the Relial region in North Africa from about 12000 years BP [69].

It is not until the Neolithic period, thousand of years later, when humans domesticate the olive tree. Traditionally, it was thought that the origin of the olive tree cultivation, as we know it today, originated at the Near and Middle East regions known as the "Fertile Crescent". However, present day olive produce are either Caucasian in origin, or come from different zones of the Upper Egypt. The expansion of olive tree cultivation was cited and regulated in the Babylonian code of Hammurabi (about 2500 years BC), and then in the XIX Egyptian Dynasty (1292-1198 BC). This explains why the first spread of cultivated olive tree was in Mesopotamia and Egypt, followed by Palestine and Crete (associated with Minoan-Cretan civilizations 5000 BC), and later to Greece and Italy. Olive trees were spread throughout North Africa, and eventually, new planting and management techniques also reached the east of the Iberian Peninsula. It is known that the olive varieties came to the Iberian Peninsula following three clear routes, and merged with varieties of olivastros that had already been established by the most ancient Iberian peoples and civilization of Tartessos. The first introduction of olive trees was by the Minoans and Phoenicians, who travelled across the Mediterranean. After the Phoenician decline, the Carthaginians followed the expansion and introduction of different varieties from the southern part of the Iberian Peninsula, and finally the Romans introduced new varieties via the north and northeast routes. The majority of them have survivied in our study area. It is important to consider the existence of Pliocene fossils (Morgandino-Italy, Tiryns, Greece, Ephesus, Turkey, Nablus, and Capsa Cirta in North Africa) as indicated by Tsouchtidi [81], confirming peri-Mediterranean expansion of the Mediterranean species [21], which is attested by the presence of debris endocarps and pollen in old stratigraphic laerys in Ugarit (Palestine), Relilai (North Africa), Crete, Rhodes and in the Argolida [52].

Mata et al. [44] studied the different types of material culture, remains of olive trees and fruits of pre-Phoenician epochs on the Iberian Peninsula, by gathering information from organic debris (namely endocarps and scrap wood) found during excavations, as well as on ceramics, sculpture, architecture, jewelry, coins and other metal objects, and by consulting classical sources. In the eastern Spanish areas, *Olea* remains have been found in at least 30 sites. Endocarps in particular, were found in 12 of these deposits, on charred olive wood or in construction debris. There are 46 recorded instances for the presence of endocarps on tools, and pollen from olive trees was identified in 6 cases. The deposits in which endocarps were present were associated with the production of olive oil, indicating that already during protohistoric periods, wild or cultivated olive trees were being used to obtain oil. The deposits in which olive residues were found are shown in the Table 1.

Table 1. East Spanish sites where olive remains were found

Levante next to the sea	Levante Interior
Cabezo Lucero (Guardamar, Alicante)	Amarejo (Bonete, Albacete)
Castellet de Bernabé (Lliria, Valencia)	Casa del Monte (Valdeganga,
Bastida de Les Alcluses (Moixent, Valencia)	Albaccte)
Cornulló dels Moros (Albocasser, Castellón)	Montón de Tierra (Griego, Teruel)
Cova de les Cendres (Teulada, Alicante)	
Cova de les Tabergues (Tirig, Castellón)	
Edeta, Tossal de San Miguel (Lliria, Valencia)	
Fonteta Raqueta (Ribarroja, Valencia)	
Hacienda Botella (Elche, Alicante)	
La Falaguera (Alcoi, Alicante)	
La Seña (Villar del Arzobispo, Valencia)	
La Vital (Gandía, Valencia)	
Kelin-Los Villares (Caudete de Las Fuentes, Valencia)	
Perengil (Vinaros, Castellón)	
Puig de la Nau (Benicarló, Castellón)	
Puntal dels Llops (Olocau, Valencia)	
Cataluña	Murcia-Andalucía
Camp de les Lloses (Tona, Barcelona)	Alhonoz (Herrera, Sevilla)
Emporion (L'Escala, Gerona)	Atayuelas (Torre del Campo, Jaén)
Mas Castellar (Pontos, Gerona)	Cero del Santuario (Baza, Granada)
Olerdola (Barcelona)	Cigarralejo (Mula, Murcia)
Reixac (Ullastre, Gerona)	

Possibly, both in Catalonia and Andalucia must have more plant material remains of olive trees but they are not mentioned in the published archaeological analysis due to the lack of sampling and excavation campaigns carried. In Andalusia there is abundant pottery of IV century BC where olive leaves have been represented, in some cases the whole tree. Coins minted in Emporio in this time were also found. The Iberians cultivated the olive tree from the Palancia River (Castellón) to the south of the peninsula, coinciding with the wild olive geographical areas. Despite the social and economic importance of olive trees, including landscape, the Iberians did not represent this tree in their pictures or in numismatics. Avieno (530 BC) cites *"fertile olive groves in Palus Naccararum"* and the existence of *Oleo Flumen* (now Ebro river) *"with abundant olive trees on its banks, and especially at its mouth"*.

The Romans and Greeks collected abundant olive images in their daily lives and well Iberia mentioned in several writings. Estrabon (29 BC) wrote *"the olive tree dominates the Iberian Coast bordering Our Sea and in the Outside sea"*, according to Hidalgo-Tablada [36] although the olive tree was known by the Tartesos, had to be introduced as a crop in the Iberian Peninsula by the Phoenicians. In areas of the western Mediterranean the first descriptions of the olive tree were made by Virgilio. Later Plinio in the year 23 AC appointed twelve kinds or species of olive and some specifications to differenciate the quality of their oils. Columella in the year 77 AC [23] discusses the differences of eleven olive oils types, indicating that some are very fine while others can be only used to illuminate, also says "where best the abundant olive trees grow in Iberia is in the hills of Baetica". Abu-Zacaria [2] only differentiates three types of olives grown in the Iberian Peninsula, one of them was brought from north Africa and other two remained from the ancient olive groves, one with punic origin and other phoenician.

The earliest references to the characteristics of the varieties, after the first basic descriptions made by the classic principles of our age, were made in the fifteenth century as it was indicated by Balaguerias [10] when he studied the Collection Oleographic of Spain that it was started few years earlier by the Royal Botanic Gardens in Madrid. But really the first classifications considered with technical criteria were due to Pitton de Tournefort who came to distinguish eighteen varieties of the genus *Olea* in his *"Institutions herbariae rei"*, and Gouan in the eighteenth century who defined and classified twelve varieties of olive in his work *"Flora Monspeliaca"*. Alonso de Herrera [4] mentions the cultivation of olives, but indicates very little about the different varieties and their oils. He basically differences varieties of rounded and elongated olives, and does reference to their sizes "thicker" and "minor". The Gouan works were completed and subsequently published by Rozier as it was indicated by Patac et al. [54]. In 1819 Tavantini [77] proposed a classification of olive varieties almost exclusively based on the morphology of the endocarps as these authors indicate, described fourteen varieties of olives for oil, and emphasized that the "Cordovi" is different to the group of "Manzanilla". He also recommended "Francesilla" mentioning the olives "Sevillana", "Verdial" and "Zorzaleño". Rojas Clemente [60] described eleven varieties from Andalusia, and some located on the central coast of the Iberian Peninsula. These varieties were fitted to the previous nomenclatures established by Couture [25].

Table 2a. Fatty acid composition of varietal oils Eastern Spain (I)

	Variety	SFA						MUFA		PUFA	
		Palmitic ac. (%)	Heptadecanoic ac. (%)	Estearic ac. (%)	Eicosanoic ac. (%)	Decosanoico ac. (%)	Tetracosanico ac. (%)	Oleic ac. (%)	Palmitoleic ac. (%)	Linoleic ac. (%)	Linolenic ac. (%)
1	Alfafara	15.05±1.06	0.19±0.03	1.73±0.07	0.32±0.01	0.10±0.01	0.05±0.01	70.57±2.05	1.26±0.15	9.48±1.12	0.72±0.06
2	Aguilar	13.43±0.05	0.22±0.06	2.25±0.06	0.40±0.01	0.14±0.01	0.06±0.01	74.38±0.1	1.41±0.04	7.08±0.05	0.65±0.01
3	Arbequina	15.63±0.69	0.11±0.12	1.73±0.12	0.34±0.03	0.09±0.01	0.05±0.01	77.99±6.76	1.66±0.07	11.46±1.5	0.54±0.02
4	Blanqueta	17.79±0.34	0.19±0.03	2.02±0.22	0.36±0.01	0.11±0.01	0.05±0.01	60.15±1.71	1.57±0.09	17.20±1.74	0.73±0.03
5	Borriolenca	10.6±0.55	0.08±0.01	2.10±0.13	0.37±0.03	0.08±0.01	0.04±0.01	78.57±0.52	0.79±0.03	6.48±0.15	0.52±0.01
6	Cabaret	15.32±0.36	0.03±0.01	1.72±0.01	0.31±0.02	0.08±0.01	0.03±0.01	66.87±0.79	1.96±0.02	13.05±0.63	0.64±0.02
7	Callosina	11.12±1.04	0.18±0.06	1.96±0.12	0.33±0.02	0.09±0.01	0.06±0.01	76.08±2.72	1.26±0.31	12.17±0.85	0.71±0.08
8	Carrasqueña	8.45±1.07	0.04±0.01	2.25±0.03	0.33±0.01	0.11±0.01	0.04±0.01	82.76±0.15	0.55±0.03	4.02±0.13	0.45±0.02
9	Cornicabra	8.35±0.07	0.06±0.01	3.12±0.07	0.42±0.02	0.10±0.01	0.05±0.01	82.01±0.66	0.50±0.03	4.89±1.48	0.50±0.04
10	Cuquillo	13.54±0.73	0.05±0.01	2.27±0.04	0.34±0.01	0.11±0.01	0.05±0.01	75.59±0.54	1.32±0.05	5.88±0.45	0.84±0.05
11	Changlot Real	11.64±0.33	0.17±0.02	2.28±0.16	0.39±0.01	0.10±0.01	0.06±0.01	77.28±1.79	0.52±0.02	7.01±1.25	0.57±0.04
12	Choco	11.32±0.25	0.06±0.01	3.62±0.04	0.50±0.01	0.16±0.01	0.06±0.01	73.60±0.33	0.78±0.02	9.29±0.09	0.61±0.01
13	De la cueva	9.75±0.14	0.17±0.05	2.85±0.06	0.44±0.02	0.16±0.02	0.05±0.01	74.80±0.26	0.52±0.05	10.48±0.14	0.58±0.02
14	De la lloma	7.73±0.2	0.10±0.01	2.37±0.04	0.35±0.01	0.10±0.01	0.04±0.01	83.25±0.28	0.39±0.03	5.27±0.24	0.40±0.06
15	Del pomet	12.65±0.35	0.05±0.05	1.83±0.06	0.31±0.02	0.10±0.01	0.02±0.01	79.83±0.37	1.24±0.1	3.42±0.04	0.56±0.04
16	Egipcia	9.91±0.06	0.23±0.01	3.09±0.01	0.48±0.01	0.16±0.03	0.04±0.01	79.73±0.04	0.68±0.01	5.25±0.01	0.43±0.01
17	Empeltre	12.83±0.16	0.09±0.01	1.50±0.04	0.29±0.02	0.10±0.01	0.06±0.01	73.98±0.35	1.44±0.05	9.14±0.06	0.64±0.02
18	Farga	9.66±0.16	0.03±0.01	2.05±0.08	0.34±0.02	0.12±0.01	0.04±0.01	78.88±0.77	0.62±0.03	7.66±0.76	0.49±0.04
19	Figuereta	7.79±0.09	0.04±0.01	2.31±0.06	0.42±0.01	0.15±0.01	0.06±0.01	79.15±0.08	0.42±0.02	8.99±0.06	0.66±0.07
20	Genovesa	11.56±1.05	0.19±0.02	2.35±0.12	0.41±0.02	0.10±0.01	0.06±0.01	77.48±2.36	0.61±0.15	6.65±1.03	0.60±0.02
21	Grossal	12.47±0.04	0.25±0.01	2.36±0.02	0.44±0.02	0.16±0.02	0.07±0.01	71.88±0.06	0.58±0.01	11.13±0.03	0.66±0.02
22	Hojiblanca	11.51±0.08	0.11±0.01	2.46±0.02	0.36±0.01	0.10±0.01	0.05±0.01	77.20±0.14	0.83±0.01	6.82±0.09	0.57±0.02
23	Jandra	15.31±0.26	0.08±0.01	1.64±0.02	0.28±0.01	0.08±0.01	0.04±0.01	69.13±0.27	1.57±0.03	10.92±0.19	0.95±0.02
24	Llumero	10.33±1.03	0.06±0.01	1.80±0.07	0.35±0.01	0.12±0.01	0.05±0.01	80.24±1.26	0.60±0.05	5.66±0.43	0.78±0.09

Table 2b. Fatty acid composition of varietal oils Eastern Spain (II)

	Variety	SFA						MUFA		PUFA	
		Palmitic ac. (%)	Heptadecanoic ac. (%)	Estearic ac. (%)	Eicosanoic ac. (%)	Docosanoic ac. (%)	Tetracosanic ac. (%)	Oleic ac. (%)	Palmitoleic ac. (%)	Linoleic ac. (%)	Linolenic ac. (%)
25	Manzanilla	12.68±0.14	0.15±0.01	2.49±0.08	0.34±0.02	0.09±0.01	0.06±0.01	76.03±0.53	3.31±0.03	5.68±0.45	0.63±0.03
26	Marfil	10.79±0.49	0.14±0.01	1.69±0.05	0.26±0.01	0.07±0.01	0.03±0.01	79.37±0.75	0.59±0.02	6.30±0.59	0.77±0.01
27	Mas Blanc	10.89±0.03	0.03±0.01	1.33±0.01	0.26±0.02	0.14±0.02	0.04±0.01	74.26±0.34	1.19±0.02	11.47±0.08	0.44±0.01
28	Millareja	13.38±0.43	0.13±0.01	2.38±0.09	0.36±0.01	0.09±0.01	0.04±0.01	70.15±0.49	0.97±0.05	11.72±0.52	0.79±0.04
29	Moixentina	15.70±0.33	0.05±0.01	4.35±0.07	0.45±0.01	0.11±0.01	0.06±0.01	71.38±0.06	0.45±0.02	11.59±0.03	0.71±0.02
30	Monteaguda	16.22±0.22	0.12±0.01	2.70±0.03	0.38±0.01	0.09±0.01	0.06±0.01	71.57±0.04	1.23±0.01	7.05±0.02	0.59±0.01
31	Morons	9.59±0.17	0.09±0.01	1.57±0.01	0.26±0.01	0.07±0.01	0.04±0.01	73.55±0.11	0.63±0.01	11.35±0.08	0.63±0.01
32	Morruda	12.57±0.63	0.06±0.02	2.52±0.12	0.41±0.03	0.12±0.01	0.05±0.01	72.71±1.77	0.89±0.16	10.15±1.41	0.54±0.09
33	Negra	10.87±0.06	0.06±0.01	3.82±0.01	0.43±0.01	0.11±0.01	0.03±0.01	73.14±0.06	0.48±0.01	10.51±0.04	0.58±0.02
34	Patronet	13.69±0.12	0.03±0.01	1.57±0.06	0.28±0.01	0.10±0.01	0.04±0.01	72.62±0.16	1.86±0.05	9.34±0.25	0.48±0.01
35	Picual	10.52±0.37	0.05±0.01	2.90±0.09	0.33±0.01	0.10±0.01	0.04±0.01	79.70±2.56	0.65±0.03	5.14±0.24	0.57±0.04
36	Romana	18.03±0.04	0.04±0.01	2.17±0.01	0.35±0.01	0.10±0.01	0.05±0.01	68.44±0.07	2.04±0.02	8.21±0.02	0.59±0.03
37	Rotja	11.70±0.07	0.10±0.01	2.14±0.04	0.33±0.01	0.08±0.01	0.04±0.01	70.73±0.12	1.05±0.06	9.73±0.14	0.61±0.01
38	Rotjeta	9.77±0.08	0.05±0.01	3.14±0.03	0.45±0.02	0.14±0.01	0.05±0.01	63.83±0.16	0.29±0.01	17.25±0.07	0.47±0.03
39	Rufina	12.94±0.21	0.23±0.01	2.72±0.06	0.43±0.02	0.12±0.01	0.05±0.01	76.40±0.2	1.17±0.01	5.33±0.09	0.61±0.02
40	Seniero	8.30±0.26	0.18±0.02	1.48±0.06	0.27±0.02	0.09±0.01	0.05±0.01	82.47±0.44	0.76±0.01	5.95±0.63	0.46±0.03
41	Serrana	11.98±0.27	0.23±0.02	2.03±0.06	0.35±0.02	0.12±0.01	0.05±0.01	69.59±0.75	0.62±0.12	15.75±0.31	0.84±0.04
42	Valentins	11.38±0.05	0.10±0.01	2.10±0.07	0.43±0.06	0.11±0.01	0.05±0.01	72.17±0.43	0.68±0.03	11.02±0.63	0.76±0.13
43	Vallesa	10.66±0.45	0.04±0.01	2.94±0.12	0.36±0.02	0.14±0.01	0.06±0.01	73.50±0.11	0.44±0.08	11.22±0.06	0.52±0.01
44	Vera	13.32±0.19	0.04±0.01	2.87±0.04	0.36±0.01	0.08±0.01	0.04±0.01	74.18±0.24	0.99±0.03	7.01±0.11	0.48±0.02
45	Verdiel	11.98±1.23	0.07±0.02	2.46±0.18	0.32±0.01	0.12±0.01	0.06±0.01	76.23±1.18	0.86±0.23	6.93±1.05	0.54±0.05
46	Villalonga	14.61±0.62	0.14±0.02	3.04±0.05	0.37±0.01	0.09±0.01	0.05±0.01	69.22±0.95	1.18±0.15	13.03±0.71	0.70±0.12
47	Vinagre	13.56±0.99	0.02±0.01	2.04±0.14	0.31±0.02	0.06±0.01	0.02±0.01	73.50±1.02	1.16±0.25	8.37±0.89	0.97±0.04

Table 3a. Sterol composition, α-tocopherol and polyphenols in the varietal oils Eastern Spain (I)

Variety	Esteroles β-sitosterol (%)	Campesterol (%)	Stigmasterol (%)	α-tocopherol (mg/kg)	Polyphenols (mg/kg)
Alfafara	96.37±0.38	2.97±0.32	0.62±0.08	206.76±8.54	463
Aguilar	94.61±0.25	3.12±0.13	2.25±0.13	138.71±1.15	159
Arbequina	96.70±0.35	2.88±0.29	0.41±0.11	184.09±8.12	201
Blanqueta	94.39±0.39	3.80±0.33	1.80±0.12	127.50±4.59	402
Borriolenca	91.93±0.34	4.36±0.23	3.68±0.18	60.97±0.72	116
Cabaret	93.69±0.29	3.72±0.09	2.58±0.24	91.14±1.51	304
Callosina	95.85±0.58	2.71±0.25	1.49±0.46	275.12±6.42	235
Carrasqueña	94.58±0.40	3.51±0.08	1.90±0.51	154.42±1.37	73
Cornicabra	96.57±0.17	2.79±0.15	0.65±0.09	233.32±6.55	473
Cuquillo	96.58±0.26	3.06±0.24	0.39±0.05	343.01±6.98	203
Changlot Real	95.63±0.88	2.76±0.19	1.51±0.34	183.89±5.1	309
Choco	98.41±0.12	1.24±0.65	0.50±0.09	184.17±1.02	162
De la cueva	95.89±0.19	3.19±C.18	0.92±0.05	169.71±2.43	126
De la lloma	93.25±0.56	3.79±0.28	3.02±0.44	183.38±0.52	218
Del pomet	94.93±0.30	2.66±0.13	2.41±0.23	98.53±0.57	305
Egipcia	97.39±0.16	1.72±0.07	0.89±0.16	201.10±3.37	306
Empeltre	96.28±0.21	2.43±0.13	1.29±0.11	166.64±1.83	316
Farga	95.36±0.5	2.96±0.2	1.67±0.38	240.21±12.81	243
Figuereta	95.49±0.38	3.23±0.21	1.34±0.15	161.29±1.31	181
Genovesa	93.69±0.39	3.70±0.16	2.63±0.37	70.41±2.72	164
Grossal	94.98±0.22	3.36±0.11	1.65±0.13	129.58±1.17	432
Hojiblanca	96.61±0.22	2.94±0.09	0.46±0.18	178.46±2.85	418
Jandra	94.40±0.18	2.65±0.07	2.94±0.21	231.19±3.19	173

Note: β-sitosterol, Campesterol and Stigmasterol content are expresed in % over total sterols content.

Table 3b. Sterol composition, α-tocopherol and polyphenols in the varietal oils Eastern Spain (II)

Variety	Sterols				α-tocopherol (mg/kg)	Polyphenols (mg/kg)
	β-sitosterol (%)	Campesterol (%)	Stigmasterol (%)			
Llumero	95.42±0.63	3.47±0.37	1.11±0.38		139.21±4.54	318
Manzanilla	95.68±0.63	2.80±0.20	1.54±0.37		184.00±1.57	169
Marfil	96.48±0.35	3.03±0.37	0.45±0.46		270.14±0.94	321
Mas Blanc	96.99±0.12	2.58±0.08	0.43±0.35		86.84±0.56	427
Millareja	95.60±0.59	3.52±0.31	0.88±0.34		181.77±1.69	373
Moixentina	93.57±0.43	3.59±0.39	2.84±0.17		138.95±2.07	473
Monteaguda	96.84±0.14	2.71±0.19	0.45±0.23		236.39±3.22	381
Morons	95.29±0.14	3.81±0.11	0.90±0.16		167.88±1.16	198
Morruda	95.47±0.54	2.91±0.49	1.62±0.25		132.16±1.51	489
Negra	96.19±0.57	2.96±0.1	0.96±0.26		250.89±1.04	552
Patronet	94.65±0.28	3.80±0.23	1.51±0.23		113.75±0.71	173
Picual	96.39±0.99	2.93±0.3	0.35±0.15		132.51±0.79	347
Romana	95.54±0.38	3.77±0.14	0.69±0.25		196.49±2.24	362
Rotja	95.69±0.37	3.25±0.34	1.06±0.08		229.70±0.63	139
Rotjeta	92.35±0.22	3.92±0.1	3.73±0.13		153.46±0.74	141
Rufina	94.61±0.37	3.25±0.16	2.14±0.28		191.54±0.57	263
Seniero	95.84±0.64	2.76±0.34	1.14±0.37		136.17±2.16	303
Serrana	95.89±0.61	2.89±0.29	1.22±0.33		158.44±1.76	268
Valentins	96.04±0.32	2.31±0.21	1.66±0.18		139.20±0.48	203
Vallesa	90.95±0.57	4.65±0.25	4.37±0.33		61.20±0.83	169
Vera	96.53±0.18	2.91±0.23	0.56±0.13		288.77±0.86	283
Verdiel	93.61±1.38	2.86±0.39	0.61±0.07		201.23±4.95	261
Villalonga	94.42±1.07	3.25±0.46	2.17±0.73		148.06±9.23	296
Vinagre	94.32±0.67	3.25±0.28	2.44±0.41		114.63±0.82	143

Note: β-sitosterol, Campesterol and Stigmasterol content are expresed in % over total sterols content.

Martinez Robles [45] studied 32 varieties based on data de Rojas Clemente and adding new others. Rojo [61] described twelve basic castes olive and specifically added the "Empletre" in Aragon, the "Royal", "Vera fina" and "Herbaquin". Colmeiro [22] did a elayographic collection where described and drew 78 branches of different varieties with their fruits. Hidalgo-Tablada [36] described 21 varieties and collected data from other 29. Abella [1] described only the fifteen most widespread varieties in Andalusia. Espejo [27] attempted to standardize and catalog all olive varieties in Spain. Valenzuela [83] described and drew the fifteen varieties of olive tree that considers majority in the Iberian Peninsula and mentions many synonyms and maintains other ten denominations as different varieties. Ribera [59] described fifteen varieties and indicated some data on the quality of their oils. Patac et al. [54] cited more than 200 varieties of olive, described 40 varieties being intensely in the first work, and 52 in another work. He established numerous synonyms for the varieties described. Ortega-Nieto [49] described a very complete pomology of 24 varieties which was also characterized morphometrically. Pagnol [53] divided the varieties grown in Spain in two groups: those were appropriate for oil and those that are preferred for eating. Barranco and Rallo [11] studied and described 197 denominations in Andalusia, grouping the prospected materials in 156 different varieties. Tous [78] characterized 17 varieties grown in the Mediterranean coast in Tarragona. Salazar [64] following the UPOV (International Union for the protection of new Varieties of Plants) rules, also using different complementary biometric data and some data about oils typified 19 varieties from central-west Mediterranean Sea. Fontanazza [30] considered the sizes of olive tree, fruit shape and characters of the endocarp as stable characters for the determination of the varieties, being excluded the chemical composition of the oils. Tous and Romero [79], after a prospection of 117 trees, identified 65 different denominations grouped into 40 varieties, being the first in studying the basic composition of their oils. Barranco and Rallo [11] established the basic pomology of the major varieties grown in Spain, which were classified into varieties for oil and dual-purpose (oil and eating), mentioning their synonyms and grouped them into four categories called primary, secondary, broadcast and local. These categories had been already previously used by other authors. They did pomological descriptions but only mentioned of some basic characteristics of their oils without consider its composition. Salazar et al. [65], [66] conducted a pomological study of the 29 varieties considered mainstream or indigenous to the central and western Mediterranean, their pomological sheets being publicated following the UPOV standards including in some cases detailed description of the composition of their oils. Barranco et al. [12] after establishing a protocol for the pomological characterization of the olive varieties sponsored by the IOC (International Olive Council), conducted a global catalog of olive varieties including 139 varieties of 23 countries. From them, 47 were associated to Spain, 18 of them varieties are considered in this study. Iñiguez et al. [38] reported 136 denominations and did pomological descriptions of 75 of these materials, considering 27 characters, but none of them includes the composoición of oils. Barranco et al. [13] studied the homonyms of the main varieties of olive grown in Spain with 511 denominations. Caballero et al. [20] related the 358 vegetal materials included in the genebank of the Alameda del Obispo in Cordoba (Spain), which it has continued to expand later.

The first chemical characterizations of the varieties that included the chemical composition of fatty acids and other unsaponifiable compounds were done by Panelli et al. [51] who studied fifteen varieties majority in Umbria (Italy). At present there are still few modern studies on the full composition of the oils of olive varieties. One of them was done by El Antari et al. [26]. After years of study of the composition of monovarietal olive oils, in this work the mean values of the oil composition of the varieties of east of Spain have collected, being studied the interannual variations and the impact of various factors in the fatty acids content and the most important unsaponifiable components, being shown in Table 2.

Photography 1. Trichomes of Serrana variety.

From the morphological point of view (using the descriptor TG99/4 UPOV), of the 32 basic characters defined by the IOC, the most stable descriptors for pomological characterization of the varieties were the endocarp (if we except its size and weight) and the characteristics of the olive (without considering size, weight and features of the lenticels). The study of the lenticels to establish its taxonomy has been addressed by electron microscopy, as well as the role of trichomes and other phytoliths of olive leaf, but it requeires more studies, not only consider its size. Germplasm analysis using eighteen materials indicated a higher polymorphism than the pointed by Fabbri et al. [28] and less than the established by Weisman et al. [87]. This researcher recognized that RAPDSs techniques identified very well the varieties of Valencia through UPGMA clustering dendrogram considering Nei index. On the one hand, the north varieties were differenciated, on other hand the south and center vaieties, and in a third group was the variety Serrana. Something similar had been observed by Trujillo et al. [80] in a wider area and with less specificity but that seemed to indicate that the varieties of the Iberian Peninsula can be grouped fairly well as their areas of origin. From our results it is deduced two groups that seem to confirm the different origin of olive varieties from East of Spain, a first group possibly introduced by the Phoenicians, Carthaginians and Greco-Roman civilization by the north (Farga, Canetera, Empeltre, Rufina, Borriolenca, Valentins), another group of varieties introduced in the

Muslim age, coming from North Africa and they are probably occupying the south and center of our study area (Morrut, Villalonga, Grossal, Alfafara, Changlot Real, Cabaret, Callosina). A third independent group comprises the Blanqueta, the Serrana and Cuquillo.

Various scientific papers have shown that the monovarietal olive oils have contents slightly different of polyphenols, types of sterols, and there is less variability in fatty acids. It has also been shown that the composition of the oil in these substances is influenced by the climate where the crop was grown, together the agronomic techniques used (Godoy-Caballero et al.) [34]. The different composition of monovarietal oils allows us to classify them according to their nutritional and therapeutic properties. The characterization of varietal oils enables better targeting diet according to the interests to be pursued, and the differences between them and other oils offer an opportunity to promote them commercially. As the composition is influenced by the weather, as well as cultural practices, whether varietal oils of a certain area can be grouped according to their composition characteristics, and therefore its dietary and terapeutic characteristics, they may be marketed under the same coverage, giving them certain commercial advantages.

In this chapter a compilation of the characteristics of the most important varietal oils from Eastern Spain has been made according to their composition in polyphenols, sterols and different fatty acids. A classification is presented under several criteria that can be considered in dietary recommendations for the alleged effects that a particular composition has in the metabolism or therapeutic.

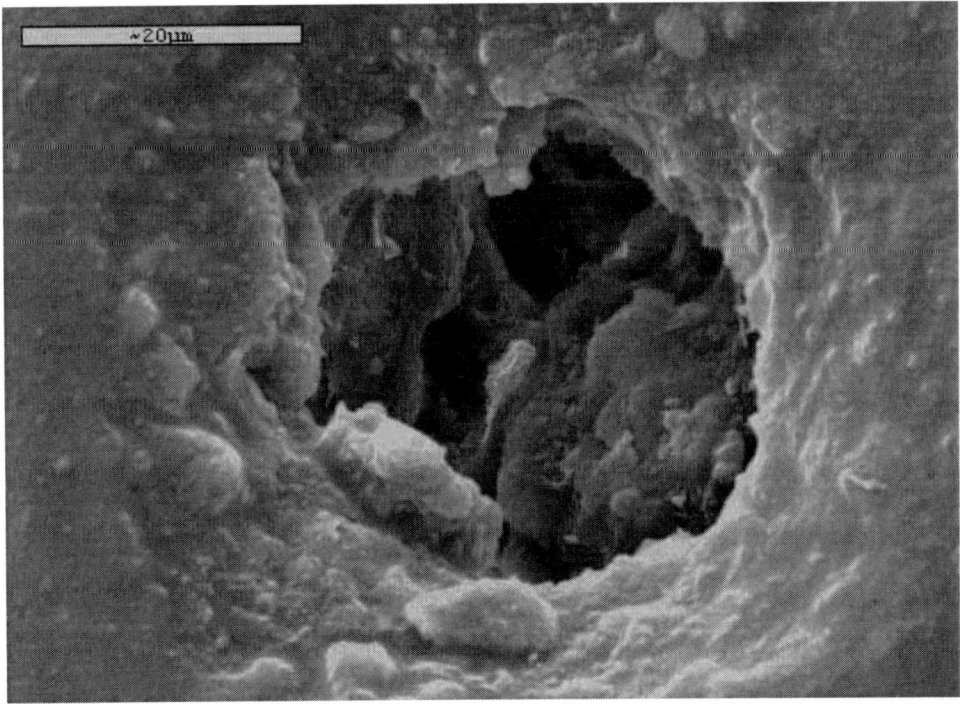

Photography 2. Lenticels of Blanqueta variety.

OLIVE OIL COMPONENTS AND THEIR INFLUENCE ON HEALTH

The components of olive oils are classified into two groups: saponifiable and unsaponifiable. Saponifiable portion occupies 98-99% and it is formed by the glycerides and fatty acids, primarily triglycerides. The unsaponifiable part do not exceed between 1 and 1.5% but it is formed by components of enormous importance in the oil properties, being mainly sterols, triterpene alcohols, squalene, chlorophyll and carotene among others. The composition of the unsaponifiable fraction of oil varies according to variety, althoug it is also influenced by the weather and soil type, which causes variations between areas. The IOC sets limits on the fatty acid composition of olive oil. The main fatty acids that compound the glycerides or are free are oleic which occupies between 55 and 83% of the total, palmitic between 7.5-20 %, linoleic acid with 3.5-21 %, stearic acid with 0.5-5%, palmitoleic acid 0.3-3.5% and linolenic acid 0.1-1.5%. The other acids have less presence. The unsaponifiable part represents a very small fraction but has great importance in the health effects. Sterols occupy an average of 0.16% of oil weight, polyphenols 0.03% of oil weight, tocopherols only 0.02%. Terpenes occupy 0.5% of the oil weight. About 230 minority components represent about 2% oil. These are primarily chlorophylls, alcohols, cetones, esters and furamic derivatives among others. A usual composition of majority and minority substances in olive oils is shown in Table 4.

Table 4. Average amounts of components in olive oil (Monzo et al. [47])

Fatty acids	Oleic acid (C18:1)	56-84%
98 g/100 g oil	Palmitic acid (C16:0)	7-20%
	Linoleic acid (C18:2)	3-21%
	Stearic acid (C18:0)	1.4-4%
	Palmitoleic acid (C16:1)	0.3-3.5%
	Linolenic acid (C18:3)	0.2-1.5%
	Myristic acid (C14:0)	0-0.1%
	Other fatty acids	2.5-3%
Sterols	Sitosterol (C29 H50 0)	65-88.5%
80 and 260 mg/100 g oil	Campesterol (C28 H48 0)	2-4%
	Stigmasterol (C29 H48 0)	6.3-2.5%
	Avenasterol (C29 H48 0)	5-31%
Phenols	50 to 500 mg/kg oil	
Carotenoids and tocopherols	3 to 35 mg/kg oil	
Carotenoids	0.5 to 10 mg/kg oil	
Phospholipids	45 to 150 mg/kg oil	
Chlorophyll	1 to 10 mg / kg oil	

In Table 2, it can be noted a higher content of oleic acid, which is monounsaturated, compared palmitic and stearic acids, which are saturated fatty acids, and a discreet polyunsaturated fatty acid content, such as linoleic and linolenic acids. This fact add special interest oil dietary health [71]. The amount of unsaturated components makes the extra virgin olive oil may suffer oxidations, but its anti-oxidants content among the minor components (α-tocopherol and phenols) makes it a particularly stable product if it is handled properly. The high content of α-tocopherol, which is typically over 90% of the tocopherol content is the most important element for oil stabilization [88].

Use olive oil in traditional medicine especially in the Mediterranean is recognized from early age to cure numerous ailments and consiguir the balance of our body, especially the endocrine system. This culture and its use in pharmacology is held in the Middle Ages. In the Islamic civilization is used as medicine, as food and food preservative. All information about olive oil used as medicine in Alexandria and Rome was writen by Gaius Cornelius Celsus and Pliny the Elder (first century).

The publication "Medicinal Plants Renovated Dioscorides" [31], after differentiating between the wild olive tree and olive-growing, discusses the properties and composition of their leaves, specifying that they contain 0.7% of oleuropein, which has positive effects in the treatment of elevated blood pressure, febrifuge, and high levels of cholesterol, and it is also recommended as a diuretic. The same author speaks of the virtues of the olive as a tonic and appetite stimulant and recommends the use of olive oil in pharmacological preparations such as excipient for ointments and liniments because it takes long before it becomes stale. Once neutralized with lime water, it can be used as an anti-scald, healing, emollient laxative product, to facilitate the expulsion of intestinal worms.

Avila [7] mentions the beneficial health effects of olive oil (extra virgin), describing its beneficial role in the prevention of cellular aging, on the circulatory system, the digestive system, the urinary system, the endocrine system, on joints, bone tissue, in the ears, skin and scalp, and as a preventative for at least prostate and breast cancer [89].

At the International Congress on the Mediterranean Diet [7] in one of the conclusions that this author transcribes, it is specified that: "The extra virgin olive oil is an essential component in the Mediterranean diet due to its antioxidant polyphenols, sterols and unsaturated fatty acids, since the consumption of extra virgin olive oil has a positive effect on disorders as diverse as excess weight, diabetes type 1 and type 2, rheumatoid arthritis, several types of cancer (prostate and breast among others), cognitive impairment, hypertension, aging, psoriasis and fundamentally on cardiovascular diseases"

Phenols

Many of the benefits olive oil has on health are due to its content of phenols [56], [62]. These are natural antioxidants. Polyphenols are used for measuring the stability of oil, i.e., its resistance to oxidation,"staling", depends on its polyphenol content. This content is highly variable depending on the climate and the origin of the oil, the cultivation techniques, water availability of the olive groves, the level of maturation of the olives and the technology used in the conservation of the oil. They can currently be used, therefore, to differentiate some varieties, although not individually but in connection with other types of compounds.

There are more than 4000 different phenolic compounds in oil [72], [18], [34]. The ones with the greatest influence on the organoleptic characteristics of the monovarietal oils and their ability to improve health are simple phenols such as hydroxytyrosol and tyrosol, compound phenols such as ligustrosids and especially oleuropein, as well as aromatic compounds. Polyphenols as antioxidants help prevent damage caused by molecules known as "free radicals" in the body tissues. Free radicals are charged molecules that readily react with living cells causing a change in their performance and therefore tissue degradation. These free radicals can be formed in the human body in certain metabolic processes, and because of environmental pollution in the air, water and food. Free radicals have been linked to heart disease, cancer and aging [50], [72]. Free radicals play an important role in the arteries, especially when they react with cholesterol and low density lipoprotein LDL, and may get attached to the walls of blood vessels impeding bloodstream with the consequent risk of having a heart attack or a stroke [37]. Extra virgin olive oil, with its contribution in polyphenols, works as a natural antioxidant and helps prevent this oxidation, it therefore acts as an anticoagulant. [56], it shows the effects of olive oil polyphenols on the oxidative damage in endothelial dysfunction. Yang et al. [90] analyzed its effect on inflammation and its antithrombotic effect

Generally, phenols are considered to be protective compounds of the endothelial system of blood vessels. The hidrotirosol is specifically considered an inhibitor of platelet disaggregation and thromboxane synthesis, which also intervenes in the regulation of the production of leukotriene B-4 which is an inducer of inflammatory processes. Hydroxytyrosol is then considered a potent anti-inflammatory and antithrombotic agent. The europeina is considered an endotelial cell protector and a strong vasodilating substance.

Sterols

Sterols (phytosterols) are plant origin steroids, chemically very similar to cholesterol, which have lower reactivity with free radicals in the human body. Moreover, they decrease the absorption of cholesterol from the small intestine into the bloodstream because they use the same transfer paths as the latter compound. One of the targets in the prevention of arteriosclerosis and other cardiovascular diseases is the reduction of cholesterol circulating in blood (total cholesterol) [37]. Cholesterol is transported in the blood locked to low density lipoprotein LDL. It has been proved that there is a relationship between the concentration of plasma cholesterol and LDL-cholesterol accumulation in the artery walls. LDL carries about two-thirds of cholesterol in plasma. This can infiltrate into the arterial walls and attract macrophages, smooth muscle cells and endothelial cells.

Once embedded in the interior of the vessel, LDL oxidises forming oxLDL, which causes the formation of a thick foam which reduces the section, so that the reduction in LDL cholesterol delays and even reverses the increase in the formation of atherosclerotic plaques. Therefore, the consumption of phytosterols has a positive effect on reducing the total concentration of coresterol and specifically LDL- cholesterol in the blood [75], thereby decreasing the risk of suffering cardiovascular diseases [46]. Moreover, fitosterores have anticancer, anti-inflammatory and atherogenicity effects [17], this may be due to the fact that they affect the stiffness of the membranes of tumor and non-tumor cells, reducing their flexibility, making them more fragile [9]. The most abundant sterols in olive oil are sitosterol, campesterol and stigmasterol. These are structurally similar to cholesterol, without the methyl group on the fourth carbon atom [33].

Campesterol

Several studies have found that campesterol, like sitosterol, decreases LDL cholesterol levels. That leads some scientists to consider campesterol as a valuable element for the control of cholesterol, reducing the risk of heart diseases. Campesterol is sometimes also used to treat specific prostate conditions. However, it is important to note that very high levels of campesterol may have different effects to those intended, it may increase the risk of cardiovascular diseases in some individuals. Their presence in olive oil is lower in proprtion than sistosterol, therefore their presence is considered, in principle, positive.

Stigmasterol

Stigmasterol shares the properties of other sterols in the regulation of the quantities of cholesterol in the blood. However, it plays an important role in reducing inflammation, because it may be a precursor of chemical compounds with these effects.

β-Sitosterol

The β-sitosterol is particularly abundant among the sterols contained in olive oil. The β-sitosterol is considered a beneficial component in the composition of olive oil. It is the most abundant sterol. Awad et al. [9] noted that the β-sitosterol that passes to the blood reduces the speed of growth of the plaque which produces arteriosclerosis.

Moreover, numerous studies have reported that β-sitosterol reduces symptoms resulting from benign prostatic hyperplasia. These intestinal conveyors return cholesterol and sitosterol to the intestine preventing its excessive blood concentration. This can be regarded as positive, since an increased sitosterol diet is generally associated with an eventual reduction of cholesterol intake. Berge et al. [16] reported much more effective in the case of sitosterol, restoring much of this sterol back into the intestine for its removal specific transporters it can cause very high levels of sitosterol in blood plasma, causing a disease known as sitosterolemia.

α-Tocopherol

The α-tocopherol, known as vitamin E, is an antioxidant which has the property of protecting polyunsaturated fatty acids in membranes and other cellular structures of lipid peroxidation [19], [6]. This fact, powers many health benefits, such as skin protection by blocking free radicals. Its strong inhibition of oxidation prevents the development of degenerative diseases like cancer, particularly cancers of the mouth and lung in smokers, colo-rectal cancer, polyps, prevention of stomach cancer, pancreas and prostate [35]. It also causes inhicicion of platelet aggregation [86]. This prevents blood clots, preventing diseases such as deep vein thrombosis, atherosclerosis or heart attack. It globally protects the cardiovascular system preventing clogged arteries due to the effect of cholesterol. The α-tocopherol has anti-inflammatory [58] effects, prevents gynecological pathologies in pregnant women such as complications of preeclampsia, premenstrual syndrome, painful periods, hot flashes. On the other hand, it prevents the development of cataracts.

Fatty Acid Profile

Fatty acid composition may vary slightly, for the same variety in very different locations. These variations are for example higher for Arbequina than for Empeltre and Picual [78]. In our case, the studied plant materials are set in two collections, located in two different locations, but very similar in soil, climate and cultivation. Studies have not detected significant differences between the two locations.

We cannot fail to note the great significance of some of the fatty acids studied. The presence of a high content of linoleic acid involves low stability of the oil, it was already mentioned by Tous [79]. This author considered the Blanqueta, Morruda, Farga, and Arbequina varieties as more unstable than other varieties, by measuring the stability caused by the effect of high temperatures (120 °C) during a long period of time.

The sum of saturated acids (palmitic and stearic) is important in each one of the varietal oils, since it is a parameter to define virgin olive oil. The legislation specifies that this content must be less than 1.5 per cent of total fatty acids. As Cortesi et al. [24] showed studying the triglyceride composition of varietal oils; this value cannot be met in the oils of some cultivars, as Fedeli also collects [29].

The beneficial effects of the fatty acids in olive oils are mainly caused because they are made of monounsaturated (MUFA) and poliisaturated (PUFA) fats. MUFAs and PUFAs ratios consumed through olive oil influence the content of lipids and lipoproteins in blood plasma [32]. MUFAs are more important than PUFAs. For example, LDL is less susceptible to oxidation by free radicals in a diet enriched by MUFAs, which are more stable and resistant to oxidation than PUFAs. So a higher content of PUFAs in relation to MUFAs and saturated fatty acids (SFA) leads to lower ateroscleromas condition [39]. Especially oleic and linoleic acids have been associated with a decrease in the plasma levels of human LDL and an increase in HDL in serum [3]. Oleic acid consumption reduces the sensitivity and platelet aggregation and reduces the levels of coagulation factor VII (FVII), and increases fibrinolysis [40]. Moreover, the least healthy fatty acid is the palmitic of the saturated fraction (SFA) as it is the one that most increases cholesterol levels in the blood, so it is the most atherogenic. However, the preventive effect of extra virgin olive oil on asteriosclerosis, and hence coronary diseases is attributed not only to the high level of monounsaturated fatty acids but also to the presence of antioxidants such as phenolic compounds and tocopherols, already mentioned above.

The high content of monounsaturated fatty acids and low levels of saturated in Extra Virgin Olive Oils make it possible to use them in moderating the effect of getting type II diabetes, causing an increase in insulin sensitivity and a decrease in the level of glycolysis hemoglobin, contributing to a progressive improvement of glycemic control. Besides, the lipid profile of diabetic patients improved significantly.

Overall, it can be summarized that the benefits traditionally attributed to olive oil are:

- Reducing blood cholesterol, especially LDL (low- density) lipoproteins and increasing or maintaining HDL)
- Avoiding or preventing the formation of kidney stones
- Enhancing the venous circulation reducing the possibility of the occurrence of phlebitis
- Regulating intestinal absorption and correcting chronic and transient transit constipation, further facilitating the absorption of minerals like calcium, iron, magnesium and phosphorus, facilitating the development of bone structure and bone mineralization.

- Improving type II diabetes by regulating gastric emptying and getting a slower absorption of glucose, decreasing blood glucose levels and reducing the dose of insulin needed to control this disease.
- It is considered a mild laxative
- It is credited with the power to slow the deterioration of some brain functions and tissue aging by mitigating cellular damage
- It is healing, it avoids skin irritation and relieves eczema, it is emollient and keratolytic, it is a crack corrector and regenerator

COMPOSITION OF MONOVARIETAL OILS

It should be noted that there is a great variability of varietal olives names picked in the olive-growing areas of the Spanish Mediterranean, reaching over 1300. Many of them synonym with several names to designate the same variety. The studied varieties were prospected throughout the Western Mediterranean area and were characterized by following three technologies, firstly by the UPOV (TG 99/4) standards, on the other hand, morphometrically, following the studies of Pérez-Camacho and Rallo [55] and finally by the chemical analysis of the oils [42], [43].

Trujillo et al. [80] and Said et al. [63] analyzed and grouped by isoenzyme techniques many of the varieties in Andalusia. Most olive varieties in the Spanish Mediterranean were also studied and grouped by using the RAPD technologies of Pontikis et al. [57] and Belaj et al. [15]. Later, Sanz -Cortés et al. [68] validated the discrimination achieved by application of the UPOV technical standards and specifications following AFLP Angiolillo et al. [5]. This confirmed the possible grouping and origins of most olive varieties studied in this chapter [48].

The results obtained in three successive olive campaigns in which the phenol content, types of sterols and fatty acid types were studied, are presented in this chapter.

Total phenolics were estimated according to Folin-Ciocalteu spectrophotometric method [73] after incubation for 30 min at room temperature, the absorbance at 765 nm was measured and expressed as mg/kg of caffeic acid. Determination of tocopherols and sterols was carried out by gas chromatography using the method established by Slover et al. [74]. This method was used because it allows simultaneous determination of tocopherols and sterols in the same sample. This procedure is applied for the determination of both tocopherols and sterols in oils and in other fatty oils and lipids of different foods.

The study of β-sitosterol content, in addition to serving on the characterization of oils along with other sterols, is a suitable parameter to defend the genuineness of the oil and prevent adulteration [29]. The β-sitosterol is easily detectable because it always has a high percentage within sterols, although as shown by mass spectrometry is not a single isomer, as this chromatographic peak includes δ-7-stigmasterol and other isomers, as well as several stigmastadienes [24], [29].

The model gas chromatograph used was a Hewlett- Packard model 5840, equipped with a FID 300 detector operating at a temperature of 290 °C and temperature of injecting the sample 280 °C, the process was isothermal at 265 ° C during the 45 minutes that lasts the elucidation of the compounds to study, the flow is 1 mL per minute and using helium as a carrier gas. The peak areas of α–cholestane, sterols and tocopherol were calculated using an integrator. The response factor for 5α–cholestane must be considered as the response pattern in the final graphics. Sterols and tocopherols were calculated expressed in mg/100 g of fat.

The usual varieties that are more or less considered to be the majority and marketed are Alfafara varietal oils, Arbequina, Blanqueta, Cornicabra, Cuquillo, Empeltre, Farga, Genovesa, Grossal, Hojiblanca, Manzanilla, Morruda, Picual, Serrana, Villalonga. The composition of fatty acids in the studied oil varieties is presented in Table 2. In Table 3 polyphenol, α-tocopherol and sterols contents are shown. Figure 1 shows that all the major oils have a quite similar percentage of oleic and palmitoleic acid (MUFAs), about 70% of total fatty acids. However, there is a wide variability in the content of α-tocopherol. Cuquillo, Cornicabra and Farga oils are the ones with the higher content of oleic, palmitoleic and α-tocopherol components, of great health importance. The α-tocopherols are considered by Reiter et al. [58] as reducers of the risk of coronary heart diseases like MUFAs. Arbequina, Manzanilla, Hojiblanca, Alfafara, Empeltre have also high contents, plus they are considered excellent oils from a taste point of view. These oils have always obtained a high valuation in the tasting panels in both technical and tasting panels.

In Figure 2, a large variability of minority oils is observed from the viewpoint of its composition. Vera, Callosina, Marfil, Negra, Rotja, Jandra distinguished by its high content of α-tocopherol oils and are therefore more antioxidant and more stable. Moreover, on the opposite side, Borriolenca, Vallesa, Mas Blanc, Pomet and Cabaret are varieties whose oils have less α-tocopherol and are considered more unstable.

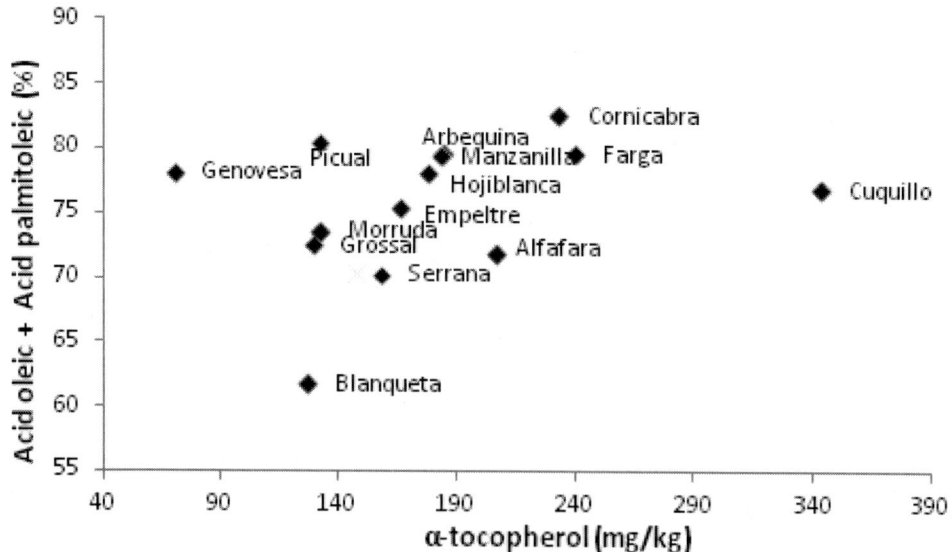

Figure 1. Distribution of majority varietal oils of Eastern Spain according to their content in palmitoleic and oleic acid and α-tocopherol.

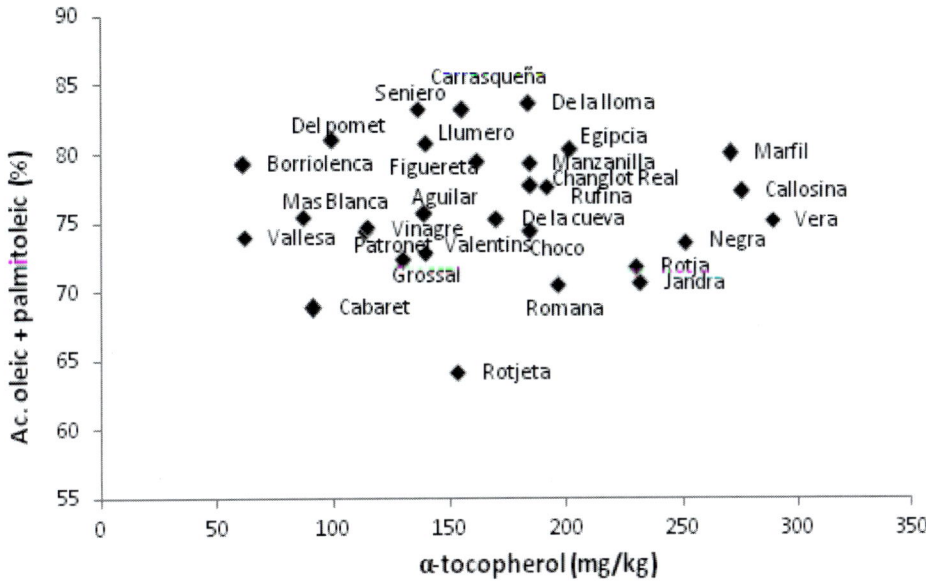

Figure 2. Distribution of minority varietal oils Eastern Spain according to their content in palmitoleic and oleic acid and α-tocopherol.

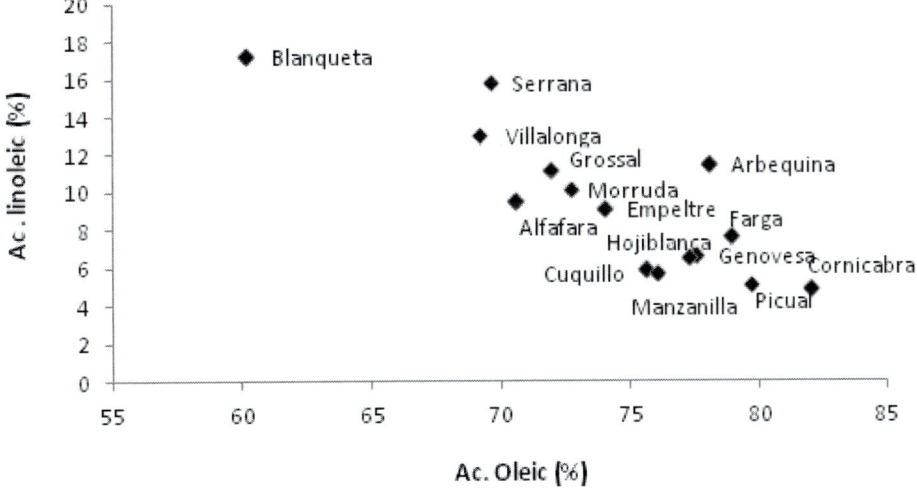

Figure 3. Distribution of majority varietal oils of Eastern Spain according to their content in linoleic and oleic acid.

In Figure 3 the main varieties according to their content of oleic and linoleic acid are shown. Blanqueta, Serrana, Villalonga, Grossal and Arbequina varieties have a low level of oleic, which in principle may seem inconvenient. However, they are high in linoleic acid (omega 6) serving as an improver of the lipid profile, getting a lower rate of triglycerides and faster disappearance of these in blood plasma according to Avila et al. [7]. The same happens in some minority varieties (Figure 4). Rotjeta, Cabaret and Callosina present low level of oleic acid and high linoleic acid. This composition can be used as a system of varietal differentiation in both groups of varieties.

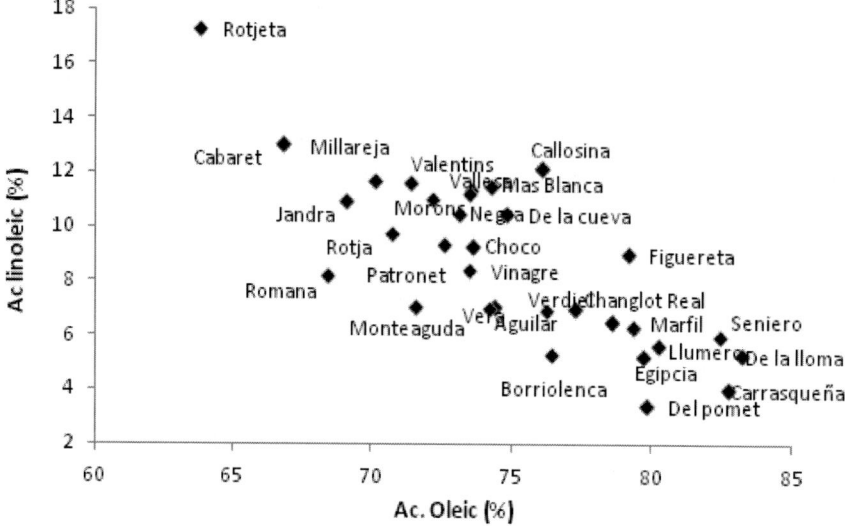

Figure 4. Distribution of minority varietal oils Eastern Spain according to their content in linoleic and oleic acid.

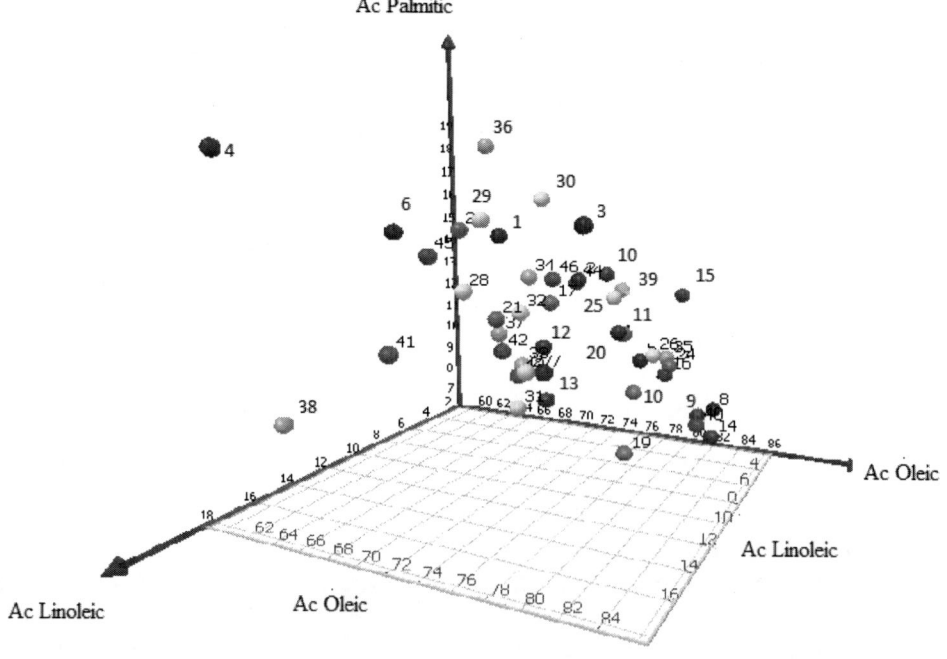

Figure 5. Positioning with respect to oleic, linoleic and palmitic acids. Oils are numbered according to table 2.

Figure 5 shows that the distribution of oils according to their content of oleic acid, palmitic and linoleic acids forms a specific spatial distribution which allows to differentiate most variaties very well, as is the case of the Arbequina (3), Picual (5), etc. We observed in this spatial distribution how the varieties with high oleic and high linoleic have low values of palmitic. It is observed that groups of varietal oils formed in the graph could be associated with a common territorial origin with native plant material.

Figures 6 and 7 show the PCA of the oils of the majority and minority varieties. In them, the importance of composition in monounsaturated in the taxonomic differentiation of varieties that are highly correlated with Factor 1 (Table 5) is confirmed. Factor 2 relates to polyphenols, sterols and tocopherols. The analysis of principal components clearly separates the varieties based on the components studied. The grouping is based on similarities as previously indicated in Figures 1, 2, 3 and 4.

We conclude with the study of our PCA, that the most balanced oils are Cornicabra, Hojiblanca and Empeltre because of their MUFAs content, mainly represented by F1, and sterols, polyphenols and α-tocopherol, represented mainly in F2, so they are more stable and of high health and diet concern.

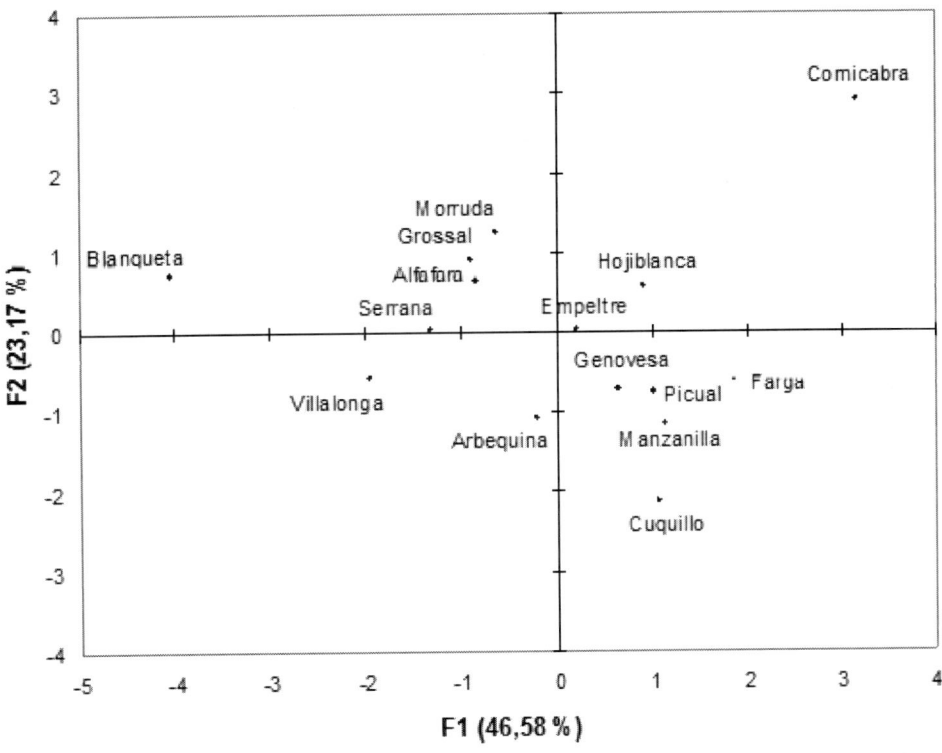

Figure 6. Principal components analysis of mayority varieties.

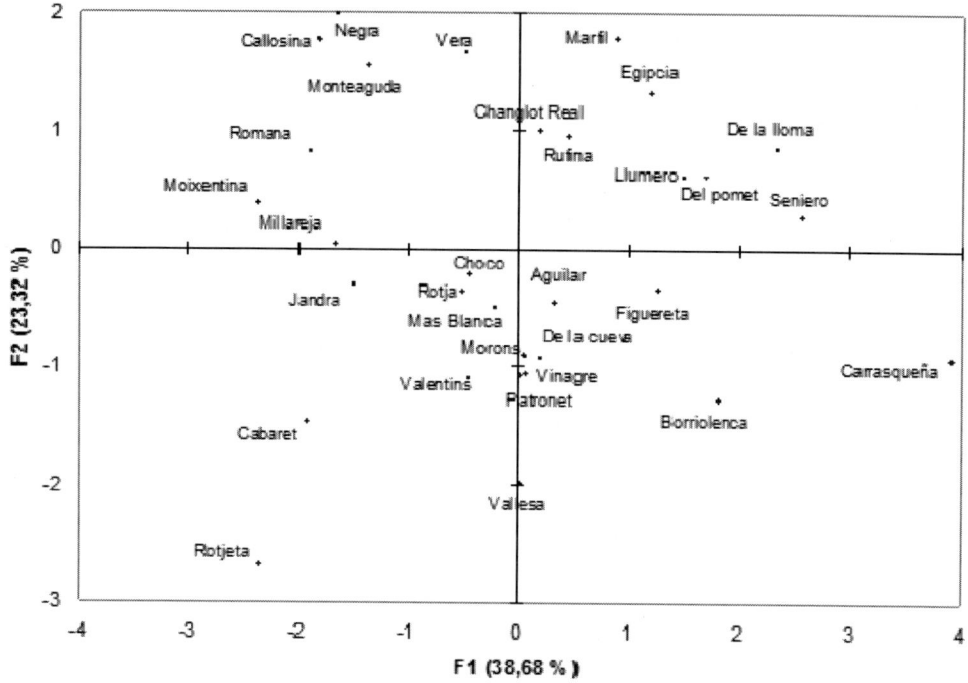

Figure 7. Principal components analysis of minority varieties.

Table 5. Correlations between variables and factors

	Mayoritarias		Minoritarias	
	F1	F2	F1	F2
SFA	-0,845	-0,201	-0,691	0,180
MUFA	0,943	-0,139	0,889	0,380
PUFA	-0,903	0,103	-0,775	-0,475
Sterols	0,388	0,747	-0,495	0,333
tocopherol	0,447	-0,217	-0,265	0,711
Polyphenols	-0,164	0,845	-0,370	0,617

It is clear that the importance of extra virgin olive oil lies mainly in its balanced composition of monounsaturated fatty acids, its high content of polyphenols and α-tocopherol for their role as antioxidants. The content of β-sitosterol, linoleic acid (omega 6) and linolenic (omega 3) are also highly interesting. As shown in Figure 7 the majority of the oils studied are well balanced in these contents.

There is a clear variability of oils and a tendency for some varieties to be much more interesting than others according to their composition, for example Cornicabra is high in monounsaturated and polyphenols, with a low content of polyunsaturated. However, other varieties like Blanqueta have low contents of monounsaturated, but high polyunsaturated and polyphenols.

We must consider that both the fatty acid content of the esterifiable and their contents in sterols, α-tocopherol, and other minor components are able to group extra varietal virgin olive oils and occasionally allow their clear and unequivocal differentiation, especially the content in fatty acid. Although the fatty acid composition has certain variability by the region of growth, it well reflects the varietal characteristics and area of origin. The fatty acid composition is considered to be a good varietal indicator in oils, especially the contents in major fatty acids: oleic, linoleic and palmitic, and oleic/palmitic ratio. Other more global relationships such as unsaturated/saturated as well as monounsaturated/polyunsaturated relationships may also be indicative of some groups of varieties.

REFERENCES

[1]   Abella I (1878). La magia de los árboles. Integral. RBA Ed. *Barcelona. España.* (Fascímil, 2000).

[2]   Abu-Zacaria (1269). Libro de agricultua. MAPA. Junta de Andalucía. *Sevilla. España* (Facsimil 2003). 1435 pp.

[3]   Aguilera C, Ramirez-Tortosa M, & Mesa M (2002). Sunflower, virgin-olive and fish oils differentially affect the progression of aortic lesions in rabbits with experimental atherosclerosis. *Atherosclerosis.* 162, 335–344.

[4]   Alonso de Herrera G (1513). Agricultura General. Real Sociedad Económica Matritiense. *Terrón Ed MAPA.* Madrid. 445 pp.

[5]   Angiolillo A, Mencuccin I, & Baldoni L (1999). Olive genetic diversity assessed using amplified fragment lengeht polymorphisms. *Theor. Appl. Genet.* 98, 411-421.

[6]   Atkinson J, Epand R.F, & Epand R.M (2008). Tocopherols and tocotrienols in membranes: A critical review. *Free Radical Biology & Medicine.* 44, 739–764.

[7]   Avila J (2002). El poder curativo del aceite de oliva. *Robin BooK Ed. Barcelona. España.* 202 pp.

[8]   Awad A B, & Fink C S (2000). Phytosterols as anticancer dietary components: Evidence and mechanism of action. *J Nutr.* 130, 2127-2130.

[9]   Awad AB, Smith A J, & Fink C S (2001). Plant sterols regulate rat vascular smooth muscle cell growth and prostacyclin release in culture. *Prostaglandins Leukot Essent Fatty Acids.* 64, 323-330.

[10]   Balagueiras E (1993). La colección oleográfica de España del Real Jardín Botánico de Madrid. *Las ciencias.* 4. 12 pp.

[11]   Barranco D & Rallo L (1984). Las variedades de olivo cultivados en Andalucía. Junta de Andalucía. Ed. *Madrid. España.* 387 pp.

[12]   Barranco D, Cimato A, Fiorino P, Touzani A, Castañeda C, Serafini F, & Trujillo I, (2000). *Catálogo mundial de variedades de olivo.* COI Ed. 360 pp.

[13]   Barranco D, Trujillo I, & Rallo L (2005). Elaiografía Hispánica, En: Variedades de olivo en España. Rallo L (Ed). MAPA. Mundiprensa. Eds. Madrid. España. 47-76.

[14]   Barton R N E, Currant A P, Fernández-Jalvo Y, Finlayson J C, Goldberg P, Macphail R, Pettitt P B, & Stringer, C B, (1999). Gibraltar Neanderthals and results of recent excavations in Gorham's, Vanguard and Ibex caves. *Antiquity.* 73, 13-23.

[15]   Belaj A, Satovic Z, Rallo L, & Trujillo I (2004). Optimal use of RAPD markers for identifying varieties in olive (*Olea europea* L) germplasm colecctions. *J. Amer. Soc. Hort. Sci.* 129, (2). 266-270.

[16] Berge K E, Tian H, Graf G A, Yu L, Grishin N V, Schultz J, Kwiterovich P, Shan B, Barnes R, Hobbs H H. (2000). Accumulation of dietary cholesterol in sitosterolemia caused by mutations in adjacent ABC transportes. *Science*. 2090, 1771-1775.

[17] Berger, A, Jones, P J H, & Abumweis, S S, (2004). Plant sterols: factors affecting their efficacy and safety as functional food ingredients. *Lipid Health Disease*. 3, 1–19.

[18] Bernal J, Mendiola J A, Ibañez E, Cifuentes, A (2011). Advanced analysis of nutraceuticals. *Journal of Pharmaceutical and Biomedical Analysis*. 55 (4), 758-74.

[19] Burton G W, Joyce A, & Ingold K U (1983). Is vitamin E the only lipid-soluble, chain breaking antioxidant in human blood plasma and erythrocyte membrane? *Arch. Biochem. Biophys*. 221, 281–290.

[20] Caballero J M, del Rio C, Navarro C, Garcia-Fernandez M D, Moraleles Hermoso M, Del Olmo L A, López F, & Cera F, Rúiz G (2005). Ensayos comparativos de olivo en Andalucia, En: Variedades de olivo en España. Rallo L En: *Variedades de olivo en España*. Rallo L (Ed). MAPA. Mundiprensa. Eds. Madrid. España.

[21] Carrión Y, Ntimon M, & Badal E (2010). *Olea Europaea* L. in the North Mediterrranean Basing during the pleniglacial and the early-middle holeocene. *Quaternary Sciency Reviews*. 29, 952-958.

[22] Colmeiro M (1865). Colección elayográfica. Láminas y herbario. Archivo del Real Jardín Botánico de Madrid. España.

[23] Columela L M (s. I). De Re Rústica (Libro XII). Holgado A Ed. (1988). MAPA. Siglo XXI Eds. Madrid. España. 339 pp.

[24] Cortesi N, Rovelli P, & Fedeli E (1992). Valuazione globale di qualità di oli e grassi. 1. Studio della componente trigliceridica. *Riv. Italiana delli Sust. Grasse*. 69. 1-3.

[25] Couture J (1786). Traité de l'Olivier. Rediviva Ed. Nimes. Francia. 460 pp. Facsimil (1996).

[26] El Antari A, El Moudini A, & Ajana H (2003). Comparación de la calidad y la composición acídica del aceite de oliva de algunas variedades mediterráneas cultivadas. *Olivae*. 95, 26-31.

[27] Espejo Z (1898). *Cultivo del olivo*. Hijos M.G. Hernández Ed. Madrid. España.

[28] Fabbri A, Hornazza J, & Polito V S (1995). Random amplified polymorphy DNA analysis of olive *(Olea europea* L.) cultivars. *J Amer. Soc. Hort. Sci*. 120 (3), 538-542.

[29] Fedeli E (2002). Un quarantennio di recherché e di lotta alla sofisticazione. *Olivo e Olio*. 5, (1/2). 26-31.

[30] Fontanazza G (1993). Olivicultura intensiva mecanizzata. *Edagricole Ed. Bolonia* 312 pp.

[31] Font-Quer P (1999). *Plantas medicinales. Dioscórides renovado*. Labor. Ed. Barcelona. España. 1033 pp.

[32] Kratz M, Cullen P, & Kannenberg F (2002). Effects of dietary fatty acids on the composition and oxidizability of low-density lipoprotein. *Eur J Clin Nutr*. 56, 72–81.

[33] Giuffrè A M, & Louadj L (2013). Influence of crop season and cultivar on sterol composition of monovarietal olive oils in Reggio Calabria (Italy). *Czech J. Food Sci*. 31, 256–263.

[34] Godoy-Caballero M P, Acedo-Valenzuela M I, & Galeano-Díaz T. (2012). Simple quantification of phenolic compounds present in the minor fraction of virgin olive oil by LC–DAD–FLD. *Talanta*. 101, 479–487.

[35] Hensley K, Benaksas E J, Bolli R, Comp P, Grammas P, Hamdheydari, L, Mou S, Pye, Q N, Stoddard, M F, & Wallis G (2004). New perspectives on vitamin E: α-tocopherol and carboxyethylhydroxychroman metabolites in biology and medicine. *Free Radic. Biol. Med.* 36, 1–15.

[36] Hidalgo-Tablada J (1870). Tratado del cultivo del olivo en España y modo de mejorarlo. Cuesta Ed. Madrid. España. 323 pp.

[37] Huang C L, Sumpio B E (2008). Olive oil, the mediterranean diet, and cardiovascular health. *Journal of the American College of Surgeons.* 207(3), 407-416.

[38] Iñiguez A, Paz S, & Illa F (2001). Variedades de olivo cultivadas en la Comunidad Valenciana. Generalitat Valenciana Ed Valencia. España. 267 pp.

[39] Jossa F, & Manzini M. (1998). Alimentación longevidad y enfermedades cardiovasculares. *Olivae.* 73, 10-17.

[40] Larsen L, Jespersen J, & Marckmann R (1999). Are olive oil diets antithrombotic? Diets enriched with olive, rapeseed, or sunflower oil affect postprandial factor VII differently. *Am. J. Clin. Nutr. 70*, 976–982.

[41] Lev Efraim, Mordechai E K, & Ofer Bar-Y. (2005). Mousterian vegetal food in Kebara cave, Mt. Carmel. *Journal of Archaeological Science.* 32. (3) 475-484.

[42] López-Cortés I, & Salazar D M. (2006). Variedades de olivo y composición de sus aceites en el oeste del Maditerránaeo. *Phytoma Ed. Valencia. España.* 443 pp.

[43] Loukas M, & Krimbas C, (1983) Hystory of olive cultivars base don their genetic distances. *J. Hort. Sci.* 58. 121-127.

[44] Mata C, Badal P, Colado E, & Ripollés P. (2010). Flora ibérica. SIP. Diputación Valencia Ed. Valencia. *España.* 184 pp.

[45] Martínez Robles F (1833). Ensayo sobre castas de olivos en Andalucía. En: *Tratado sobre el movimiento y aplicaciones de las aguas.* Vallejo J.M. (Ed) Miguel Burgos Ed Madrid. España.

[46] McNamara D J (2000). Dietary cholesterol and atherosclerosis. *Biochim Biophys Acta* 1529, 310-320.

[47] Monzo Ch, Oreggia M, Tiliacos C (2003). *Olio extra vergine d'oliva.* Nardini Ed. Florencia. Italia. 185pp.

[48] Nei M, & Li W H (1979). Mathematical model for studing genetic variation in terms of restriction endonucleases. *Prod. Acad Soc of Usa.* 76, 5269-5273.

[49] Ortega-Nieto J M. (1955). Las variedades de olivo cultivadas en España. *INIA-Min. Agricultura.* Eds. Madrid. España. 73 pp.

[50] Owen R W, Giacosa A, Hull W E, Haubner R, Spiegelhalder B, &Bartsch H, (2000). The antioxidant/anticancer potential of phenolic compounds isolated from olive oil. *European Journal of Cancer* 36 (10), 1235-47.

[51] Pannelli G, Alfeti B, Ambrosio A, Rosati S, & Famiani F (2000). *Varietá di olivo in Umbria.* Pliniana Ed. Perugia. Italia. 95 pp.

[52] Papajatsin N (1999). *Argolida. La cultura de los palacios. Clio Ed. Heraclion. Creta.* 151 pp.

[53] Pagnol J. (1975). L´Olivier. Aubanel Ed. *Avignon.* Francia. 180 pp.

[54] Patac L, Cadahia P, & Del Campo E (1954). *Tratado de olivicultura.* Sindicato Nac. Del Olivo Ed. Madrid. España. 646 pp.

[55] Perez-Camacho F, & Rallo L (1979). Selección de caracteres morfológicos cunatitativos de cultivares de olivo *(Olea europea* L.*)* a efectos de clasificación. *An. INIA. Prod. Veg.* 10.213-231.

[56] Perona J S, Cabello-Moruno R, & Ruiz-Gutierrez V (2006). The role of virgin olive oil components in the modulation of endothelial function. *Journal of Nutritional Biochemistry.* 17, 429–445.

[57] Pontikis C A, Loukas M, & Kousinis G (1980). The use of biochemical markers to distinguish olive cultivar. *Journal of Horticultural Science.* 55, (4). 333-343.

[58] Reiter E, Jiang Q, & Christen S (2007). Anti-inflammatory properties of a- and c-tocopherol. *Molecular Aspects of Medicine.* 28, 668–691.

[59] Ribera J (1946). *Agricultura y zootecnia. Tratado teórico-pratico.* Soler Ed. Barcelona. España. 411 pp.

[60] Rojas Clemente S (1815). Lecciones de Agricultura Sandalio Arias Ed. *Real Jardín Botánico.* Madrid. España.

[61] Rojo C (1840). Arte de cultivar el olivo. *Cabrerizo El Olivo Ed. Valencia. Jaen.* 320 pp. (Facsimil, 2001).

[62] Rosignoli P, Fuccelli R, Fabiani R, Servili M, & Morozzi G (2013). Effect of olive oil phenols on the production of inflammatory mediators in freshly isolated human monocytes. *Journal of Nutritional Biochemistry.* 24 (8), 1513-1519.

[63] Said H, Nassima G, & Beltrassem B (1995). Caracterización biométrica y enzimática de algunas variedades de olivo pertenecientes a la colección mediterránea. *Olivae.* 55, 31-54.

[64] Salazar D M (1990). Pomological typication of olive tree in Valencia. *Acta Hortculturae.* 286, 101-104.

[65] Salazar D M, López-Cortés I, Fernandez M A, Carot R, (1997). El olivo en Valencia. Materiales vegetales (I). *Fruticultura profesional.* 88, 20-27.

[66] Salazar D M, López-Cortés I, & Carot R, (2001). Nuevas prospecciones, caracterización y tipificación de variedades y aceites monovarietales en la Comunidad Valenciana. *Fruticultura profesional.* 120, 13-17.

[67] Salazar-García D C, Power R C, Sanchis-Serra A, Villaverde V, Walker M J, Henry A G. (2013). Neanderthal diets in central and southeastern Mediterranean Iberia. *Quaternary International.* http://dx.doi.org/10.1016/j.quaint.2013.06.007.

[68] Sanz-Cortes F, Badenes M, Paz S, Iñiguez A, & LLacer G. (2001) Molecular Characterization of olive cultivars using RAPD markers. *J Amer. Soc. Hort. Sci.* 126, 1. 7-12.

[69] Schäfer-Schuchardt H. (1988). L'Oliva: la grande storia di un piccolo frutto. *Arti Graffavia.* Bari, Italy.

[70] Solecki R. (1971). Shanidar The First Flower People. Knopf. Columbia. New York.

[71] Stanton R (1996). Perspectivas de las ventajas nutricionales del aceite de oliva. 345-381 En: *Enciclopedia Mundial del Olivo.* COI. Plaza y Janés Ed. Barcelona. España. 247-248.

[72] Servili M, Baldioli M, Selvaggini R, Macchioni A, & Montedoro G (1999). High-performance liquid chromatography evaluation of phenols in olive fruit, virgin olive oil, vegetation waters, and pomace and 1d- and 2d-nuclear magnetic resonance characterization. *Journal of the American Oil Chemists.* 76 (7), 873-882.

[73] Singleton V L & Rossi J (1965). Colorunetry of total phenolics with phosphomolybdic-phosphotungstic acid reagents. *Amer. J. Enol. Viticult.* 16, 144-58.

['/4] Slover H T, Thompson R II, & Mcrola G V (1983). Determinationof tocopherols and sterols by capillary gas chromatography. *J. Assoc. Of. Chem. Sci.* 80, (8) 1524-1528.

[75] St-Onge M P, Lamarche B, Mauger J F, & Jones P J H (2003). Consumption of a functional oil rich in phytosterols and medium-chain triglyceride oil improves plasma lipid profiles in men. *Journal of Nutrition.* 133, 1815–1820.

[76] Suc J P. Origin and evolution of the Mediterranean vegetation and climate in Europe. *Nature*, 307 pp.

[77] Tavantinia G (1819). *Trattato teorico-pratico completo sull-olivo.* Florencia. Italia. (Facsimil).

[78] Tous J (1990). El Olivo, situación y perspectivas en Tarragona. Diputación Prov. Tarragona Ed. *Tarragona. España.* 376 pp.

[79] Tous J, & Romero A (1993). Variedades de olivo con especial referencia a Cataaluña. La Caixa-Aedos Eds. *Barcelona. España.* 172 pp.

[80] Trujillo I, Rallo L, & Arus P (1995). Identifying olive cultivars by isozyme analysis. *J. Amer. Soc. Hort. Sci.* 123. 2, 318-324.

[81] Tsouchtidi I (2003). Olive and oil. *Toubi Ed Koropi. Grecia.* 127 pp.

[82] UPOV. International Union for the protectionof new Varieties of Plants. Norma TG 99/4.

[83] Valenzuela (1940). Curso de olivicultura e industrias derivadas. *J. Hays Bell Ed.* Buenos Aires. Argentina. 160 pp.

[84] Vergari G, Patumi M, & Fontanazza G. (1996). Use of RAPD markers in the Characterization of olive germoplasm. *Olivae.* 60, 19-22.

[85] Visioli F, & Galli C (2001). Antiatherogenic compoomnets of olive oil. *Current Atheroscleorsis Reports.* 3, 64-67.

[86] Visioli F, Ieri F, Mulinacci N, Vincieri F F, & Romani A (2008). Olive-oil phenolics and health: Potential biological properties. *Natural Product Communications* 3 (12), 2085-2088.

[87] Weisman Z, Avidan N, Lavee S, & Quebedeaux B (1998). Molecular Characterization of common olive cultivars in the West Bank (Israel) using randomly amplified polymorphic DNA (RAPD) markers. *J. Amer. Soc. Hort. Sci.* 123. 5, 837-841.

[88] Willet W C, Sacks F, & Trichopoulou A (1995). Mediterraneon diet pyramid a cultural model for healthy eating. *Amer. J. Clin. Nutri.* 61, 1402-1406.

[89] Willet W C (1997). Especific fatty acids and risk of breast prostate cancer: dietary intake *Amer. J. Clin. Nutri.* 66, (6). 1557-1563.

[90] Yang D, Kong D, & Zhang H (2007). Multiple pharmacological effects of olive oil phenols. *Food Chemistry.* 104, 1269–1271.

In: Virgin Olive Oil
Editor: Antonella De Leonardis

ISBN: 978-1-63117-656-2
© 2014 Nova Science Publishers, Inc.

Chapter 5

EFFECT OF THE GROWING AREA AND CULTIVAR ON PHENOLIC CONTENT AND VOLATILES COMPOUNDS OF RELATED PRODUCTS OF SELECTED TUNISIAN OLIVE VARIETIES

Faten Brahmi[1]*, Guido Flamini[2], Beligh Mechri[1]
and Mohamed Hammami[1]

[1]Laboratory of Biochemistry, UR "Human Nutrition and Metabolic Disorder"
Faculty of Medicine, Tunisia
[2]Dipartimento di Chimica Bioorganica e Biofarmacia via Bonanno Pisa, Italy

ABSTRACT

This chapter reports a comparative study based on the volatile compounds, polyphenols, orthodiphenols and flavonoides content from the northern Tunisian cultivars; *chemlali* and *neb jmel* and from the southern Tunisian cultivar; *chemchali*. There were differences between the leaves, fruits and stems from the cultivars when grown in the different environments. Volatiles oils were also influenced by the pedoclimatic conditions and the organ; hence, the leaves and stems of the southern varieties seem to be richer in total alcohols and total C6 alcohols than the northern variety. The sesquiterpene hydrocarbons, present in all the volatiles oils studied was more abundant in leaves volatile oil from the northern varieties (14.7% and 15.9% for *chemlali* and *neb jmel*, respectively) than in the southern variety. Results presented suggest the assumption that accumulation of polyhenols is most active in the fruits of the southern cultivar and generally more active in the leaves and stems of the northern cultivars than southern cultivars. In general, the genetic (varietal) and organs studied may be used to discriminate and to characterise the location conditions.

Keywords: Tunisian olive varieties, location, organs, volatiles compounds, total phenols

1. INTRODUCTION

In Tunisia, *Olea europaea* has a wide distribution, with a cultivated area of 1.6 million ha, predominantly in the center and south areas of the country. It represents an important economical and environmental species, making Tunisia the fourth major olive producing country in the world, accounting for 210,000 tons of olive oil production per year [1]. Tunisian olivicoltura constitutes one of the principal economical and agricultural strategic sectors that are known for their richness of varieties [2]. More than 50 different cultivars are found throughout Tunisia. The olive-growing areas spread from the northern to the southern regions, where a wide range of edaphic-climatic conditions are prevailing, from mild semi-arid to arid conditions [3]. On the other hand, this crop is endowed with adaptation flexibility to different environments and cropping methods.

The concentrations of volatile compounds of olive are a frequent objective of recent studies. Their levels depend on the level and activity of enzymes which are genetically determinant. In fact, many papers on volatiles from olives and olive oil can be found in the literature. Some of them describe the relationships between volatile compounds and virgin olive oil odor notes [4-6]. These chemicals have been also used to characterize the aroma of the oil obtained from the cultivars. A direct comparison with literature data is not possible because of the great variability of the volatiles composition with reference to the different ripeness stages of olives, extraction techniques, and analytical methods;

Polyphenols belong to the category of natural antioxidants and are the most abundant antioxidants in our diet [7]. They play an important role in human nutrition as preventative agents against several diseases, protecting the body tissues against oxidative stress. Many studies indicate an antioxidant capacity of these polyphenols with respect to the oxidation of low-density lipoproteins [8] and oxidative alterations due to free radical and other reactive species [9]. Polyphenols, although their minor concentrations in olive oil, have shown to be effective antioxidants. These antioxidant properties arise from their high reactivity as hydrogen or electron donors, and from the ability of the polyphenol derived radical to stabilize and delocalize the unpaired electron (chainbreaking function), and from their ability to chelate transition metal ions (termination of the Fenton reaction) [10-11]. The present chapter aimed to study the constituents of the volatile oils of leaves, fruits and stems from the Tunisian cultivars (*chemlali* and *neb jmel*) planted in north Tunisia and (*chemchali*) planted in south Tunisia and to evaluate their polyphenols, orthodiphenols and flavonoids content.

2. MATERIALS AND METHODS

2.1. Sample Preparation

The aerial part (leaves, fruits and stems) of the Tunisian cultivars (*chemlali* and *neb jmel*) was collected from the coastal region of Mahdia—the centre of Tunisia and (*chemchali*) was collected from Gafsa—the south of Tunisia.

2.2. Extraction Method

Each sample (500 g) of the fresh leaves, fruits and stems was subjected to hydrodistillation for 3 h using a Clevenger- type apparatus. The volatiles obtained after trapping in diethyl ether were dried over anhydrous sodium sulfate, evaporated and concentrated under a gentle stream of N2 and stored at 4°C until tested and chemically analyzed.

2.2.1. Methanol Extract

The leaves and fruits samples (5 g) were extracted with 100 ml of methanol solvent in an orbital shaker (Eyela Model MMS-300, Tokyo Rikakikai Co., Ltd., Japan) at 200 rpm, at room temperature and after centrifugation at 1,000g for 15 min, supernatant was decanted and the pellets were extracted under identical conditions. The stems were manually ground to a powder and then extracted with absolute methanol, in a 1:10 (w/v) ratio of herb to solvent, for 4 h under a continuous reflux set up in a Soxhlet extractor.

2.3. Identification of the Volatile Constituents

GC analyses were accomplished with an HP-5890 series II instrument equipped with a HP-5 capillary column (30 m 9 0.25 mm, 0.25 μm film thickness), working with the following temperature program: 60°C for 10 min, ramp of 5°C/min to 220°C; injector and detector temperatures, 250°C; carrier gas, nitrogen (2 ml/min); detector, dual FID; split ratio, 1:30; injection, 0.5 μl.

The identification of the components was performed, for both columns, by comparison of their retention times with those of pure authentic samples and by means of their linear retention indices (LRI) relative to the series of n-hydrocarbons.

Gas chromatography–electron impact mass spectrometry (GC–EIMS) analyses were performed with a Varian CP 3800 gas chromatograph (Varain, Inc. Palo Alto, CA) equipped with a DB-5 capillary column (Agilent Technologies Hewlett-Packard, Waldbronn, Germany; 30 m = 0.25 mm; coating thickness × 0.25 mm) and a Varian Saturn 2000 ion trap mass detector. Analytical conditions were as follows: injector and transfer line temperature at 250 and 240°C, respectively; oven temperature was programmed from 60 to 240°C at 3°C/min; carrier gas, helium at 1 ml/min; splitless injector. Identification of the constituents was based on comparison of the retention times with those of the authentic samples, comparing their LRI relative to the series of n-hydrocarbons, and on computer matching against commercial [NIST 98 (U.S. National Institute of Standards and Technology) and homemade library mass spectra built from pure substances and components of known samples and MS literature data [12-13].

Moreover, the molecular weights of all the identified substances were confirmed by gas chromatography–chemical ionization mass spectrometry (GC–CIMS), using methanol as chemical ionization gas.

2.4. Total phenols and O-Diphenols

Total phenolic and o-diphenols contents of fractions were determined according to the method of Montedoro et al. (1992) [14] with minor modifications.

2.5. Determination of Total Flavonoids

Total flavonoid contents (TF) of the extracts were determined according to the colorimetric assay developed by Zhishen et al. (1999) [15].

2.6. Statistical Analyses

All extractions and determinations were conducted in triplicates and results were expressed on the basis of dry matter weight. Data were recorded as mean ± standard deviations.

3. RESULTS

The major constituents of the volatile oil of the leaves of *chemlali* were 3-ethenylpyridine (18.1%), (E)-3-hexenol (16%) and (E)-ß-damascenone (14.7%). From the leaves of *neb jemel* (E)-3-hexenol (15.9%), (E)-β-damascenone (12.0%), and 3-ethenylpyridine (10.9%) were characterized as main constituents, while the volatile oil from the leaves of *chemchali* was found to be rich in (E)-3-hexenol (28%), phenylethyl alcohol (14.9%), and 1-hexanol (11.0%). On the other hand, the main constituents of the volatile oil from the fruits of *chemlali* were (E,E)-2,4-decadienal (23%), (E,Z)- 2,4-decadienal (14.9%), nonanal (6.7%), (E)-2-decenal (4.7%) and benzaldehyde (4%) whereas the most abundant components of *neb jmel* olives were 3-ethenylpyridine (15.5%), (E)-2-decenal (14.4%), (E)-2-undecenal (7%), (E,Z)-2,4-hexadienal (5.6%), (Z)-2-heptenal (5%), Benzaldehyde (4.5%) and (E)-2-octenal (4%). In the case of volatile oil from the fruits of *chemchali.*, the main components of the volatile oil were phenylethyl alcohol (15.1%), nonanal (10.0%), and benzyl alcohol (7.4%). In the case of stems, the major constituents of the volatile stems of *chemlali* composition were phenylethyl alcohol, benzyl alcohol, methyl salicylate and 3-ethenylpyridine with, respectively, 14.0, 11.0, 10.6 and 9.2 %, while the volatile oil from the stems of *neb jmel* was dominated by phenylethyl alcohol 12.9 %, nonanal (8.8 %), 3-ethenylpyridine (7.4 %) and benzyl alcohol (6.7 %). In fact, the volatile oil from the stems of *chemchali* was found to be rich in nonanal, (E)-2-decenal and benzyl alcohol were the main components with respectively 9,9%, 9,6% and 9,0%.

According to Figure 1, considerable amounts of various chemical classes of total alcohols, total aldehydes, total C6 alcohols and total C6 aldehydes compounds were found in leaves, fruits and stems examined in this study.

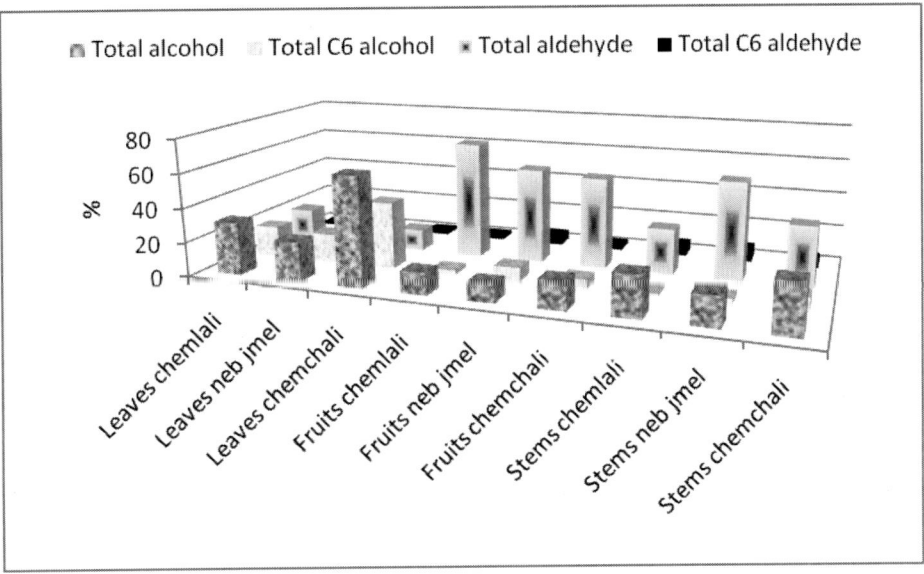

Figure 1. Content of total alcohol, total aldehyde, total C6 alcohol and total C6 aldehyde compounds in olive leaves, fruit and stems from the Tunisian cultivars.

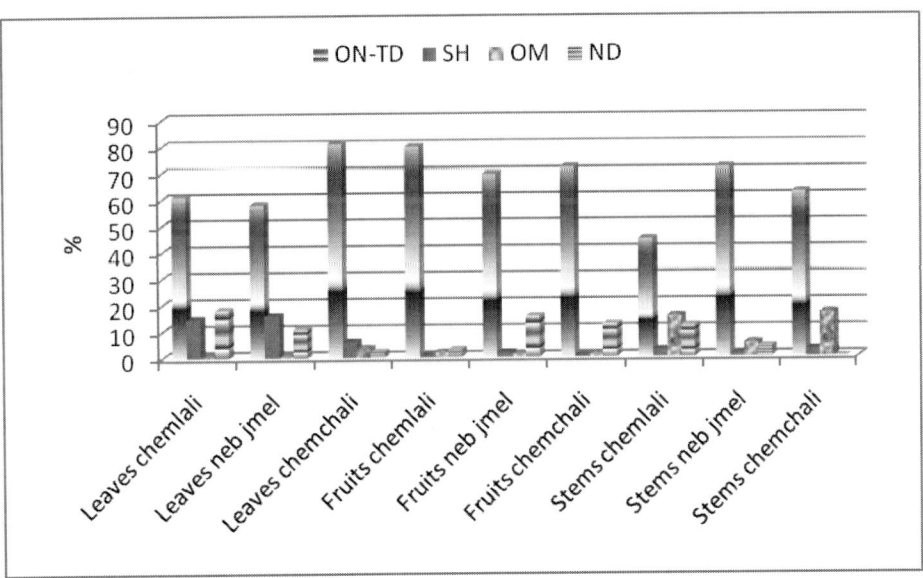

Figure 2. Composition according to compound families of the volatiles oils obtained from leaves, fruits and stems of O. europaea L. chemlali, neb jmel and chemchali. OM, oxygenated monoterpenes; SH, sesquiterpene hydrocarbons; ON-TD, oxygenated non terpene derivatives; ND, nitrogen derivatives.

The leaves and stems volatile oils were characterised by a high content of total alcohols with 22.6% to 63.3% and with 17.1% to 28.4% for leaves and stems, respectively (Figure 1). In fact, the fruits of three cultivars studied were more representative by total aldehydes than the leaves and stems studied. On the hand, the amounts of C6 alcohols were clearly higher in the leaves of the northern and southern varieties than in the fruits and stems. On the other hand, the leaves and stems of the southern varieties seem to be richer in total alcohols and total C6 alcohols than the northern variety. Furthermore, generally the amount of total aldehydes and total C6 aldehydes is greater in the fruits and stems of the varieties *chemlali* and *neb jmel* than fruits and stems of the variety *chemchali*.

Oxygenated non terpene derivatives showed the highest levels of volatiles oils in all leaves, fruits and stems from the northern and southern varieties analyzed, being generally superior in the leaves and fruits, with global average content ranging from 58.1 to 81.1 % and from 69.3 to 80.0 % in leaves and fruits, respectively (Figure 2). It is important to note, that the sesquiterpene hydrocarbons, present in all the oils studied was significantly more abundant in leaves volatile oil from the northern varieties (14.7% and 15.9% for *chemlali* and *neb jmel*, respectively) than in the southern variety (6.1% for *chemchali*). Consequently, Nitrogen derivatives had an intermediate place, since they represented generally the highest level in the organ of the northern varieties (Figure 2). The fruits of olives had the highest level of total polyphenols while the stems had the lowest amount (Figure 3). The leaves of olive cultivars had a medium value. In the case of leaves, the extract of *chemlali* showed higher phenolic content (219.85 mg/100g) than extract obtained from *neb jmel* and *chemchali*. For the fruits, the phenol levels in the extract from the south (642.89 mg/100g) were about 2 times greater than those in the extract from the north (484.1 mg/100g and 443.27 mg/100g for *chemlali* and *neb jmel*, respectively). However, for stems, the methanolic extract from *neb jemel* had higher total phenols phenols value (78.26 mg/100g), compared to the another northern variety (40.87 mg/100g) and southern variety (47.0 mg/100g). Concerning total orthodiphenols contents, the fruit of the cultivars studied was detected to have the highest orthodiphenol contents followed by leaves and stems (Figure 3).

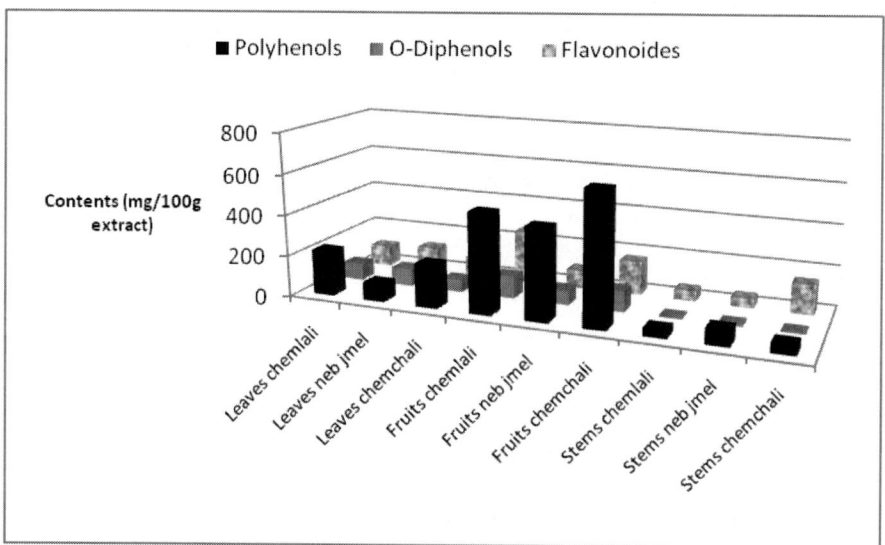

Figure 3. Total polyphenols, o-diphenols and flavonoids contents of different methanolic extracts of Olea europaea L. leaves, fruits and stems from the Tunisian cultivars. Total phenol and total o-diphenols contents were determined as hydroxytyrosol equivalents in milligrams per 100 g of dry weight; total flavonoid content was determined as mg catechin equivalents CEQ per 100 g of dry weight (DW).

The cultivars *chemlali* and *neb jmel* have the highest concentration of orthodiphenol; hence leaves planted in north produced orthodiphenol in greater amounts than those planted in south. It was noticeable that fruits of the varieties *chemlali*, *neb jmel* and *chemchali* produced important levels of the orthodiphenol content (Figure 3). However, for the stems, the three varieties contain poor level of total orthodiphenol. In fact, no remarkable variation was found in o-diphenols contents for the stems (Figure 3). The total flavonoids contents varied between the extracts (Figure 3). As a comparison, leaves of *chemlali* and *neb jmel* had higher total flavonoids contents than leaves of *chemchali*. On the contrary, concerning the stems, the southern variety; *chemchali* was detected to have the highest flavonoids contents followed by the varieties *chemlali* and *neb jmel* (Figure 3).

4. DISCUSSION

Comparing the three cultivars and organs collected for olive in the same period, a great qualitative and quantitative difference in the constituents of the volatile oils may be observed. Therefore, the volatiles oils obtained from olive leaves, fruits and stems from southern and northern varieties showed different volatiles constituents that permitted to distinguish among them. This supports the results reported by Flamini et al. (2003) [16] in the leaves, olive paste and virgin olive oil. Although there are surely different biogenesis pathways of volatile compounds between leaves, fruits and stems, it is indeed important to consider some enzymatic pathways which occur in olive oil flavor compounds biosynthesis. Taking into consideration the effect of the growing area conditions, it is important to note that organs of the variety grown at low altitude and high temperature (south) had the highest level of total alcohols; hence total alcohols may be used as markers to differentiate the organs of olive of different geographical origins [17]. These results are in agreement with the study of Vichi et al (2003) [18], which found that levels of alcohols in olive oil showed a strong dependence on geographical origin. Therefore, the amounts of C6 alcohols clearly higher in the leaves of the northern and southern varieties may be explained by a higher activity of the enzymes involved in their production. In fact, the richest of leaves and stems of the southern varieties in total C6 alcohols than total C6 aldehydes could be attributed either to the differential activity of the enzyme alcohol dehydrogenase (ADH), which reduces the C6 aldehyde compounds to the corresponding alcohols [19] (Figure 4).

In fact, C6 aldehydes and alcohols and their corresponding esters are the most volatile compounds present in olive oils and are formed by polyunsaturated fatty acids through a sequence of enzymatic reactions collectively known as the LOX pathway [20]. In addition, the important amounts of total aldehydes and total C6 aldehydes in fruits and stems from the three cultivars studied and generally their higher content in fruits and stems of the northern varieties than southern varieties is believed to depend on location. Consequently, the amounts of C6 aldeydes and their acetate (LOX products) are mainly related to the different olive cultivar, organ and location. In the studied chapter, the environmental factors, cultivar kind and different organs of olive appeared to influence the formation of C6-compound, probably because of changes in LOX activity.

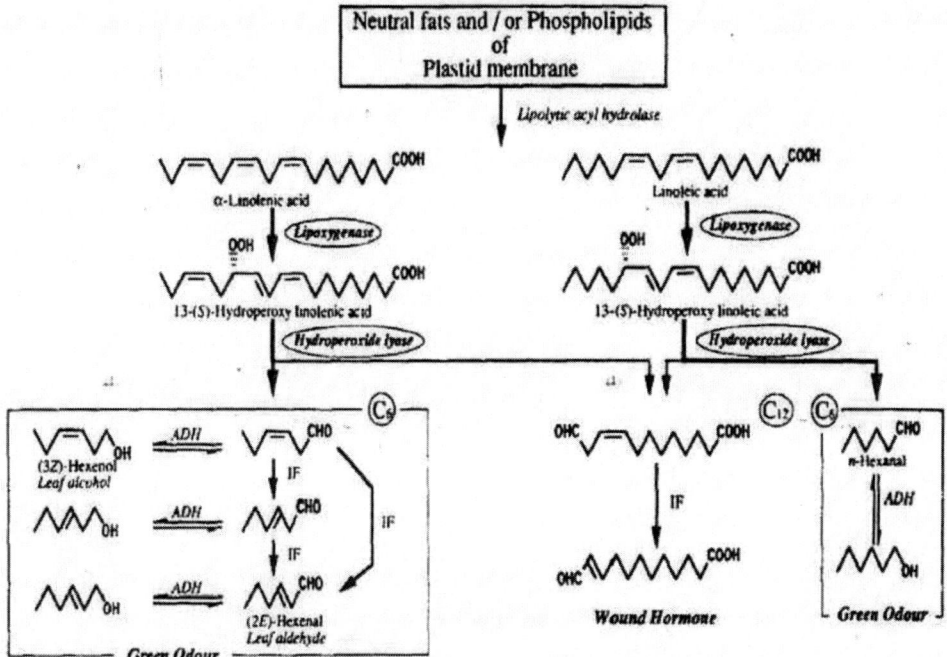

Figure 4. Lipoxygenase pathways.

These results confirmed that each organ and cultivar of each region can be distinguishable from the other. The families compounds accumulate differently depending on the cultivar and organ that their metabolites accumulation depends from the genetically determined enzyme composition. These results proved the findings of Angerosa et al. (1999) [21]. These elaborated results indicated a strict dependence of different compounds families (sesquiterpene hydrocarbons and nitrogen derivatives) on the enzymatic store, which generally is more active in the different organ from the northern cultivars than the organ from the southern cultivars. In the literature only very few papers [22-25] report the presence of the sesquiterpene hydrocarbons, which could play a very important role in the fragrance of the olive.

Different responses to the location were observed for the cultivars. The elaborated result is not in accordance with Romani et al. (2004) [26] which found that the Myrtle stem had higher total phenols than leaf. These results suggest the assumption that accumulation of polyhenols is most active in the fruits of the southern cultivar and generally more active in the leaves and stems of the northern cultivars than southern cultivars. Differences were also found in o-diphenols contents. This result showed that the location (north) affect the accumulation of total orthodiphenols and total flavonoids in leaves tissues from the cultivars *chemlali* and *neb jmel*. Furthermore, the richness of the stems from the south by total flavonoids, in comparison to the stems from the north of Tunisia seem to support the idea that phenolic accumulation can be a organ dependent regulated process. Apart from the geographic growing area conditions, genetic (varietal) and organs studied may also be critical in this respect.

CONCLUSION

In conclusion, the results of this chapter demonstrated that the organs of the variety grown at low altitude and high temperature (south) had the highest level of total alcohols. Therefore, these elaborated results indicated that such compounds families (sesquiterpene hydrocarbons and nitrogen derivatives) are higher in the different organ from the northern cultivars than the organ from the southern cultivars. Moreover, it can be concluded that the location (north) affect the accumulation of total orthodiphenols and total flavonoids in leaves tissues. However, concerning the stems, the southern variety; *chemchali* was detected to have the highest content of flavonoids activity, indicating an important impact of the region and organ on their phenolic content.

REFERENCES

[1]   Hannachi H, Msallem M, Ben Elhadj S, & El Gazzah M (2007) Influence du site géographique sur les potentialités agronomiques et technologiques de l'olivier (*Olea europea* L.) en Tunisie. *Comptes Rendus. Biologies,* 330, 135–142.

[2]   Abaza L, Daoud D, Msallem M, & Zarrouk M (2001) Evaluation biochimique des huiles de sept variétés d'Olivier cultivées en Tunisie. *Oléa Corps Gras Lipides*, 9, 174–179.

[3]   Issaoui M, Ben Hassine K, Flamini G, Brahmi F, Chehab H, & Aouni Y (2009) Discrimination of some monovarietal olive oils according to their oxidative stability, volatiles compounds and sensory analysis. *Journal of Food Lipids*, 16, 164–186.

[4]   Angerosa F, Mostallino R, Basti C, & Vito R (2000) Virgin olive oil odour notes: their relationships with volatile compounds from the lipoxygenase pathway and secoiridoid compounds. *Food Chemistry*, 68, 283-287.

[5]   Blekas G, Guth H, & Grosch W (1994) Changes in the levels of olive oil odorants during ripening of the fruits. In *Trends in FlaVour Research*; Maarse H, van der Heij DG, Eds.; Elsevier: Amsterdam, The Netherlands, pp 499-502.

[6]   Guth H, & Grosch W (1991) A Comparative study of the potent odorants of different virgin olive oils. *Fat Science Technology*, 93, 335.

[7]   Boskou D, & Visioli F (2003) Biophenols in table olives. In M. P. Vaquero, T. Garcia-Arias, & A. Garbajal (Eds.), Bioavailability of micronutrients and minor dietary compounds. Metabolic and technical aspects. *Research Signpost*.

[8]   Andrikopoulos NK, Antonopoulou S, & Kaliora AC (2002) Oleuropein inhibits LDL oxidation induced by cooking oil frying by products and platelet aggregation induced by platelet-activating factor. *Lebensmittel Wissenschaft und Technologie*, 35, 479–484.

[9]   Soler-Rivas C, Espı´n J. C, & Wichers HJ (2000) Oleuropein and related compounds. *Journal of the Science of Food and Agriculture*, 80, 1013–1023.

[10]   Rice-Evans CA, Miller NJ, & Paganga G (1997) Antioxidant properties of phenolic compounds. Reviews. *Trends in Plant Science*, 2, 152-159.

[11]   Covas MI, Ruiz-Gutiérrez V, De la Torre R, Kafatos A, Lamuela- Raventos RM, Osada J, Owen RW, & Visioli F (2006) Minor components of olive Oil: evidence to date of health benefits in humans. *Nutrition Reviews*, 64(9),20-30.

[12] Davies NW (1990) Gas chromatographic retention indexes of monoterpenes and sesquiterpenes on methyl silicone and Carbowax 20M phases. *Journal of Chromatography*, 503, 1-24.

[13] Adams RP (1995) Identification of Essential Oil Components by Gas Chromatography/Mass Spectroscopy; Allured: Carol Stream, IL.

[14] Montedoro GF, Servili M, Baldioli M, & Miniati E (1992) Simple and hydrolysable compounds in virgin olive oil. Their extraction, separation and quantitative and semiquantitative evaluation by HPLC. *Journal of Agricultural and Food Chemistry,* 40,1571–1576.

[15] Zhishen J, Mengcheng T, & Jianming W (1999) The determination of flavonoid contents in mulberry and their scavenging effects on superoxide radicals. *Food Chemistry,* 64,555–559.

[16] Flamini G, Cioni P, & Morelli I (2003) Volatiles from Leaves, Fruits, and Virgin Oil from *Olea europaea* Cv. Olivastra Seggianese from Italy. *Journal of Agricultural and Food Chemistry,* 51,1382–1386.

[17] Issaoui M, Flamini G, Brahmi F, Dabbou S, Ben Hassine K, Taamali A, Chehab H, Ellouz M, Zarrouk M, & Hammami M (2010) Effect of the growing area conditions on differentiation between Chemlali and Chetoui olive oils. *Food Chemistry,* 119,220–225.

[18] Vichi S, Pizzale L, Conte LS, Buxaderas S, & López-Tamames E (2003) Solidphase microextraction in the analysis of virgin olive oil volatiles fraction: Characterisation of virgin olive oils from two distinct geographical areas of northern Italy. *Journal of Agricultural and Food Chemistry,* 57, 6572–6577.

[19] Angerosa F, & Basti C (2001) Olive oil volatile compounds from the lipoxygenase pathway in relation to fruit ripeness. *Italian Journal of Food Science,* 13,421–428.

[20] Feussner I, & Wasternack C (2002) The lipoxygenase pathway. *Annual Review of Plant Biology,* 53, 275–297.

[21] Angerosa F, Basti C, & Vito R (1999) Virgin olive oil volatile compounds from lipoxygenase pathway and characterization of some Italian cultivars. *Journal of Agricultural and Food Chemistry,* 47, 836–839.

[22] Flath RA, Forrey RR, & Guadagni D G (1973) Aroma components of olive oil. *Journal of Agricultural and Food Chemistry,* 21, 948-952.

[23] Fedeli E (1977) Caratteristiche organolettiche dell'olio di oliva. *Rivista Italiana Delle Sostanze Grasse,* 54, 202-205.

[24] Bentivenga G, D'Auria M, De Luca E, De Bona A, & Mauriello G (2001) The use of SPME-GC-MS in the analysis of flavor of virgin olive oil. *Rivista Italiana Delle Sostanze Grasse,* 78, 157- 162.

[25] Bortolomeazzi R, Berno P, Pizzale L, & Conte LS (2001) Sesquiterpene, alkene, and alkane hydrocarbons in virgin oilve oils of different varieties and geographical origins. *Journal of Agricultural and Food Chemistry.* 49, 3278-3283.

[26] Romani A, Coinu R, Carta S, Pinelli P, Galardi C, Vincieri FF, & Franconi F (2004) Evaluation of antioxidant effect of different extracts of *Myrtus communis* L. *Free Radical Research,* 38,97–103.

In: Virgin Olive Oil
Editor: Antonella De Leonardis

ISBN: 978-1-63117-656-2
© 2014 Nova Science Publishers, Inc.

Chapter 6

INVESTIGATIONS ON THE QUALITY OF ITALIAN AND GREEK OLIVE OILS IN TERMS OF VARIETAL AND GEOGRAPHICAL ORIGIN BY USING DIFFERENT ANALYTICAL METHODS

Francesco Longobardi*, Grazia Casiello, Daniela Sacco,
Andrea Ventrella, Antonio Sacco and Michael G. Kontominas
Dipartimento di Chimica, Università di Bari "Aldo Moro", Bari (Italy)
Laboratory of Food Chemistry and Technology, Department of Chemistry,
University of Ioannina, Ioannina, Greece

ABSTRACT

The European Union provides important guidelines for maintaining the Protected Designation of Origin of numerous food products, as in case of the olive oil, since the origin marks certify a high quality level of food, conferring, as a consequence, prestige and economic value. Therefore, in this context it is extremely important to consider the topic of the characterization of foods based on variety (cultivar) and geographical origin, as this may be used as a criterion for determining authenticity and quality.

This chapter constitutes an overview of the results obtained by the authors through studies aiming at the differentiation of the geographic and varietal origin of olive oils, dealing with the classification of Italian olive cultivars based on compositional data and Nuclear Magnetic Resonance determinations, and with the characterization of the geographical origin of Western Greek and Italian virgin olive oils based on instrumental and multivariate statistical analysis.

Keywords: Olive oil, geographical origin, varietal origin, chemometrics, Nuclear Magnetic Resonance, conventional analyses

---

* Corresponding author: Dr. Francesco Longobardi e-mail: francesco.longobardi@uniba.it.

LIST OF ABBREVIATIONS

ANOVA: Analysis of Variance
CA: Canonical Analysis
CV: Cross Validation
DA: Discriminant Analysis
DF: Discriminant Function
FA: Fatty acid
FAME: Fatty acid methyl ester
FT-IR: Fourier transform infrared
GC: Gas Chromatography
IRMS: Isotopic Ratios Mass Spectrometry
HPLC: High Performance Liquid Chromatography
LDA: Linear Discriminant Analysis
LOO: Leave one out
MCCV: Monte Carlo embedded cross validation
MS: Mass Spectrometry
NCM: Nearest class mean
NMR: Nuclear Magnetic Resonance
PC: Principal Component
PCA: Principal Component Analysis
PLS-DA: Partial Least Squares Discriminant Analysis
SIMCA: Soft Independent Modeling of Class Analogies
TAG: Triacylglycerol
UV: Ultraviolet
VIP: Variable importance in projection
VOO: Virgin olive oil

1. INTRODUCTION

The general concept of food quality covers a very wide range of aspects, such as nutrition, organoleptic properties, geographical origin, socio-economic topics etc. Indeed, the unique and peculiar balance of such factors has a strong influence on the final food product characteristics.

This occurs also in the case of olive oil, whose high quality aspects, such as prestigious nutritional and health benefits, are strictly linked to the variety of olives employed, to the particular geographic area of olive cultivation, to the techniques and production processes [1-5].

For all the above considerations, the European Union with the aim to guarantee the consumers about the declared characteristics of food products, has set fundamental acts of regulation, such as the directives EC 510/2006 and EC 1898/2006, establishing the rules for conferring the "protected designation of origin" (PDO), "protected geographical indication" (PGI) and "traditional specialty guaranteed" (TSG) labels. In the specific case of olive oil, new restrictions have been established by the EC Regulation 182/2009.

The EC regulations provide guidelines also for maintaining the PDO of olive oil and other foods; this includes characterization of foods based on cultivar and geographical origin, as they are used as indicators of authenticity and quality. Therefore, there is an economic basis for identifying characteristics that differentiate olive oils according to their geographical and varietal origins.

This has been tackled by many research groups, and has led to the application of different methods to authenticate olive oil origin. In particular, a widely applied operational scheme in numerous studies involved in performing conventional and/or innovative instrumental analyses on a large number of olive oil samples, applying multivariate statistical techniques with the purpose to classify these according to their geographical or varietal origins. The employment of chemometrics resulted to be essential, due to the huge number of chemical-physical variables that had to be employed simultaneously.

Various routine techniques for oil analyses have been used to verify olive oil traceability and authenticity in combination with chemometrics tools, as in the case of fatty acids [6-12], triacylglycerols [13], fatty acids and triacylglycerols [14, 15], sterols [16], fatty acid and sterol composition etc. [17].

On the other hand, several innovative instrumental techniques, such as FT-IR spectroscopy [18], MS [19], IRMS [20, 21], and ^{1}H- or ^{13}C-NMR [22-24], have been employed with the purpose to study olive oil signals, finding the ones mainly useful for determining the product origin. The use of ^{1}H-NMR to determine olive oil geographical origin has been proven particularly interesting, especially if used in the so called fingerprinting approach [25], where the olive oil spectra are employed as a whole without giving particular attention to the assignment of specific signals, and thus gaining an overall description of each sample by its spectral overview.

With regard to the multivariate statistical techniques, the previously cited examples adopted various approaches, that included descriptive statistics, exploratory unsupervised techniques such as PCA, supervised classification techniques such as LDA, PLS-DA, SIMCA etc. [5, 26-31].

Up to date, this type of investigation is still in progress. Therefore, the aim of this chapter is to give an overview of the results obtained by its authors in recent studies regarding the differentiation of olive oil geographical and varietal origins employing conventional analytical parameters, ^{1}H and ^{13}C-NMR, in combination with suitable multivariate statistical tools [32-36].

2. METHODS AND TECHNIQUES

The methods and the techniques employed for the physico-chemical analyses are listed below. All the detailed descriptions can be found in the literature cited [32-36].

2.1. Conventional Quality Indices

Free acidity, expressed as percentage of oleic acid, peroxide value, expressed as milliequivalents of active oxygen per kilogram of oil, and K232 and K270 extinction coefficients, calculated from absorption at 232 and 270 nm, were measured as described in European Commission Regulation EEC 2568/91 and later amendments [37].

2.2. Chlorophylls

Chlorophyll content was calculated from the absorption value of an olive oil solution (obtained by diluting an aliquot of sample with a suitable volume of cyclohexane) at 670 nm and specific coefficient for pheophytin a (E_0 =613).

2.3. Fatty Acids

The analytical method for the determination of the FA composition was performed via trans-esterification into FAME, and subsequent GC analysis [37, 38]. Quantification was achieved using a FAME standard mixture and FA composition was expressed as percentage.

FA Nomenclature. C14:0, myristic acid (tetradecanoic acid); C16:0, palmitic acid (Hexadecanoic acid); C16:1n-7, palmitoleic acid (9- hexadecenoic acid); C17:0, margaric acid (heptadecenoic acid); C17:1n-8, margaroleic acid (9-heptadecenoic acid); C18:1n-9, oleic acid (9-octadecenoic acid); C18:1n-7, cis-vaccenic acid (11-octadecenoic acid); C18:2n-6, linoleic acid (9,12-octadecenoic acid); C18:3n-3, linolenic acid (9,12,15-octadecenoic acid); C20:0, arachidic acid (eicosanoic acid); C20:1n-9, gondoic acid (9-eicosenoic acid); C22:0, behenic acid (docosanoic acid); C24:0, lignoceric acid (tetracosanoic acid).

2.4. Triacylglycerols

For TAG analysis an aliquot of olive oil samples was diluted with a suitable volume of acetone and injected into a HPLC system. Identification of TAGs was carried out as reported by Ollivier et al. [14].

TAG nomenclature. The TAGs are designated by letters corresponding to abbreviated names of fatty acid carbon chain that are fixed on the glycerol molecule. The abbreviations of fatty acid names are as follows: P, palmitoyl, Po, palmitoleyl S, stearoyl; O, oleoyl; L, linoleoyl; Ln, linoleolenyl; and A, arachidoyl.

2.5. Phenolic Compounds

The extraction of phenolic compounds was performed by using pure methanol; the extracts were dried, diluted with acetonitrile, and the resulting solution, after a washing step with hexane, was evaporated giving a residue that was dissolved in a suitable volume of a mixture of methanol/water (1:1, v/v).

The concentration of total polyphenols was determined using the Folin-Ciocalteau method.

The concentration of total o-diphenols was estimated by adding 1 mL of a 5% solution of sodium molybdate dehydrate in ethanol/water (1:1, v/v) to the filtered aqueous-methanolic extract, and by measuring the absorbance at 370 nm.

2.6. Sterols

The olive oil was saponified with potassium hydroxide in ethanolic solution and the unsaponifiable fraction was then extracted with ethyl ether. The bands corresponding to the sterol fraction were separated by thin-layer chromatography on a basic silica gel plate. The sterols recovered from the plate were transformed into trimethylsilyl ethers. The mixture was analyzed by GC.

2.7. NMR Spectra

For ^{13}C-NMR analysis of olive oils a suitable volume of oil was dried and placed in an ultra-sonic bath; after filtration, oil was diluted with a suitable volume of $CDCl_3$ and ^{13}C-NMR spectra were recorded on a 500 MHz spectrometer.

For ^{1}H-NMR of phenolic compounds a suitable volume of methanol was used to extract phenolic compounds and, after concentration, the residue was resuspended in CH_3CN, and washed with hexane. After solvent evaporation, the extracts were dissolved in DMSO-d_6 and ^{1}H-NMR spectra were recorded on two high-resolution spectrometers (400 and 500 MHz).

^{1}H-NMR fingerprinting of olive oils was obtained by dissolving oil in a suitable volume of $CDCl_3$ using a 600 MHz spectrometer. Two different ^{1}H-NMR experiments were performed for each sample: a standard single pulse experiment, using 90^n pulse angle, and a one-dimensional ^{1}H-NMR pulse sequence with suppression of the strong lipid signals.

2.8. Chemometrics

The statistics and chemometrics data were obtained by using Statistica software (StatSoft Italia srl, Padova, Italy), V-Parvus release 2010 (http://www.parvus.unige.it, Genova, Italy), SIMCA-P+ Version 12 (Umetrics, Ume, Sweden), and MatLab 7.6 (MathWorks, Natick, Massachusetts, USA).

3. RESULTS AND DISCUSSION

3.1. Classification of Olive Oils According to Their Cultivars

For this study thirty-seven Apulian VOOs were considered and they included the following cultivars: Coratina (7 samples), Leccino (10), Peranzana (7), and Oliarola (13).

All samples were analyzed by means of conventional techniques (FA and TAG) and ^{13}C-NMR. Considering FA and TAG, with the aim to eliminate the parameters that did not provide relevant information for cultivar discrimination and to reduce the dimensionality of data, first a selection of variables was carried out performing ANOVA ($p \leq 0.05$), which selected 22 variables, and then PCA was applied. Five PCs were extracted, covering 82% of the total variance. Plotting the first two PCs a good grouping of VOOs was observed according to the cultivars except for a partial overlapping of the Peranzana and Coratina clusters.

The examination of the loadings associated to each PC allowed singling out the most important variables useful to distinguish the 4 cultivars. In particular, samples of the Oliarola cultivar were correlated with their content of OLLn, POL, PPL and linoleic acid while samples of Peranzana and Coratina cultivars had a high content of GaOO, AOO, oleic and linolenic acids. Leccino oils were better correlated with PoPO and palmitoleic acid.

Subsequently, to discriminate samples according to the cultivars, LDA was used. In order to obtain reliable LDA results, avoiding risks of overfitting, it is well documented that the number of variables should not be too high with respect to the samples; as a general rule, the variable/sample ratio should be $\leq 1:3$ [39]. For this reason, in this study, LDA was performed on the five PCs previously selected obtaining a classification rate of 95%. All the samples were correctly classified except one Coratina, that fell within the Peranzana group and one Leccino, assigned to the Oliarola group. Finally, to test the predictability of the model, samples were randomly split into a training set to develop a discriminant model and a validation set (2 from the Coratina, 4 from the Oliarola, 4 from the Leccino and 2 from the Peranzana cultivar) on which the model could be tested. The prediction ability for the validation set was 92%, demonstrating the validity of the model.

With regard to the NMR data, after a suitable selection of the signals in the ^{13}C-NMR spectrum, PCA was employed and PC1 vs. PC2 score plot showed that samples of Oliarola and Leccino cultivars were differentiated from the others, while there was a partial overlapping between samples of Peranzana and Coratina cultivars. By using LDA, the classification and prediction abilities of the model were 95% and 88%, respectively.

Results showed that the two approaches were substantially in agreement and therefore they could be used both to differentiate oil cultivars. FA and TAG resulted to be most successful to distinguish olive oil cultivars. Moreover, chromatographic data gave more detailed information on the composition, while the ^{13}C-NMR experiments presented advantages in terms of sample preparation.

In a similar study, classification of Apulian VOOs, according to cultivar, was obtained by using ^{1}H-NMR spectra of the olive oil phenolic extracts. Phenolic amount, strongly affecting the variety, geographical origin, and maturity degree of olives [40-42], is a very important parameter for the evaluation of virgin olive oil quality since phenols are strictly related to the typical bitter taste of olive oil and to the oil resistance to oxidation.

Multivariate analysis was performed to assess the relationship between phenolic compounds and olive varieties. For this purpose, ^{1}H-NMR spectra of the phenolic extracts were normalized to the ^{1}H signal of the solvent peak (DMSO-d_6). Variable selection was applied to select the resonances having the highest power to distinguish between different oils. The selected variables were the intensities of peaks due to aldehyde protons, vinyl protons, and aromatic protons.

Also in this case, a good classification of the samples with LDA, according to their variety, was attained with a classification rate of 100%.

3.2. Classification of Olive Oils According to Their Geographical Origin

3.2.1. Western Greek Virgin Olive Oils

This work had the purpose to employ the determination of free acidity, peroxide value, UV spectrometric parameters, phenols, chlorophylls, FAs, TAGs and sterols to find models for predicting the geographical origin of VOO samples from the Ionian Islands region of Greece. In particular the samples were from Kefalonia (KEF), Kerkyra (KER), Lefkada (LEF), and Zakynthos (ZAK). All the analytical results are reported in Table 1.

The different typologies of analytical determinations were used separately or together, evaluating which approach led to better prediction results.

Starting by using all the variables together, in view of the use of discriminant techniques, a feature selection method was employed to reduce the number of variables, that were too many if compared to samples, eliminating the parameters that did not provide relevant information for geographical discrimination. In particular, the variable selection was performed by using ANOVA, highlighting the 26 analytical parameters (free acidity, peroxide value, K232, K270, chlorophylls, C14:0, C17:1, C18:0, C20:0, OLLn + PoLL, PLLn, PLL, PLnO + PPoL + PPoPo, EeOO, SOL + POO, POP + PSL, SOO + OLA, POS + SLS, AOO, SOS + POA, campesterol, β–sitosterol, sitostanol, Δ-5-avenasterol, total sterols, o-diphenols) that showed statistically significant differences ($p<0.01$) in the means among the VOOs under study (see Table 1).

Therefore, the data were subjected to a LDA and a complete separation of the four geographical groups was achieved projecting the data points in the space of the first two discriminant functions (Figure 1), obtaining a recognition ability of 100% for each class. The discrimination between KER and the other three groups was clearly displayed along DF1, while DF2 led to the separation of LEF from the other islands.

Samples from KEF and ZAK, seemingly close in Figure 1, were well separated along DF3. A LOO CV procedure showed a prediction ability of about 81% for the four classes and after the examination of the standardized coefficients, FAs and TAGs resulted to be the most important variables for classification.

Interestingly, high significant correlations ($p<0.01$) were found by plotting the canonical scores as a function of the relevant latitudes and longitudes, with coefficients of determination (r^2) of 0.9230 and 0.8640, respectively, confirming that the analytical determinations studied, besides giving important information on olive oil quality, were useful to determine the VOO geographical origin. Moreover, since DF1 correlated better to latitude than longitude, it was deduced that the group discrimination along DF1 was mainly due to a climatic effect.

With the purpose to perform less analytical determinations, reducing time and costs of analysis, four sub-datasets were extracted from the total one, containing respectively FAs, TAGs, sterols, and all the remaining parameters (called "conventional"). LDA was applied to each sub-dataset, after a variable reduction based on ANOVA was applied whenever necessary. In particular, by applying LDA only to conventional variables, the classification and prediction abilities resulted to be about 79% and 67%, respectively; the same results were obtained by using sterols as variables.

Table 1. Means and standard deviations (SD) of analytical parameters of Greek virgin olive oils [a]

Reprinted from Food Chemistry, 133, Longobardi F, Ventrella A, Casiello G, Sacco D, Tasioula-Margari M, Kiritsakis AK, & Kontominas MG, Characterisation of the geographical origin of Western Greek virgin olive oils based on instrumental and multivariate statistical analysis, pp. 169–175, Copyright (2012) with permission from Elsevier

Geographical Origin	ZAK			KEF			LEF			KER		
Number of samples	14			14			8			7		
	Mean		SD	Mean		SD	Mean		SD	Mean		SD
Free acidity (oleic acid %)	0.34	a	0.10	0.22	a	0.10	0.42	a	0.22	0.80	b	0.62
Peroxide value (meq O_2 kg^{-1} oil)	9.66	a	1.27	9.30	a	1.84	8.77	ab	1.74	6.99	b	1.67
K232	1.36	a	0.09	1.31	a	0.06	1.40	a	0.09	1.76	b	0.31
K270	0.11	a	0.01	0.09	a	0.01	0.11	a	0.01	0.16	b	0.06
Chlorophyll content (mg kg^{-1})	5.71	b	1.20	3.86	a	1.34	3.06	a	0.51	2.79	a	1.13
Total phenols (mg gallic ac. kg^{-1})	212.59	b	95.34	124.79	a	31.33	186.07	ab	86.59	109.94	a	58.08
o-Diphenols (mg gallic ac. kg^{-1})	78.03	b	41.70	38.49	a	12.91	66.35	ab	31.95	58.20	ab	19.69
C14:0	0.01	b	0.01	0.01	a	0.01	0.01	ab	0.00	0.00	a	0.00
C16:0	12.89	ab	1.00	13.04	a	1.19	11.70	b	0.50	13.47	a	1.35
C16:1	1.11	ab	0.22	1.13	ab	0.18	0.91	a	0.10	1.27	b	0.26
C17:0	0.07	a	0.05	0.16	a	0.13	0.24	a	0.13	0.24	a	0.36
C17:1	0.08	a	0.00	0.14	b	0.06	0.08	a	0.01	0.10	ab	0.01
C18:0	2.77	a	0.19	2.74	a	0.20	2.74	a	0.33	2.02	b	0.15
C18:1	73.09	a	2.19	72.89	a	2.66	73.32	a	2.62	71.76	a	2.47
C18:2	8.01	a	1.33	7.95	a	1.70	8.95	a	2.30	9.21	a	1.70
C18:3	0.75	ab	0.08	0.70	ab	0.12	0.66	a	0.07	0.80	b	0.04
C20:0	0.46	a	0.05	0.45	a	0.04	0.42	ab	0.03	0.39	b	0.02
C20:1	0.26	a	0.05	0.28	a	0.03	0.25	a	0.05	0.30	a	0.02
C22:0	0.31	a	0.19	0.24	a	0.18	0.33	a	0.30	0.13	a	0.15
C24:0	0.14	a	0.03	0.14	a	0.02	0.12	a	0.03	0.14	a	0.02
LLL	0.12	b	0.06	0.08	ab	0.04	0.09	ab	0.04	0.06	a	0.03
OLLn + PoLL	0.27	a	0.03	0.25	a	0.05	0.28	ab	0.05	0.34	b	0.07
PLLn	0.19	ab	0.02	0.18	a	0.02	0.22	bc	0.04	0.26	c	0.03
OLL + PoOL	1.79	a	0.66	1.63	a	0.61	2.22	a	0.91	2.06	a	0.65
OOLn + PoPoO	1.91	ab	0.14	1.88	ab	0.34	1.73	a	0.09	2.10	b	0.17

Geographical Origin	ZAK			KEF			LEF			KER		
Number of samples	14			14			8			7		
PLL	0.66	a	0.06	0.64	a	0.13	0.51	b	0.09	0.80	c	0.10
PLnO + PPoL + PPoPo	0.08	ab	0.01	0.09	b	0.02	0.07	a	0.01	0.13	c	0.02
OOL + PPLn + PoOO	12.52	a	1.15	12.24	a	1.55	14.21	a	2.64	14.19	a	1.85
POL + SLL	6.70	a	1.15	6.83	a	1.39	6.69	a	1.37	7.99	a	1.38
PPL	0.66	a	0.18	0.71	a	0.24	0.63	a	0.18	0.89	a	0.23
EeOO	0.11	a	0.02	0.18	b	0.09	0.09	a	0.06	0.14	ab	0.03
OOO	39.45	a	3.47	40.10	a	4.04	40.80	a	4.25	38.01	a	3.49
SOL + POO	24.59	a	0.87	24.35	a	1.34	22.48	b	1.21	23.84	ab	1.79
POP + PSL	3.43	a	0.48	3.44	a	0.56	2.81	b	0.23	3.71	a	0.54
GOO	0.34	a	0.07	0.34	a	0.07	0.31	a	0.13	0.36	a	0.05
SOO + OLA	4.89	a	0.43	4.85	a	0.49	4.83	a	0.46	3.47	b	0.40
POS + SLS	1.25	a	0.06	1.20	a	0.14	1.15	a	0.20	0.88	b	0.10
AOO	0.76	a	0.10	0.74	a	0.10	0.65	ab	0.11	0.59	b	0.08
SOS + POA	0.29	a	0.02	0.28	a	0.03	0.26	a	0.04	0.18	b	0.02
Campesterol	3.53	bc	0.28	3.90	c	0.29	2.94	a	0.70	3.26	ab	0.47
Campestanol	0.14	a	0.09	0.21	a	0.18	0.08	a	0.13	0.05	a	0.06
Stigmasterol	0.68	a	0.17	0.73	a	0.11	0.84	a	0.23	0.68	a	0.20
Δ-7-Campesterol	0.46	a	0.35	0.34	a	0.31	0.29	a	0.31	0.19	a	0.23
Δ-5,23-Stigmastadienol	0.34	a	0.42	0.26	a	0.39	0.12	a	0.20	0.07	a	0.12
Clerosterol	1.04	a	0.54	0.90	a	0.11	0.76	a	0.15	0.62	a	0.28
β-Sitosterol	75.55	c	3.74	82.07	a	2.74	85.62	ab	2.28	86.08	b	2.47
Sitostanol	0.30	a	0.39	0.74	b	0.49	0.06	a	0.16	0.10	a	0.28
Δ-5-Avenasterol	16.46	b	3.74	9.72	a	3.22	7.87	a	1.51	7.64	a	2.12
Δ-5,24-Stigmastadienol	0.73	a	0.30	0.56	a	0.14	0.74	a	0.19	0.56	a	0.06
Δ-7-Stigmastenol	0.29	a	0.09	0.31	a	0.38	0.30	a	0.05	0.31	a	0.06
Δ-7-Avenasterol	0.49	a	0.25	0.39	a	0.19	0.66	a	0.25	0.56	a	0.23
Total sitosterols (%)	94.49	a	0.80	94.15	a	0.73	95.01	a	0.77	95.02	a	0.68
Total sterols (ppm)	1312.64	ab	274.89	1065.43	a	169.38	1689.88	c	352.17	1431.00	bc	163.41

[a]Results for geographical origin of HSD for unequal N Tukey test are reported: groups of one row with different letters are statistically different (p<0.01), a>b>c (= significantly different contents).

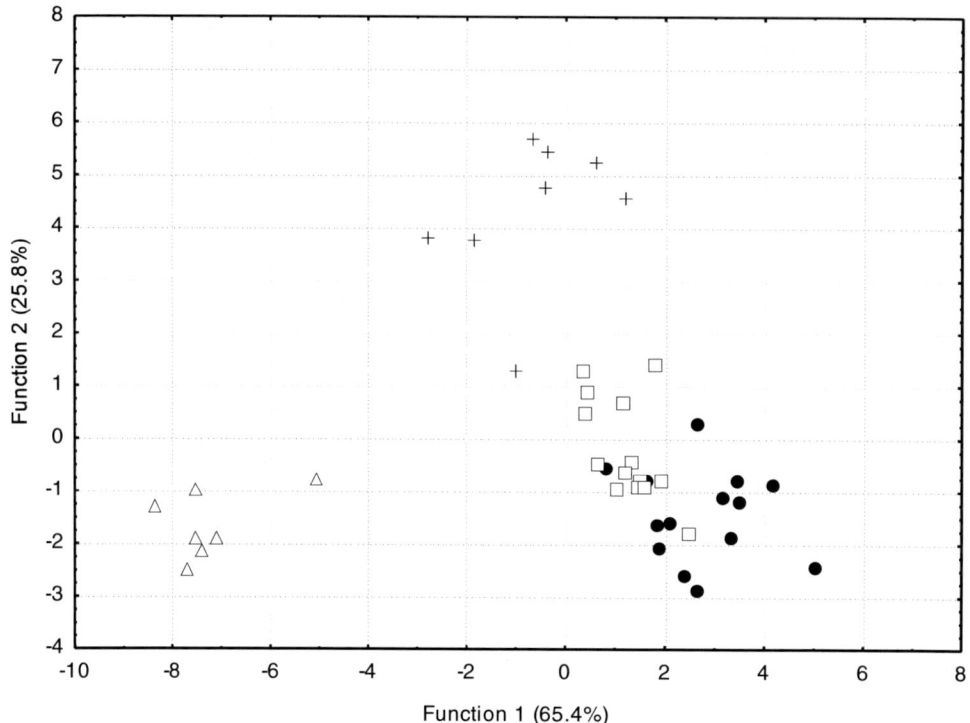

Figure 1. A graphical representation of the discriminant function analysis of Greek virgin olive oil samples coming from four different geographical origins, i.e. Zakynthos (•), Kefalonia (□), Lefkada (+), Kerkyra (Δ). Reprinted from Food Chemistry, 133, Longobardi F, Ventrella A, Casiello G, Sacco D, Tasioula-Margari M, Kiritsakis AK, & Kontominas MG, Characterisation of the geographical origin of Western Greek virgin olive oils based on instrumental and multivariate statistical analysis, pp. 169–175, Copyright (2012) with permission from Elsevier.

For FAs, the classification and the prediction abilities were found to be equal to 77% and 72%, respectively, while the LDA on TAGs gave 91% and 70%. By comparing the prediction results, among the different types of determinations carried out, FAs and TAGs were the most appropriate analytical parameters to discriminate VOO geographical origin, as observed by applying LDA to all variables simultaneously. Nevertheless, it was noticed that the model performance achieved by using sub-datasets was worse than that obtained by using all determinations together.

3.2.2. Apulian Virgin Olive Oils

This work was strictly linked to the previous one. The main aim was to employ data of free acidity, peroxide value, spectrophotometric indices, chlorophyll content, sterol, FA and TAG compositions, in combination with chemometrics, in order to predict the geographical origin of Southern Italian VOOs. In particular, Apulian VOOs were collected and analyzed (Table2) and, taking into account the areas of production, grouped into three classes of geographic origins identified as "North", "Centre" and "South", respectively.

In contrast to the study performed on Greek VOOs, the analytical determinations carried out in this study were considered uniquely as a whole dataset, but different classification techniques were employed and compared in order to test which one could be more suitable to discriminate the olive oil samples on the basis of their geographical origin.

Table 2. Means and standard deviations (SD) of analytical parameters of Apulian VOOs. Free acidity is expressed as oleic acid percentage, PVs are expressed as meq O$_2$ kg^{-1} of oil, chlorophyll content, FAs, TAGs, sterols and eritrodiol + uvaol are expressed as percentages, while total sterols are expressed as mg/100g. Reprinted from Food Chemistry, 133, Longobardi F, Ventrella A, Casiello G, Sacco D, Catucci L, Agostiano A, Kontominas MG, Instrumental and multivariate statistical analyses for the characterisation of the geographical origin of Apulian virgin olive oils, pp. 579–584, Copyright (2012) with permission from Elsevier

Geographical origin	North		Centre		South	
Number of samples	15		31		11	
	Mean	SD	Mean	SD	Mean	SD
Free acidity	0.33	0.08	0.28	0.12	0.65	0.47
PV	11.29	5.33	10.75	2.84	12.05	2.72
K232	2.02	0.41	1.94	0.22	2.09	0.19
K270	0.12	0.03	0.12	0.02	0.13	0.02
ΔK	0.005	0.002	0.006	0.003	0.003	0.003
Chlorophyll	4.40	1.80	4.02	1.50	3.42	1.08
C16:0	12.39	1.13	11.65	1.13	13.33	1.28
C16:1	0.96	0.36	0.97	0.59	1.58	0.44
C17:0	0.26	0.08	0.22	0.09	0.22	0.10
C18:0	2.14	0.13	2.39	0.41	2.05	0.46
C18:1W11	1.62	0.32	1.63	0.53	1.56	0.50
C18:1W9	72.96	2.31	73.15	1.99	70.13	2.90
C18:2W9-12	8.46	1.26	8.53	1.39	9.88	1.35
C18:3W9-12-15	0.64	0.12	0.70	0.13	0.65	0.17
C20:0	0.31	0.07	0.41	0.07	0.33	0.08
C20:1	0.26	0.11	0.36	0.08	0.27	0.09
LLL	0.16	0.12	0.30	0.46	0.20	0.16
PLLn	0.10	0.09	0.11	0.19	0.06	0.07
OLL + PoOL	1.94	1.17	1.88	1.26	2.79	1.63
OOLn + PoPoO	1.70	0.45	1.78	0.82	1.46	1.49
PLL	0.58	0.24	0.56	0.31	0.65	0.12
PLnO + PPoL + PPoPo	0.05	0.04	0.06	0.09	0.03	0.04
OOL + PPLn + PoOO	13.10	1.93	13.40	2.16	14.64	1.52
POL + SLL	7.13	1.61	6.55	2.04	7.90	1.45
PPL	0.68	0.31	0.63	0.43	0.77	0.43
OOO	38.90	3.68	40.03	8.12	37.15	3.98
SOL + POO	24.04	6.17	24.71	3.52	24.14	1.44
POP + PSL	4.88	5.34	3.25	1.44	3.72	0.64
GOO	0.42	0.36	0.56	0.47	0.58	0.54
SOO + OLA	4.05	0.50	4.32	1.25	3.77	0.54
POS + SLS	1.09	0.35	0.94	0.55	1.21	0.92
AOO	0.97	0.88	0.67	0.59	0.59	0.51

Table 2. (Continued)

Geographical origin	North		Centre		South	
Number of samples	15		31		11	
	Mean	SD	Mean	SD	Mean	SD
SOS + POA	0.11	0.22	0.16	0.39	0.17	0.34
Colesterol	0.21	0.12	0.25	0.13	0.17	0.06
Brassicasterol	0.10	0.00	0.10	0.01	0.10	0.02
24-Metylene colesterol	0.11	0.11	0.17	0.11	0.12	0.07
Campesterol	2.75	0.11	2.83	0.34	3.06	0.25
Campestanol	0.27	0.08	0.36	0.14	0.64	0.22
Clerosterol	0.81	0.017	0.80	0.16	0.78	0.05
β-Sitosterol	83.54	0.57	83.26	0.59	82.01	1.71
Sitostanol	0.63	0.07	0.81	0.17	1.29	0.52
Δ-5-Avenasterol	9.66	0.21	9.21	0.73	9.53	0.47
Δ-5,24-Stigmastadienol	0.33	0.14	0.43	0.15	0.55	0.21
Δ-7-Stigmastenol	0.43	0.14	0.26	0.07	0.27	0.11
Δ-7-Avenasterol	0.28	0.20	0.49	0.08	0.60	0.26
Total β-sitosterol	94.88	0.38	94.46	0.52	94.26	0.57
Total sterols	181.30	53.92	143.13	40.63	118.41	18.81
Eritrodiol + uvaol	1.95	0.53	1.86	0.71	1.62	0.21

As a consequence, three different statistical pattern recognition techniques, i.e. LDA, PLS-DA and SIMCA were applied to the data. The best results were obtained by applying LDA and PLS-DA and are commented below.

For the application of LDA, the choice of the variables for the method building process was carried out by calculating, for any possible combination of variables from a pre-reduced set, the relevant Wilks' Lambda (Λ), an inverse index of the goodness of the separation among classes. The obtained best subset of six variables consisting of FFA, ΔK, C20, campesterol, campestanol, and Δ7-stigmastenol. When applying LDA on such variables, recognition abilities of about 80%, 90% and 73% for North, Centre and South were obtained, respectively, with an average recognition ability equal to 84%. A LOO CV procedure led to an average prediction ability of 82%. In particular, the prediction ability was 80% for North, 90% for Centre and 64% for South. By evaluating the standardized coefficients: campestanol, ΔK and FFA proved to be the most important original variables for DF1, while for the DF2 high values were observed in the cases of Δ7-stigmastenol and C20.

As a second method, PLS-DA was applied [43] to all the original variables. The optimal complexity of the model was found to include 2 latent variables: t[1] showed ability in discriminating South class, while t[2] was able in separating North samples. The recognition abilities with PLS-DA resulted to be 80% for North samples, 97% for Centre samples and 73% for South samples with an average recognition ability of 88%. The examination of the VIP scores showed that the original variables with the highest discriminating power were Δ7-stigmastenol, C20, campestanol, Δ7-avenasterol, sitostanol and C20:1. Such information was confirmed by studying the loading values [43]: in particular, Δ7-stigmastenol was found to be particularly useful in the discrimination of the North samples, while C20 showed great loadings on both t[1] and t[2], being useful in classifying Centre samples; sitostanol, campestanol, C18:1W9 and β-sitosterol seemed to be important for the discrimination of South samples. The PLS-DA model was then cross-validated by a LOO procedure resulting to the prediction abilities of 60%, 87% and 73% for North, Centre and South samples, respectively. The average prediction ability was found to be 77%.

3.2.3. Western Greek and Apulian Olive Oils

In this study the olive oil samples considered in the two previous studies (sections 3.2.1 and 3.2.2) were pooled all together, facing the ambitious challenge to distinguish VOO samples coming from even 7 origins (three different areas of Apulia and four different Greek Islands) by using only a single analytical technique. To reach this target, the ^{1}H-NMR spectroscopy, used in the fingerprinting approach, in combination with Multivariate Statistical Analysis, was chosen as the most suitable and promising technique, having the potential to obtain information about a great number of constituents by means of short time experiments.

In fact, in order to exploit not only the dominating lipid signals, but also the signals from less abundant compounds, both a simple one pulse experiment (ZG1H) and an experiment with multiple saturation (NOESYGPPS) of the lipid signals were applied to each sample.

As a result, the fast and fully automated combination of both experiments involved a largely enhanced dynamic range and a substantially improved NMR sensitivity giving information on both the main and the minor constituents of the olive oil samples, these latter being generally hidden by the major ones in a standard proton spectrum. This is a remarkable point since the minor components could play an important role for authenticity assessment purposes being more difficult to be adulterated.

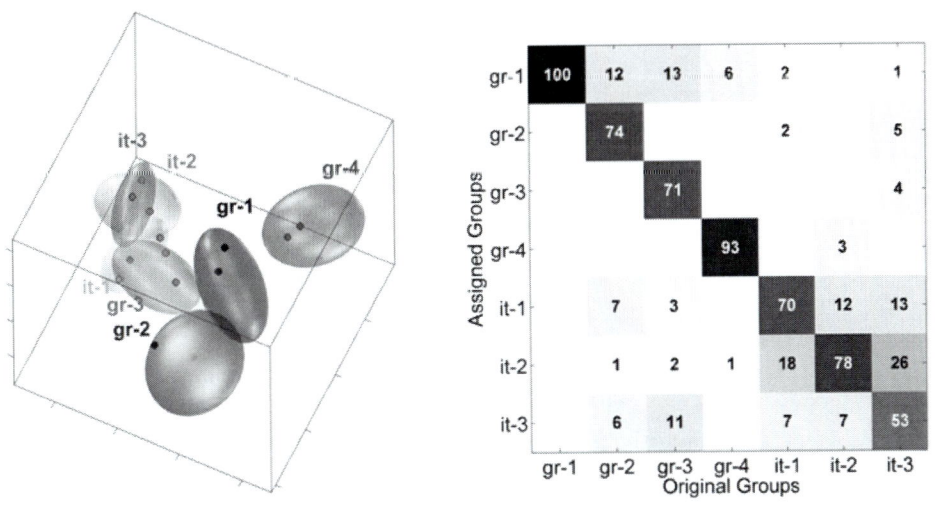

Figure 2. A graphical representation of the PCA/CA/NCM – classifier approach for the prediction of the origin of the Western Greek (4) and Apulian (3) olive oils from the ^{1}H-NMR spectra. The model set of each class is symbolized by its confidence-ellipsoid and the test sets are drawn as circles. The right illustration shows the confusion-matrix of the Monte-Carlo prediction. Reprinted from Food Chemistry, 130, Longobardi F, Ventrella A, Napoli C, Humpfer E, Schütz B, Schäfer H, Kontominas MG, & Sacco A, Classification of olive oils according to geographical origin by using ^{1}H NMR fingerprinting combined with multivariate analysis, pp. 177–183, Copyright (2011) with permission from Elsevier.

For data treatment, three different data matrices were generated from the NMR spectra: matrix 1 obtained from ZG1H spectra, matrix 2 derived from NOESYGPPS spectra, and matrix 3 obtained by combining the two previous ones.

Prediction ability was tested via MCCV on PCA/CA/NCM classification on all three data sets: in particular, PCA was used for the dimension reduction, CA was applied to PCA scores to identify the multivariate subspace for maximum group separation (Figure 2), and CV was carried out with the purpose to provide prediction ability subdividing the samples in training set (used to build the model) and test set (used to test the prediction ability).

The assignment of samples contained in the test sets to the geographical origin classes was carried out by comparing distances between test objects and class means using the NCM method.

For matrices 1, 2, and 3 the obtained average correct prediction probabilities were 74%, 65% and 78%, respectively; such results were considered very interesting since they were obtained by a procedure that did not require extraction/purification steps, so minimizing potential interferents and reducing time of analyses, and since a great number of classes were taken into account.

In order to test the reliability and the validity of the classification models, artificial randomization of origin-assignment (class labels) was performed to obtain unreal random data set for comparison with those produced by NMR (matrix 3). The random correct prediction rates ranged from 4 to 19%. It was deduced that only a small contribution of noise was modelled, whilst a remarkable part of the variance in the final model was effectively due to the different geographical origins.

CONCLUSION

Food authenticity is a topic of worldwide interest and regards the control of the entire chain of food production and marketing, leading to trace the food through each single step of its production back to its origin. The verification of food traceability, and of course of olive oil, is essential for the prevention of mislabelling and allows ensuring public health.

In this context, the identification of the geographical origin and of cultivar is of considerable importance for authentication, in order to guarantee consumers and producers of high-quality food products, as in the case of high quality olive oils.

The works commented in this chapter are all linked one to each other and fundamentally deal with the use of the so called "conventional" analytical methods and "innovative" instrumental methods, with the help of the highly useful multivariate statistical techniques, for the differentiation of varietal and geographical origins of olive oils.

In particular, with regard to the classification of the olive oil cultivars, results showed that both the FA and TAG composition and ^{13}C-NMR information can be used to reach that target as evidenced by prediction results, i.e. 92% and 88%, respectively. These results are to be considered valuable considering that four classes were taken into account.

Concerning the topic of the geographical origin determination of olive oil, considerably interesting information was obtained. The use of the conventional and instrumental analytical methods and of LDA on a considerable set of Greek olive oils, coming from four different Ionian islands, allowed obtaining prediction ability approximately equal to 81%. Worse prediction abilities were obtained by excluding some classes of determinations during the construction of the final LDA classification model, highlighting that the multidisciplinary approach was highly competitive.

A similar study allowed classifying Apulian olive oil samples, from three different geographic areas, by considering conventional data but focusing the attention on the use of different classification techniques. The best results were obtained by using LDA (prediction ability of about 83%) confirming its suitability for discrimination purposes.

Finally, fingerprinting ^{1}H-NMR in combination with multivariate statistics was successfully used to distinguish olive oils coming from seven different geographical origins. This ambitious aim became possible (mean prediction rate of 78%) thanks to the use of two different NMR sequences that allowed gathering information regarding major and minor olive oil components.

With all the above considered, it is possible to assert that the potential of conventional and/or innovative analytical tools combined with chemometrics in the geographical and varietal origin assessment of olive oil is clearly demonstrated, although there are issues that still have to be dealt with. In particular, validation and harmonization of the developed methods should be carried out with the purpose of using them in a regulatory framework relevant to olive oil quality.

REFERENCES

[1] Diraman H, Saygi H, & Hisil Y (2010) Relationship Between Geographical Origin and Fatty Acid Composition of Turkish Virgin Olive Oils for Two Harvest Years. *Journal of the American Oil Chemists' Society, 87*(7), 781-789.

[2] Fontanazza G, Patumi M, Solinas M, & Serraiocco A (1993) Influence of cultivars of the composition and quality of olive oil. *Acta Hortic, 356*, 358-361.

[3] Solinas M (1987) HRGC analysis of virgin olive oil phenolic compounds in relation to olive ripening degree and variety. *Riv. Ital. Sostanze Grasse, 64*, 255-262.

[4] Vasquez Roncero A (1978) Les Poliphénoles de l'Huile et leur Influence sur les Caractéristiques de l'Huile. *Revue Française des Corps Gras, 25*, 21-26.

[5] Alonso-Salces RM, Héberger K, Holland MV, Moreno-Rojas JM, Mariani C, Bellan G, Reniero F, & Guillou C (2010). Multivariate analysis of NMR fingerprint of the unsaponifiable fraction of virgin olive oils for authentication purposes. *Food Chemistry,* 118, 956–965.

[6] Alonso Garcia MV, & Aparicio Lopez R (1993). Charaterization of european virgin olive oils using fatty acids. *Grasas y Aceites, 44*, 18–24.

[7] Bucci R, Magri AD, Magri AL, Marini D, & Marini F (2002). Chemical authentification of extra virgin olive oil varieties by supervised chemometric procedures. *Journal of Agricultural and Food Chemistry, 50*, 413–418.

[8] Forina M, & Tiscornia E (1982). Pattern recognition methods in the prediction of italian olive oil origin by their fatty acid content. *Annali di Chimica,72*, 143–155.

[9] Stefanoudaki E, Kotsifaki F, & Koutsaftakis A (1999). Classification of virgin olive oils of the two major cretan cultivars based on their fatty acid composition. *Journal of the American Oils Chemist's Society, 76*, 623–626.

[10] Tsimidou M, & Karakostas KX (1993) Geographical Characterization of Greek Virgin Olive Oil by Non-Parametric Multivariate Evaluation of Fatty Acid Composition. *Sci. Food Agric.* 62, 253-257.

[11] Lanza CM, Russo C, & Tomaselli F (1998) Relationship between Geographical Origin and Fatty Acid Composition of Extra-Virgin Olive Oils Produced in Three Areas of Eastern Sicily. *Ital. J. Food Sci.* 4, 359-366.

[12] Sacco A, Brescia MA, Liuzzi V, Reniero F, Guillou C. Ghelli S, & Van Der Meer P (2000) Characterization of Italian Olive Oils Based on Analytical and Nuclear Magnetic Resonance Determinations. *J. Am. Oil Chem. Soc.* 77, 619-625.

[13] Gigliotti C, Daghetta A, & Sidoli A (1993) Indagine Conoscitiva sul Contenuto Trigliceridico di Oli Extra Vergini di Oliva di Varia Provenienza. *Riv. Ital. Sostanze Grasse* 70, 483-489.

[14] Ollivier D, Artaud J, Pinatel C, Durbec JP, & Guérère M (2006). Differentiation of French virgin olive oil RDOs by sensory characteristics, fatty acid and triacylglycerol compositions and chemometrics. *Food Chemistry*, 97, 382–393.

[15] Aranda F, Gòmez-Alonso S, Rivera del Alamo RM, Salvador MD, & Fregapane G (2004). Triglyceride, total and 2-position fatty acid composition of Cornicabra virgin olive oil: comparison with other Spanish cultivars. *Food Chemistry*, 86, 485–492.

[16] Leardi R, & Paganuzzi V (1987). Characterization of the origin of extra virgin olive oils by chemometric methods applied to the sterols fraction. *Rivista Italiana delle Sostanze Grasse*, 64, 131–136.

[17] Armanino C, Leardi R, & Lanteri S (1989) *Chemometric Analysis of Tuscan Olive Oils. Chemom. and Intell. lab. Sys.* 5, 343-354.

[18] Bertran E, Blanco M, Coello J, Iturriaga H, Maspoch S, & Montolin I (2000). Near-infrared spectrometry and pattern recognition as screening methods for the authentification of virgin olive oils of very close geographical origin. *Journal of Near Infrared Spectroscopy*, 8(1), 45–52.

[19] Cajka T, Riddellova K, Klimankova E, Cerna M, Pudil F, & Hajslova J (2010). Traceability of olive oil based on volatiles pattern and multivariate analysis. *Food Chemistry*, 121, 282–289.

[20] Angerosa F, Breas O, Contento S, Guillou C, Reniero F, & Sada E (1999). Application of stable isotope ratio analysis to the characterization of the geographical origine of olive oils. *Journal of Agricultural and Food Chemistry*, 47, 1013–1017.

[21] Bianchi G, Angerosa F, Camera L, Reiniero F, & Anglani C (1993). Stable carbon isotope ratio 13C/12C of olive oil components. *Journal of Agricultural and Food Chemistry*, 41, 1936–1940.

[22] Mannina L, Patumi M, Proietti N, Bassi D, & Segre AL (2001). Geographical characterization of Italian extra virgin olive oil using high field 1H NMR spectroscopy. *Journal of Agricultural and Food Chemistry*, 49, 2687–2696.

[23] Mannina L, Marini F, Gobbino M, Sobolev AP, Capitani D (2010) NMR and chemometrics in tracing European olive oils: The case study of Ligurian samples. *Talanta*, 80, 2141–2148.

[24] Vlahov G, Del Re P, & Simone N (2003). Determination of geographical origin of olive oils using 13C nuclear magnetic resonance spectroscopy. I - Classification of olive oils of the Puglia region with denomination of protected origin. *Journal of Agricultural and Food Chemistry*, 51, 5612–5615.

[25] Fiehn O (2001). Combining genomics, metabolome analysis, and biochemical modelling to understand metabolic networks. *Comp. Funct Genomics*, 2, 155-168.

[26] Petrakis PV, Agiomyrgianaki A, Christophoridou S, Spyros A, & Dais P (2008) Geographical characterization of greek virgin olive oils (cv. Koroneiki) using ^{1}H and ^{31}P NMR fingerprinting with canonical discriminant analysis and classification binary trees. *Journal of Agricultural and Food Chemistry*, 56, 3200-3207.

[27] Rezzi S, Axelson DE, Héberger K, Reniero F, Mariani C, & Guillou C (2005). Classification of olive oils using high throughput flow 1H NMR fingerprinting with principal component analysis, linear discriminant analysis and probabilistic neural networks. *Analytica Chimica Acta*, 552, 13–24.

[28] Aparicio R, Ferreiro L, Cert A, & Lanzon A (1990) Caraterización de aceites de oliva vírgenes andaluces. *Grasas y aceites*, 41, 23-39.

[29] Sacchi R, Mannina L, Fiordiponti P, Barone P, Paolillo L, Patumi M, & Segre A (1998) Characterization of Italian extra virgin olive oils using ^{1}H-NMR spectroscopy. *J. Agric. Food Chem.* 46, 3947-3951.

[30] Shaw AD, Di Camillo A, Vlahov G, Jones A, Bianchi G, Rowland J, & Kell DB (1997) Discrimination of the variety and region of origin of extra virgin olive oils using ^{13}C-NMR and multivariate calibration with variable reduction. *Analytica Chimica Acta* 348, 357-374.

[31] Spugnoli P, Parenti A, Cardini D, Modi G, & Caselli S (1998) Caratterizzazione di oli extra vergini monovarietali di tre cultivar toscane. *La Rivista Italiana delle Sostanze Grasse* 75, 227-233.

[32] Brescia MA, Alviti G, Liuzzi V, & Sacco A (2003) Chemometric classification of olive cultivars based on compositional data of oils. *Journal of the American Oil Chemists' Society* 80, 945-950.

[33] Sacco A, Brescia MA, Liuzzi V, Reniero F, Guillou G, Ghelli S, & van der Meer P (2000) Characterization of italian olive oils based on analytical and nuclear magnetic resonance determinations. *Journal of the American Oil Chemists' Society* 77, 619-625.

[34] Longobardi F, Ventrella A, Napoli C, Humpfer E, Schütz B, Schäfer H, Kontominas MG, & Sacco A (2012) Classification of olive oils according to geographical origin by using ^{1}H NMR fingerprinting combined with multivariate analysis. *Food Chemistry*, 130, 177–183.

[35] Longobardi F, Ventrella A, Casiello G, Sacco D, Tasioula-Margari M, Kiritsakis AK, Kontominas MG (2012) Characterisation of the geographical origin of Western Greek virgin olive oils based on instrumental and multivariate statistical analysis. *Food Chemistry*, 133, 169–175.

[36] Longobardi F, Ventrella A, Casiello G, Sacco D, Catucci L, Agostiano A, Kontominas MG (2012) Instrumental and multivariate statistical analyses for the characterisation of the geographical origin of Apulian virgin olive oils. *Food Chemistry*, 133, 579–584.

[37] European Communities. Regulation 2568 (1991) *Off. J. Eur. Communities*, L 248.

[38] Morrison WR., & Smith LH (1964) Preparation of fatty acid methyl esters and dimethylacetals from lipids with boron fluoride-methano. *Journal of lipid Research* 64, 373-376.

[39] Defernez M, & Kemsley EK (1997) The Use and Misuse of Chemometrics for Treating Classification Problems. *Trends in analytical chemistry* 16, 216-221.

[40] Solinas M, Di Giovacchino L, & Mascolo A (1978) The polyphenols of olives and olive oil. Note III: influence of temperature and kneading time on the oil polyphenol content. *Riv. Ital. Sostanze Grasse* 55, 19-23.

[41] Montedoro GF, & Garofolo L (1984) The qualitative characteristics of virgin olive oils. The influence of variables such as variety, environment, preservation, extraction, conditioning of the finished product. *Riv. Ital. Sostanze Grasse* 61, 157-168.

[42] Amiot MT, Fleuriet A, & Macheix JT (1986) Importance and evolution of phenolic compounds in olive during growth and maturation. *J. Agric. Food Chem.* 34, 823-826.

[43] Wold S, Sjöström M, & Eriksson L (2001) PLS-regression: a basic tool of chemometrics. *Chemometrics and Intelligent Laboratory Systems*, 58, 109–130.

In: Virgin Olive Oil
Editor: Antonella De Leonardis

ISBN: 978-1-63117-656-2
© 2014 Nova Science Publishers, Inc.

Chapter 7

OLIVE VARIETY SUITABILITY AND TRAINING SYSTEM FOR MODERN OLIVE GROWING: PLANT GROWTH AND YIELD COMPONENTS

Mouna Aïachi Mezghani[*]

Institut de l'olivier, Station Régionale de Sousse, Sousse, Tunisie

ABSTRACT

In Tunisia, new olive orchards are planted at higher densities. Therefore, there has been a significant increase in the number of intensively managed orchards. The olive orchard like other fruits crops' orchards, have two principal aims: early bearing and regular production. Therefore, a comparative trial was set up in 2002 to study the responses of different olive table varieties, like *Picholine, Meski, Manzanilla* and *Ascolana,* and oil varieties, like *Chemluli, Chetoui, Koroneiki* and *Coratina,* to different tree-trainings and pruning system (central leader form, open vase form and free form). The study was carried out at the *Experimental orchard of 'Taoues',* central Tunisia (34°N, 10°E). Trees were planted at the density of 204 trees ha^{-1}. Plant growth parameters, yield and fruit characteristics were measured regularly each year. The results showed that free form trees have had an important increase of their canopy volume comparing to the two others form. It varied from 6296 m^3 ha^{-1} (*Meski*) to 12073 m^3 ha^{-1} (*Koroneiki*). On the two other systems (central leader form and open vase form), trees showed an important reduction of their canopy, ranging between a minimum of 3001 m^3 ha^{-1} and a maximum of 9201 m^3 ha^{-1}. In 2012, trees reached an average height which varied from 3.4 m (*Chemlali*) to 4.5m (*Picholine*), from 3.8 m (*Chemlali*) to 4.5 m (*Chetoui*) and from 3.7 m (*Chemlali*) to 4.7 m (*Chetoui*), respectively for the open vase form, the central leader form and the free form. The cumulative yield during the eight production years, from the first in 2005 to the last in 2012, was the highest on the free form. For this form, *Koroneiki* noted the highest yield equal to 42177 kg ha^{-1}. The most important cumulative yield was noted for *Koroneiki* trained on the open vase form (22213 kg ha^{-1}) and for *Chemlali* trained on central leader axis (16131 kg ha^{-1}). The alternate bearing index varied according to the variety and the management of the

---

[*] Corresponding author: Mme Mouna Aïachi Mezghani, Institut de l'olivier, Station Régionale de Sousse, B.P. 14, 4061 Sousse, Tunisie. Email: ayachimouna@yahoo.fr.

adopted orchard management. *Koroneiki* noted the lowest criterion which was equal to 0.39 and 0.26 for the open vase form and the central leader form, respectively. *Meski* showed more important values which were equal to 0.90 and 0.83 for the open vase form and the central leader, respectively. For the free form, varieties like *Chemlali, Ascolana, Meski* and *Koroneiki*, showed values higher than 0.6. To conclude, we can say that the free form and the open vase form systems are highly productive, easy to develop (received little pruning or no pruning at all) and the most economical training system for intensive tree conditions. However, *Koroneiki, Chemlali, Coratina* (oil varieties), and *Picholine, Meski, Manzanilla and Ascolana* (olive table varieties), showed a good suitability mainly due to their growth habit, the high yield and the less alternate bearing. Moreover, *Chetoui* seems to be not adapted to the Southern Tunisian Region, even under intensive conditions.

Keywords: Varieties, intensive conditions, tree-trainings, plant growth, yield

INTRODUCTION

The olive tree is considered a heady species giving satisfactory yields in a variety of areas. In addition, the consumption of olive oil and table olives has increased, even in countries that do not have the tradition of consuming it. Also, the nutritional value of the Mediterranean diet was emphasized [23, 34]. Olive orchards have been established in new areas with less favourable soil and climatic conditions, to obtain higher yield responding to the increasing demand. Also, intensive olive cultivation was introduced to increase production and to limit costs. New varieties were used in this type of orchard while local varieties were also used with higher densities with little information about orchard management (training system, adapted pruning, quantity of irrigation ...) and cultivar architecture characteristics (branching, flowering frequency, fruit location ...). The cultivar suitability for these orchards is not clear. Therefore, good knowledge of vegetative and productive properties of the olive tree was required. As in the case of other fruit crops, the olive orchard will incorporate three main aims: adapted tree habit, early entry into bearing and regular production with good fruit quality [8, 22, 23, 29]. These objectives stimulated researches to better understand the tree comportment and to find solutions to problems appearing with higher densities.

However, the increase of density (200-300 trees ha^{-1}) and the introduction of more intense cultural techniques today, are considered useful to enhance the production and the profit of the growers. The adoption of an appropriate technological package for each planting system (choice of variety, training system, pruning, irrigation, fertilization) is necessary. Wide range of types could be classified as semi-intensive with low inputs (100-200 trees ha^{-1}) and high inputs (120-280 trees ha^{-1}), owing to different combinations of their basic elements: tree age, density, irrigation and tillage. In European country, the intensive olive groves are characterized by a tree density of about 250 trees ha^{-1}, yearly fertilization and pruning, several sprays for pest control, soil tillage once to three per year and irrigation up to 2700 m^3 ha^{-1} year^{-1}. Intensive management results in high yields of 3600-6500 kg ha^{-1}, but also in higher labor costs [27, 28, 35, 43].

The type of pruning and the training system are two of the major factors in successful tree performance and the orchard productivity. The best system is the one that allows high yield of excellent quality and can be obtained at minimum cost without negative side effects on plant performance and orchard management [26, 36, 43]. It would be important that olive trees can express along the year, a high photosynthesis and utilize assimilates mainly for fruit production. All the processes must be optimized so that the productive cycle could be completed without anomalies and disorders which would heavily compromise the yield and the planting life [12, 26, 18, 43].

Variety is a key factor regarding the adaptability to different training system. The varieties diffused in the most important areas, showed some positive attributes appreciated by growers. These characteristics are high oil quality and quantity, soil and climate adaptability and low requirement on specific cultivation techniques. In order to obtain the best conditions to express good production, trees were trained to adequate form, put at right density and under standard norms of irrigation, fertilization and soil management [26, 30, 40, 43].

If plants are to be harvested manually for oil production, the pruning should be as simple and rapid as possible. The most frequent training systems in modern olive growing are the vase form, the free form and the central leader form all compatible with minimum pruning strategies. The vase and the central leader form, are equally suitable for manual and mechanical harvesting. A Study [38] reported the effect of training system on growth, yield and oil quality. They showed that there were no differences in fruit yield between trees of *Frantoio* and *Leccino* trained to both central leader form and vase form, but trees of *Maurino* trained to the central leader form, yielded 40% less than those trained to a vase form. Oil quality was unaffected by the training system. A recent work [31] compared three training systems in a 5m x 5m trial of three cultivars under dry land cultivation. Trees trained to a vase form produced more than central leader form of *Leccino*, but no differences were found between the two systems in *Frantoio* and *Maraiolo*. Recent studies have showed that there was no clear relation between higher productivity, growth, yield efficiency and mechanical harvesting [9, 31, 38, 43]. In a few cases, central leader form was even less productive than other training systems. A recent work [39] studied the variability of tree architecture of three varieties (e.g., *Arbequina*, *Arbosana* and *Koroneiki*) used in high density olive orchards. This work reported that high branching and small diameters are important to increase yield efficiency and affect the cultivar suitability for super high density orchards.

In Tunisia, the geographical distribution of the olive tree is determined by the soil and climatic factors. This is why more than 97% of olive groves are conducted under rain-fed and only 3% of olive groves is irrigated. There are 63 million trees consisting of 60 million for olive oil and 3 million for olive table. To overcome this situation, and particularly the fluctuating productions, and to meet the important amount of olive oil exported to the EU, the Tunisian government, through its financial assistance program, tried since the nineties to encourage farmers to plant olive trees at higher densities. However, and because the increase of tree density enhances competition for water and nutrients, trees should be supplied by water and fertilizers [2, 5, 6]. Irrigation has to be applied as a compulsory practice during the critical stages of shoot and fruit development, i.e., when the tree is the most responsive to water. The irrigated plantations constituted 50000 hectares of which 79% is for the production of olive oil and the rest produce table olives. The orchards can be classified in three categories: traditional orchards with a density about 14-100 trees ha^{-1} and low productivity (650 kg ha^{-1} to 950 kg ha^{-1}) and intensive plantations with a density of 200-300 trees ha^{-1}, trained mainly with the open vase form or sometimes on the central leader form (2000 to 3000 kg ha^{-1}). Super intensive plantations about 1500 ha with a density ranging between 1250 and 1666 trees ha^{-1}, were introduced since the year 2000 [19, 20, 21]. Varieties employed in these orchards, are foreign varieties.

The collection of olive varieties in Tunisia is very rich and most of them were strongly adopted in their original sites [41]. Different foreign varieties were introduced in Tunisia. Recent works studied the behavior of different varieties planted at the *Experimental orchard of 'Taoues'*, central Tunisia (34°N, 10°E) at a density of 204 trees ha^{-1} [2, 3, 13]. *Chemlali, Koroneiki* and *Arbequina* noted the most important yield per tree, varying between 11 and 15 kg tree^{-1}. A comparative trial was established to evaluate the suitability of four olive cultivars (*Arbosana, Arbequina, Chemlali* and *Chetoui*) to a high planting density of 1250 trees ha^{-1} [19, 20]. These works showed that *Chetoui* and *Chemlali* are not adapted to this planting system due to their high vigor, their lower productivity and their highest alternate bearing indexes; while *Arbosana* and *Arbequina* were the most adapted to this new planting density system. There is a little known about the behavior of varieties under Southern climatic conditions of Tunisia and their reaction to different tree training systems. It is important to evaluate the reaction of Tunisian and foreign varieties to different trainings and tree-pruning form.

Therefore, the aim of this study, was to study the behavior of local and foreign varieties grown under different tree-trainings (the central leader form, the open vase form and the free form), in an intensive olive orchard located at the *Experimental orchard of olive tree Institute*, Southern Tunisia (34°N, 10°E) planted at a density of 204 trees ha^{-1}. Plant growth parameters and yield characteristics were studied from their entry into bearing production in 2005 to the last harvesting in 2012.

MATERIALS AND METHODS

Research Site and Plant Material

The trial was carried at the *Research Station of Taoues*, about 40 km far from Sfax, Southern Tunisia (34°N, 21°E). In this region, the climate is semi-arid with yearly averages of 250 mm of rainfall and 1400 mm of reference evapotranspiration (ET$_0$); it has been dry and hot from May to October. Absolute summer temperature often exceeds 35°C with maximum in August and minimum in January (11°C). Rainfall series showed large variations between years and months with minimum values recorded in July and maximum values in December. There was no rainfall during summer months.

Trees were planted in 2002 at 204 trees ha^{-1} on silty (76% S, 14% L, 10% C), non-calcareous and alkaline soil. It was poor in organic matter (0.32%), with a pH of 8.7, an actual density of 1.7 g cm^{-3} and a porosity of 34%. The volumetric soil water content was measured in the laboratory at field capacity (10.2%) and at the wilting point (6.6%).

The experimental set-up was based on eight varieties: *Chemlali, Chetoui, Koroneiki, Coratina* (olive oil cultivar), *Manzanilla, Meski* and *Ascolana* (table olives cultivar) and *Picholine* (a double aptitude cultivar). *Chemlali, Chetoui* and *Meski* cvs., are the most important Tunisian olive varieties. *Coratina* and *Ascolana* are Italian varieties introduced in Tunisia twenty years ago as well as *Picholine* and *Manzanilla* which are originate from Spain and usually cultivated in Tunisia under irrigated conditions. *Koroneiki* is a Greek variety which is frequently cultivated in Tunisia because it showed a high yield even under rain-fed conditions [2, 3, 14]. All plants were issued from semi hardwood cuttings.

Three different tree-trainings and tree shapes were developed and kept through the study for eight years. The chosen treatments were as follows: (1) Central leader form-Trees trained to a single trunk with primary branches arranged in an helicoidal fashion along the central axis for maximum occupation of space and minimum overlapping. The primary branch showed decreasing length from the base to the top of the tree which assumes a conical shape, (2) Open vase form-Trees are formed of a single trunk who trunk height was equal to 0.7 meters, 3 to 5 branches are oriented in different directions. Most of the fruiting shoots are born on secondary and tertiary branches and (3) Free form-It's a free-canopy system which is obtained with minimum pruning, primary branches originating directly from the trunk of the tree or inserted directly from the base of the tree. The canopy is allowed to grow freely.

Crop management practices were carried out in the orchard, i.e., tilling, irrigation, fertilize and pest management practices were similar to those applied in the intensive orchards. Soil was tilled annually in late winter, in summer (June) and in autumn (September).

Irrigation was applied from early spring (April) to harvest (September). Olive trees were irrigated twice a week, by a localized system with four emitters per tree. Crop evapotranspiration (ET_c) was determined following the formulae developed for the non-standard conditions [4] as: $ET_c = ET_0 \times K_c \times K_r$ with values of 0.5 for K_c following the tree age and 0.7 for K_r determined on the base of the tree down projection canopy flat area. Irrigation was applied for the last five reported years, and varied between 400 m^3 ha^{-1} and 570 m^3 ha^{-1}. Pruning was practiced annually from the second year after plantation. The different tree trainings for selecting the main axis and the lateral branches were gradually established. The fertilization program was based essentially on N, P and K, and has been applied since the third year after planting.

The Experiment was designed as a randomized complete block with seven replications (variety). Each block is a different variety and composed of two rows of ten trees each. Eight repetitions (trees) were selected for each form.

Plant Growth Parameters and Yield Characteristics

Tree height, trunk and canopy diameter were monitored annually after harvest (on January) from 2005 until 2012. Measurements were performed for each tree of the different tree form. Canopy volume of each tree was calculated by using the formula:

Canopy volume = 1/6 ($\pi \times D^2 \times H$)

where H is the height of the tree and D is the width of the canopy.

From the third to the twelfth years after planting, all trees were harvested separately and production was weighed individually. Olives used for conserves (*Meski, Ascolana, Picholine* and *Manzanilla*) were harvested on October, while oil fruits (*Chetoui, Coratina, Koroneiki* and *Chemlali*) were collected later, during the winter.

Average yield was determined for each training tree-from and variety. Yield per hectare was determined by multiplying the yield per tree density.

The alternate bearing criterion (K) was evaluated [44]. It was calculated according the formula.

$$K = \sqrt{\frac{1}{n-1}\sum_{i=2}^{N}\left(\frac{Pi - P(i-1)}{Pi + P(i+1)}\right)^2}$$

where n is the harvest number years; Pi, P(i+1) and P(i-1) are the crop yield of the three successive crops.

RESULTS

Plant Growth

Vigour Parameters

Annual height and diameters of the trunk were analysed for the eight years of experiment, from 2005 to 2012. Cumulative values were measured on all varieties of the three training form.

Average height trunk of the different varieties was represented in Figure 1. This one increased rapidly for the first three years of the monitoring period (2005-2007). At the end of the growing season (2007), the trunk height varied between 34.0 cm (*Chemlali*) and 81.6 cm (*Manzanilla*) for the open vase form and increased until 42.5 cm (*Manzanilla*) to 86.2 cm (*Ascolana*) for the central leader form.

At the end of 2007, the trunk seemed to be formed indicating then a slowdown in the increase of its length. In 2012, the final height of trunk ranged between 39.2 cm and 83.7 cm for the open vase form (Figure 1a) and between 43.1 cm and 93.3 cm for the central leader form (Figure 1b). For the free form, primary branches are inserted directly on the base of the trunk. This is why the trunk was absent or very short (Figure 1c).

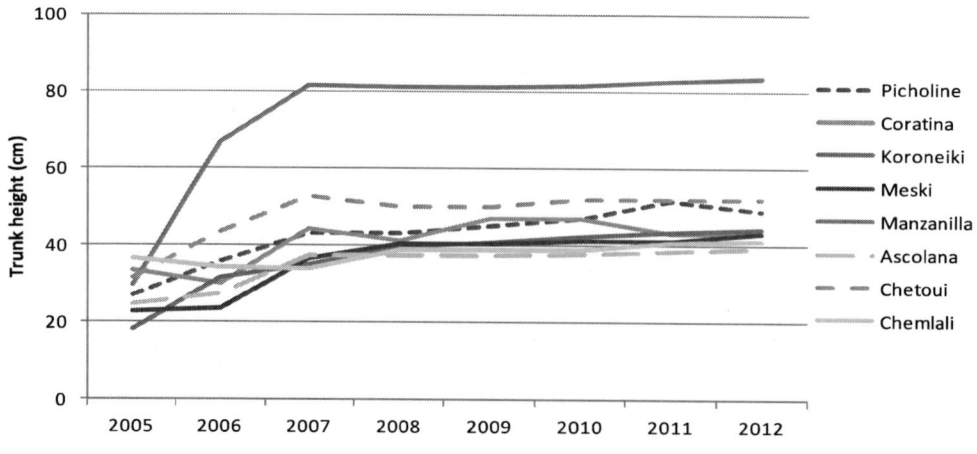

a

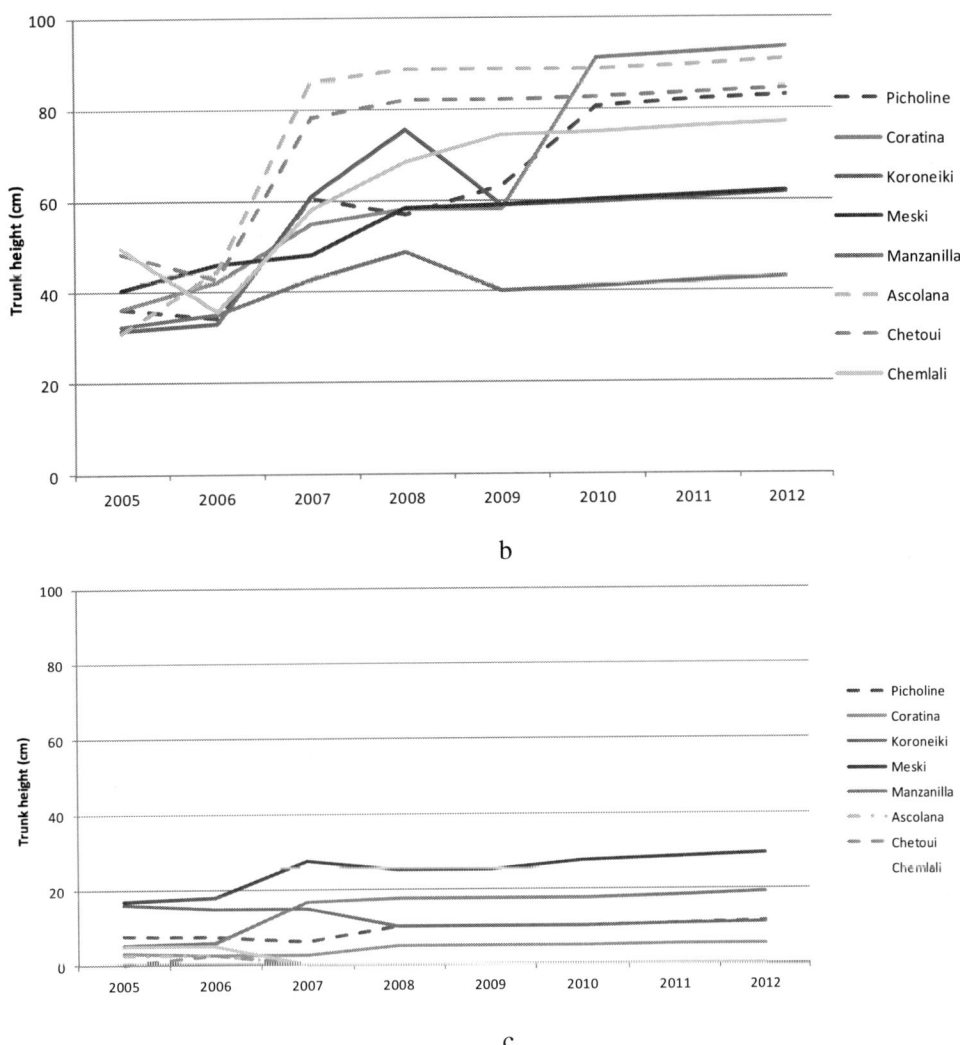

Figure 1. Average trunk height made on basis of field measurements carried out on the eight studied cultivars, under the three different tree-training- and for the experimental period (2005-2012). a-Open vase form, b-Central leader form and c-Free form. Each value is an average of 8 observations (tree).

The trunk diameter grew regularly with increasing rates (Figure 2). Differences were noted between varieties according to form. The growth in the thickness of the trunk was very active and continued throughout the studied period.

At the end of the experiment, trees reached a maximum diameter of 18.87 cm and 16.79 cm for the open vase form and the central leader axis respectively, while lower diameters were measured on *Chetoui* variety; e.g., 12.78 cm for open vase form and on *Chemlali*, e.g., 13.65 cm for central leader form.

Tree Parameters

Regular measurements of tree height recorded during the eight experimental years (2005-2012) for each variety and training system are reported in Figure 3.

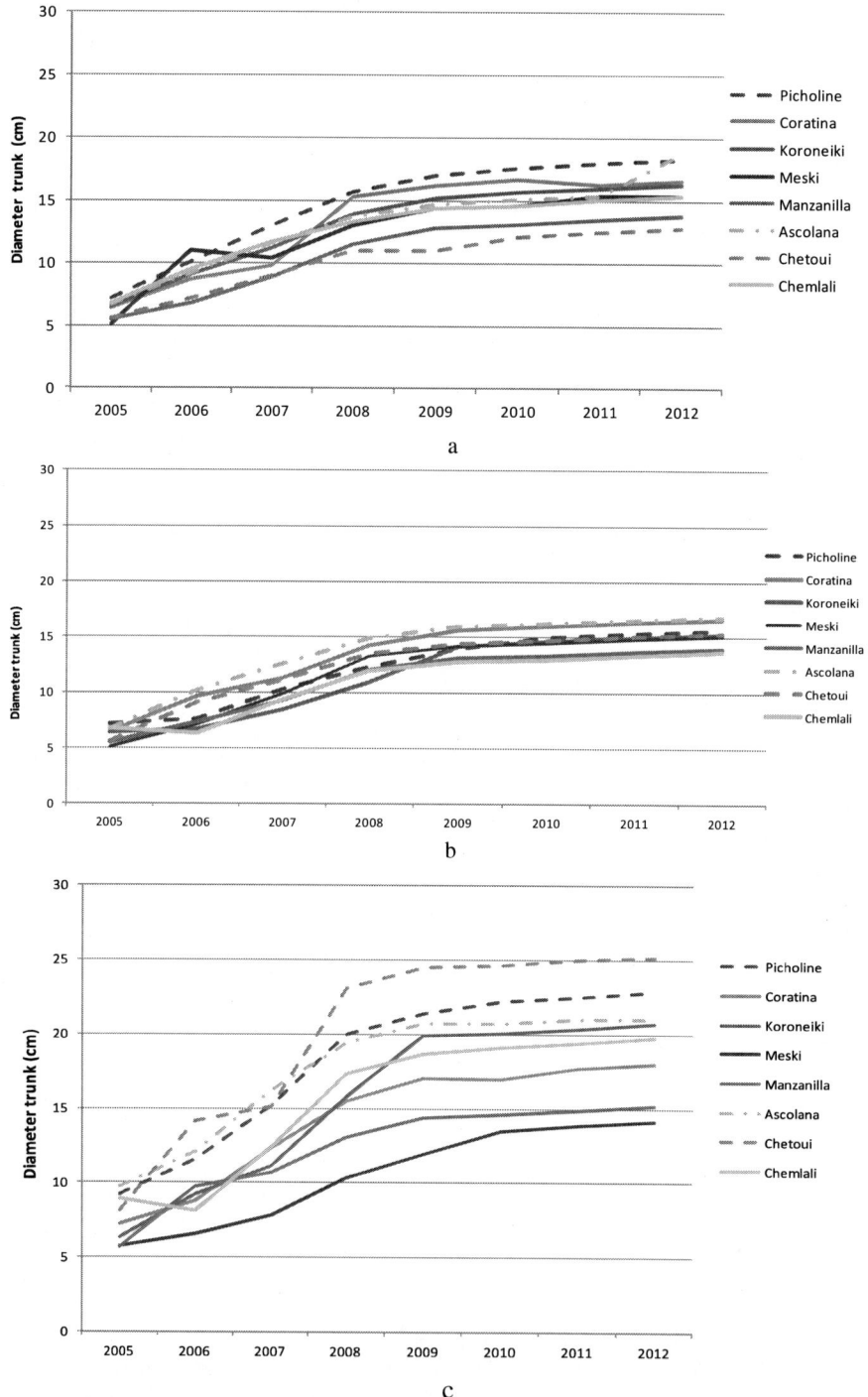

Figure 2. Average trunk diameter made on basis of field measurements carried out on the eight studied cultivars under the three different tree-trainings and during the experimental period (2005-2012). a-Open vase form, b-Central leader form and c-Free form Each value is an average of 8 observations (tree).

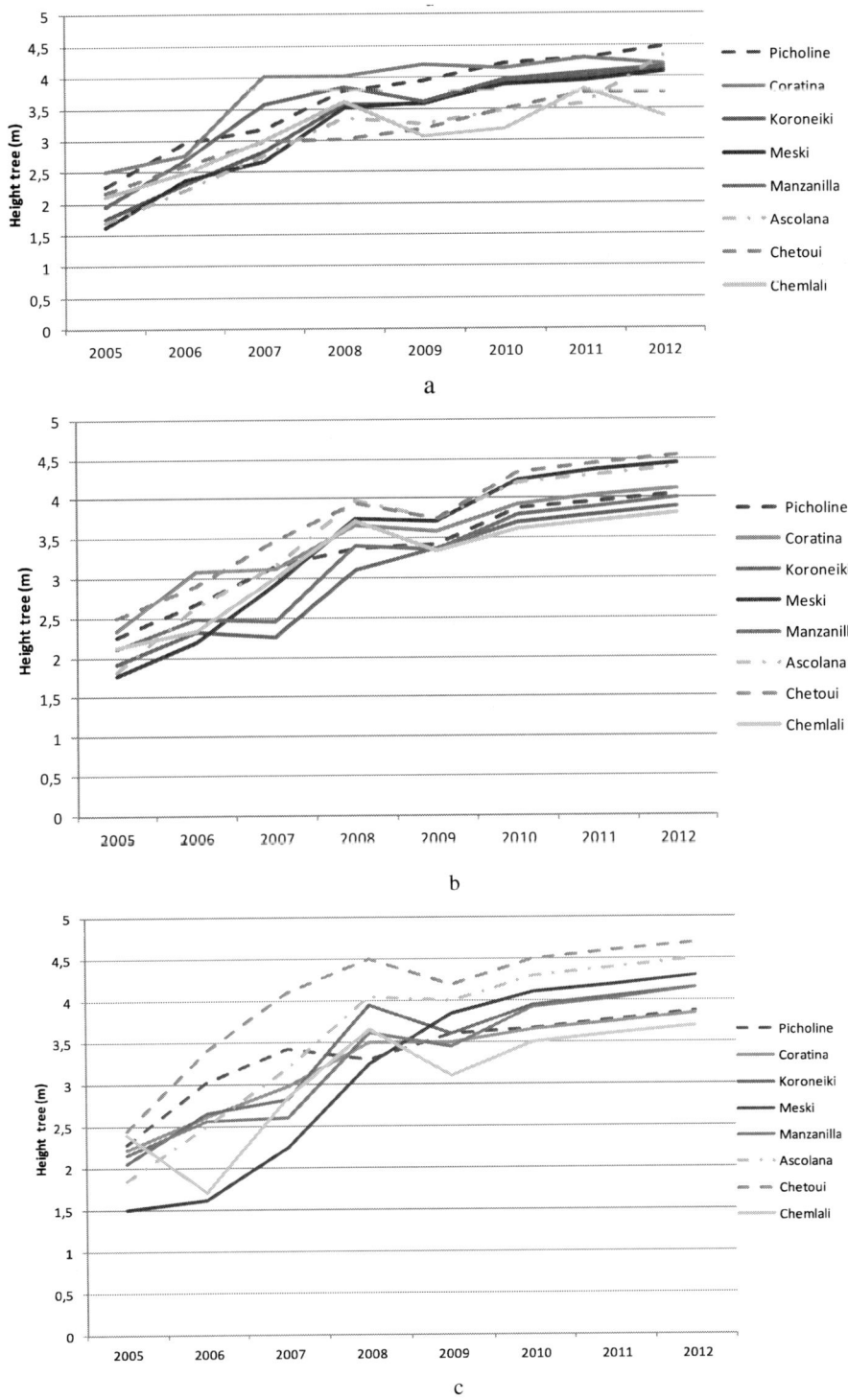

Figure 3. Evolution of height tree (m) measured on the eight studied cultivars under the three different tree-trainings and for the experimental period (2005-2012). a-Open vase form, b-Central leader form and c-Free form. Each value is an average of 8 observations (tree).

The evolution of the tree height followed an exponential curve, with increasing differences between varieties (Figure 3). It is important to mention that the height of the tree, increased rapidly on the first five studied years, *i.e.* from 2003 to 2008. In the fifth year of growth, *i.e.,* 2008, more than 80% of the final height was accumulated for the different varieties cultivated under the three training systems. From this date, the tree height growth began to be slower. At the end of the studied period, trees reached an average height varying from 3.4 m (*Chemlali*) to 4.5 m (*Picholine*), from 3.8 m (*Chemlali*) to 4.5 m (*Chetoui*) and from 3.7 m (*Chemlali*) to 4.7 m (*Chetoui*), respectively for the open vase form, the central leader form and the free form (Table 1). It appeared that trees issued from *Chemlali,* appeared to be the smallest considering the three tree-trainings.

Table 1. Average tree height (m) measured on all studied varieties under the three tree-trainings and at the end of the experimental period, i.e., over the eight years of monitoring

Varieties	Open vase form	Central leader form	Free form
Meski	4.1 ± 0.2	4.4 ± 0.6	4.3 ± 0.8
Manzanilla	4.2 ± 0.8	4.0 ± 0.7	4.2 ± 0.6
Ascolana	4.3 ± 0.1	4.4 ± 0.1	4.5 ± 0.1
Picholine	4.5 ± 0.3	4.0 ± 0.4	3.9 ± 0.4
Chetoui	3.7 ± 0.4	4.5 ± 0.1	4.7 ± 0.1
Chemlali	3.4 ± 0.2	3.8 ± 0.6	3.7 ± 0.4
Coratina	4.2 ± 0.4	4.1 ± 0.4	3.8 ±0.1
Koroneiki	4.1 ± 0.7	3.9 ± 0.7	4.1 ±0.3

Canopy width evolution showed significant differences among varieties, over the studied period, for the three tree training form (Figure 4). For the open vase form, *Chetoui* showed the lowest canopy diameter (2.9 m), while *Picholine* noted the highest one (4.7 m).

For the central leader axis, the canopy width ranged between 3.1 m for *Koroneiki* and 3.7 m for *Ascolana* and *Picholine*. Trees with free-canopy system noted a more important development of the crown, with minimum and maximum values recorded for cultivars *Chemlali* (4.3 m) and *Koroneiki* (5.3 m), respectively. Final canopy volume was calculated per tree and then reported per hectare by multiplying it to number of trees, according the three different training systems (Table 2). Significant differences were observed between the three forms. The canopy volume varied from 3646.1 to 6389.9 m^3 ha^{-1}, from 3001.7 to 9201.1 m^3 ha^{-1} and from 6296.1 m^3 ha^{-1} to 12073.4 m^3 ha^{-1} respectively, for the central leader form, the open vase form and the free form. Additionally, the ground cover index was more important for all varieties trained on free form. In this case, it ranged between 29.8% (*Chemlali*) and 43% (*Koroneiki*). The lowest soil coverage was noted in trees conducted under the central leader form and this result was verified for all varieties.

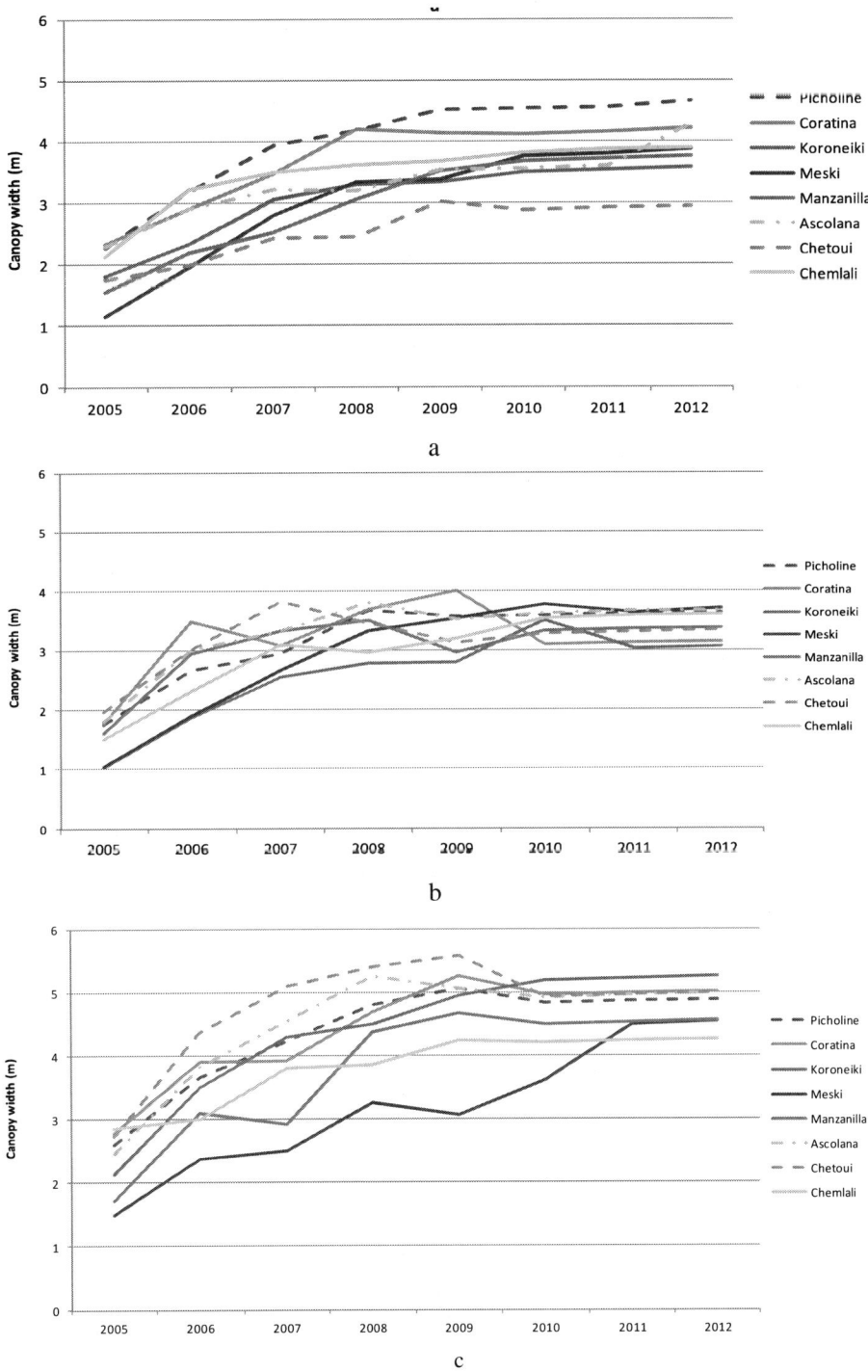

Figure 4. Evolution of width canopy (m) measured on the eight studied cultivars under the three different tree-trainings and for the experimental period (2005-2012). a-Open vase form, b-Central leader form and c-Free form. Each value is an average of 8 observations (tree).

Table 2. Canopy volume and ground cover index measured on all studied varieties at the end of the experimental period, i.e., over the eight years of monitoring (in 2012), under the three different tree-trainings

Varieties	Open vase form		Central leader form		Free form	
	Canopy volume $(m^3 ha^{-1})$	Ground cover index (%)	Canopy volume $(m^3 ha^{-1})$	Ground cover index (%)	Canopy volume $(m^3 ha^{-1})$	Ground cover index (%)
Meski	6274.1	24.3	6296.3	22.2	6296	34.3
Manzanilla	6512.4	23.5	4843.9	18.8	9007.1	33.8
Ascolana	8110.3	29.2	6389.3	21.6	11796.8	39.7
Picholine	9201.1	34.7	5270.1	21.5	9370.8	38.2
Chetoui	3001.7	14.7	4817.7	17.8	11111.2	39.7
Chemlali	5197.4	24.4	5237.6	21.2	6488.1	29.8
Coratina	7807.3	28.5	4193.5	15.9	12057.4	40.1
Koroneiki	5304.5	34.7	3646.1	15.1	12073.5	43.9

The ground cover index ranged between 15.1 and 22.2 %. For the open vase form, *Chetoui* showed the lowest value (3001.7 m^3 ha^{-1}), while varieties like *Picholine, Ascolana* and *Coratina*, noted the highest volume of the canopy which was equal to 9201.1, 8110, 7807 m^3 ha^{-1} respectively. For the central leader form, all varieties noted a reduction of their canopy volume, compared to the two others systems. For the free forms, all varieties seem to be more vigorous and developed more important canopy volume. The crown volume varied from 6296 m^3 ha^{-1} (*Meski*) to 12073 m^3 ha^{-1} (*Koroneiki*).

Yield Components

The cumulative yields of the different varieties over a period of eight years from 2005 (entry into production) to 2012, were recorded and calculated on the basis of individual trees and then per hectare on the three training-tree systems (Table 3).

Trees trained on the central leader axis showed a total yield per tree which varied from 34.9 kg tree^{-1} (*Chetoui*) to 79.4 kg (*Chemlali*), while trees trained on the open vase form showed yield ranging between 21.3 kg tree^{-1} (*Chetoui*) and 110 kg tree^{-1} (*Koroneiki*). All varieties trained on the free form, showed an increase of the yield. For the olive table variety, *Manzanilla* noted the highest total yield per tree and per hectare which was equal to 110.4 kg tree^{-1} and 22406.6 kg ha^{-1}. For the oil variety *Koroneiki*, the total yield was equal to 208.1 kg tree^{-1} and 42177 kg ha^{-1}.

The annual mean yield was higher on the free form. The highest yield was recorded for *Koroneiki* equal to 26 kg tree^{-1}year^{-1}and 5272 kg ha^{-1} year^{-1}. For varieties like *Manzanilla, Ascolana, Coratina* and *Chemlali*, the yield was higher than 2000 kg ha^{-1} year^{-1} and fluctuated between 2200 to 2800 kg ha^{-1} year^{-1}. *Meski* noted the lowest yield which was only equal to 1573 kg ha^{-1} year^{-1}. The annual mean yield varied between a minimum noted on *Chetoui* of 541 kg ha^{-1} year^{-1}and a maximum of 2776 kg ha^{-1} year^{-1} for the open vase form. It ranged between 891 kg ha^{-1} year^{-1}on *Chetoui* and 2016 kg ha^{-1} year^{-1}on *Chemlali* for the central leader axis.

Table 3. The total yield per tree and per hectare during the first 8 years of production and mean annual fruit yield per tree and hectare of the studied varieties under the three different tree-trainings

Tree form	Varieties	Yield per total period		Annual mean yield	
		kg tree^{-1}	kg ha^{-1}	kg tree^{-1}	kg ha^{-1}
Open vase form	Meski	68.0	13781.4	8.5	1722.7
	Manzanilla	63.6	12896.3	7.9	1612.0
	Ascolana	46.9	9476.9	5.8	1184.6
	Picholine	82.7	16743.1	10.3	2092.8
	Chetoui	21.3	4335.0	2.6	541.9
	Chemlali	78.2	15887.7	9.7	1985.9
	Coratina	47.7	9631.9	5.9	1203.9
	Koroneiki	109.5	22213.0	16.4	2776.6
Central leader form	Meski	75.1	15324.4	9.3	1912.5
	Manzanilla	52.2	10653.7	6.5	1331.7
	Ascolana	53.1	10847.7	6.6	1355.9
	Picholine	44.0	8984.1	5.5	1123.0
	Chetoui	34.9	7134.6	4.3	891.8
	Chemlali	79.4	16131.3	9.9	2016.4
	Coratina	40.4	8242.6	5.0	1030.3
	Koroneiki	72.8	14860.3	9.1	1857.5
Free form	Meski	62.2	12587.8	7.8	1573.5
	Manzanilla	110.4	22406.6	13.8	2800.8
	Ascolana	103.8	21112.6	12.9	2639.1
	Picholine	71.6	14494.4	8.9	1811.8
	Chetoui	65.1	13376.5	8.1	1672.1
	Chemlali	79.0	17840.0	9.9	2230.0
	Coratina	101.4	20512.1	12.7	2564.1
	Koroneiki	208.0	42177.1	26.0	5272.1

The alternate bearing index varied according to the variety and the adopted orchard management (Table 4). Regarding the open vase form and the central leader form, *Koroneiki* noted the lowest criterion which was equal to 0.39 and 0.26, respectively. In contrast, this criterion on *Meski* was more important and reached value equals to 0.90 for the open vase form and 0.83 for the central leader axis. For the free form, varieties like *Chemlali, Ascolana, Meski* and *Koroneiki*, showed values higher than 0.6. It is important to mention that *Koroneiki* noted a tendency to alternate bearing (k = 0.67) when it was conducted on the free form, but the cumulative yield was still important. This can be explained by highly productive On year causing an extreme fruiting alternate bearing in the trees issued of this variety. The yield per tree reached 42, 51 and 55 kg tree^{-1} respectively, for 2007, 2009 and 2011, while it was equal to 7.8, 11.5 and 0 kg tree^{-1} for 2006, 2008 and 2010 (Figure 5).

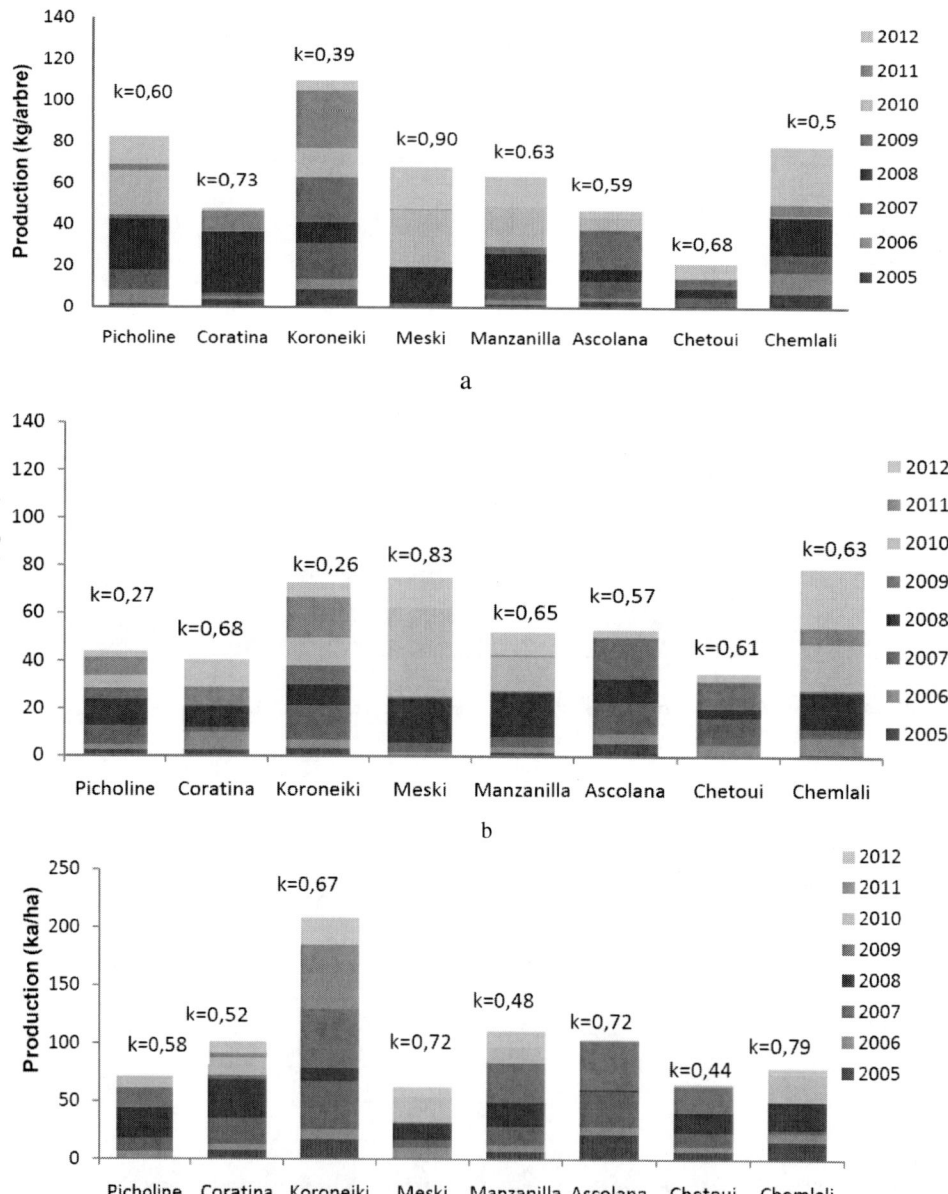

Figure 5. Yield per tree during the first 8 years of production an alternate bearing criterion of the studied varieties under the three different tree trainings. a-Open vase form, b-Central leader form and c-Free form.

Table 4. The alternate bearing criterion calculated on all studied varieties over the experimental period, i.e. over the eight years of monitoring

	Open vase form	Central leader form	Free form
Picholine	0.60	0.27	0.58
Coratina	0.73	0.68	0.52
Koroneiki	0.39	0.26	0.67
Meski	0.90	0.83	0.72
Manzanilla	0.63	0.65	0.48
Ascolana	0.59	0.57	0.72
Chetoui	0.68	0.61	0.44
Chemlali	0.52	0.63	0.79

Finally, the highest value of alternate bearing criterion was observed on *Chemlali* which was equal to 0.79. This result can be explained by the fact that the yield was very low during Off years, e.g. 2006, 2007, 2009, 2011 and 2012, which varied between 0 and 7 kg tree^{-1}. For this same variety, yield was equal to 16, 24.5 kg tree^{-1} for 2005, 2009 and 2010, respectively.

Particular interest was focused on the annual yield of the different varieties. The number and frequency of On and Off year has been determined (Table 5). The production was considered On year when the annual production exceeded 10 kg tree^{-1} or +/- 2000 ha^{-1}. Generally, the frequency of On year was the most important on all varieties trained on the free form; it varied from 25% to 75% or from 2 to 6 years out of 8. *Koroneiki* and *Manzanilla* variety, showed the most important On year number; 6 of 8 years were On. In this case, the yield per tree for *Koroneiki*, was equal to 24.8 and 3.9 kg tree^{-1} for On and Off year, respectively. For *Manzanilla*, the yield was equal to 16.3 kg tree^{-1} for On year and 4.2 kg tree^{-1} for Off year. The sum of the yield during the On years, varied from 59% (*Meski*) to 96% (*Koroneiki*) of the total yield during the studied period.

Number and frequency of Off year were more significant on the open vase form and the central leader axis, with values ranging between 50 and 100% and from 62.5 to 87.5%, respectively. On the open vase form, *Coratina, Ascolana* and *Chetoui*, noted higher number of year when the yield was very low. The frequency varied between 7 and 8 Off year. *Chetoui* variety did not show any fruit production. The mean yield of On year, varied between 16.2 kg tree^{-1} (*Koroneiki*) and 30.1 kg tree^{-1} (*Coratina*) for the open vase form and between 11 kg tree^{-1} (*Picholine* and *Chetoui*) and 22.9 kg tree^{-1} (*Meski*) for the central leader form.

During the first year of entry into bearing (2005-2007), the free tree form showed a higher crop efficiency (Figure 6). Maximum values were noted on *Koroneiki* in 2005 (3.63 kg m^{-3}), on *Chemlali* in 2006 (3.42 kg m^{-3}) and on *Meski* in 2007 (2.6 kg m^{-3}). From 2008, the crop efficiency has been less important.

Maximum values ranged between 0.26 kg m^{-3} (for *Chetoui* in 2008) and 1.12 kg m^{-3} (for *Koroneiki* in 2009). For the two other training systems, the crop efficiency varied depending on the variety and the year. For the open vase form, maximum values varied between 0.79 kg m^{-3} (*Picholine* in 2009) and 2.79 kg m^{-3} (*Koroneiki* in 2005). For the central leader form, crop efficiency values were the lowest. Maximum values ranged between 0.58 kg m^{-3} (*Picholine* in 2005) and 1.5 kg m^{-3} (*Chetoui* in 2007).

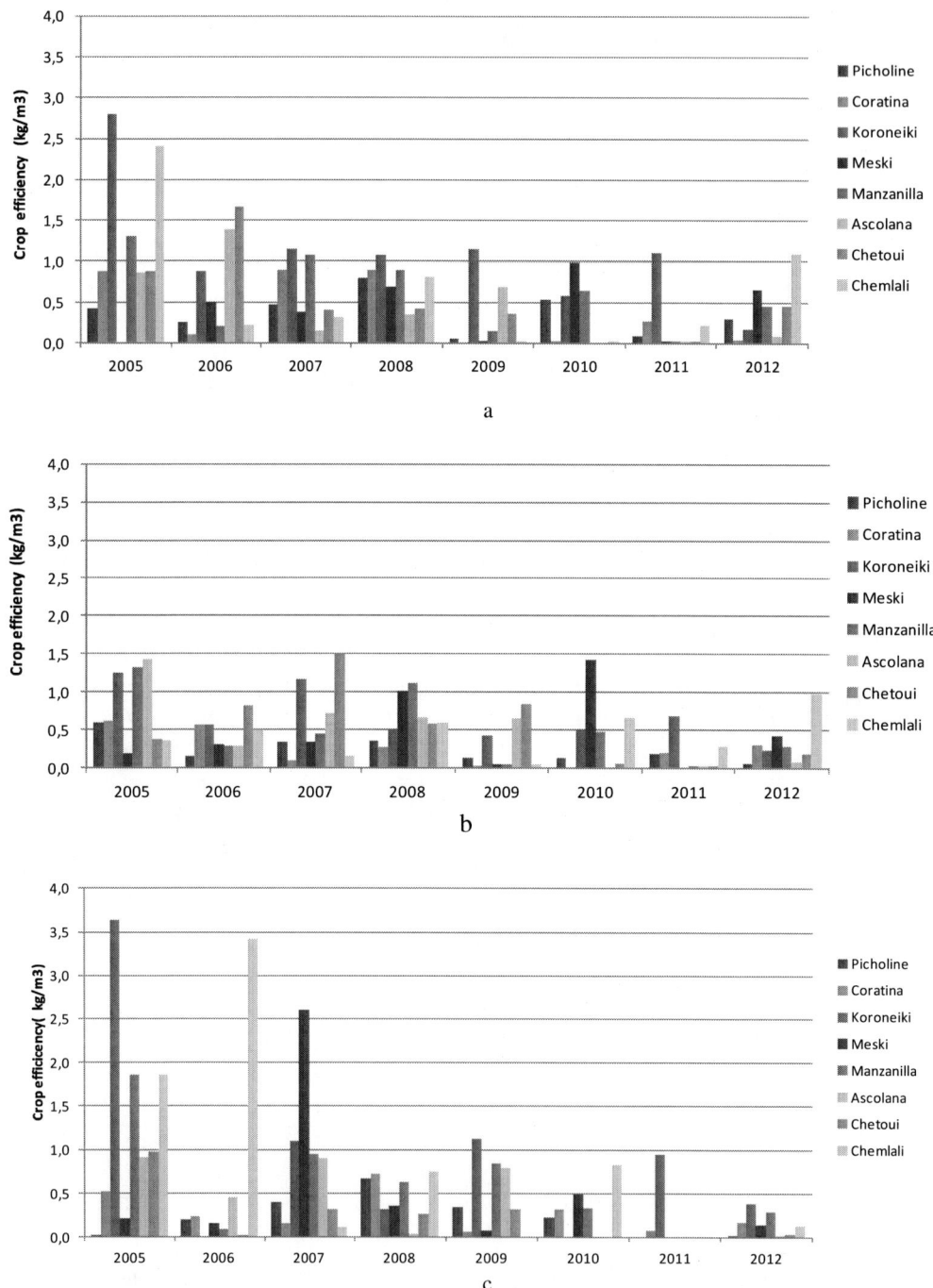

Figure 6. Crop efficiency (kg m^{-3}) estimated on the eight studied cultivars under the three different tree-trainings and for the experimental period (2005-2012). Each value is an average of 8 observations (tree). a-Open vase form, b-Central leader form and c-Free form.

Table 5. Number and frequency of On and Off years and the mean yield per tree during the On and Off years of the different varieties under the three different tree-trainings

Varieties	Number		Frequency		Annual means	
	On years	Off years	On years	Off years	On years	Off years
Open vase form						
Meski	3	5	37.5	62.5	21.7	0.6
Manzanilla	3	5	37.5	62.5	17.1	1.7
Ascolana	1	7	12.5	87.5	18.5	4.0
Picholine	3	5	37.5	62.5	19.9	4.6
Chetoui	0	8	0	100	0	2.6
Chemlali	2	6	25	75	23.1	5.3
Coratina	1	7	12.5	87.5	30.1	2.5
Koroneiki	4	4	50	50	16.2	7.1
Central leader form						
Meski	3	5	37.5	62.5	22.9	1.2
Manzanilla	2	6	25	75	16.4	3.2
Ascolana	1	7	12.5	87.5	17.4	5.1
Picholine	1	7	12.5	87.5	11	4.7
Chetoui	2	6	25	75	11.1	2.1
Chemlali	3	5	37.5	62.5	19.8	3.9
Coratina	1	7	12.5	87.5	11.25	4.1
Koroneiki	3	5	37.5	62.5	14.4	5.9
Free form						
Meski	2	6	25	72	18.5	2.7
Manzanilla	6	2	75	25	16.3	4.2
Ascolana	3	5	37.5	62.5	31.6	1.8
Picholine	3	5	37.5	62.5	18.2	3.4
Chetoui	3	5	37.5	62.5	17.3	2.6
Chemlali	3	5	37.5	62.5	21.7	2.8
Coratina	3	5	37.5	62.5	20.4	4.9
Koroneiki	6	2	75	25	24.8	3.9

DISCUSSION

The aim of this Chapter is to study the reaction of different table varieties, like *Picholine, Meski, Manzanilla* and *Ascolana* and oil varieties, like *Chemlali, Chetoui, Koroneiki* and *Coratina*, to the different tree-trainings and pruning system (the central leader, the open vase form and the free form). This study was carried out at the *Experimental orchard of 'Taoues'*, central Tunisia (34°N, 10°E). Plant growth parameters and yield components were monitored yearly for the eight years of experiment, from 2005 (entry into production) to 2012 (last harvest).

According to the vigour, the height of the trunk increased rapidly for the first three years of the monitoring period (2005-2007) for the open vase form and the central leader axis. At the end of 2007 (five year of cultivation), the trunk seemed to be formed, indicating a slowdown in the increase of its length. For the free form, primary branches are inserted directly on the base of the trunk. This is why the trunk was absent or very short. The growth thickness of the trunk was very active and continued throughout the observation period. These results are in accordance with those reported on olive tree which indicated that during the planting year (unproductive year) and the low fruit load years, vegetative growth was important for all growth parameters producing high amounts of assimilates and low C/N ratio [2,3,26]. Similar patterns of growth were reported for almonds [1], showing that optimum increase was noted four years after plantation. High variation from year to year was in response to the changing conditions and the adaptation of varieties to the environmental factors and their changes.

The *Chemlali* variety showed trees much smaller than others under the three tree-trainings. However, *Chetoui* conducted on the free form and on the central leader axis, noted higher trees. This can be explained by the natural habit of the variety. *Chemlali* showed an open growth habit while *Chetoui* indicated erect-year growth type [21, 41].

The final canopy volume reported per hectare showed significant differences between the three forms of tree. Canopy volume varied from 3646.1 to 6389.9 m^3 ha^{-1}, from 3001.7 to 9201.1 m^3 ha^{-1} and from 6296.1 m^3 ha^{-1} to 12073.5 m^3 ha^{-1}, respectively for the central leader form, the open vase form and the free form. Additionally, the ground cover index was more important for all varieties trained on the free form. In this case, it ranged between 29.8% for *Chemlali* and 43% for *Koroneiki*. For the free form, all varieties seem to be more vigorous and developed more important canopy volume. The central leader axis and the open vase form, showed less important canopy volume because they require more severe pruning to develop the canopy during the training period. The tree trained on the free form noted a more important canopy volume because no pruning was used. The crown volume varied from 6296 m^3 ha^{-1} (*Meski*) to 12073 m^3 ha^{-1} (*Koroneiki*). This variation between can be explained by both the endogenous factor within the same species which determined the plant vigor and the tree architectural characteristics distinguished between varieties. Variability of the tree architecture was noted between three varieties (i.e. *Arbequina*, *Arbosana* and *Koroneiki*) cultivated in high density olive orchards [22, 23, 29, 39].

The total volume reached 12000 m^3 ha^{-1} in the intensive conditions of our experiment. This result corroborates previous works indicating that the total volume of trees in the region of Andalucia, in non-irrigated land reached 8000 to 10000 m^3 ha^{-1}. In irrigated land, it ranged between 12000 to 15000 m^3 ha^{-1} [32, 33]. It is important to mention that the assimilate production depends on the leaf area and its light exposure, expressed by canopy volume, by leaf density, by light interception and exposure. Trees that showed different crops, without supplementary irrigation, are more efficient. They had volumes of 10-12000 m^3 ha^{-1}, a leaf area index around 6, and a canopy density of about 2 m^2 of leaves per cubic meter. The interception of radiant energy that fell over the canopy was 90% [7, 43].

The annual mean yield was higher on the free form than on the ones of the open vase form and the central leader axis. On the free form, the highest yield was recorded for *Koroneiki* which was equal to 5272 kg ha^{-1}. For varieties like *Manzanilla, Ascolana, Coratina* and *Chemlali*, the yield was higher than 2000 kg ha^{-1} and fluctuated between 2200 to 2800 kg ha^{-1}. All these varieties were highly productive and easy to conduct under the free form tree.

For the open vase form and the central leader axis, the yield doesn't exceed 2800 kg ha^{-1}. As the varieties were planted all at uniform distance and density, it could be explained that the main factor affecting the yield in this case was the training tree form and pruning of the olive tree on one hand and the degree of light interception on the other [15, 24, 36, 37]. Furthermore, the severe pruning required to develop the open vase form and the central leader axis, delayed these tree to reach their full fruiting potential, while the free form tree developed a more important canopy and consequently more fruits. Under the open vase form, *Picholine* (table variety) and *Koroneiki* (oil variety) were more productive. The low yield of *Chetoui* may be explained by the fact that this variety is not adapted to the Southern Tunisian region, as it is not its original area [3, 20, 21, 41].

The alternate bearing index varied according to the variety and the adopted orchard management. Regarding the open vase form and the central leader axis, *Koroneiki* showed the lowest criterion. In contrast, this criterion on *Meski* was more important and reached values equal to 0.90 and 0.83 for the open vase form and the central leader axis, respectively. For the free form, varieties like *Chemlali, Ascolana, Meski* and *Koroneiki*, showed values higher than 0.6. On the free form, *Koroneiki* noted a tendency to alternate bearing (k = 0.67), but the cumulative yield was still important. This can be explained by highly productivity in On year causing an extreme fruiting alternate bearing in the trees issued of this variety. The frequency of On year was more important for all varieties trained on the free form; it varied from 25% to 75% or from 2 from 8 years to 6 from 8 years. *Koroneiki* variety showed the most important On year number. Six of 8 years were On. However, the number and frequency of Off year was more significant on the open vase form and the central leader axis which varied between 6 and 8 year. *Chetoui* variety did not show any On year.

The olive tree is a well-known alternate-bearing species [11, 12, 25, 30, 44], this is why a high yield year is generally followed by a low yield year even under optimal conditions of cultivation. The vegetative growth is reduced during the On year due to competition for assimilates between new shoots and growing fruits, which become the main sinks after the fruit set. The bearing condition of the tree influences consistently both the current-year shoots' number and their growth [10, 11, 12, 25, 36]. This behavior is not only specific for olives. A consistent reduction of shoot growth was reported for apple trees [17, 18] and peach trees [16] during the high fruit load years.

Indeed, it is clear that the level of the alternate bearing varied considerably between the variety, the tree training system and the year. It is important to mention that only the alternate bearing index was sufficient to have an idea on the yield over a long period. The frequency of the On year is also important to describe the evolution of the yield. Furthermore, the free form was more productive and showed a higher frequency of On years which is beneficial for grower. These results were not in agreement with those [24], for which, the degree of alternate bearing was only slightly affected by the different tree shapes.

Still, we concluded that the free form is highly productive and easy to develop. It can be used in intensive growing conditions, even in southern of Tunisia. Varieties like *Koroneiki, Manzanilla, Ascolana, Picholine, Coratina* and *Chemlali* showed a good suitability for this tree-training form, mainly due to their adapted canopy, high crop efficiency and less alternate bearing. However it is important to establish the technical package accompanying this type of training, in terms of irrigation, fertilization and chemical treatments. However, further research should be carried to study variety architectural differences, and their affect on the plant growth and the distribution of fruiting.

ACKNOWLEDGMENTS

Author thanks Mr. F. Labidi, K. Meddeb L. Attia and A. Ouled Amor for their assistance during the field experimental work.

REFERENCES

[1]   Abdel-Rhahmane A.A & Sharkawi H.M (1974). Responses of olive and almonds orchards to partial irrigation under dry farming practices in semi-arid regions. Water relations in olive during the growing season. *Plant and soil*, 41, 13-31.

[2]   Aïachi Mezghani M, Sahli A, Grati N, Ben Amar F, Ben Ali S, Ben Amor R, Labidi F & Ouled Amor A (2009). Olive trees characteristics (*Olea europaea* L.) under intensive condition cultivated in Tunisia (Sfax site). In Proceedings of ''For a renovate, profitable and competitive Mediterranean olive growing sector''. Olivebioteq 2009, Sfax, Tunisie, 15-19 Décembre 2009, 443-447.

[3]   Aïachi Mezghani M, Sahli A, Labidi F, Khairi M & Ouled Amor A (2012). Study of the Behaviour of Olive Varieties cultivated under Different Tree-trainings: Vegetative and Productive Characteristics. VI-International Symposium on olive growing, Book Series, *Acta Horticulturae*, 949, 439-446.

[4]   Allen G, Pereira S, Raes D & Smith M (1998). Crop evapotranspiration-Guideline for computing crop water requirements. Irrigation and drainage, paper No 56, FAO, Rome.

[5]   Ben Ahmed C, Ben Rouina B & Boukhris M.M (2007). Effects of water deficit on olive trees cv. Chemlali under field conditions in arid region in Tunisia. *Scientia Horticulturae*, 113, 267-277.

[6]   Berman M.E & Dejong T.M (2003). Seasonal patterns of vegetative growth and competition with reproductive sinks in peach (*Prunus Persica*). *Journal of Horticultural Science & Biotechnology*, 78, 303-309.

[7]   Bongi G & Palliotti A (1994). Olive In. B Schaffer et PC Anderson (eds). Handbook of environmental physiology of fruit crop. CRC Press Inc, USA, 165-182.

[8]   Camposeo S, Ferrara G, Palasciano M & Godini A (2008). Varietal behaviour according to the superintensive olive culture training system. *Acta Horticulturae*, 791, 271-274.

[9]   Cantini C & Sillari B (1998). Esperienza toscane nell' intensificazioni colturale dell'olivo. In Seminari di Olivicolture, Academia Nazionale dell'olivo, Italy, 127-139.

[10]   Castillo-Llanque F.J, Rapoport H.F& Navarro C (2005). Caracterizacion de la ramification del olivo: influencia de differentes fechas de recolleccion y estados de carga. In: Congreso Iberico de Ciencias Horticolas, May 2005, Porto, Portugal, 1-7.

[11]   Castillo-Llanque F.J, Rapoport H.F & Navarro C (2008). Interaction between shoot growth and reproductive behavior in olive trees. *Acta Horticulturae*, 791, 453-457.

[12]   Castillo-Llanque F.J & Rapoport H (2011). Relation between reproductive behavior and new shoot development in 5-year-old branches of olive tree (*Olea europaea* L.). *Trees*, 25, 823-832.

[13]   Chaari Rkhiss A, Gueriani L, Kammoun N, Ouled Amor A & Maalej M (2009). Comportement de six variétés d'oliviers à huile dans le biotope de Taous (Sfax,

Tunisie) : Résultats de 4 campagnes de suivi. In Proceedings of "For a renovate, profitable and competitive Mediterranean olive growing sector". Olivebioteq 2009, Sfax, Tunisie, 15 19 Décembre 2009, 45-50.

[14] COI (2000). Catalogue mondial des variétés d'olivier. Conseil oléicole International, Madrid, 360 p.

[15] Connors D.J & Fereres E (2005). The physiology of adaptation and yield expression in olive. *Horticultural Reviews*, 31.

[16] DeJong T.M & Goudrian J (1989). Modelling peach fruit growth and carbohydrate requirements: reevaluation of the double-sigmoid growth pattern. *Journal American Society Horticultural Sciences*, 114, 800-804.

[17] Erf J.A & Proctor J.T.A (1987). Changes in apple leaf water status and vegetative growth as influenced by crop load. *Journal American Society Horticultural Sciences*, 112, 617-620.

[18] Forshey C.G & Elfving D.C (1989). The relationship between vegetative growth and fruiting in apple trees. *Horticultural Review*, 11, 229-87.

[19] Jardak T (2006). The olive industry in Tunisia. In Proceedings of Olivebioteq. Volume Special seminars and invited lectures, Marza del Vallo, Italy 5-10 November, 35-46.

[20] Larbi A, Ayadi M, Ben Mabrouk M, Kharroubi M, Kammoun N & Msallem M (2006). Agronomic and oil characteristics of some olive varieties cultivated under high-density planting conditions. In Proceedings Olivebioteq. Marza del Vallo, Italy 5-10 November, Vol II, 135-138.

[21] Larbi A, Ayadi M, Ben Dhiab A & Msallem M (2011). Olive cultivars suitability for high-density orchards. *Spanish Journal of agricultural Research*, 9 (4).

[22] Lauri P.E, Moutier N& Garcia G (1998). Architecture et conduite de l'olivier (Olea europaea L.) - Repères bibliographiques et perspectives de recherche. In Architecture et modélisation en arboriculture fruitière. 11^{éme} colloque sur les recherches fruitières, Mars 1998, 5-6, Montpellier, France, Montpellier, France : INRA-CITFL, 132-141.

[23] Lavee S (1997). Biologie et physiologie de l'olivier. In: Plaza, S., Janes, A. (Eds.). Encyclopédie Mondiale de l'Olivier. Servers Editorials Estudi Balm, Barcelona, Spain, 61-110.

[24] Lavee S, Haskal A & Avidan B (2012). The effect of planting distances and tree shape on yield and harverst. *Scientia Horticulturae*, 142, 166-173.

[25] Martin-Vertedor A.I, Perez Rodriguez J.M, Prieto Losada M.H & Fereres Castiel E (2011). Interactive responses to water deficits and crop load in olive (*Olea Europaea* L., cv.Morisca). I-Growth and water relations. *Agricultural Water Management*, 98, 941-949.

[26] Masmoudi-Charfi C (2013). Growth of young olive trees: water requirements in relation to canopy and root development. *American Journal of Plant Sciences*, 4, 1316-1344.

[27] Metzidakis I.T& Koubouris G.C (2006). Olive cultivation and industry in Greece. In Proceedings Olivebioteq. Volume Special seminars and invited lectures, Marza del Vallo, Italy 5-10 November, 133-150.

[28] Metzidakis I.T, Martinez-Vilela A, Castro Nieto G and Basso B (2008). Intensive olive orchards on slopping land: good water and pest management are essential. *Journal of Environnemental management*, 89, 120-128.

[29] Moutier N, Garcia G et Lauri P.E (2004). Shoot architecture of the olive tree: effects of cultivar on the number and distribution of vegetative and reproductive organs on branches. *Acta Horticulturae*, 636, 689-694.

[30] Palese A.M, Nuzzo V, Favati F, Pietrafesa A, Celano G & Xiloyannis C, 2010. Effects of water deficit on the vegetative response, yield and oil quality of olive trees (*Olea europaea* L.,cv.. Coratina) grown under intensive cultivation. *Scientia Horticulturae*, 125, 222-229.

[31] Parlati MV, Iannotta N & Pandolfi S (1996). Studio dell' influenza della forma di allevamento sul comportamento vegetativo e productive dell' olivo. In Atti del convegno "il olivicoltura Mediterranea; stato e prospettive della coltura e della ricerca" la grafica Commerciale, Cosenza, Italy, 343-353.

[32] Pastor M (1983). Plantation density. Proceedings of International Course of FAO on fertilization and intensification of olive cultivation, Cordoba, 160-176.

[33] Pastor M & Humanes J (1990). Plantation density experiments of non-irrigated groves in Andalucia. *Acta Horticulturae*, 286, 287-289

[34] Patumi M, D'Andria R, Marsilio V, Fontanazza G, Morelli G & Lanza B (2002). Olive and olive oil quality after intensive monocone olive growing (*Olea europaea* L., cv. Kalamata) in different irrigation regime. *Food Chemistry*, 77, 27-34.

[35] Pinhiero A.C (2006). Olive farming in Portugal. An overview. In Proceedings Olivebioteq. Volume Special seminars and invited lectures, Marza del Vallo, Italy 5-10 November, 163-171

[36] Proietti P, Palliotti A & Preziosi P (1995). Light distribution and yield efficiency in olive trees trained to monocone and vase systems. *Olea*, 23, 8 (abstract).

[37] Proietti P& Tombesi A (1996). Translocation of assimilates and source -sink influences on productive characteristics of the olive tree. *Advanced Horticultural Sciences*, 10, 11-14.

[38] Proietti P, Famiani F & Antagnozzi E, 1998. Confronto tra le forme di allavamento a monocono e a vaso in diverse cultivar d'olivo. *Rivista frutticoltura*, 60, 7-8, 69-72.

[39] Rosati A, Paoletti A, Caporali S & Perri E (2013). The role of tree architecture in super high density olive orchards. *Scientia Horticulturae*, 161, 24-29.

[40] Therios I (2009). Olives. Crop production in horticulture, 18, 409p.

[41] Trigui, A & Msallem M (2002). Oliviers de Tunisie, Catalogue des Variétés Autochtones et types locaux, volume 1. Institution de la Recherche et de l'Enseignement Supérieur Agricoles, Tunis.

[42] Tombesi A (1996). La raccolta meccanica delle olive. *Rivista Fructticulltora*, 58, 2, 31-35.

[43] Tombesi A (2006). Planting systems, canopy management and Mechanical harversting. In Proceedings of Olivebioteq. Volume Special seminars and invited lectures, Marza del Vallo, Italy 5-10 November, 307-316.

[44] Villemur P & Delmas J. M (1981). A propos de quelques facteurs du rendement en culture intensive de l'olivier. Proc. Séminaire international sur la culture intensive de l'olivier. Marrackech, Maroc, Octobre 1981,115-128.

In: Virgin Olive Oil
Editor: Antonella De Leonardis

ISBN: 978-1-63117-656-2
© 2014 Nova Science Publishers, Inc.

Chapter 8

HOW AGRONOMIC FACTORS AFFECTS OLIVE OIL COMPOSITION AND QUALITY

Ricardo Malheiro[1,2], Susana Casal[2], Paula Baptista[1] and José Alberto Pereira[1]*

[1]Mountain Research Centre (CIMO), School of Agriculture,
Polytechnic Institute of Bragança, Campus de Santa Apolónia, Bragança, Portugal
[2]REQUIMTE/Laboratório de Bromatologia e Hidrologia, Faculdade de Farmácia,
Universidade do Porto, Rua de Jorge Viterbo Ferreira, Porto, Portugal

ABSTRACT

Olive oil is one of the most popular vegetable oils worldwide but several factors might affect its quality and composition, from the tree to the spoon. Olive oil quality and composition is mainly influenced by olive fruit characteristics, and therefore all aspects that influence their development have a crucial effect on olive products. Those factors include the selection of olive cultivar, its cultivation, degree of crop intensification and production systems, agricultural practices, including irrigation and fertilization, olive pests and diseases management, all these factors clearly defining the composition of olive fruits and the inherent quality and properties of olive products.

In the last decades, huge modifications in olive tree cultivation have been observed, related essentially with two great factors: development of olive cultivations in new producing areas and crop intensification in traditional producing areas. Generally, most agronomic factors, including crop density, farming system, irrigation and fertilization, have no substantial effects on fresh olive oil quality parameters and classification. Nevertheless, a considerable incidence of olive pests and diseases can easily take fresh olive oils to the lampante category. In opposition, all agronomic factors seem to influence olive oil composition. Antioxidants are the main affected components, with a crucial effect on olive oil sensorial attributes, bioactive and nutritional properties, as well as its oxidative stability.

In present chapter the influence of diverse agronomic factors on olive fruits and olive oils production, composition and quality, is reviewed and discussed, giving special

---

* Email: jpereira@ipb.pt.

importance to olive farming-systems, fertilization and irrigation, as well as the incidence of olive pests and diseases.

Keywords: Olive oil, agronomic factors, composition, quality, olive-farming systems, fertilization, irrigation, olive pests and diseases

INTRODUCTION

Olive oil is a premium vegetable oil that, contrary to the majority of commercial vegetable oils, can be consumed in its crude form, without refining, maintaining and preserving its composition and potentially all associated beneficial properties. Olive oil is the 9th most produced vegetable oil worldwide [1] and it is gaining importance and relevance comparatively to other vegetable oils. Indeed, in the last two decades both production and consumption figures increased considerably (Figure 1).

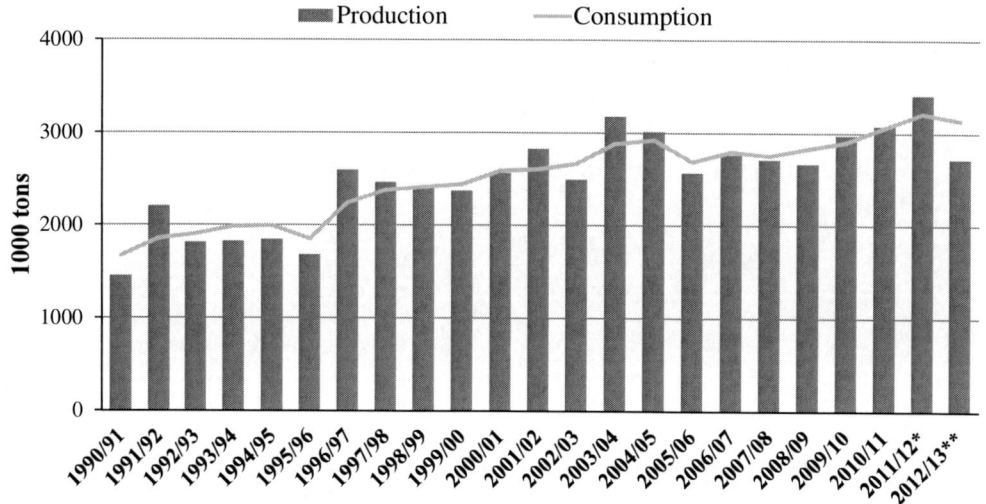

Figure 1. Worldwide olive oil production and consumption data (1000 tons) since 1990/91 (IOC 2012a; IOC 2012b). *Provisional data; **Predicted data.

Since the early 90's, olive oil production increased about 50% while its consumption increased 65% [2, 3]. Such increase is not only associated to global population growth, but mostly with the popularity that this vegetable oil gained over the last decades. Originally restricted almost to the Mediterranean region, olive orchards are now planted in regions where olives were absent or had a small representation. In particular, its cultivation is increasing in America (in the North mainly in California; Central America, mainly in Mexico; and in South America in Argentina, Chile and Brazil), South Africa and Australia. Olive oil is also being introduced in new potential markets, namely in Brazil, China and Russia, which also foment the consumption of this premium vegetable oil.

The inclusion of olive oil and related olive products in the daily diet is associated to innumerous health benefits, a direct consequence of its characteristic chemical composition.

Among those components, monounsaturated fatty acids, pigments (chlorophylls and carotenoids with β-carotene as a very active component), tocopherols (mainly α-tocopherol), sterols (mainly β-sitosterol), and diverse phenolic compounds (hydroxytyrosol; tyrosol; oleuropein and derivatives) are recognized as the most active compounds (Figure 2).

In respect to major components, fatty acids, olive oil is rich in monounsaturated fatty acids (MUFA), mainly oleic acid ($C_{18:1}$), and has reduced amounts of both saturated (SFA) and polyunsaturated fatty acids (PUFA) but provides adequate amounts of essential fatty acids (linoleic and linolenic acids) [4]. This entails a high antiatherosclerotic potential, and a lower risk of cardiovascular diseases [5].

Concerning pigments, β-carotene is one of the most abundant carotenoids in olive oils [6] and, together with lutein, has antioxidant properties [7], inhibiting olive oil photoxidation by acting as singlet oxygen quenchers [8].

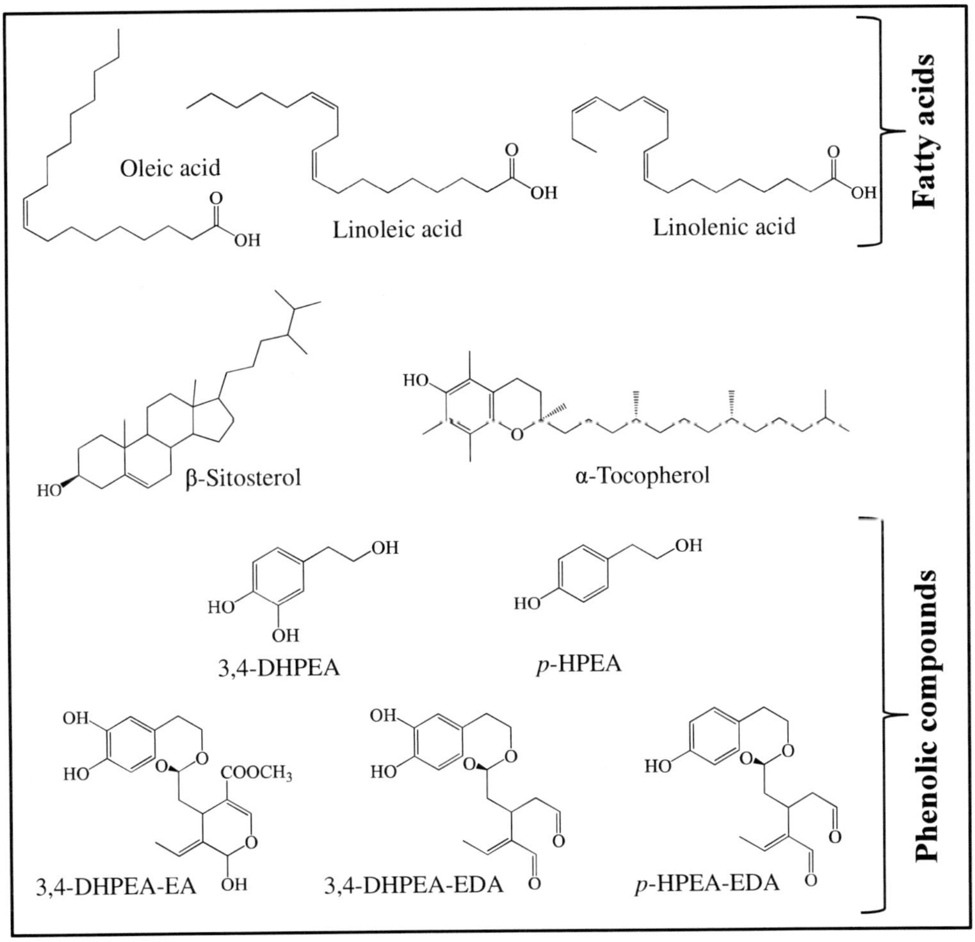

Figure 2. Olive oil components with nutritional and bioactive potential. (3,4-DHPEA – 3,4-dihydroxyphenylethanol (hydroxytyrosol); ρ-HPEA – ρ-hydroxyphenylethanol (tyrosol); 3,4-DHPEA-EA – oleuropein aglycon; 3,4-DHPEA-EDA – Dialdehydic form of decarboxymethyl elenolic acid linked to hydroxytyrosol; ρ-HPEA-EDA – Dialdehydic form of decarboxymethyl elenolic acid linked to tyrosol).

Tocopherols, especially α-tocopherol, display a dual activity since they are antioxidant compounds and exert a vitaminic action (vitamin E) contributing considerably to olive oils stability [9, 10] and consumer's health. Additionally, ingestion of α-tocopherol appears to have a preventive effect on the some of the most important diseases of the modern society, as cancer and Alzheimer [11, 12].

Olive oil phenolic compounds, also with recognized antioxidant capacity, are associated to innumerous healthy properties and benefits to consumers' health. There are several reports and literature revisions that highlight the *in vitro* [13-15] and *in vivo* [16-18] activities of these compounds and their pharmacological properties [19]. Phenolic compounds also have a marked influence on the olive oil sensorial attributes, particularly the spicy, astringent and pungent sensations, as well as on the olive oil global quality and stability.

Olive oil composition, however, is the results of a diversified array of agronomic and technological factors. On this chapter, the impact of the application of fertilizers in olive orchards, the new planting systems linked to irrigation methods, together with the incidence of olive pests and diseases on the production, on the olive oil composition and quality will be reviewed.

PLANTING SYSTEMS AND IRRIGATION

Olive trees are cultivated in three main olive-farming systems: i) traditional/low density olive groves (Figure 3A); ii) intensive olive groves (Figure 3B); iii) and high density or super intensive olive groves, also known as hedgerow olive orchards (Figure 3C) [20].

Traditional olive groves have usually 100 to 200 trees ha^{-1} (Figure 3A), are rain fed and, in some cases, harvest is still manual. These groves are usually composed by large and old trees, with several decades and, in some cases, even centenary olive trees, making mechanical harvest difficult. In some regions this kind of olive-farming system is being gradually abandoned or replaced by the two other more modern olive-farming systems. Intensive plantations are mainly irrigated, have between 200 and 550 olive trees ha^{-1}, and harvest is usually made mechanically (Figure 3B). In hedgerow olive orchards olives are exclusively collected by mechanical means to turn harvest as profitable as possible, being the olive trees irrigated following deficit irrigation programmes. These kind of olive orchards contain more than 1500 olive tree ha^{-1} (Figure 3C) [21]. Additionally, olive groves can be classified according to their production system: conventional olive groves and organic olive groves.

Olive-farming systems

Figure 3. Different olive-farming systems. A – Traditional/low density olive grove; B – intensive olive grove; C – super intensive or hedgerow olive grove.

Planting System and Olive Oils Quality and Composition

Studies comparing the influence of different olive-farming systems on the olive oil composition and quality are scarce. These studies are difficult to implement since many variables interfere in the outline, starting from olive varieties, irrigation regimes, soil composition and the standardization of olive trees density. Therefore, most studies on olive-farming systems focus mainly in trials to test the adaptation of different olive cultivars to high density olive orchards, irrigation and fertilization optimization, as well as to observe production yields and the quality and composition of the obtained olive oil from specific olive cultivars.

By studying low density plantations (from 51 to 156 trees ha^{-1}) in traditional olive groves of cv. Chemlali, Guerfel et al., [22] verified that planting density is a key aspect to be considered for the quality and composition of olive oils in arid areas. These authors verified that tree densities from 100 to 156 trees ha^{-1} result in olive oils with higher oxidative stability, as observed in Figure 4. Authors verified higher content of chlorophylls and carotenoids, oleic acid and total phenols in densities of 100 trees per ha, contributing to the observed effects, but the specific coefficients of extinction at 232 and 270 nm (K$_{232}$ and K$_{270}$ respectively) increased, eventually threatening extra-virgin olive oil classification.

Regarding intensive olive groves, adaptation of cv. Arbequina in different planting densities (from 179 to 385 trees ha^{-1}) was studied by Tous et al., [23]. These authors verified that, at full production, an average of more than 5000 kg of olives per ha were obtained, with the highest economic return with 312 trees ha^{-1}. The same group also studied adaptation of different olive cultivars (Arbequina-i 18, Arbosana, Canetera, Joanenca, Koroneiki, and Fs-17) to super intensive olive orchards with a plantation density near 2500 trees ha^{-1} [24]. The results clearly demonstrated that Arbequina-i 18 is the most adapted to this olive-farming system, producing nearly 24000 kg of olives per ha, followed by Arbosana and Canetera [24]. According to these authors, an increased plant density is mainly affected by light competition: being cultivated closely light competition is high, reducing crop yield [23].

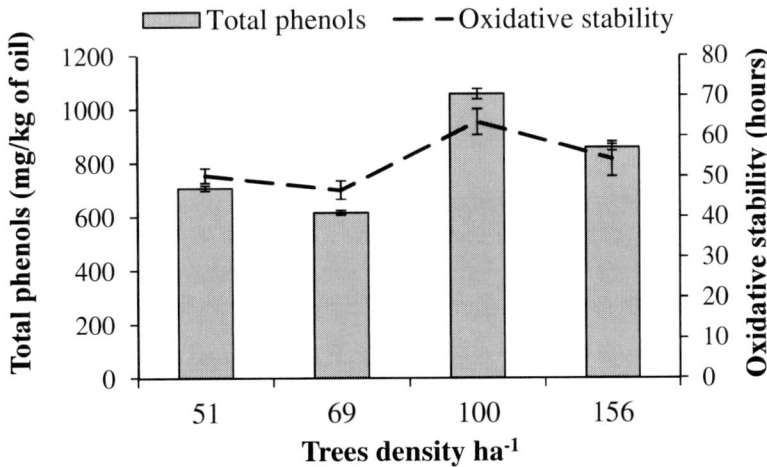

Figure 4. Oxidative stability (hours) and total phenols content of olive oils from cv. Chemlali cultivated in traditional olive groves under different olive tree densities (data updated from Guerfel et al., 2010).

In high density olive orchards (1250 trees ha[-1]) Allalout et al., [25] studied olive oil quality and composition of several olive cultivars (Arbequina, Arbequina-i 18, Arbosana, and Koroneiki). The authors verified that Arbequina was the cultivar with lower oil content, while in the extreme opposite was Koroneiki. Nevertheless, all cultivars reported good quality indices, being classified as extra-virgin olive oils.

Plant density seems to have a crucial effect on olive oil yield but it appears to have no influence in the olive oils quality. Meanwhile, some aspects need to be taken in consideration because if plant densities are extremely high it can affect plants physiological development. The competition for soil mineral nutrients as well as for light exposure can influence negatively the production of olive fruits, affecting yield, and possibly having repercussions at quality and composition of the obtained olive products. Furthermore high density olive-farming plantations are irrigated, which also influence productivity, quality and composition of olive oil, as we discuss ahead in this chapter.

Organic vs. Conventional Olive Oil: Quality and Composition

Organic agriculture is increasing in all crops worldwide, mainly due to consumers' awareness about pesticides residues in food products. Indeed, conventional agriculture, due to the application of phytosanitary products, leads to pesticide residues in olives [26], passing them to olive products [27, 28]. In 2010, nearly half a million ha were already dedicated to organic olives production [29]. A consumers' satisfaction study in Greece demonstrated that 78% of consumers were satisfied with organic olive oil, describing its beneficial health aspects as the most competitive advantage [30].

Comparing productivities of conventional and organic olive groves, higher productions are usually observed in conventional groves. Ninfali et al., [31] verified that both Leccino and Frantoio olive varieties produce more in conventional groves (5514 and 4721 kg/ha, respectively) than in organic ones (4125 and 3494 kg/ha). The same authors also verified higher pest incidence in organic olive groves (10%) than in conventional ones (4%).

In which respect to olive oil quality from organic and conventional olive groves, data reported in literature is sometimes contradictory. Gutiérrez et al., [32] reported that organic olive oils possess higher quality when compared to conventional ones, while Ninfali et al., [31] were unable to found the same trend during a 3-year consecutive study. In the study performed by Gutiérrez et al., [32] the olive oil from cv. Picual displayed better results in all quality indices evaluated (free acidity, peroxide value and K_{232}). The differences found between both cultural practices were more pronounced with olives maturation, with higher degradation in conventional olive oils. Sensorial evaluation was also better scored in organic olive oil than in conventional ones, in particular the positive attributes green fruit, mature fruit, bitter and spicy [32]. In which respect to Frantoio and Leccino olive cultivars [31], the same quality parameters were not consistent since in some years conventional olive oils display significant higher quality, and in other years the same trend is not observed. In the sensorial evaluation, no marked differences were observed between both agricultural practices in cv. Frantoio. However, conventional olive oils from cv. Leccino were fruitier, pungent, bitter, and exhaled higher cut grass sensations than olive oils extracted from organic olive fruits [31].

Regarding olive oils composition, and according to Gutiérrez et al., [32], organic olive oils are richer in antioxidant compounds, such as α-tocopherol, o-diphenols and phenolic compounds, while Ninfali et al., [31] did not found significant differences in the same components when the three years of study were compared. The same authors' found no differences in the antioxidant activity displayed by organic and conventional olive oils while organic cv. Picual olive oils were reported as having significantly higher oxidative resistance than conventional olive oils, partially attributed by the high content of antioxidant compounds, by Gutiérrez et al., [32]. Sterolic fractions of both organic and conventional olive oils were similar, while the fatty acids profiles differ mainly in oleic and linoleic acids contents. Oleic acid content was higher in organic olive oils while linoleic acid content was superior in conventional olive oils [32, 33].

Irrigation and Olive Oils Quality and Composition

Irrigation is already a common practice in intensive, super intensive or hedgerow olive orchards and, in minor proportion, in traditional olive groves. It is clear that irrigation practices increment considerably olive fruits as well as olive oil production yields [34-41]. But do irrigation practices influence olive oils quality and composition? One approach to context this question could be given in Figure 5, which schematizes the general effect of olive trees irrigation on quality, composition, stability and bioactivity of olive oils.

As illustrated in Figure 5, olive trees irrigation considerably affects olive oils composition, sensorial component and, in minor extension, olive oils quality. Starting from quality, several studies highlight slight increases in free acidity and higher peroxide formation in olive oils extracted from irrigated olives, but without statistical meaning [37, 41-43]. Regarding specific extinction coefficients at 232 and 270 nm (K_{232} and K_{270}, respectively) some studies report a slight decrease in these quality parameters while others report no significant differences with different irrigation regimes. However, olive oil classification is not endangered by irrigation practices since all olive oils continue within the legal limits of extra-virgin olive oils [44]. Regarding sensorial characteristics, olive oils extracted from irrigated olives are less fruity, pungent, and bitter, and the higher the quantity of water applied to olive trees, the lower the score of those positive attributes [41-45]. By contrast, the same studies report that olive oil becomes sweeter due to the lower grade of bitterness and pungency of the oils obtained from irrigated trees. The bitterness index (K_{225}) is in accordance to these observations, being negatively correlated with the amount of water applied to olive trees, a fact also linked to the phenolic composition of olive oils. Almost all studies which conducted trials with different irrigation regimes determined the phenolic content of the extracted olive oils. The results are clear and concise: irrigation reduces drastically the phenolic compounds in olive oils [35, 39, 41-45]. These results were also corroborated by Romero et al., [46] and Servili et al., [47] that verified richer qualitative and quantitative profiles of phenolic compounds in olives from different regulated deficit irrigation programs. Besides phenolic compounds, chlorophylls, carotenoids, tocopherols [43], sterolic fraction [41], and volatile composition [41, 47] are reduced in olive oils extracted from irrigated olives. Changes in the fatty acids profile were also observed, generally with higher contents of oleic acid being reported in olive oils from irrigated trees and higher linoleic content in olive oils from non-irrigated olive orchards [43, 47].

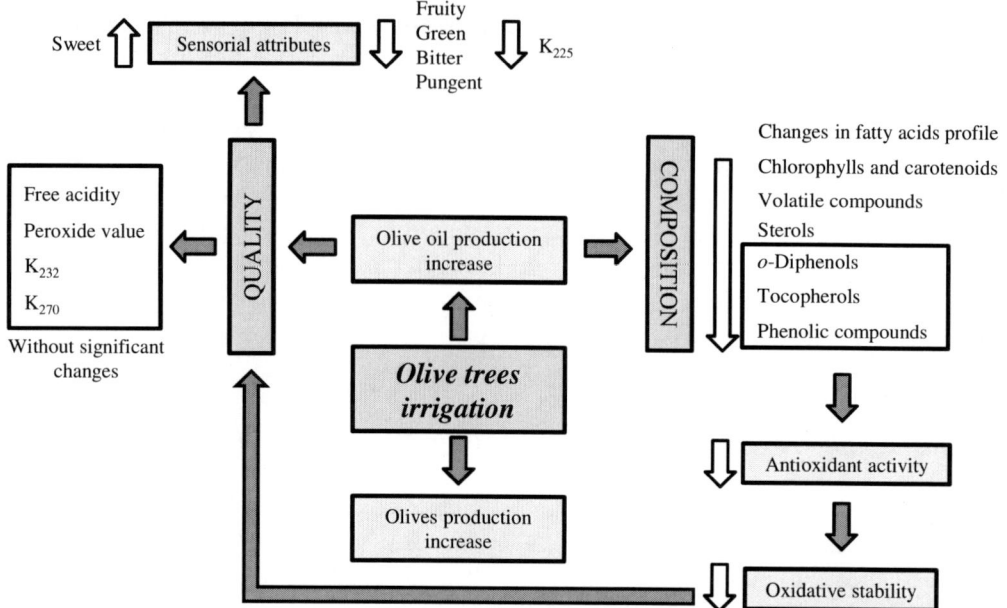

Figure 5. Global impact of irrigation practices on olives and olive oil productivity, quality and composition.

The changes inflicted by irrigation practices in the composition of olive oils interfere with their bioactivity, reducing the antioxidant potential, since olive oils from irrigated olives lose important antioxidant compounds, including phenolics, o-diphenols and tocopherols. This loss of antioxidant potential by olive oils is critical, affecting their stability and resistance to oxidation.

Indeed, oxidative stability loss is directly correlated with the amount of water applied in the irrigation regimes, with lower oxidation resistance as water increases. Berenguer et al., [42] reported oxidative stability losses between 10% and 71% in Arbequina olive oils with an irrigation treatment of 140% and 104% of ETc (olive tree evapotranspiration), respectively. Other authors reported, for the same olive cultivar, losses between 17% [43] and 27% [46] in studies of linear irrigation strategies and deficit irrigation strategies, respectively. Gómez-Rico et al., [44] also reported oxidative stability losses between 4% and 36% in Cornicabra olive oils.

Overall, olive oils extracted from irrigated olive fruits display good quality after extraction, allowing them to be classified as extra-virgin olive oils. However, since olive oil composition is severely affected, particularly being depleted of antioxidant compounds, their storage stability could be reduced. This is a very important aspect since, once stored and bottled, olive oils from irrigated olives may pass through preservation deficiencies and may arrive to consumers already degraded and rancid due to autoxidation [48]. Furthermore, these aspects may raise authentication concerns, since labelled extra-virgin olive oils may arrive to consumers with sensorial defects, being the product inconsistent with the denomination labelled.

FERTILIZATION

One of the goals of olive trees fertilization is improving and increasing olives production and yields. In each campaign, substantial amounts of mineral nutrients are lost in olive groves due to soil lixiviation and erosion, irrigation, tillage practices, fruits removal, branches removal by pruning practices, and natural drop of olive leaves. Therefore, it is essential to provide to the crop the necessary mineral nutrients in order to assure adequate growth and yields in the following years. The most modern olive-farming systems with intensive production foment an increased consumption of mineral nutrients, needed to ensure a proper production. According to Rodrigues et al., [49], olive tree responds markedly to the application of nitrogen (N), mainly in low fertility soils. Also, the application of potassium (K) leads to an increase in olive trees growth and olives yield in fertilized olive groves. Indeed, 60% of K is located in olive fruits [50], which are annually removed with the harvest, being its reposition essential. Phosphorous (P) is another mineral nutrient applied to olive trees, being essential to root growth and plant tissues. Boron (B) and magnesium (Mg) are other two mineral nutrients applied to olive trees [51, 52].

Nevertheless the application of fertilizers is not free of consequences to olive production and the quality and composition of the obtained olive oils. The most related cases are nitrogen over-fertilization [53], as excessive doses reduce flowering, flower quality and ovule longevity [54] as well as fruit set, reducing fruit load [50] and consequently also olive oil production. Excess of nitrogen has also influence in olive oils quality and composition [55, 56] as described in the next section of the present chapter. Therefore, a proper diagnostic of olive tree mineral status is essential, including foliar diagnosis [57], as well as an optimization in the application of fertilizers according to the olive tree needs [58].

Fertilization and Olive Oil Quality and Composition

Most studies regarding fertilization effects in olive oils quality seem to be in agreement as to no major interferences are observed. A good example is that the application of different doses of N-P-K (4N-1P-3K complex fertilizer) in olive orchards submitted to irrigation did not cause quality changes in olive oils from cv. Manzanilla de Sevilla [59]. Quality parameters (free acidity, peroxide value, K_{232} and K_{270}) remained low with the increasing dose of fertilizers. However, these observations do not extendable to the olive oil composition, with differences observed at the molecular level. The main components affected are the phenolic compounds [55, 56, 59, 60]. According to the different treatments applied to olive trees in different years, cv. Manzanilla olive oils lost between 16 and 32% of phenolic compounds comparatively to control treatments without application of fertilizers [59, 60]. Regarding conditions of N over-fertilization, olive oils from cv. Picual lost between 16 and 51% of phenolic compounds [55, 56]. With a considerable decrease in phenolic compounds, several aspects of olive oil are expectedly affected: being strong antioxidant compounds, olive oil becomes less protected from oxidative agents. This hypothesis was confirmed in the above mentioned studies, since the oxidative stability in both olive oils from cvs. Manzanilla and Picual was considerably reduced [55, 56, 59, 60]. Another aspect directly related with the reduction of phenolic compounds is the olive oil natural bitterness (measured by K_{225}). K_{225}

parameter reduces with fertilization, therefore the sensorial score is compromised, with positive attributes like bitterness, spicy and pungent sensations being considerably reduced in olive oils due to phenolics loss.

Tocopherols (mainly α-tocopherol) and pigments (chlorophylls and carotenoids) appear to have an opposite tendency. Nitrogen application significantly increases the amounts of α-tocopherol in olive oils, while for β- and γ-tocopherols no significant changes are observed [56]. Fernández-Escobar et al., [56] also verified an increment in chlorophylls and carotenoids with increasing N doses, but without statistical meaning. Regarding fatty acids composition of olive oils, the changes observed due to N application appears to be related to olive cultivar as only merely small changes were observed in cv. Picual [56] but some important modifications were reported in cv. Manzanilla olive oils with different fertilization treatments [59]. The PUFA fraction increased significantly with the application of higher doses of N-P-K, mainly due to the increase in linoleic acid, while MUFA decreased significantly due to the reduction in oleic acid representation. These observations were accompanied during two consecutive years, which corroborate the influence of fertilization in the fatty acids profile of olive oils.

In a long term, just as discussed in the section "Irrigation and olive oils quality and composition" of the present chapter, the changes inflicted by fertilization in olive oils composition may be critical for its preservation. Oxidative stability is reduced due to antioxidant compounds loss, and the unsaturation degree increases, which turn them more prone to oxidation.

OLIVE PESTS AND DISEASES

Olive pests and diseases are responsible for serious crop losses and quality degradation of olive products, including olive oil. According to Bueno and Jones [61] 15% of olives production is lost each year due to pest incidence. Nowadays, due to the increase of olives cultivation and their expansion in new areas of the globe, olive pests spread across olive producing areas, raising damages. On this chapter we will focus on two of the most important olive pests: the olive fruit fly (*Bactrocera oleae* (Rossi); Diptera: Tephritidae), and the olive moth (*Prays oleae* Bern.; Lepidoptera, Plutellidae), and in two of the main diseases that affect olive tree: the olive anthracnose (*Colletotrichum* sp.) and verticillium wilt (*Verticillium dahliae* Kleb.) (Figure 6).

Olives Production Losses Caused by Olive Pests and Diseases

Olive moth has important economic effects on olives production, affecting therefore olive oil quality and amounts. Olive moth cause high levels of fruit drop due to burrowing into fruits in the early stages of development. The larvae of antophagous generation feed on the seed inside olive stone and then exit the fruit, increasing fruit drop, mainly in the month of September, but also during fruit maturation. In a long-term study, Ramos et al., [62] verified the consequences of different levels of incidence of olive moth in southern Spanish olive groves. These authors observed that high levels of olive moth attack (more than 40% of fruit drop) occurs approximately every three years and cause an average loss of 131792 tons of olives. These losses entailed an economic damage of about 79.5 million € at that time. With moderate levels of attack (average of 29% of fruit drop), 50332 tons of olives were lost, with a damage of 30.4 million €. With low level of attack (average of 10% of fruit drop) 18623 tons of olives were lost with an economic prejudice of 11.2 million € [62]. These results highlight the importance and threat that this olive pest represents to olive crop.

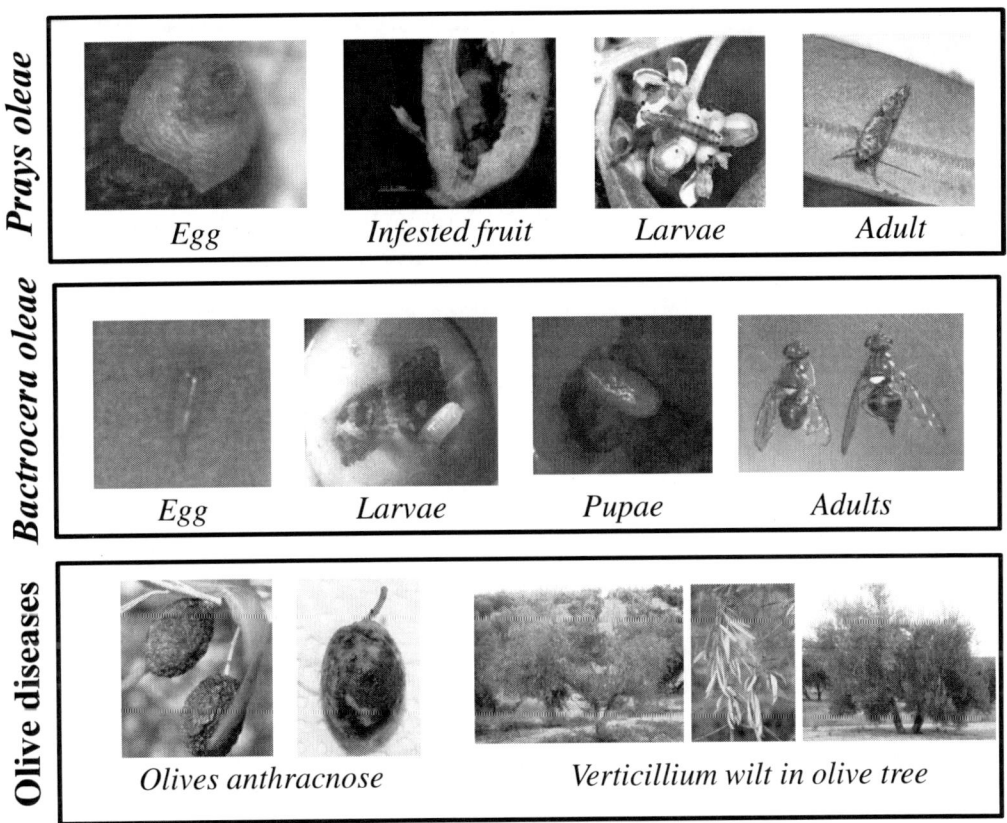

Figure 6. Main olive pests and diseases that affect *Olea europaea* L..

Even regarding olives production losses, after olive moth attack, olives are susceptible to olive fly infestation. The adult female of this dipteran lay her eggs on olive fruits, and the larvae feed on olive pulp, consuming from 3 to 20% of pulp according to the olive variety [63, 64], and creating galleries inside the fruit. When the larvae is ready to pupate, that can occur inside the fruit or in the soil, larvae opens an exit hole in fruit for larvae or adults leave. Similarly to what happens with olive moth, olive fly infestation also causes fruit drop from the tree but in lower amounts comparatively to olive moth, as witnessed by Bento et al., [65]. In no chemical sprayed olive groves, these authors report a fruit drop of about 19%of olives per tree. The most critical data was checked when authors assessed the infestation level of olives at the harvest moment, and verified that 84% of the fruits were infested.

Concerning olive diseases, anthracnose is considered the most destructive disease worldwide in olive crop. Firstly reported in Portugal [66] this disease rapidly spread to all olive producing areas in the globe. Olive anthracnose is caused by several causal agents, including *Colletotrichum acutatum*, more prevalent and aggressive than other species, as example *C. gloeosporioides* [67, 68]. Anthracnose causes fruits dehydration, fruit rot and mummification [69] as observed in Figure 6. This pest also aids in the spread and entrance of the causal agent in olive fruits. In years when *B. oleae* populations and infestation levels are high, olive anthracnose can cause up to 100% of fruit production losses [70, 71]. According to Moral et al., [72], only in Spain, about 70 million € are lost each year due to anthracnose prevalence.

Other disease with global importance is verticillium wilt. This is a vascular disease caused by the soil-borne fungus *Verticillium dahliae* Kleb. [73]. Similarly to olive anthracnose, verticillium wilt can be found in almost all olive producing regions in the world (Table 1).

Table 1. Verticillium wilt incidence in olive groves and disease incidence in olive trees cultivated in different olive producing countries

Country	Incidence in olive orchards	Disease incidence in olive trees	References
Algeria	90%	12%	[74]
Greece	-	2-3%	[75]
Israel	-	12-50%	[76]
Italy	6.2-35.8%	-	[77]
Morroco	60%	10-30%	[78]
Spain	39.5%	-	[79]
Syria	-	0.85-4.5%	[80]
Turkey	35%	3.1%	[81]

This is a very dangerous disease since it can attack partially the trees or it can kill the entire olive tree. Another worrying aspect about this disease is the high infestation levels recorded at nurseries (about 50%), as witnessed in Italian olive nurseries by Nigro et al., [77]. Its symptoms are partial leave defoliations, inflorescences necrosis and branches dieback. This disease influence considerably olive trees development and growth, as well as fruit production affecting considerably olive oil production.

Olive Pests and Diseases on Olive Oil Quality

Studies regarding the effect of olive moth on olive oil quality are inexistent, once attacked fruits drop to the soil and this fruits are not recommended to be used for oil extraction. However, literature highlights accordingly the nefarious effect of olive fruit fly infestation in olive oil quality. Olive oils quality degradation is directly proportional to the amount of olive fruits infested by olive fly and mainly olive fruits with exit holes [82-85]. Two main chemical processes occur in infested olives, hydrolysis and oxidation. During its development, olive fly larvae consume considerable amounts of olive pulp, creating galleries in the fruit. This consumption and tissue destruction lead to enzymatic reactions between lipases and triglycerides, increasing the amounts of free fatty acids in the pulp and therefore olive oils free acidity (FA). By other hand olive pulp become oxidized, by the entrance of exogenous elements as air, cold/heat, water and several types of microorganisms, mainly bacteria and fungi that provoke fruit rot, through the hole created in the fruits. Olive oils extracted from infested olives report higher peroxide value (PV), due to the compounds formed during primary oxidation, mainly hydroperoxides [48] when compared with olive oils extracted with healthy fruits. Others quality parameters that allow monitoring oxidation occurrence also report higher values in olive oils extracted from infested fruits: K_{232} and K_{270} (which measure respectively the presence of primary and secondary products of oxidation).

As already mentioned, exit holes created by olive fly are an infectious window, since many pathogenic agents enter by those sites. *Camarosporium dalmaticum* (Thüm.) Zachos & Tzav.-Klon, and *Botryosphaeria dothidea* are some examples of pathogenic agents that cause fruit rot and that are correlated with olive fly infestation [86, 87]. *C. acutatum* incidence increase in years of high olive fly populations and infestation levels. Olive oil extracted from olives with anthracnose reports lower quality compared to healthy olives and lower quality than olives infested by olive fly [88]. Furthermore, an increase in bacteria, yeasts and moulds is observed in olives infested by pests leading to an increase in the free acidity of olive oils especially if olives are stored prior to oil's extraction [89].

Sensorial characterization is a very important component of olive oils quality and classification. Regarding sensorial component, olive oils from infested fruits by pests and diseases have lower fruity, green, bitter and pungent sensorial attributes and increased defects are noted, mainly fusty, musty, winey, grubby and many times rancid [83, 84, 90]. Many defects arise from degradation and fermentative processes of olives. The perceived sensorial component is mainly affected by the changes observed in the volatile compounds released by the olive oil. In fact, olive oils from fruits infested by pests and diseases report lower green and cut-grass sensations due to loss in (*E*)-2-hexenal, one of the main volatiles responsible for those notes [91, 92], and one of the most abundant volatile compounds in olive oils [93]. The increase in fusty and musty defects is due to microbial contamination of the olive fruits and winey defect due to fermentative processes that release high amounts of alcohols (methanol, ethanol and isoamylic) [94] and acetic acid [95].

The conjunction of chemical degradation due to pests and diseases incidence together with the sensorial component gives us an idea of the changes inflicted in olive oils and the overall quality that they display. Olive oil quality is severely affected and fresh olive oils cannot be classified as extra-virgin olive oils, being considered virgin, or even lampante olive oils [96, 97] due to both quality and sensorial defects.

Olive Pests and Diseases on Olive Oil Composition, Stability and Bioactivity

Olive oil minor components are directly implicated in the olive oil quality and its properties as well. A very important factor that enhances the deleterious effects of pests and diseases in olive oils composition is the maturation process. Several authors studied the composition of olive oils extracted from infested fruits (mainly olive fly) and they highlight the fact that mature olives are characterized by a poorer composition in several olive oil components than greener ones and the intrinsically related quality is also severely affected [83, 84, 95]. Some of the most affected compounds include the fatty acids, mainly unsaturated fatty acids, sterols (β-sitosterol), tocopherols (α-tocopherol) and, in great extent, phenolic compounds (Figure 2).

In this section we will focus mainly in olive fly due to the scarce and many times inexistent information about the impact of olive moth, anthracnose and verticillium wilt in the composition of olive oils. Therefore, and starting by olive oil pigments, both chlorophylls and carotenoids are severely affected by olive fly infestation. Olive oils extracted from green cv. Chemlali olive fruits lost about 73.8% and 39.2% of chlorophylls and carotenoids, respectively, when 100% of the fruits are infested [84]. Similar remarks were observed by Tamendjari et al., [95] when studying Algerian olive cultivars. This is an aspect that not only influences the composition of olive oil, but can easily have repercussions at the consumers' preference, since olive oils become less green and more yellow, an aspect which isn't appreciated by regular consumers.

Regarding fatty acids profile, the information collected from literature is not consistent. Some authors [84, 85, 95] studied the effect of olive fly in fatty acids of olive oils without noticeable t changes in the profile. However, by studying three Algerian olive cultivars (cvs. Chemlal, Azzeradj and Bouchouk), Tamendjari et al., [97] established negative correlations between olive infestation level and unsaturated fatty acids, mainly oleic and linoleic acids. The same authors report a positive correlation with saturated fatty acids according to the infestation level of olive fly. These data highlights the oxidative processes suffered by the olive fruits during infestation, which leads to a lower amount of unsaturated fatty acids in the olive oil. Furthermore, unsaturated fatty acids are proner to oxidative processes [98] than saturated ones.

Sterolic fraction (mainly β-sitosterol) and the triterpenic alcohol uvaol are affected by olive fly infestation [99]. These compounds have their content reduced with olive fly infestation level.

Besides being important minor components of olive oil, tocopherols are scarcely studied regarding the effect of pests and diseases incidence on their contents in olive oil. When studying Portuguese olive cultivars (cvs. Cobrançosa, Madural and Verdeal Transmontana), Pereira et al., [85] verified that the β-tocopherol and γ-tocopherol content in the olive oils was not affected by the infestation level. However, α-tocopherol content decreased with the olive fly infestation, with a characteristic response according to the olive cultivar, as witnessed in Figure 7.

Cv. Cobrançosa was the one that lost lower amounts of α-tocopherol, only 6% at 100% infestation level. Meanwhile cvs. Madural and Verdeal Transmontana reported losses of about 22 and 38%, respectively, a lost that clearly compromises the composition and quality of the olive oils obtained from these olive cultivars. The trend observed in Figure 7 for α-tocopherol is the same observed for total tocopherols, since α-tocopherol is the most abundant tocopherol in olive oils. Therefore, olive fly influences the amount of a powerful antioxidant compound with impact on the olive oil stability and therefore, as already mentioned, in the quality of olive oils.

In which concerns to phenolic compounds, literature is consistent: olive oils extracted from olive fly infested fruits are poor in phenolic compounds [82-84, 97]. Gucci et al., [82] verified that all phenolic compounds represented in Figure 2, namely hydroxytyrosol, tyrosol, 3,4-DHPEA-EA, 3,4-DHPEA-EDA, and p-HPEA-EDA, are negatively correlated with the percentage of olive fruits with exit holes. Only (+)-pinoresinol and (+)-1-acetoxypinoresinol maintain their contents with the increasing percentage of exit holes [82].

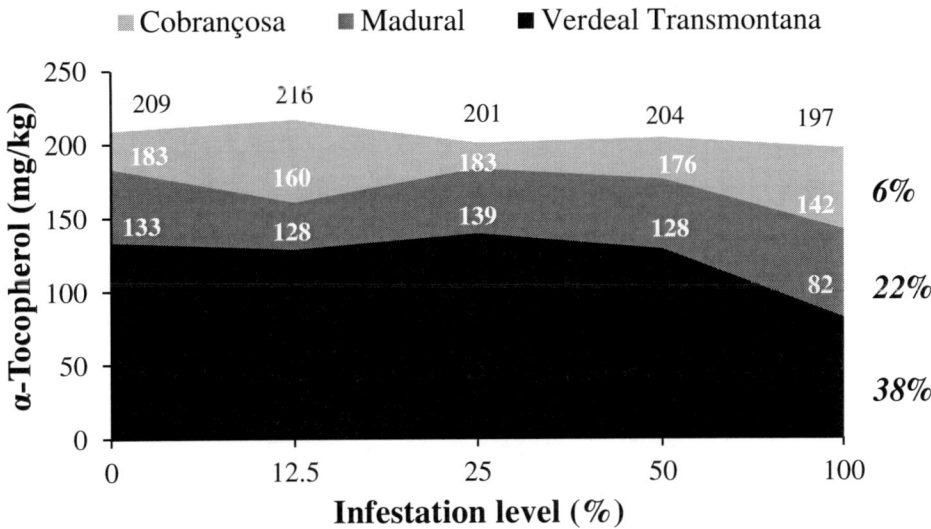

Figure 7. Changes recorded in α-tocopherol amounts in olive oils from cvs. Cobrançosa, Madural and Verdeal Transmontana extracted from olive fruits with different infestation levels (in bold and italic is presented the percentage of loss of α-tocopherol between 0 and 100% infestation level). Data from Pereira et al., (2004).

Regarding total phenols content, Gucci et al., [82] report a loss of about 75% of phenolic compounds when comparing cv. Frantoio olive oils from healthy olives and those extracted from olive fruits with 100% exit holes. Similar results were verified in other olive varieties: 21% and 50% respectively in Croatian olive cultivars Istarska bjelica and Buža [83]; 83% in Tunisian olive cultivar Chemlali [84]; and between 60-68% in Algerian olive cultivars Azzeradj, Chemlal and Bouchouk [97]. The loss of phenolic compounds is induced by the olive fly larvae. Part of the phenolic compounds lost are ingested by the larvae during its development and pulp consumption, while other part of phenols are oxidized in the olive fruit and degraded by microorganisms and enzymes present in the fruit or that contaminate the fruit after adult emergence.

With all the described changes suffered by olive fruits components due to olive fly infestation, the oxidative stability of olive oils as well as its bioactivity are compromised. According to Gómez-Caravaca et al., [100] and Mraicha et al., [84], olive oil antioxidant properties are reduced with olive fly infestation, as reported in Figure 8A.

Antioxidant potential is reduced (Figure 8A) due to the loss of antioxidant compounds as it is the case of tocopherols, namely α-tocopherol (Figure 7), and, in higher extent, in the content of phenolic compounds as already witnessed and reported in Figure 8B. Olive oils extracted from olives with an infestation level below 30% report nearly three times more phenolic compounds than olive oils extracted from olives above the mentioned infestation level [100]. This scenario leads, in a final stage, to loss of oxidative stability by olive oils. Higher infestation levels are related with lower resistance to oxidation (Figure 8C) since lower amounts of antioxidants are available to protect fatty acids from oxidative agents, specially unsaturated ones, therefore the quality of the obtained olive oils is reduced. Below 30% of olive fly infestation, the oxidative stability is comparatively 10 hours higher than from olive oils extracted from olives with an infestation level superior than 30% [100].

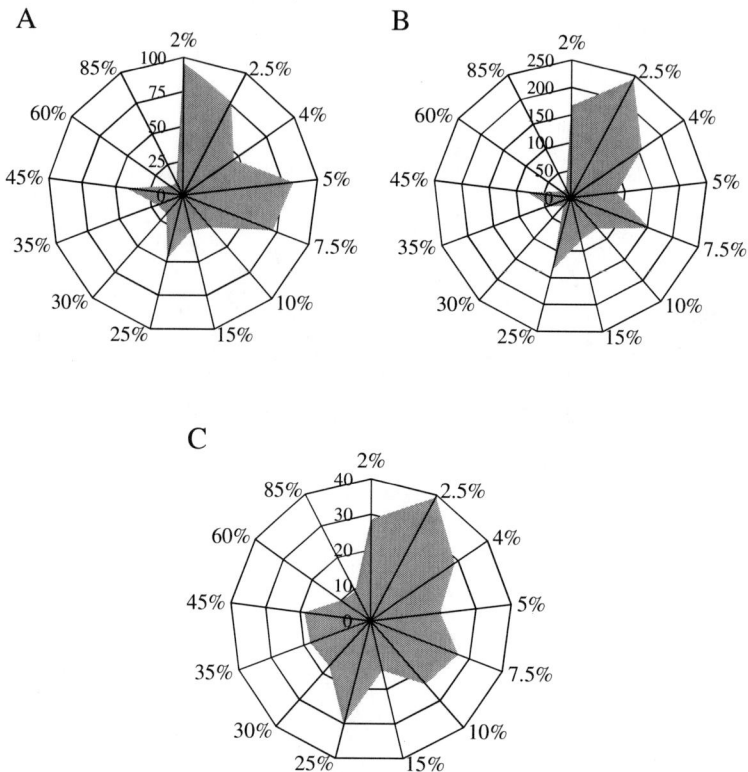

Figure 8. Antioxidant activity (Figure 8A) (µg quercetin/mL of extract), phenolic content (sum of simple phenols, lignans and secoiridoids; Figure 8B) (mg/kg of oil) and oxidative stability (Figure 8C) (hours), of olive oils extracted from olive fruits with different infestation levels. Data updated from Gómez-Caravaca et al., (2008).

CONCLUSION

Agronomic factors influence considerably olive oils quality and composition. While irrigation and fertilization applied in proper amounts, according to the plant needs, does not affect olive oils quality significantly, it considerably affects olive oil composition. For both factors this entails important issues regarding olive oils stability and preservation during storage, reducing significantly olive oils shelf-life. Furthermore, a diversified array of functions and health properties are lost in olive oils from over-irrigated and over-fertilized olives, due to the considerable reduction in antioxidant compounds (mainly phenols, *o*-diphenols, α-tocopherol).

Regarding olive pests and diseases, olive oil is primarily affected on the economic field, since significant losses are entailed each year in olive fruits production. Olive oils quality and composition is significantly changed by olive fly, a fact that leads to reduced quality, and stability. Olive fly actions are so severe that olive oils are frequently classified as lampante olive oils.

Therefore, proper optimization of irrigation and fertilization programs need to be carried out according to the olive varieties, the geographical region, climate conditions, soil properties and most of all, plant status and needs. Regarding olive pests diseases, effective control programs and phytosanitary programs need to be implemented to ensure olive oils quality, composition, stability and properties.

REFERENCES

[1] FAOSTAT (2013). Available at http://faostat.fao.org/site/636/default.aspx#ancor; Accessed 24th October 2013.

[2] International Olive Council (IOC) (2012a) World Olive Oil Figures – Production. Available at: http://www.internationaloliveoil.org/estaticos/view/131-world-olive-oil-figures; Accessed 24th October 2013.

[3] International Olive Council (IOC) (2012b) World Olive Oil Figures – Consumption. Available at: http://www.internationaloliveoil.org/estaticos/view/131-world-olive-oil-figures; Accessed 24th October 2013.

[4] Simopoulos AP (2002) Omega-3 fatty acids in inflammation and autoimmune diseases. *Journal of the American College of Nutrition*, 21, 495-505.

[5] Huang CL, & Sumpio BE (2008) Olive oil, the Mediterranean diet, and cardiovascular health. *Journal of the American College of Surgeons,* 207, 407-416.

[6] Gandul-Rojas B, & Minguez-Mosquera MI (1996) Chlorophyll and carotenoid composition in virgin olive oils from various Spanish olive varieties. *Journal of the Science of Food and Agricuture,* 72, 31-39.

[7] Sies H, & Stahl W (1995) Vitamin E and C, beta-carotene, and other carotenoids as antioxidants. *American Journal of Clinical Nutrition*, 62, 1315S-1321S.

[8] Velasco J, & Dobarganes C (2002) Oxidative stability of virgin olive oil. *European Journal of Lipid Science and Technology,* 104, 661-676.

[9] Blekas G, Tsimidou M, & Boskou D (1995) Contribution of α-tocopherol to olive oil stability. *Food Chemistry,* 52, 289-294.

[10] Aparicio R, Roda L, Albi MA, & Gutiérrez F (1999) Effect of various compounds on virgin olive oil stability measured by Rancimat. *Journal of Agricultural and Food Chemistry,* 47, 4150-4155.

[11] Albanes D, Malila N, Taylor PR, Huttunen JK, Virtamo J, Edwards BK, Rautalahti M, Hartman AM, Barrett MJ, Pietinen P, Hartman TJ, Sipponen P, Lewin K, Teerenhovi L, Hietanen P, Tangrea JA, Virtanen M, & Heinonen OP (2000) Effects of supplemental α-tocopherol and β-carotene on colorectal cancer: results from a controlled trial (Finland). *Cancer Causes & Control* 11, 197-205.

[12] Dysken MW, Sano M, Asthana S, Vertrees JE, Pallaki M, Llorente M, Love S, Schellenberg GD, McCarten R, Malphurs J, Prieto S, Chen P, Loreck DJ, Trapp G, Bakshi RS, Mintzer JE, Heidebrink JL, Vidal-Cardona A, Arroyo LM, Cruz AR, Zachariah S, Kowall NW, Chopra MP, Craft S, Thielke S, Turvey CL, Woodman C, Monnel KA, Gordon K, Tomaska J, Segal Y, Peduzzi PN, & Guarino PD (2014) Effect of vitamin E and memantine on functional decline in Alzheime disease. The TEAM-AD

VA Cooperative Randomized Trial. *The Journal of the American Medical Association,* 311, 33-44.

[13] Andrikopoulos NK, Kaliora AC, Assimopoulou AN, & Papageorgiou VP (2002) Inhibitory activity of minor polyphenolic and nonpolyphenolic constituents of olive oil against in vitro low-density lipoprotein oxidation. *Journal of Medicinal Food,* 5, 1-7.

[14] Malheiro R, Sousa A, Casal S, Bento A, & Pereira JA (2011) Cultivar effect on the phenolic composition and antioxidant potential of stoned table olives. *Food and Chemical Toxicology,* 49, 450-457.

[15] Owen RW, Giacosa A, Hull WE, Haubner R, Spiegelhalder B, & Bartsch H (2000) The antioxidant/anticancer potential of phenolic compounds isolated from olive oil. *European Journal of Cancer,* 36, 1235-1247.

[16] Covas MI, Ruiz-Gutiérrez V, de la Torre R, Kafatos A, Lamuela-Raventós RM, Osada J, Owen RW, & Visioli F (2006a) Minor components of olive oil: evidence to date of health benefits in humans. *Nutrition Reviews,* 64, S20-S30.

[17] Covas MI, Nyyssönen K, Poulsen HE, Kaikkonen J, Zunft H-JF, Kiesewetter H, Gaddi A, de la Torre R, Mursu J, Bäumler H, Nascetti S, Salonen JT, Fitó M, Virtanen J, & Marrugat J (2006b) The effect of polyphenols in olive oil on heart disease risk factors: a randomized trial. *Annals of Internal Medicine,* 145, 333-341.

[18] Konstantinidou V, Covas MI, Muñoz-Aguayo D, Khymenets O, de la Torre R, Saez G, Tormos MC, Toledo E, Marti A, Ruiz-Gutiérrez V, Mendez MVR, & Fito M (2010) *In vivo* nutrigenomics effects of virgin olive oil polyphenols within the frame of the Mediterranean diet: a randomized controlled trial. *The FASEB Journal,* 24, 2546-2557.

[19] Obied HK, Prenzler PD, Omar SH, Ismael R, Servili M, Esposto S, Taticchi A, Selvaggini R, & Urbani S (2012) Pharmacology of olive biophenols. In: Fishbein JC, & Heilman JM (Eds.) *Advances in Molecular Toxicology,* 6, 195-242.

[20] Loumou A, & Giourga C (2003) Olive groves: "The life and identity of the Mediterranean". *Agriculture and Human Values,* 20, 87-95.

[21] Rallo L (2009) Iberian olive growing in a time of change. *Chronica Horticulturae,* 49, 15-17.

[22] Guerfel M, Zaghdoud C, Jebahi K, Boujnah D, & Zarrouk M (2010) Effects of the planting density on virgin olive oil quality of "Chemlali" olive trees (*Olea europaea* L.). *Journal of Agricultural and Food Chemistry,* 58, 12469-12472.

[23] Tous J, Romero A, Plana J, & Baiges F (1999) Planting density trial with 'Arbequina' olive cultivar in Catalonia (Spain). *Acta Horticulturae,* 474, 177-180.

[24] Tous J, Romero A, Plana J, & Hermoso JF (2008) Olive oil cultivars suitable for very-high density planting conditions. *Acta Horticulturae,* 791, 403-408.

[25] Allalout A, Krichène D, Methenni K, Taamalli A, Oueslati I, Daoud D, & Zarrouk M (2009) Characterization of virgin olive oil from super intensive Spanish and Greek varieties grown in northern Tunisia. *Scientia Horticulturae,* 120, 77-83.

[26] Cunha SC, Fernandes JO, & Oliveira MBPP (2007a) Determination of phosmet and its metabolites in olives by matrix solid-phase dispersion and gas chromatography-mass spectrometry. *Talanta,* 73, 514-522.

[27] Cunha SC, Fernandes JO, & Oliveira MBPP (2007b) Comparison of matrix solid-phase dispersion and liquid-liquid extraction for the chromatographic determination of

fenthion and its metabolites in olives and olive oils. *Food Additives and Contaminants,* 24, 156-164.

[28] Cunha SC, Lehotay SJ, Mastovska K, Fernandes JO, & Oliveira MBPP (2007c) Evaluation of the QuEChERS sample preparation approach for the analysis of pesticide residues in olives. *Journal of Separation Science,* 30, 620-632.

[29] Willer H, & Lukas K (2012) The World of Organic Agriculture – Statistics and Emerging Trends 2012.

[30] Sandalidou E, Baourakis G, & Siskos Y (2002) Customers' perspectives on the quality of organic olive oil in Greece: a satisfaction evaluation approach. *British Food Journal,* 104, 391-406.

[31] Ninfali P, Bacchiocca M, Biagiotti E, Esposto S, Servili M, Rosati A, & Montedoro G (2008) A 3-year study on quality, nutritional and organoleptic evaluation of organic and conventional extra-virgin olive oils. *Journal of the American Oil Chemists' Society,* 85, 151-158.

[32] Gutiérrez F, Arnaud T, & Albi MA (1999) Influence of ecological cultivation on virgin olive oil quality. *Journal of the American Oil Chemists' Society,* 76, 617-621.

[33] Samman S, Chow JWY, Foster MJ, Ahmad ZI, Phuyal JL, & Petocz P (2008) Fatty acid composition of edible oils derived from certified organic and conventional agricultural methods. *Food Chemistry,* 109, 670-674.

[34] Dabbou S, Chehab H, Faten B, Dabbou S, Esposto S, Selvaggini R, Taticchi A, Servili M, Montedoro GF, & Hammami M (2010) Effect of three irrigation regimes on Arbequina olive oil produced under Tunisian growing conditions. *Agricultural Water Management,* 97, 763-768.

[35] d'Andria R, Morelli G, Giorgio P, Patumi M, Vergari G, & Fontanazza G (1999) Yield and oil quality of young olive trees grown under different irrigation regimes. *Acta Horticulturae,* 474, 185-188.

[36] Inglese P, Barone E, & Gullo G (1996) The effect of complementary irrigation on fruit growth, ripening pattern and oil characteristics of olive (*Olea europaea* L.) cv. Carolea. *Journal of Horticultural Science,* 71, 257-263.

[37] Ismail AS, Stavroulakis G, & Metzidakis J (1999) Effect of irrigation on the quality characteristics of organic olive oil. *Acta Horticulturae,* 474, 687-690.

[38] Palese AM, Nuzzo V, Favati F, Pietrafesa A, Celano G, & Xiloyannis C (2010) Effects of water deficit on the vegetative response, yield and oil quality of olive trees (*Olea europaea* L., cv Coratina) grown under intensive cultivation. *Scientia Horticulturae,* 125, 222-229.

[39] Patumi M, d'Andria R, Fontanazza G, Moreli G, Giorgio P, & Sorrentino G (1999) Yield and oil quality of intensively trained trees of three cultivars of olive (*Olea europaea* L.) under different irrigation regimes. *Journal of Horticultural Science and Biotechnology,* 74, 729-737.

[40] Patumi M, d'Andria R, Marsilio V, Fontanazza G, Moreli G, & Lanza B (2002) Olive and olive oil quality after intensive monocone olive growing (*Olea europaea.,* cv. Kalamata) in different irrigation regimes. *Food Chemistry,* 77, 27-34.

[41] Stefanoudaki E, Williams M, Chartzoulakis K, & Harwood J (2009) Effect of irrigation on quality attributes of olive oil. *Journal of Agricultural and Food Chemistry,* 57, 7048-7055.

[42] Berenguer MJ, Vossen PM, Grattan SR, Connell JH, & Polito VS (2006) Tree irrigation levels for optimum chemical and sensory properties of olive oil. *HortScience, 41*, 427-432.

[43] Tovar MJ, Motilva MJ, Luna M, Girona J, & Romero MP (2001a) Analytical characteristics of virgin olive oil from young trees (Arbequina cultivar) growing linear irrigation strategies. *Journal of the American Oil Chemists' Society, 78*, 843-849.

[44] Gómez-Rico A, Salvador MD, Moriana A, Pérez D, Olmedilla N, Ribas F, & Fregapane G (2007) Influence of different irrigation strategies in a traditional Cornicabra cv. olive orchard on virgin olive oil composition and quality. *Food Chemistry, 100*, 568-578.

[45] Tovar MJ, Motilva MJ, & Romero M (2001b) Changes in the phenolic composition of virgin olive oil from young trees (*Olea europaea* L. cv. Arbequina) grown under linear irrigation strategies. *Journal of Agricultural and Food Chemistry, 49*, 5502-5508.

[46] Romero MP, Tovar MJ, Girona J, & Motilva MJ (2002) Changes in the HPLC phenolic profile of virgin olive oil from young trees (*Olea europaea* L. cv. Arbequina) grown under different deficit irrigation strategies. *Journal of Agricultural and Food Chemistry, 50*, 5349-5354.

[47] Servili M, Esposto S, Lodolini E, Selvaggini R, Taticchi A, Urbani S, Montedoro G, Serravalle M, & Gucci R (2007) Irrigation effects on quality, phenolic composition, and selected volatiles of virgin olive oils cv. Leccino. *Journal of Agricultural and Food Chemistry, 55*, 6609-6618.

[48] Laguerre M, Lecomte J, & Villeneuve P (2007) Evaluation of the ability of antioxidants to counteract lipid oxidation: Existing methods, new trends and challenges. *Progress in Lipid Research, 46*, 244-282.

[49] Rodrigues MA, Ferreira IQ, Claro AM, & Arrobas M (2012) Fertilizer recommendations for olive based upon nutrients removed in crop and pruning. *Scientia Horticulturae, 142*, 205-211.

[50] Erel R, Dag A, Ben-Gal A, Schwartz A, & Yermiyahu U (2008) Flowering and fruit set of olive trees in response to nitrogen, phosphorous and potassium. *Journal of the American Society for Horticultural Science, 133*, 639-647.

[51] Marcelo ME, Jordão PV, Matias H, & Rogado B (2010) Influence of nitrogen and magnesium fertilization of olive tree 'Picual' on yield and olive oil quality. *Acta Horticulturae, 868*, 445-450.

[52] Rodrigues MA, Pavão F, Lopes JI, Gomes V, Arrobas M, Moutinho-Pereira J, Ruivo S, Cabanas JE, & Correia CM (2011) Olive yields and tree nutritional status during a four-year period without nitrogen and boron fertilization. *Communications in Soil Science and Plant Analysis, 42*, 803-814.

[53] Fernández-Escobar R (2011) Use and abuse of nitrogen in olive fertilization. *Acta Horticulturae, 888*, 249-258.

[54] Fernández-Escobar R, Ortiz-Urquiza A, Prado M, & Rapoport HF (2008) Nitrogen status influence on olive tree flower quality and ovule longevity. *Environmental and Experimental Botany, 64*, 113-119.

[55] Fernández-Escobar R, Sánchez-Zamora MA, Uceda M, & Beltrán G (2002) The effect of nitrogen overfertilization on olive tree growth and oil quality. *Acta Horticulturae, 586*, 429-431.

[56] Fernández-Escobar R, Beltrán G, Sánchez-Zamora MA, García-Novelo J, Aguilera MP, & Uceda M (2006) Olive oil quality decreases with nitrogen over-fertilization. *HortScience*, 41, 215-219.

[57] Fernández-Escobar R, Parra MA, Navarro C, & Arquero O (2009) Foliar diagnosis as a guide to olive fertilization. *Spanish Journal of Agricultural Research*, 7, 212-223.

[58] Marín L, & Férnandez-Escobar R (1997) Optimization of nitrogen fertilization in olive orchards. *Acta Horticulturae*, 448, 411-414.

[59] Morales-Sillero A, Jiménez R, Fernández JE, Troncoso A, & Beltrán G (2007) Influence of fertigation in 'Manzanilla de Sevilla' olive oil quality. *HortScience*, 42, 1157-1162.

[60] Tognetti R, Morales-Sillero A, d'Andria R, Fernández JE, Lavini A, Sebastiani L, & Troncoso A, (2008) Deficit irrigation and fertigation practices in olive growing: convergences and divergences in two case studies. *Plant Biosystems*, 142, 138-148.

[61] Bueno AM, & Jones O (2002) Alternative methods for controlling the olive fly, *Bactrocera oleae*, involving semiochemicals, *IOBC wprs Bulletin*, 25, 1-11.

[62] Ramos P, Campos M, & Campos JM (1998) Long-term study on the evaluation of yield and economic losses caused by *Prays oleae* Bern. in the olive crop of Granada (southern Spain). *Crop Protection*, 17, 645-647.

[63] Neuenschwander P, & Michelakis S (1978) The infestation of *Dacus oleae* (Gmel.) (Diptera, Tephritidae) at harvest time and its influence on yield and quality of olive oil in Crete. *Journal of Applied Entomology*, 86, 420-433.

[64] Neuenschwander P, & Michelakis S (1981) Olive fruit drop caused by *Dacus oleae* (Gmel.) (Dipt. Tephritidae). *Journal of Applied Entomology*, 91, 193-205.

[65] Bento A, Torres L, & Sismeiro JLR (1999) A contribution to the knowledge of *Bactrocera oleae* (Gmel.) in Trás-os-Montes region (Northeastern Portugal): phenology, losses and control. *Acta Horticulturae*, 474, 541-544.

[66] de Almeida MJV (1899) La gaffa des olives en Portugal. *Bulletin de la Société Mycologique de France*, 15, 90-94.

[67] Talhinhas P, Neves-Martins J, Oliveira H, & Sreenivasaprasad S (2009) The distinctive population structure of *Colletotrichum* species associated with olive anthracnose in the Algarve region of Portugal reflects a host-pathogen diversity hot spot. *FEMS Microbiology Letters*, 296, 31-38.

[68] Talhinhas P, Mota-Capitão C, Martins S, Ramos AP, Neves-Martins J, Guerra-Guimarães L, Várzea V, Silva MC, Sreenivasaprasad S, & Oliveira H (2011) Epidemiology, histopathology and aetiology of olive anthracnose caused by *Colletotrichum acutatum* and *C. gloeosporioides* in Portugal. *Plant Pathology*, 60, 483-495.

[69] Cacciola SO, Faedda R, Sinatra F, Agosteo GE, Schena L, Frisullo S, & San Lio GM (2012) Olive anthracnose. *Journal of Plant Pathology*, 94, 29-44.

[70] Moral J, Bouhmidi K, & Trapero A (2008) Influence of fruit maturity, cultivar susceptibility, and inoculation method on infection of olive fruit by *Colletotrichum acutatum*. *Plant Disease*, 92, 1421-1426.

[71] Moral J, & Trapero A (2009) Assessing the susceptibility of olive cultivars to anthracnose caused by *Colletotrichum acutatum*. *Plant Disease*, 93, 1028-1036.

[72] Moral J, Oliveira R, & Trapero A (2009) Elucidation of disease cycle of olive anthracnose caused by *Colletotrichum acutatum*. *Phytopathology*, 99, 548-556.

[73] López-Escudero FJ, & Mercado-Blanco J (2011) Verticillium wilt of olive: a case study to implement an integrated strategy to control a soil-borne pathogen. *Plant Soil*, 344, 1-50.

[74] Bellahcene M, Fortas Z, Geiger JP, Matallah A, & Henni D (2000) Verticillium wilt in olive in Algeria: geographical distribution and extent of the disease. *Olivae*, 82, 41-43.

[75] Thanassoulopoulos CC, Biris DA, & Tjamos EC (1979) Survey of Verticillium wilt of olive trees in Greece. *Plant Disease Report*, 63, 936-940.

[76] Levin AG, Lavee S, & Tsror L (2003) Epidemiology of *Verticillium dahliae* on olive (cv. Picual) and its effect on yield under saline conditions. *Plant Pathology*, 52, 212-218.

[77] Nigro F, Gallone P, Romanazzi G, Schena L, Ippolito A, & Salemo MG (2005) Incidence of Verticillium wilt on olive in Apulia and genetic diversity of *Verticillium dahliae* isolates from infected trees. *Journal of Plant Pathology*, 87, 13-23.

[78] Serrhini MN, & Zeroual A (1995) Verticillium wilt in Morroco. *Olivae*, 58, 58-61.

[79] Sánchez-Hernández ME, Ruiz-Dávila A, Pére de Algaba A, Blanco-López MA, & Trapero-Casas A (1998) Occurrence and aetiology of death of young olive trees in Southern Spain. *European Journal of Plant Pathology*, 104, 347-357.

[80] Al-Ahmad MA, & Mosli MN (1993) Verticillium wilt of olive in Syria. *Bulletin OEPP/EPPO Bulletin*, 23, 521-529.

[81] Dervis S, Mercado-Blanco J, Erten L, Valverde-Corredor A, & Pérez-Artés E (2010) Verticillium wilt of olive in Turkey: a survey on disease importance, pathogen diversity and susceptibility of relevant olive cultivars. *European Journal of Plant Pathology*, 127, 287-301.

[82] Gucci R, Caruso G, Canale A, Loni A, Raspi A, Urbani S, Taticchi A, Esposto S, & Servili M, (2012) Qualitative changes of olive oils obtained from fruits damaged by *Bactrocera oleae* (Rossi). *HortScience*, 47, 301-306.

[83] Koprivnjak O, Dminić I, Kosić U, Majetić V, Godena S, & Valenčič V (2010) Dynamics of oil quality parameters changes related to olive fruit fly attack. *European Journal of Lipid Science and Technology*, 112, 1033-1040.

[84] Mraicha F, Ksantini M, Zouch O, Ayadi M, & Sayadi S (2010) Effect of olive fruit fly infestation on the quality of olive oil from Chemlali cultivar during ripening. *Food and Chemical Toxicology*, 48, 3235-3241.

[85] Pereira JA, Alves MR, Casal S, & Oliveira MBPP (2004) Effect of olive fruit fly infestation on the quality of olive oil from cultivars Cobrançosa, Madural, and Verdeal Transmontana. *Italian Journal of Food Science*, 16, 355-365.

[86] Iannotta N, Noce ME, Ripa V, Scalercio S, & Vizzarri V (2007) Assessment of susceptibility of olive cultivars to the *Bactrocera oleae* (Gmelin, 1790) and *Camarosporium dalmaticum* (Thüm.) Zachos & Tzav.-Klon. attacks in Calabria (Southern Italy). *Journal of Environmental Science and Health Part B*, 42, 789-793.

[87] Latinović J, Mazzaglia A, Latinović N, Ivanović M, & Gleason ML (2013) Resistance of olive cultivars to *Botryosphaeria dothidea*, causal agent of olive fruit rot in Montenegro. *Crop Protection*, 48, 35-40.

[88] Sousa A, Pereira JA, Casal S, Oliveira B, & Bento A (2005) Effect of the olive fruit fly and the olive antrachnose on oil quality of some Portuguese cultivars. 2nd Meeting of the IOBC/WPRS WG Integrated Protection of Olive Crops. Florence, Italy, book of abstracts: 35pp.

[89] Torres-Vila LM, Rodríguez-Molina MC, & Martínez JA (2003) Olive fly damage and olive storage effects on paste microflora and virgin olive oil acidity. *Grasas y Aceites,* 54, 285-294.

[90] Angerosa F, Di Giacinto L, & Solinas M (1992) Influence of *Dacus oleae* infestation on flavor of oils, extracted from attacked olive fruits, by HPLC and HRGC analyses of volatile compounds. *Grasas y Aceites,* 43, 134-142.

[91] Angerosa F, Mostallino R, Basti C, & Vito R (2000) Virgin olive oil odour notes: their relationships with volatile compounds from the lipoxygenase pathway and secoiridoid compounds. *Food Chemistry,* 283-287.

[92] Angerosa F, Servili M, Selvaggini R, Taticchi A, Esposto S, & Montedoro G (2004) Volatile compounds in virgin olive oil: occurrence and their relationship with the quality. *Journal of Chromatography A,* 1054, 17-31.

[93] Kalua CM, Allen MS, Bedgood DR, Bishop AG, Prenzler PD, & Robards K (2007) Olive oil volatile compounds, flavor development and quality: A critical review. *Food Chemistry,* 100, 273-286.

[94] Bendini A, Cerretani L, Cichelli A, & Lercker G (2008) Come l'infestazione da *Bactrocera oleae* può causare variazioni nel profilo aromatico di oli vergini da olive. *Rivista Italiana di Sostanze Grasse,* 86, 167-177.

[95] Tamendjari A, Angerosa F, & Bellal MM (2004) Influence of *Bactrocera oleae* infestation on olive oil quality during ripening of Chemlal olives. *Italian Journal of Food Science,* 16, 343-354.

[96] Tamendjari A, Laribi R, & Bellal MM (2011) Effect of attack of *Bactrocera oleae* on olive oil by the quality of the volatile fraction of oil from two varieties Algerian. *Rivista Italiana di Sostanze Grasse,* 88, 114-122.

[97] Tamendjari A, Sahnoune M, Mettouchi S, & Angerosa F (2009) Effect of *Bactrocera oleae* infestation on the olive oil quality of three Algerian varieties: Chemlal, Azzeradj and Bouchouk. *Rivista Italiana di Sostanze Grasse,* 86, 103-111.

[98] Kamal-Eldin A (2006) Effect of fatty acids and tocopherols on the oxidative stability of vegetable oils. *European Journal of Lipid Science and Technology* 58, 1051-1061.

[99] Parlati MV, Petruccioli G, & Pandolfi S (1990) Effects of the *Dacus* infestation on the oil quality. *Acta Horticulturae,* 286, 387-390.

[100] Gómez-Caravaca AM, Cerretani L, Bendini A, Segura-Carretero A, Fernández-Gutiérrez A, Del Carlo M, Compagnone D, & Cichelli A (2008) Effects of fly attack (*Bactrocera oleae*) on the phenolic profile and selected chemical parameters of olive oil. *Journal of Agricultural and Food Chemistry,* 56, 4577-4583.

In: Virgin Olive Oil   ISBN: 978-1-63117-656-2
Editor: Antonella De Leonardis   © 2014 Nova Science Publishers, Inc.

Chapter 9

OLIVE OIL: PRODUCTION AND NUTRITIONAL PROPERTIES

María del Pilar Godoy-Caballero[1,],*
María Isabel Acedo-Valenzuela[1], Teresa Galeano-Díaz[1],
Héctor Goicoechea[2,3] and María Julia Culzoni[2,3,]*

[1]Departamento de Química Analítica,
Universidad de Extremadura, Badajoz, España
[2]Laboratorio de Desarrollo Analítico y Quimiometría (LADAQ),
Cátedra de Química Analítica I, Facultad de Bioquímica y Ciencias Biológicas,
Universidad Nacional del Litoral, Ciudad Universitaria,
Santa Fe, Argentina
[3]Consejo Nacional de Investigaciones Científicas y Técnicas (CONICET),
Buenos Aires, Argentina

ABSTRACT

Virgin olive oil (VOO) is one of the basic components of the Mediterranean diet and its importance, due fundamentally to its nutritional and sensory characteristics, is currently known. The extraordinary properties of VOO derive from its content in hydrophilic phenolic compounds, which are only present in VOO and differentiate it from all the other vegetable oils used by humans. There are many published works related to the VOO properties as well as to its content in phenolic compounds. In the present chapter, a complete review of the most interesting aspects of VOO production and its nutritional properties is presented.

On one hand, both the quantitative and qualitative composition of VOO in phenolic compounds is affected by the agronomical and technological aspects of production. The cultivar, the fruit ripeness, the climatic conditions and the agronomical techniques, such as irrigation, are the agronomical most influential factors and, as a consequence, more widely studied. Regarding the technological aspects of production, the presence of

---

[*] Corresponding author: Ph: +54 342 4575206 (190), E-mail: mculzoni@fbcb.unl.edu.ar.
[*] Corresponding author: Ph: +34 924 289300 (86194), E-mail: pgodoy@unex.es.

phenolic compounds in VOO is related to the activity of endogenous enzymes present in the olive fruit. Therefore, the extraction conditions, as well as the extraction system, greatly affect their concentration in VOO.

On the other hand, phenolic compounds are considered as a very important part of the fruit defense chemical system. Diverse functions, such as its antimicrobial activity and its protection against the oxidative damages are attributed to these compounds. In addition, they are responsible for the VOO oxidation stability and for its sensory properties, and have been identified as the main components providing antioxidant properties to this food. Besides to the sensory properties and antioxidant activity, the current interest in phenolic compounds and VOO is based on their important nutritional and biological properties.

Keywords: Virgin olive oil, nutritional properties, production, hydrophilic phenolic compounds

1. INTRODUCTION

Virgin olive oil (VOO) is a vegetable oil extracted from the olive tree (*Olea Europaea* L.) fruit, called olive. Olive oil is a very energetic food, since it provides about 9 kcal·g^{-1} from its triglycerides, which are constituted by fatty acids, being oleic acid, which is present in the higher amount in olive oil, between 68.0 and 81.5 %.

There are different olive oil categories. In this way and taking into account the art. 118 of the CE Regulation no. 1234/2007 [1], virgin olive oils are defined as olive oils which have been exclusively obtained from the olives by mechanical media and other physical procedures. Virgin olive oils are classified and defined as: extra virgin olive oil (EVOO), virgin olive oil (VOO) and lampante olive oil. Moreover, there are other olive oil categories, such as refined olive oil, olive oil (which exclusively contains refined olive oils and virgin olive oils), crude orujo olive oil, refined orujo olive oil and orujo olive oil.

Olive oil is considered a monounsaturated fat (given it high content in oleic acid) and a functional food [2], with positive health effects for their consumers. Despite the chemical composition of olive oil depends on different factors, such as agronomical conditions and technological production characteristics, the olive variety and fruit ripening, in general terms, it is composed of major and minor components. The major components represent more than 98% of the total oil weight and are also called the saponifiable fraction. This fraction is mainly composed by triglycerides, although there are also present diglycerides, monoglycerides and free fatty acids in a lower proportion. Regarding the minor fraction, it represents about the 2% of the total olive oil weight and includes more than 230 different components. The determination of each component in this fraction with high precision is very difficult, due to its complex nature and the low amount in which they are present. There are two different classes of compounds in this fraction: (a) fatty acid derivatives, such as the phospholipids, waxes and steroid esters, and (b) compounds which are not chemically related to the fatty acids: hydrocarbons, fatty alcohols, free alcohols, free sterols, tocopherols, pigments (chlorophylls, pheophytins and carotenoids) and phenolic compounds.

Phenolic compounds, together with tocopherols and carotenoids, are the most important antioxidants in olive oil. While the lipophilic phenols, such as the tocopherols, can be found in other vegetable oils, hydrophilic phenols of VOO are not generally present in other oils and fats, since the technological process, for example the refining, suppresses them [2, 3, 4]. The hydrophilic fraction of olive oil, from now on phenolic compounds, constitutes a heterogeneous mixture of compounds with different structures, which are present in the olive mesocarp and which traditionally have been extracted from olive oil using hydromethanolic mixtures.

The number of publications related to these compounds has increased so much in the last years for several reasons. Phenolic compounds are related not only to the olive oil stability, but also to its healthy properties. Regarding this, many phenolic compounds from VOO, such as hydroxytyrosol and their derivatives are being thoroughly researched nowadays with the goal to establish relationships between the olive oil consumption and the risk of suffering cardiovascular diseases or cancer.

Finally, it is important to indicate that olive oil also contains a great amount of volatile aromatic compounds from diverse nature, which provides a particular aroma to the different edible oils. These compounds are generated by enzyme action in the olive crushing and they are incorporated in the oily phase as aldehydes and alcohols. While phenolic compounds contribute to the olive oil bitter, the volatile compounds are the main compounds responsible for the greenness attribute of the VOO. The concentration range of the components in the VOO sample is very broad and can be used for variety and geographical origin identification, since there are variations associated to these factors, and also on the fruit degree of ripeness. There have been identified more than one hundred compounds, such as hydrocarbons, alcohols, aldehydes, esters, phenols, terpenes and furan derivatives [2].

2. COMPOSITION AND NUTRITIONAL PROPERTIES OF VIRGIN OLIVE OIL

2.1. Virgin Olive Oil Composition

More than 40 years ago, Cantarelli showed the occurrence of natural antioxidants in VOO obtained by pressure to explain the difference in oxidative stability between virgin and other vegetable oils [5], carrying out the extraction and colorimetric evaluation of total phenols in several Italian VOO obtained by pressing the olives. The results were compared with the phenolic composition of refined olive oils, and a strong discrimination in phenolic concentration between the two groups of oils was shown [5, 6]. Since them, systematic studies in relation to the presence of polar phenolic compounds in VOO have increased. The obtained results presented in a great variety of publications show that the peculiar composition of VOO in relation to its content in phenolic compounds has not been found in any other vegetable oil [3].

VOO contains different classes of phenolic compounds: (1) phenolic acids and derivatives: gallic acid, 4-hydroxybenzoic acid, vanillic acid, caffeic acid, syringic acid, *p*-coumaric acid, *o*-coumaric acid, ferulic acid, cinnamic acid, protocatechuic acid, 4-(acetoxyethyl)-1,2-dihydroxybenzene, benzoic acid and hydroxy-isocromans, (2) phenolic alcohols: hydroxytyrosol or 3,4-(dihydroxyphenyl)ethanol (HYTY or 3,4-DHPEA), tyrosol or *p*-(hydroxyphenyl)ethanol (TH or *p*-HPEA) and 3,4-DHPEA glucoside and acetate, between others, (3) secoiridoids: dialdehydic form of decarboxymethyl elenolic acid linked to 3,4-

DHPEA also called dialdehydic form of decarboximethyl oleuropein aglycon (3,4-DHPEA-EDA), dialdehydic form of decarboxymethyl elenolic acid linked to *p*-HPEA also called dialdehydic form of decarboxymethyl ligstroside aglycon or oleocanthal (*p*-HPEA-EDA), aldehydic form of elenolic acid linked to 3,4-DHPEA also called aldehydic form of oleuropein aglycon (3,4-DHPEA-EA), aldehydic form of elenolic acid linked to *p*-HPEA also called aldehydic form of ligstroside aglycon (*p*-HPEA-EA); other secoiridoids derivatives of *p*-HPEA, oleuropein (OI), the dialdehydic form of oleuropein aglycon and the dialdehydic form of ligstroside aglycon, which are isomers of the 3,4-DHPEA-EA and *p*-HPEA-EA, respectively, (4) lignans: (+)-1-acetoxypinoresinol and (+)-pinoresinol, and (5) flavonoids, such as luteolin and apigenin.

The chemical structures of some of these compounds are shown in Figure 1.

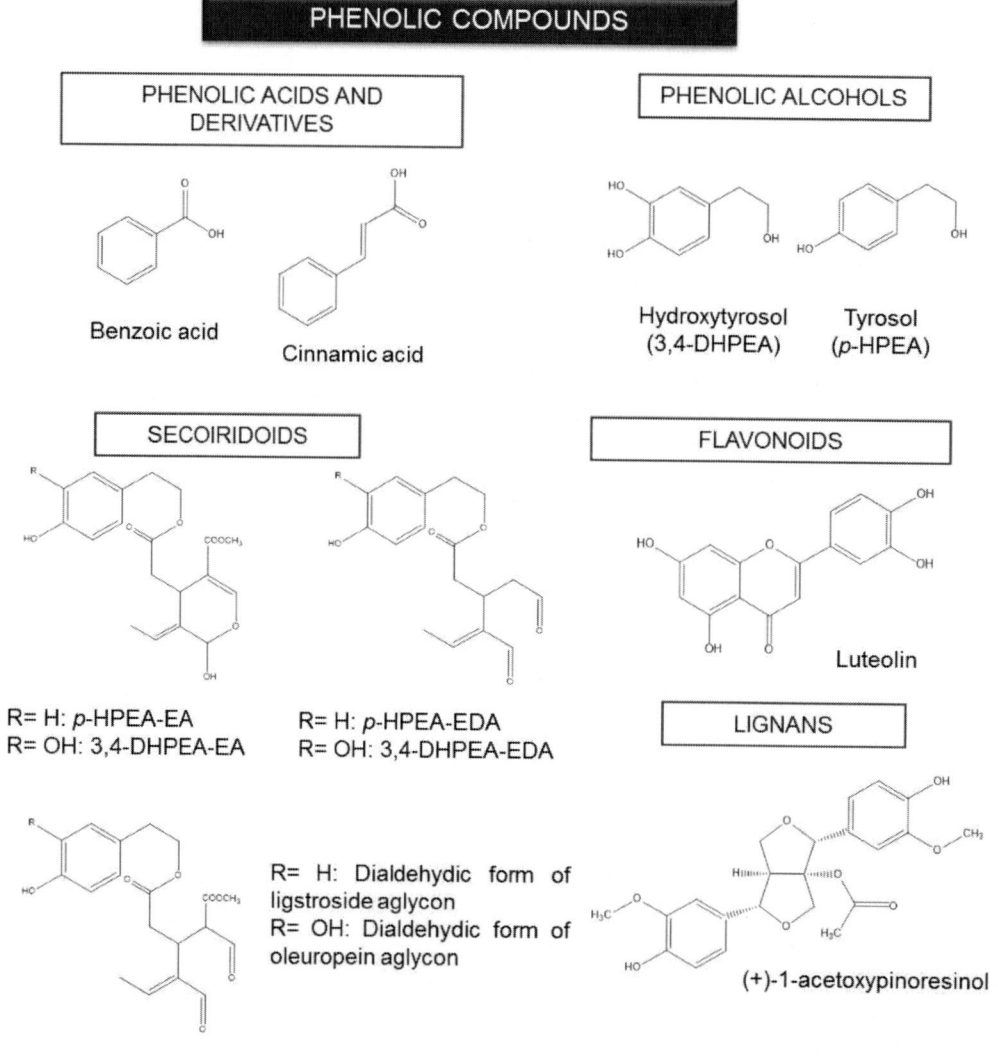

Figure 1. Phenolic compounds in VOO. As example, the chemical structures of some compounds of each family are shown.

Despite they are present in small amounts in VOO, phenolic acids were the first group of phenolic compounds found in VOO together with phenyl-alcohols, hydroxy-isochromans and flavonoids [7, 8, 9], while secoiridoids and lignans are its main phenolic compounds. Phenolic acids in VOO can be divided in two groups: benzoic acids (with basic chemical structure C6-C1) and cinnamic acids (C6-C3). Benzoic, protocatechuic, gallic, vanillic and syringic acids are benzoic acids, while sinapic, cinnamic, caffeic, ferulic, *p*-coumaric and *o*-coumaric acids are cinnamic acids.

The mechanisms through which phenolic compounds are part of the chemical composition of olive oil are largely unknown, despite their importance. Some compounds, such as the secoiridoids, are derivatives of secoiridoid glucosides present in the olive fruit that are released in the oil during the mechanical extraction process [10, 11]. The main classes of phenols in olive fruit are phenolic acids, phenolic alcohols, flavonoids and secoiridoids. The 3,4-DHPEA and *p*-HPEA are the most abundant phenolic alcohols in the olive fruit [12, 13]. The flavonoids include flavonol glycosides such as luteolin-7-glucoside and rutin [13], and anthocyanins, cyanidin and delphinidin glycosides, in particular [14, 15, 16]. While phenolic acids, phenolic alcohols and flavonoids occur in many fruits and vegetables belonging to various botanical families, secoiridoids, on the contrary, are exclusively present in the family of *Oleaceae* that includes *Olea Europaea* L. Olives and VOO are the main products obtained from this specie used in human nutrition.

Despite the fact that phenolic acids were the first phenolic compounds found in VOO and that they have been found in the olive fruit by different authors [17, 18, 19], in general terms, they are present in VOO in concentrations lower than 1 mg·kg^{-1}, while secoiridoids are the most abundant phenolic compounds in VOO with concentrations ranging between 27 and 32 mg·kg^{-1}. The secoiridoids more abundant in VOO are mainly derivatives of oleuropein, demethyloleuropein and ligstroside, which are present in high amounts in the olive fruits (Figure 2). They are the 3,4-DHPEA-EDA, the *p*-HPEA-EDA and the isomer of the oleuropein aglycon 3,4-DHPEA-EA. These compounds were found, for the first time, by Montedoro et al. [18, 20], whom also assigned their chemical structures [21] that were later confirmed by other authors [22, 23, 24].

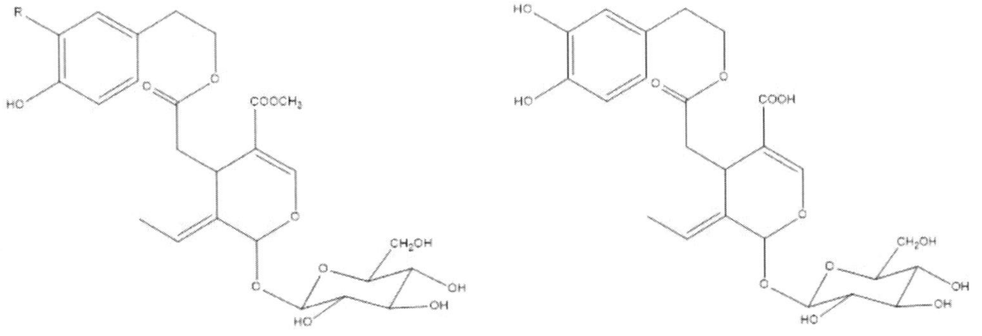

R= OH: Oleuropein
R= H: Ligstroside

Demethyloleuropein

Figure 2. Chemical structures of some secoiridoid glucosides found in olives.

These compounds are characterized by the presence of either elenolic acid or elenolic acid derivatives in their molecular structure and come from the secondary terpene metabolism. They consist of a phenyl ethyl alcohol (3,4-DHPEA or p-HPEA), elenolic acid and sometimes a glucosidic residue (generally, they are glucosilated in plants).

Oleuropein and ligstroside aglycon and their dialdehydic forms were also detected as minor hydrophilic phenols of VOO [24, 25, 26]. These compounds are intermediate structures of the biochemical transformation of secoiridoid glucosides in olive fruit such as oleuropein, demethyloleuropein and ligstroside in the final aglycon derivatives corresponding to the 3,4-DHPEA-EDA from oleuropein and demethyloleuropein and p-HPEA-EDA from ligstroside, respectively [26].

The 3,4-DHPEA and p-HPEA are the main phenolic alcohols of VOO. Their concentration is generally low in fresh oils but increases during oil storage due to the hydrolysis of VOO secoiridoids such as 3,4-DHPEA-EDA, p-HPEA-EDA and 3,4-DHPEA-EA that contain 3,4-DHPEA and p-HPEA in their molecular structure [27]. For this reason, there are variations in the 3,4-DHPEA and p-HPEA concentrations found in VOO by different authors [2].

Flavonoids such as luteolin and apigenin were also reported as phenolic compounds of VOO [28]. Flavonoids present aromatic rings in their structures: one of them and the lateral chain with three carbon atoms come from L-phenylalanine, while the rest comes from the acetyl-CoA. Flavonoids are divided in flavones, flavonols, flavanones and flavonols. Luteolin and apigenin are flavones.

The lignans were the last phenolic compounds found in VOO. Brenes et al. confirmed the occurrence of (+)-1-acetoxypinoresinol and (+)-pinoresinol in Spanish VOO [29]. These authors also reported that the concentrations of lignans could also be employed to discriminate the oils produced from Picual from the other VOOs extracted from the Hojiblanca, Cornicabra and Arbequina varieties [30].

Due to the agronomical and technological aspects of olive oil production, that strongly affect their occurrence, is rather difficult to establish an average concentration of hydrophilic phenols in VOO [3]. Their concentration can vary between 40 and 900 mg·kg^{-1}, although there are oils with contents even higher than 1000 mg·kg^{-1} [18, 31]. In addition, the chromatographic profiles of VOO phenolic compounds show strong differences that may be related to the agronomical and technological aspects of production. In this context, the relationships between the phenolic profile of oils and their genetic or geographical origin should be better investigated.

2.2. Properties of the Phenolic Compounds

Phenolic compounds are considered as a very important part of the chemical defense system of the fruit [2]. They present diverse functions, such as their antimicrobial activity and protection against the oxidative damage, since they limit the UV light effects. In addition, they contribute to the stability of the olive oil against oxidation and to its sensorial characteristics, such as flavor and aroma, and they have been identified as the main responsible compounds for the antioxidant properties of VOO.

In addition to the antioxidant activity and sensorial properties, phenolic compounds present important healthy properties, which are mainly responsible for the great interest that these compounds have nowadays [32, 33].

Phenolic acids, for example, have been associated with sensorial and organoleptic (flavor, astringency and hardness), as well as the antioxidant properties. In addition, through the ingestion of fruits and vegetable these compounds provide a protective function against illnesses that can be related to oxidative damage (coronary hearth and cerebrovascular diseases, and cancer) [34, 35, 36].

2.2.1. Sensory Properties

The phenolic composition has a great influence in the sensory properties of VOO. In particular, these compounds have been associated to the bitter and pungent sensory notes of oil [3, 35, 37, 38, 39]. In fact, there are several papers focusing on the study of the relationship between the "bitter" and the "pungent" taste of VOO and the total concentration of phenols [40, 41, 42], although the relationship between individual hydrophilic phenols of VOO and its sensory characteristics are not clearly defined [3]. In general, it has been assumed that the main responsible compounds for the sensory properties are 3,4-DHPEA, *p*-HPEA and their derivatives [39]. Some authors point out that 3,4-DHPEA-EDA and 3,4-DHPEA-EA contribute more to the bitter than other compounds [43, 44]. However, another authors state that there is an important correlation between the sensory properties of bitter and astringency and the content in ligstroside derivatives, such as the *p*-HPEA-EDA [45]. The pungent sensory note present in oils obtained at the beginning of the season has been also associated with the content in ligstroside derivatives [46] and the content of the aldehydic form of oleuropein [43].

García et al. [43] indicate that the VOO bitter, sensory evaluated and chemically estimated by the sum of the content of secoiridoids derivatives of 3,4-DHPEA in olives, considerably decreases with the increase of the temperature used in the heating of the olives. In this way, the bitter intensity was objectively estimated by the content of the secoiridoids derivatives of 3,4-DHPEA in oil. For each variety, an adequate relationship between the content of secoiridoids derivatives of 3,4-DHPEA and their bitter was found.

In last years, many researchers have found correlations between the sensory properties of VOO and its content in phenolic compounds, but only few have examined correlations with the lignan fraction. On the other hand, most of the studies show statistical correlations between some organoleptic properties and the total concentration of secoiridoids, as it has been previously indicated. However, there are few studies related to the correlation between these properties and the chemical structures of the secoiridoids.

Gutiérrez-Rosales et al. [47] isolated each secoiridoid derivative by preparative liquid chromatography for the first time, with the aim of evaluating the bitter intensity in 20 different VOO samples and they observed a very strong correlation between the 3,4-DHPEA-EDA, *p*-HPEA-EDA and 3,4-DHPEA-EA content and the bitter. Finally, they concluded that these three compounds are the main responsible for the bitter flavor of VOO.

The first paper that established a relationship between the chemical structure of secoiridoids, such as *p*-HPEA-EDA and 3,4-DHPEA-EDA, with the pungent taste of olive oil was published in 2003 by Andrewes et al. [48]. The isolation of phenolic compounds was carried out by a previous liquid-liquid extraction coupled to reversed-phase liquid chromatography with fraction collection. After several assays using taste panels, it was concluded that *p*-HPEA-EDA is mainly responsible for this pungent taste in VOO, since any other phenolic fraction of the analyzed VOO did not produce such an intense taste.

These results were then highlighted by Beauchamp et al. [49], whom, after the *p*-HPEA-EDA synthesis, studied the sensory properties of this compound dissolved in corn oil (which did not own such properties). The result, which is probably due to the presence of two dialdehydic groups in the *p*-HPEA-EDA molecule, was a pungent taste similar to VOO.

2.2.2. Antioxidant Activity

Oxidation is an unavoidable process that starts after VOO extraction and lead to its damage, which can become more pronounced due to storage. Phenolic compounds can inhibit the oxidation through different mechanisms [50, 51, 52, 53], being the most common the known as radical scavenging. According to this mechanism, their antioxidant activity can be explained taking into account that phenolic compounds own the ability to capture free radicals inhibiting the oxidation procedure by the offshoring of an electron and the stabilization by resonance in the aromatic nucleus, which avoids the continuation of the chain reactions of free radicals.

The antioxidant activity of phenolic compounds in VOO has been thoroughly studied [3, 35, 39, 54, 55, 56, 57]. In fact, correlations among total phenols, evaluated by colorimetric method on the methanolic extract of VOO, their antioxidant activity, expressed using the oxygen radical absorbance capacity (ORAC) test, and the shelf life of oil, evaluated by the Rancimat method, have been found [55, 58, 59].

Baldioli et al. [31] studied the antioxidant activity of several secoiridoid derivatives, isolated from VOO and dissolved in purified olive oil, and found that the *o*-diphenols, such as 3,4-DHPEA, 3,4-DHPEA-EDA and 3,4-DHPEA-EA, possess a much greater antioxidant activity than *p*-HPEA and α-tocopherol [31, 54]. These authors also proved that 3,4-DHPEA and the other secoiridoids containing these compounds in their molecular structure (3,4-DHPEA-EDA and 3,4-DHPEA-EA) are the natural antioxidants of VOO with the highest antioxidant power. In addition, the results have confirmed that the antioxidant activity depends on the concentration of these phenols in VOO. Afterwards, other authors have confirmed that, in general, the major antioxidant effects are observed in presence of compounds with 3,4-dihydroxy, 3,4,5-trihydroxy structures linked to an aromatic ring, such as oleuropein, 3,4-DHPEA-EDA and 3,4-DHPEA-EA [60].

In a relatively recent research, Carrasco-Pancorbo et al. [61] have studied the antioxidant activity of different simple phenolic compounds of VOO by chemical methodologies, accelerate oxidation in a lipid model and an electrochemical method. They have demonstrated the ability of these compounds to act as hydrogen donors and inhibit the oxidation, due to the high number of hydroxyl group in their molecules. In particular, the compounds with hydroxyl groups in *ortho* position showed higher antioxidant capacity due to the formation of intermolecular hydrogen bonds in the reaction with the free radicals. In addition, the substituents in *ortho* position, which act as electron donors, tend to weaken the phenol O-H bond and promote the higher stability of the phenoxyl radical. The obtained results showed again that the 3,4-DHPEA, the 3,4-DHPEA-EDA and the 3,4-DHPEA-EA are the stronger antioxidant and, in addition, it was confirmed that the presence of a simple hydroxyl group confer more limited antioxidant activity. It has been also demonstrated that the presence of –$COOCH_3$ in the molecule tends to decrease the antioxidant power, since it is not an electron donor group.

Several simulations in relation to the VOO behavior in the frying and other cooking processes have been also carried out. The results in relation to the stability of phenols during these processes and the microwave cooking confirm the strong effect of some oleuropein derivatives, such as the 3,4-DHPEA-EDA and the 3,4-DHPEA-EA in the oil stability: the concentration of these compounds sharply decrease in the oil heating process to preserve it from oxidation [62, 63].

Other authors have demonstrated that 3,4-DHPEA, elenolic acid, 3,4-DHPEA-EDA and 3,4-DHPEA-EA experiment a reduction in their concentration more quickly than other phenolic compounds present in olive oil, such as 3,4-DHPEA acetate and p-HPEA-EA, while lignans were confirmed as the more resistant phenolic compounds to thermal treatments and, therefore, with lower antioxidant capacity [64].

2.2.3. Biological Activity and Beneficial Effects on Human Health

The antioxidant activity of VOO compounds has attracted a great interest in last years, since it is related to the protection against important chronic and degenerative disorders such as coronary heart diseases, neurodegenerative disorders and tumors in different localizations [3, 32, 33, 35, 39, 65, 66].

The reactive oxygen species (ROS) are originated as a consequence of the oxidative stress, and they are involved in many chronic and degenerative disorders [67, 68, 69] through several partially known mechanisms. The oxidative stress is also related to the aging [70] and, in addition, with the progress of neurodegenerative disorders, such as Parkinson [71].

Natural antioxidants play an important role in the prevention of this kind of pathologies. As a consequence, in the last years, foods with high levels of these compounds have been object of an increased interest. In this context, the effect of olive oil against cancer has been considered by several researchers. Martin-Moreno et al. [72], Trichopoulou et al. [73] and La Vecchia et al. [74] demonstrated that the frequent intake of olive oil reduces the risk of cancer. These findings were further confirmed by other authors [75] obtaining similar conclusions for tumors in pancreas [76], oral cavity [77], esophagus [78], right and colon [79], prostate [80, 81] and lung [82]. In addition, the effect of olive oil against UV radiation damage onto the skin was also demonstrated [83].

On the other hand, 3,4-DHPEA and the methanolic complexes of VOO containing phenolic compounds and other purified compounds (3,4-DHPEA-EDA and p-HPEA-EDA) are capable of producing apoptosis in different tumor cells as was confirmed by several authors [84, 85, 86]. Interestingly, while 3,4-DHPEA strongly induces the apoptosis, p-HPEA has not action at the same concentration (100 µM). In addition, p-HPEA-EDA is more active than 3,4-DHPEA-EDA. These results suggest that the different parts of the molecular structure of a given compound can get diverse biological activity.

One of the most important aspects regarding the interaction between the different parts of a molecule and the healthy properties is the antioxidant activity, which, as it was previously indicated, is directly related to the ability of these compounds to transfer the hydroxyl hydrogen to free radicals. As a result, the high antioxidant capacity of phenolic compounds with hydroxyl groups in *ortho* position is confirmed. In this context, a recent study carried out by Fabiani et al. [87] reveals that both 3,4-DHPEA and 3,4-DHPEA-EDA are more able to prevent the oxidative damages in the DNA than p-HPEA and p-HPEA-EDA.

On the other hand, there are evidences supporting the fact that the protector effect against chronic and degenerative diseases is related to the amounts of phenolic compounds in olive oil, being 3,4-DHPEA more important than unsaturated fatty acids [88, 89]. In this sense, the peroxidation of phospholipids in liposomes [90], the protection against oxidation of low density lipoproteins [91, 92], the decrease in the erythrocyte oxidative damage by 3,4-DHPEA [93], the inhibition of platelet aggregation by 3,4-DHPEA and its role in the synthesis of thromboxanes in human cells [94], and the inhibition in DNA base changes by peroxynitrites [95], can be cited. In addition, it has been also showed that olive oil phenolic compounds have protective effects regarding inflammation in animals [96, 97].

Phenolic compounds present in olive oil can be easily available for humans, even at low doses [98]. 3,4-DHPEA and *p*-HPEA, the two major VOO phenolic alcohols, are absorbed as a function of the olive oil intake [98, 99]. Consequently, it was suggested that monitoring their concentrations could be useful for clinical studies and acts as an indicator of the olive oil consumption [65]. 3,4-DHPEA and *p*-HPEA are present in plasma and urine in their conjugated forms (ca. 98%), fact that suggests an intestinal/hepatic metabolism [100, 101].

It should be also commented that most of the beneficial concerns about the olive consumption when considering cardiovascular disorders are associated with the vasodilatory, anti-inflammatory, antioxidant, antimicrobial and platelet anti-aggregation properties related to the oleuropein content [102, 103]. This compound and vanillic and *p*-coumarinic acids can also inhibit the in-vitro growth of some bacteria as *Escherichia coli*, *Klebsiella peneumoniae* and *Bacilluscereus* [104].

Nevertheless, more studies are necessary to illumine on the influence that the VOO phenolic compounds exert on health. Undoubtedly, this research would be better carried out if the commercial standards of the mentioned compounds would be available to the analytical chemists.

3. Olive Oil Production

3.1. Agronomical and Technological Aspects

The quali- and quantitative content of phenolic compounds in VOO is affected by the agronomical and technological aspects of production.

The cultivar, the fruit ripeness, the climatic conditions and the agronomical techniques, such as irrigation, are the agronomical factors with more influence and, as a consequence, the more studied [45, 105, 106, 107]. As it was pointed out by several authors, the olive composition is quantitatively influenced by the olive variety [30, 108, 109, 110, 111], although the fruit ripening has also effect on it, i.e. the oleuropein content decreases, while the demethyloleuropein content increases with the ripening. Nevertheless, the concentration of both compounds decreases in over ripened fruits [112]. The culture variety also affects the concentration of VOO specific phenolic compounds while the phenolic profile stands practically invariable.

It was also pointed out the relationship between the VOO quality and the season. Several studies show that the concentration of VOO phenolic compounds is affected by the availability and distribution of water in the olive tree vegetative cycle [113, 114].

Regarding the technological aspects of production, the presence of phenolic compounds in VOO is related to the activity of endogenous enzymes present in the olive fruit. This fact indicates that the extraction conditions exert high influence on the concentration of these compounds in the oil. The grinding and the malaxation are the critical steps in the oil production concerning the content of phenolic compounds [115].

Some secoiridoid derivatives like 3,4-DHPEA-EDA, p-HPEA-EDA, p-HPEA-EA and 3,4-DHPEA-EA are originated during the grinding by hydrolysis of oleuropein, demethyloleuropein and ligstroside. This is due to the action of β-glucosidase endogenous enzymes (Figure 3) [3]. This enzymatic hydrolysis has been studied by several authors using oleuropein and demethyloleuropein as substrates [116, 117], as well as the relationship between the enzymatic hydrolysis of secoiridoids and the presence of some of these derivatives of secoiridoid aglycones (3,4-DHPEA-EDA, p-HPEA-EDA and 3,4-DHPEA-EA) in VOO [118]. Although the mechanism of 3,4-DHPEA-EDA as a final product of the enzymatic hydrolysis of the olive demethyloleuropein is known [118], the formation mechanism of this compound and the corresponding one to p-HPEA-EDA, both coming from oleuropein and ligstroside, respectively, is currently under study.

Nevertheless, owing to oleuropein, demethyloleuropein and ligstroside concentrations in olives depend on the fruit variety [112, 119, 120], the VOO phenolic compounds from a given olive variety should be related to the concentration of oleuropein, demethyloleuropein and ligstroside in the olive of the same variety. Thus, the mechanisms in Figure 3 can be assumed due to the relations found between concentration of VOO final derivatives and the original olive glucoside for a given variety [119]. In this way, it is assumed the enzymatic transformation of oleuropein in 3,4-DHPEA-EDA, which also includes the enzymatic activity of methylesterase according with the proposed mechanism [3].

It has been proved that the concentration of derivatives of secoiridoid aglycons and phenolic alcohols decreases both in the olive paste and olive oils during the malaxation process along with time and temperature increment in the processing [10, 121, 122, 123]. The higher solubility of the phenolic compounds in the aqueous phase than in the oil is not the only mechanism that diminishes their final concentration in the oil, but the oxidation reactions catalyzed by endogenous oxidoreductases that promote the phenolic compounds oxidation during this process [10, 124, 125]. The use of new technologies that allow, for example, the extraction of boneless olive paste with increased concentrations of phenolic acids confirm the relationship between the oxidation reactions during the processing and the concentration of phenolic compounds in the oil [3].

On the other hand, interactions between polysaccharides and phenolic compounds present in the olive paste also influence their lost during the processing. Throughout grinding and malaxation processes, polysaccharides can join phenolic compounds, and thus their total concentration in olive oil is reduced [126].

Another technological aspect that has also influence in the phenolic composition of olive oil is the extraction system. There are two types of decanters that are used to separate the olive oil from the rest of the aqueous phase through centrifugation: the traditional system including three phases and the most recent one consisting in two phases.

In the three-phase system water is added in high quantities (10-100 L/100 kg of olive oil paste) before centrifuging, with the aim of decreasing the paste viscosity and improving the olive oil separation. However, the addition of water modifies the distribution of phenolic compounds between the oil and the aqueous phase, favoring their migration to the latter.

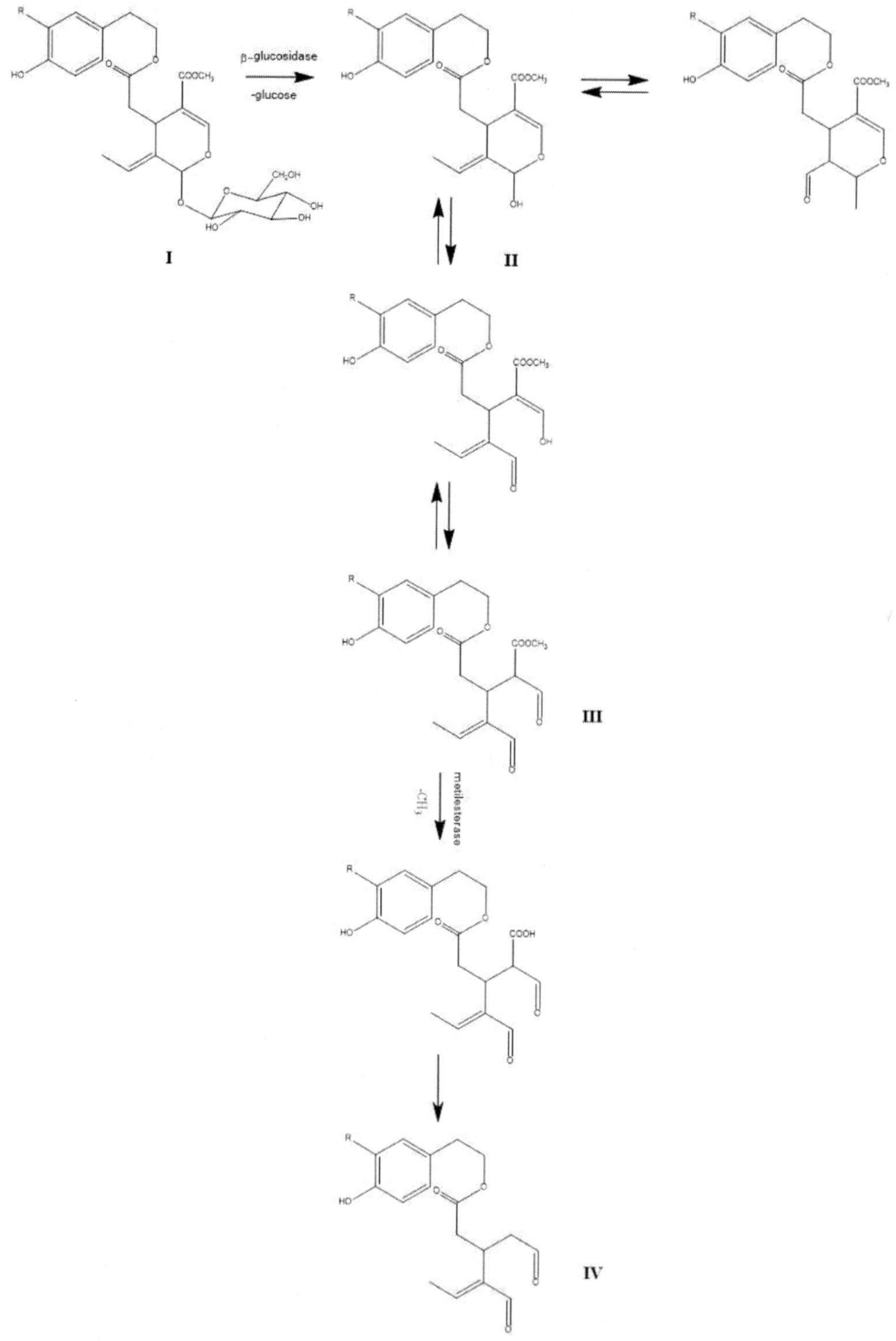

Figure 3. Proposed mechanism for the production of derivative secoiridoids. (I) R= H: ligstroside; R= OH: oleuropein; (II) R= H: *p*-HPEA-EA; R= OH: 3,4-DHPEA-EA; (III) Dialdehydic forms of the ligstroside and oleuropein aglycons; (IV) R= H: *p*-HPEA-EDA; R= OH: 3,4-HPEA-EDA.

The three-phase name is due to the fact that three products are obtained through this process, i.e. olive oil, a liquid phase (olive mill or *alpechín*) and a solid phase (*orujo*). In these systems separation is performed in two steps; the first one consists in separate the liquid phase from the solid phase (*orujo*), and the second one in spliting the oil from the *alpechín*.

It has been proved that with the most recent two-phase system, called after this name due to the generation of oil and a unique residue called *alperujo* (mixture of *orujo* and *alpechín*), olive oils with higher concentrations of phenolic compounds than with the traditional three-phase system can be obtained [127, 128, 129]. Besides, it has been obtained a higher concentration of phenolic compounds in oils using the three-phase decanters with less water addition in comparison with those that use high water volumes [130].

CONCLUSION

In this chapter, a complete review regarding production and nutritional properties of VOO was presented. It has been proved that the unique extraordinary features of VOO rely on its content in hydrophilic phenolic compounds, which confer fundamental nutritional and sensory characteristics to the oil, such as antioxidant properties and human protection against severe diseases. Besides, these phenolic compounds take part in the fruit defense chemical system, due to their antimicrobial activity, and in the protection against oxidative damages. The agronomical and technological aspects of production have great influence on the content of hydrophilic phenolic compounds in VOO, being their concentrations related to the activity of the endogenous enzymes present in the olive fruit.

ACKNOWLEDGMENTS

M.P. Godoy-Caballero is grateful to the *Ministerio de Educación of Spain* for a *FPU* grant (Orden EDU/3083/2009, de 6 de noviembre, BOE nº 277, de 17/11/09). The authors are grateful to the *Ministerio de Ciencia e Innovación* of Spain (Project CTQ2011-25388) cofinanced by the European FEDER funds. Funding from the *Gobierno de Extremadura* and European FEDER Funds (Consolidation Project of Research Group FQM003, Project GR1003) is also acknowledged. H.G. and M.J.C. are grateful to Universidad Nacional del Litoral (Project CAI+D Nº 12-65), CONICET (Consejo Nacional de Investigaciones Científicas y Técnicas, Project PIP 455) and ANPCyT (Agencia Nacional de Promoción Científica y Tecnológica, Project PICT 2011-0005) for financial support.

REFERENCES

[1]   REGLAMENT (EC) nº 1234/2007, 22nd October 2007, DO L 299 of 16.11.2007, p. 1.

[2]   Lozano-Sánchez J., Segura-Carretero A., Fernández-Gutiérrez A. (2009) Composición química del aceite de oliva. In: *El Aceite de Oliva Virgen: Tesoro de Andalucía*. Editors: Fernández-Gutiérrez A., Segura-Carretero A., Servicio de Publicaciones de la Fundación Unicaja. Málaga, Spain, pp. 195-224.

[3]   Servili M., Selvaggini R., Esposto S., Taticchi A., Montedoro G. F., Morozzi G. (2004) Health and sensory properties of virgin olive oil hydrophilic phenols: agronomic and technological aspects of production that affect their occurrence in the oil. *J. Chromatogr. A*, 1054, 113-127.

[4]   Boskou D. (2006) Olive Oil Chemistry and Technology AOCS Press, Champaign, IL, USA.

[5]   Cantarelli C. (1961) Sui polifenolipresentinella drupa e nell'olio di oliva. *Riv. Ital. Gr.*, 38, 69-72.

[6]   Montedoro G. F., Cantarelli C. (1969) Indagine sulle sostanze fenoliche presenti nell'olio di oliva. *Riv. Ital. Sostanze Gr.*, 46, 115-124.

[7]   Montedoro G. F. (1972) I costituenti fenolici presenti negli oli vergini di oliva. Nota 1: identificazione di alcuni acidi fenolici e loro potere antiossidante. *S. T. A.*, 3, 177-186.

[8]   Vázquez Roncero A. (1978) Les polyphenols de l'huile d'olive et leur influence sur les characteristiques de l'huile. *Rev. Fr. Corps Gras*, 25, 21-26.

[9]   Solinas M., Cichelli A. (1981) Sulla deteminazione delle sostanze fenoliche dell'olio di oliva. *Riv. Ital. Sostanze Gr.*, 58, 159-164.

[10] Bianco A., Coccioli F., Guiso M., Marra C. (2001) The occurrence in olive oil of a new class of phenolic compounds: hydroxy-isochromans. *Food Chem.*, 77, 405-411.

[11] Montedoro G. F., Servili M., Baldioli M., Selvaggini R., Begliomini A. L., Taticchi A. (2002) Relationships between phenolic composition of olive fruit and olive oil; the importantce of endogenous enzymes. *Acta Horticult.*, 586, 551-556.

[12] Garrido Fernández A., Fernández Díez M. J., Adams M. R. (1997) Physical and chemical characteristics of the olive fruit. In: *Table olives*. Editors: Garrido Fernández A., Fernández Díez M. J., Adams M. R., Chapman& Hall, London, UK, pp. 67-109.

[13] Vázquez Roncero A., Graciani Constante E., Maestro Durán R. (1974) Componentes fenólicos de la aceituna. I. Polifenoles de la pulpa. *Grasas Aceites*, 25, 269-279.

[14] Maestro Durán R., Vázquez Roncero A. (1976) Colorantes anticiánicos de las aceitunas manzanillas maduras. *Grasas Aceites*, 27, 237-243.

[15] Mazza G., Miniati E. (1993) Anthocyanins in Fruits, Vegetables and Grains. CRC Press, Boca Raton, FL, USA.

[16] Marekow N. L. (1984) Polyphenols Research in Bulgaria. Bull Liaison Groupe Polyphenols, pp. 12-31.

[17] Solinas M., Cichelli A. (1982) GC and HPLC determination of phenolic substances in olive oil: Hypothetical role of tyrosol in the assessment of virgin olive oil in mixture with refined oils. *Riv. Soc. Ital. Sci. Alim.*, 11, 223-230.

[18] Montedoro G. F., Servili M., Baldioli M., Miniati E. (1992) Simple and hydrolyzable phenolic compounds in virgin olive oil. 1. Their extraction, separation, and quantitative and semiquantitative evaluation by HPLC. *J. Agric. Food Chem.*, 40, 1571-1576.

[19] Tsimidou M., Lytridou M., Boskou D., Paooa-Lousi A., Kotsifaki F., Petrakis C. (1996) On the determination of minor phenolic acids of virgin olive oil by RP-HPLC. *Grasas Aceites*, 47, 151-157.

[20] Montedoro G. F., Servili M., Baldioli M., Miniati E. (1992) Simple and hydrolyzable phenolic compounds in virgin olive oil. 2. Initial characterization of the hydrolyzable fraction. *J. Agric. Food Chem.*, 40, 1577-1576.

[21] Montedoro G. F., Servili M., Baldioli M., Selvaggini R., Miniati E., Macchioni A. (1993) Simple and hydrolyzable compounds in virgin olive oil.3. Spectroscopic characterizations of the secoiridoid derivatives. *J. Agric. Food Chem.*, 41, 2228-2234.

[22] Angerosa F., D'Alessandro N., Corana F., Mellerio G. (1995) GC-MS Evaluation of Phenolic Compounds in Virgin Olive Oil. *J. Agric. Food Chem.*, 43, 1802-1807.

[23] Angerosa F., D'Alessandro N., Konstantinou P., Di Giacinto L. (1996) Characterization of phenolic and secoiridoid aglycons present in virgin olive oil by gas chromatography-chemical ionization mass spectrometry. *J. Chromatogr. A*, 736, 195-203.

[24] Owen R. W., Mier W., Giacosa A., Hull W. E., Spiegelhalder B., Bartsch H. (2000) Phenolic compounds and squalene in olive oils: the concentration and antioxidant potential of total phenols, simple phenols, secoiridoids, lignans and squalene. *Food Chem. Toxicol.*, 38, 647-659.

[25] Perri E., Raffaelli A., Sindona G. (1999) Quantitation of Oleuropein in Virgin Olive Oil by Ionspray Mass Spectrometry—Selected Reaction Monitoring. *J. Agric. Food Chem.*, 47, 4156-4160.

[26] Rovellini P., Cortesi N. (2002) Liquid chromatography-mass spectrometry in the study of oleuropein and ligstroside aglycon in virgin olive oil: aldehydic, dialdehydic form and their oxidized products. *Riv. Ital. Sostanze Gr.*, 79, 1-14.

[27] Brenes M., García A., García P., Garrido A. (2001) Acid Hydrolysis of Secoiridoid Aglycons during Storage of Virgin Olive Oil. *J. Agric. Food Chem.*, 49, 5609-5614.

[28] Rovellini P., Cortesi N., Fedeli E. (1997) Analysis of flavonoids from *Olea europaea* by HPLC-UV and HPLC-electrospray-MS. *Riv. Ital. Sostanze Gr.*, 74, 273-279.

[29] Brenes M., Hidalgo F. J., García A., Ríos J. J., García P., Zamora R., Garrido A. (2000) Identification of lignans as major components in the phenolic fraction of olive oil. *J. Am. Oil Chem. Soc.*, 77, 715-720.

[30] Brenes M., García A., Ríos J. J., García P., Garrido A. (2002) Use of 1-acetoxypinoresinol to authenticate Picual olive oils. *Int. J. Food Sci. Technol.*, 37, 615-625.

[31] Baldioli M., Servili M., Perretti G., Montedoro G. F. (1996) Antioxidant activity of tocopherols and phenolic compounds of virgin olive oil. *J. Am. Oil Chem. Soc.*, 73, 1589-1593.

[32] Ghanbari R., Anwar F., Alkharfy K. M., Gilani A. H., Saari N. (2012) Valuable Nutrients and Functional Bioactives in Different Parts of Olive (*Olea europaea* L.)—A Review. *Int. J. Mol. Sci.*, 13, 3291-3340.

[33] Cicerale S., Lucas L., Keast R. (2010) Biological Activities of Phenolic Compounds Present in Virgin Olive Oil. *Int. J. Mol. Sci.*, 11, 458-479.

[34] Morales M. T., Angerosa F., Aparicio R. (1999) Effect of the extraction conditions of virgin olive oil on the lipoxygenase cascade: Chemical and sensory implications *Grasas Aceites*, 50, 114-121.

[35] Bendini A., Cerretani L., Carrasco-Pancorbo A., Gómez-Caravaca A. M., Segura-Carretero A., Fernández-Gutiérrez A., Lercker G. (2007) Phenolic Molecules in Virgin Olive Oils: a Survey of Their Sensory Properties, Health Effects, Antioxidant Activity and Analytical Methods. *Molecules*, 12, 1679-1719.

[36] Robbins R. J. (2003) Phenolic Acids in Foods: An Overview of Analytical Methodology. *J. Agric. Food Chem.*, 51, 2866-2887.

[37] Gutiérrez-Rosales F., Perdiguero S., Gutiérrez R., Olías J. M. (1992) 11. Evaluation of the bitter taste in virgin olive oil. *J. Am. Oil Chem. Soc.*, 69, 394-395.

[38] Gomes C. A., Girao da Cruz T., Andrade J. L., Milhazes N., Borges F., Marques. M. P. M. (2003) Anticancer Activity of Phenolic Acids of Natural or Synthetic Origin: A Structure-Activity Study. *J. Med. Chem.*, 46, 5395-5401.

[39] Servili M., Esposto S., Fabiani R., Urbani S., Taticchi A., Mariucci F., Selvaggini R., Montedoro G. F. (2009) Phenolic compounds in olive oil: antioxidant, health and organoleptic activities according to their chemical structure. *Inflammopharmacology*, 17, 76-84.

[40] Gutiérrez Rosales F., Perdiguero S., Gutiérrez R., Olias J. M. (1992) Evaluation of bitter taste in virgin olive oil. *J. Am. Oil Chem. Soc.*, 69, 394-395.

[41] Montedoro G. F., Baldioli M., Servili M. (1992) I composti fenolici dell'olio di oliva e la loro importanza sensoriale, nutrizionale e merceologica. *Giornale Ital. di Nutriz Clin. e Prev.*, 1, 19-32.

[42] Tsimidou M. (1998) Polyphenols and Quality of Virgin Olive Oil in Retrospect. *Ital. J. Food Sci.*, 10, 99-116.

[43] García J. M., Yousfi K., Mateos R., Olmo M., Cert A. (2001) Reduction of bitterness by heating of olive (*Olea europaea*). *J. Agric. Food Chem.*, 49, 4231-4235.

[44] Kiritsakis, A. K. (1998) Flavor components of olive oil. *J. Am. Oil Chem. Soc.*, 75, 673-81.

[45] Tovar M. J., Motilva M. J., Paz-Romero M. (2001) Changes in the phenolic composition of virgin olive oil from young trees (*Olea europaea* L. cv. Arbequina) grown under linear irrigation strategies. *J. Agric. Food Chem.*, 49, 5502-5508.

[46] Kiritsakis A. K. (1998) Flavor components of olive oil - A review. *J. Am. Oil Chem. Soc.*, 75, 673-681.

[47] Gutiérrez Rosales F., Rios J. J., Gomez-Rey Ma. L. (2003) Main polyphenols in the bitter taste of virgin olive oil. Structural confirmation by on-line high-performance liquid chromatography electrospray ionization mass spectrometry. *J. Agric. Food Chem.*, 51, 6021-6025.

[48] Andrewes P., Busch J. L. H. C., de Joode T., Groenewegen A., Alexandre H. (2003) Sensory properties of virgin olive oil polyphenols: identification of deacetoxy-ligstroside aglycon as a key contributor to pungency. *J. Agric. Food Chem.*, 51, 1415-1420.

[49] Beauchamp G. K., Keast R. S. J., Morel D., Lin J., Pika J., Han Q., Lee C. H., Smith A. B., Breslin P. A. S. (2005) Ibuprofen like activity in extra-virgin olive oil. *Nature*, 437, 45-46.

[50] Steinmetz K. A., Potter J. D. (1991) Vegetables, fruit, and cancer. II. Mechanisms. *Cancer Causes Control*, 2, 427-442.

[51] Cuvelier M. E., Richard H., Berset C. (1992) Comparison of the antioxidative activity of some acid-phenols: structure-activity relationship. *Biosci. Biotech. Biochem.*, 56, 324-325.

[52] Kahkonen M. P., Hopia A. I., Vuorela H. J., Rauha J. P., Pihlaja K., Kujala T. S., Heinonen M. (1999) Antioxidant Activity of Plant Extracts Containing Phenolic Compounds. *J. Agric. Food Chem.*, 47, 3954-3962.

[53] Stich H. F., Rosin M. P. (1984) Naturally occurring phenolics as antimutagenic and anticarcinogenic agents. *Adv. Exp. Med. Biol.*, 177, 1-29.

[54] Gutiérrez González-Quijano R., Janer del Valle C., Janer del Valle M. L., Gutierrez Rosales F., Vázquez Roncero A. (1977) Relationships between polyphenols content and the quality and stability of virgin olive oil. *Grasas Aceites*, 28, 101-106.

[55] Gordon M. H. (1990) The mechanism of antioxidant action in vitro, Elsevier Applied Science, London, UK.

[56] Briante R., La Cara F., Tunziello P., Febraio F., Nucci R. (2001) Antioxidant Activity of the Main Bioactive Derivatives from Oleuropein Hydrolysis by Hyperthermophilicβ-Glycosidase. *J. Agric. Food Chem.*, 49, 3198-3203.

[57] Riachy M. E., Priego-Capote F., León L., Rallo L., Luque de Castro M. D. (2011) Hydrophilic antioxidants of virgin olive oil. Part 1: Hydrophilic phenols: A key factor for virgin olive oil quality. *Eur. J. Lipid Sci. Technol.*, 113, 678-691.

[58] Tsimidou M. (1998) Polyphenols and quality of virgin olive oil in retrospect. *Ital. J. Food Sci.*, 10, 99-116.

[59] Ninfali P., Bacchiocca M., Biagiotti E., Servili M., Begliomini A. L., Montedoro G. F. (2002) Validation of the Oxygen Radical Absorbance Capacity (ORAC) parameter as a new Index of quality and stability of virgin olive oil. *J. Am. Oil Chem. Soc.*, 79, 971-976.

[60] Artajo L. S., Romero M. P., Morello J. R., Moltiva M. J. (2006) Enrichment of refined olive oil with phenolic compounds: evaluation of their antioxidant activity and their effect on the bitter index. *J. Agric. Food Chem.*, 54, 6079-6088.

[61] Carrasco-Pancorbo A., Cerretani L., Bendini A., Segura-Carretero A., Del Carlo M., Gallina-Toschi T., Lercker G., Compagnone D., Fernández-Gutiérrez A. (2005) Evaluation of the antioxidant capacity of individual phenolic compounds in virgin olive oil. *J. Agric. Food Chem.*, 53, 8918-8925.

[62] Brenes M., García A., Dobarganes C., Velasco J., Romero C. (2002) Influence of Thermal Treatments Simulating Cooking Processes on the Polyphenol Content in Virgin Olive Oil. *J. Agric. Food Chem.*, 50, 5962-5967.

[63] Gómez-Alonso S., Fregapane G., Salvador M. D., Gordon M. H. (2003) Changes in phenolic composition and antioxidant activity of virgin olive oil during frying. *J. Agric. Food Chem.*, 51, 667-672.

[64] Carrasco-Pancorbo A., Cerretani L., Bendini A., Segura-Carretero A., Lercker G., Fernández-Gutiérrez A. (2007) Evaluation of the influence of thermal oxidation on the phenolic composition and on the antioxidant activity of extra-virgin olive oils. *J. Agric. Food Chem.*, 55, 4771-4780.

[65] Covas M. I. (2007) Olive oil and the cardiovascular system-Review. *Pharm. Res.*, 55, 175-186.

[66] Visioli F., Bernardini E. (2001) Extra Virgin Olive Oil's Polyphenols: Biological Activities. *Curr. Pharm. Design*, 17, 786-804.

[67] Berliner J. A., Navab M., Fogelman A. M., Frank J. S., Demer L. L., Edwards P. A., Watson A. D., Luiss A. J. (1995) Atherosclerosis: basic mechanism-oxidation, inflammation and genetics. *Circ.*, 91, 2488-2496.

[68] Nakae D., Kobayashi Y., Akai H., Andoh N., Satoh H., Ohashi K., Tsutsumi M., Konishi Y. (1997) Involvement of 8-hydroxyguanine formation in the initiation of rat liver carcinogenesis by low dose levels of *N*-nitrosodiethylamine. *Cancer Res.*, 57, 1281-1287.

[69] Orlando R. C. (2002) Mechanisms of epithelial injury and inflammation in gastrointestinal diseases" *Rev. Gastroenterol. Disord.*, 2, S2-8.

[70] O'Donnel E., Lynch M. A. (1998) Dietary antioxidant supplementation reverses age-related neuronal changes. *Neurobiol. Aging*, 19, 461-467.

[71] Jenner P., Olanow C. W. (1996) Oxidative stress and the pathogenesis of Parkinson's disease. *Neurology*, 47, S161-70.

[72] Martin-Moreno J. M., Willet W. C., Gorgoio L., Banegas J. R., Rodriguez-Artalejo F., Fernandez-Rodriguez J. C., Maisonneuve P., Boyle P. (1994) Dietary fat, olive oil intake and breast cancer risk. *Int. J. Cancer*, 58, 774-780.

[73] Trichopoulou A., Katsouyanni K., Stuver S., Tzala L., Gnardellis C., Rimm E., Trichopoulos D. (1995) Consumption of olive oil and specific food groups in relation to breast cancer risk in Greece. *J. Natl. Cancer Inst.*, 87, 110-116.

[74] La Vecchia C., Negri E., Franceschi S., de Carli A., Giacosa A., Lipworth L. (1995) Olive oil, other dietary fats, and the risk of breast cancer (Italy). *Cancer Cause Control*, 6, 545-550.

[75] Lipworth L., Martinez M. L., Angell J., Hsieh C. C., Trichopoulos D. (1997) Olive oil and human cancer: an assessment of the evidence. *Prev. Med.*, 26, 181-190.

[76] Soler M., Chatenaud L., La Vecchia C., Franceschi S., Negri S. (1998) Diet, alcohol, coffee and pancreatic cancer: final results from an Italian study. *Eur. J. Cancer Prev.*, 7, 455-460.

[77] Franceschi S., Favero A., Conti E., Salamini R., Volpe R., Negri E., Barman L., Vecchia C. L. (1999) Food groups, oils and butter, and cancer of the oral cavity and pharynx. *Br. J. Cancer*, 80, 614–620.

[78] Bosetti C., Gallus S., Trichopoulou A., Talamini R., Franceschi S., Negri E. (2003) Influence of the Mediterranean diet on the risk of cancers of the upper aerodigestive tract. *Cancer Epidemiol. Biomarkers Prev.*, 12, 1091-1094.

[79] Stoneham M., Goldacre M., Seagroatt V., Gill L. (2000) Olive oil, diet and colorectal cancer: an ecological study and a hypothesis. *J. Epidemiol. Community Health*, 54, 756-760.

[80] Tzonou A., Signorello L. B., Lagiou P., Wuu J., Trichopoulos D., Trichopoulou A. (1999) Diet and cancer of the prostate: a case-control study in Greece. *Int. J. Cancer*, 80, 704-708.

[81] Hodge A. M., English D. R., McCredie M. R. E., Severi G., Boyle P., Hopper J. L., Giles G. G. (2004) Foods, nutrients and prostate cancer. *Cancer Causes Control*, 15, 11-20.

[82] Fortes C., Forastiere F., Farchi S., Mallone S., Trequattrinni T., Anatra F., Schmid G., Peducci C. A. (2003) The protective effect of the Mediterranean diet on lung cancer. *Nutr. Cancer*, 46, 30-37.

[83] Ichihashi M., Ueda M., Budiyanto A., Bito T., Oka M., Fukunaga M., Tsuru K., Horikawa T. (2003) UV-Induced skin Damage. *Toxicology*, 189, 21-39.

[84] Fabiani R., De Bartolomeo A., Rosignoli P., Servili M., Selvaggini R., Montedoro G. F., Di Saverio C., Morozzi G. (2006) Virgin olive oil phenols inhibit proliferation of human promyelocytic leukemia cells (HL60) by inducing apoptosis and differentiation. *J. Nutr.*, 136, 614-619.

[85] Fabiani R., De Bartolomeo A., Rosignoli P., Servili M., Montedoro G. F., Morozzi G. (2002) Cancer chemoprevention by hydroxytyrosol isolated from virgin olive oil through G1 cell cycle arrest and apoptosis. *Eur. J. Cancer Prev.*, 11, 351-358.

[86] Ragione F. D., Cucciolla V., Borriello A., Pietra V. D., Pontoni G., Racioppi L., Manna C., Galletti P., Zappia V. (2000) Hydroxytyrosol, a natural molecule occurring in olive oil, induces cytochrome c-dependent apoptosis. *Biochem. Biophys. Res. Commun.*, 278, 733-739.

[87] Fabiani R., Rosignoli P., De Bartolomeo A., Fuccelli R., Servili M., Montedoro G. F., Morozzi G. (2008) Oxidative DNA damage is prevented by extracts of olive oil, hydroxytyrosol, and other phenolic compounds in human blood mononuclear cells and HL60 cells. *J. Nutr.*, 138, 1411-1416.

[88] Ruíz-Gutierrez V., Muriana F. J., Guerrero A., Cert A. M., Villar J. (1996) Plasma lipids, erythrocyte membrane lipids and blood pressure of hypertensive women after ingestion of dietary oleic acid from two different sources. *J. Hypertens.*, 14, 1483-1490.

[89] Fitó M., Cladellas M., Torre R. D. L., Martí J., Alcántara M., Pujadas-Bastardes M., Marrugat J., Bruguera J., López-Sabater M. C., Vila J. (2005) Antioxidant effect of virgin olive oil in patients with stable coronary heart disease: A randomised, crossover, controlled, clinical trial" *Atherosclerosis*, 181, 149-158.

[90] Aeschbach R., Loliger J., Scott B. C., Murcia A., Butler J., Halliwell B., Aruoma O. I. (1994) Antioxidant actions of thymol, carvacrol, 6-gingerol, zingerone and hydroxytyrosol. *Food Chem. Toxicol*, 32, 31-36.

[91] Grignaffini P., Roma P., Galli C., Catapano A. L. (1994) Protection of low-density lipoprotein from oxidation by 3,4-dihydroxyphenylethanol. *Lancet*, 343, 1296-1297.

[92] Visioli F., Bellomo G., Montedoro G. F., Galli C. (1995) Low density lipoprotein oxidation is inhibited in vitro by olive oil constituents. *Atherosclerosis*, 117, 25-32.

[93] Manna C., Galletti V., Cucciolla P., Montedoro G. F., Zappia V. (1999) Olive oil hydroxytyrosol protects human erythrocytes against oxidative damages. *J. Nutr. Biochem.*, 10, 159-165.

[94] Petroni A., Blasevich M., Salami M., Papini N., Montedoro G. F., Galli C. (1995) Inhibition of platelet aggregation and eicosanoid production by phenolic components of olive oil. *Thromb. Res.*, 78, 151-160.

[95] Deiana M., Arouma O. I., Bianchi M. P., Spencer J. P. E., Harparkash K., Halliwell B., Haeschbach R., Banni S., Dessi M. A., Corongiu F. (1999) Inhibition of peroxinitite dependent DNA base modification and tyrocin nitration by the extra virgin olive oil derived antioxidant hydroxytyrosol. *Free Radic. Biol. Med.*, 26, 762-769.

[96] Martinez-Dominguez E., de la Puerta R., Ruiz-Gutierrez V. (2001) Protective effects upon experimental inflammation models of a polyphenol-supplemented virgin olive oil diet. *Inflamm. Res.*, 50, 102-106.

[97] Perona J. S., Cabello-Moruno R., Ruiz-Gutierrez V. (2006) The role of virgin olive oil components in the modulation of endothelial function. *J. Nutr. Biochem.*, 17, 429-445.

[98] Marrugat J., Covas M. I., Fitó M., Schröder H., Miró-Casas E., Gimeno E., López-Sabater M., Torre R., Farré M. (2004) Effects of differing phenolic content in dietary olive oils on lipids and LDL oxidation. Arandomized controlled trial. *Eur. J. Nutr.*, 43, 140-147.

[99] Visioli F., Galli C., Bornet F., Mattei A., Patelli R., Galli G., Caruso D. (2000) Olive oil phenolics are dose-dependently absorbed in humans. *FEBS Lett.*, 486, 159-160.

[100] Caruso D., Visioli F., Patelli R., Galli C., Galli G. (2001) Urinary excretion of olive oil phenols and their metabolites in humans" *Metabolism,* 50, 1426-1428.

[101] Miro-Casas E., Covas M. I., Farre M., Fito M., Ortuño J., Weinbrenner T., Roset P., Torre R. D. L. (2003) Hydroxytyrosol disposition in humans. *Clin. Chem.*, 49, 945-952.

[102] Soler-Rivas C., Epsin J. C., Wichers H. J. (2000) Oleuropein and related compounds. *J. Sci. Food Agric.*, 80, 1013-1023.

[103] Tuck K. L., Hayball P. J. (2002) Major phenolic compounds in olive oil: Metabolism and health effects. *J. Nutr. Biochem*, 13, 636-644.

[104] Aziz N. H., Farag S. E., Mousa L. A. A., Abo-Zaid M. A. (1998) Comparative antibacterial and antifungal effects of some phenolic compounds. *Microbios*, 93, 43-54.

[105] García J. M., Yousfi K., Mateos R., Olmo M., Cert. A. (2001) Reduction of oil bitterness by heating of olive (*Olea europaea*) fruits" *J. Agric. Food Chem.*, 49, 4231-4235.

[106] Montedoro G. F., Garofolo L., Bertuccioli M., Pannelli G. (1989) *Proceedings of 6th International Flavor Conference*, Rethymnon, Crete, Greece, pp. 881-891.

[107] Uceda M., Hermoso M., García-Ortiz A., Jiménez A., Beltrán G. (1999) Intraspecific variation of oil contents and their characteristics of oils in olive cultivars. *Acta Hortic.*, 474, 659-662.

[108] Ryan D., Prenzler P. D., Lavee S., Antolovich M., Robards K. (2003) Quantitative Changes in Phenolic Content during Physiological Development of the Olive (*Olea europaea*) Cultivar Hardy's Mammoth. *J. Agric. Food Chem.*, 51, 2532-2538.

[109] Gómez-Alonso S., Salvador M. D., Fregapane G. (2002) Phenolic Compounds Profile of Cornicabra Virgin Olive Oil. *J. Agric. Food Chem.*, 50, 6812-6817.

[110] Pinelli P., Galardi C., Mulinacci N., Vinceri F. F., Cimato A., Romani A. (2003) Minor polar compound and fatty acid analyses in monocultivar virgin olive oils from Tuscany. *Food Chem.*, 80, 331-336.

[111] Briante R., Patumi M., Limongelli S., Febbraio F., Vaccaio C., DiSalle A., LaCara F., Nucci R. (2002) Changes in phenolic and enzymatic activities content during fruit ripening in two Italian cultivars of *Olea europaea* L. *Plant Sci.*, 162, 791-798.

[112] Amiot M. J., Fleuriet A., Macheix J. J. (1989) Accumulation of oleuropein derivatives during olive maturation. *Phytochemistry*, 28, 67-69.

[113] Pannelli G., Servili M., Selvaggini R., Baldioli M., Montedoro G. F. (1994) Effect of agronomic and seasonal factors olive (*Olea europaea* L.) production on the qualitative characteristics of the oil. *Acta Hortic.*, 356, 239-244.

[114] Ismail A. S., Stavroulakis G., Metzidakis J. (1999) Effect of irrigation on the quality characteristics of organic olive oil. *Acta Hortic.*, 474, 687-690.

[115] Servili M., Piacquadio P., DeStefano G., Taticchi A., Sciancalepore V. (2002) Influence of a new crushing technique on the composition of the volatile compounds and related sensory quality of virgin olive oil. *Eur. J. Lipid Sci. Technol.*, 104, 483-489.

[116] Bianco A. D., Piperno A., Romeo G., Uccella N. (1999) NMR Experiments of Oleuropein Biomimetic Hydrolysis. *J. Agric. Food Chem.*, 47, 3665-3668.

[117] Lo Scalzo R., Scarpati M. L. (1993) A New Secoiridoid from Olive Wastewaters" *J. Nat. Prod.*, 56, 621-623.

[118] Servili M., Montedoro G. F. (2002) Contribution of phenolic compounds to virgin olive oil quality. *Eur. J. Lipid Sci. Technol.*, 104, 602-613.

[119] Esti M., Cinquanta L., La Notte E. (1998) Phenolic Compounds in Different Olive Varieties. *J. Agric. Food Chem.*, 46, 32-35.

[120] Amiot M. J., Fleuriet A., Macheix J. J. (1986) Importance and evolution of phenolic compounds in olive during growth and maturation. *J. Agric. Food Chem.*, 34, 823-826.

[121] Servili M., Baldioli M., Montedoro G. F. (1994) Phenolic composition of virgin olive oil in relationship to some chemical and physical aspects of malaxation. *Acta Hortic.*, 356, 331-336.

[122] Angerosa F., Mostallino R., Basti C., Vito R. (2001) Influence of malaxation temperature and time on the quality of virgin olive oils. *Food Chem.*, 72, 19-28.

[123] Di Giovacchino L., Sestili S., Di Vincenzo D. (2002) Influence of olive processing on virgin olive oil quality. *Eur. J. Lipid Sci. Technol.*, 104, 587-601.

[124] Sciancalepore V. (1985) Enzymatic Browning in Five Olive Varieties. *J. Food Sci.*, 50, 1194-1195.

[125] Servili M., Baldioli M., Selvaggini R., Mariotti F., Federici E., Montedoro G. F. (1998) Proceedings of the 13th International Symposium on Plant Lipids, Sevilla, Spain, p. 307.

[126] Montedoro G. F., Baldioli M., Servili M. (2001) Estrazione dell'olio vergine da paste denocciolate. *Olivo Olio*, 4, 28-32.

[127] García A., Brenes M., Martínez F., Alba J., García P., Garrido A. (2001) High-performance liquid chromatography evaluation of phenols in virgin olive oil during extraction at laboratory and industrial scale. *J. Am. Oil Chem. Soc.*, 78, 625-629.

[128] Di Giovacchino L., Costantini N., Serraiocco A., Surricchio G., Basti C. (2001) Natural antioxidants and volatile compounds of virgin olive oils obtained by two or three-phases centrifugal decanters. *Eur. J. Lipid Sci. Technol.*, 103, 279-285.

[129] De Stefano G., Piaquadio P., Servili M., Di Giovacchino L., Sciancalepore V. (1999) Effect of the extraction systems on the phenolic composition of virgin olive oils. *Fat/Lipids*, 101, 328 332.

[130] Amirante P., Catalano P., Amirante R., Montel G., Dugo G., LoTurco V., Baccioni L., Fazio D., Mattei A., Marotta F. (2001) Estrazione da paste denocciolate. *Olivo Olio*, 43, 48-55.

In: Virgin Olive Oil
Editor: Antonella De Leonardis

ISBN: 978-1-63117-656-2
© 2014 Nova Science Publishers, Inc.

Chapter 10

CHEMOPREVENTIVE ACTIVITIES OF HYDROXYTYROSOL: THE MAJOR PHENOL ALCOHOL OF EXTRA-VIRGIN OLIVE OIL

Patrizia Rosignoli[1,], Maria Vittoria Sepporta[1], Raffaela Fuccelli[1] and Roberto Fabiani[1]*
[1]Dipartimento di Chimica, Biologia e Biotecnologie,
University of Perugia, Italy

ABSTRACT

The beneficial role of extra-virgin olive oil on human health has recently been attributed to the presence of minor components, with a particular interest in phenolic compounds. The phenolic composition of olive oil is complex and includes hydroxytyrosol (3,4-dihydroxyphenylethanol: HT), tyrosol (hydroxyphenylethanol), the dialdehydic form of elenolic acid linked to hydroxytyrosol or tyrosol (oleocanthal), oleuropein aglycon and lignans. Olive oil phenols, in particular HT, have been carefully studied over the last few years and it appears that this phenol could help to prevent chronic-degenerative diseases such as cardiovascular diseases and cancer due to its antioxidant faculties. However, the biological effects of olive oil polyphenols are not limited to their antioxidant ability. Indeed, several researchers have recently demonstrated that olive oil polyphenols are also able to modulate both gene expression and various pathways involved in the regulation of many physiological and pathological conditions. This study illustrates our recent findings on the in-vitro chemo-preventive activities of HT. We have demonstrated that HT, at relatively high doses ($100\mu M$), is able to inhibit proliferation and to arrest cell cycle progression through the up-regulation of the cyclin-dependent, protein-kinase inhibitors p21 and p27 on HL60 promyelocytic cell line. These effects were associated to the induction of differentiation and apoptosis. The initial stress signal responsible for these effects was the HT-induced extracellular production of H_2O_2. HT was also able to inhibit the proliferation of human tumour cells from the colon, prostate and breast, indicating its important role in the prevention of the promotion phase of carcinogenesis. Moreover, we have observed through the comet assay

* E-mail: patrizia.rosignoli@unipg.it.

that at lower doses (1-10μM) HT is able to inhibit the oxidative DNA damage in human leukocytes thus suggesting a preventive mechanism in the initiation phase of carcinogenesis. These results underline that HT is able to induce either pro-oxidant or anti-oxidant effects which depend on exposition doses. Furthermore, we demonstrated that HT (50-100μM) was able to prevent LPS-mediated COX2 induction on human monocytes at both mRNA and protein levels. These effects were associated with a significant reduction of PGE2 accumulation in the culture medium. Moreover, in the same experimental conditions HT increased the release of TNFα by monocytes. All together, our in-vitro data supports the hypothesis that olive oil phenols may be an important element in the cancer prevention properties of the Mediterranean diet.

Keywords: Chemoprevention, hydroxytyrosol, olive oil

INTRODUCTION

The traditional Mediterranean diet is rich in fruit, vegetables, fish, and whole grain. It is also characterized by a high intake of olive oil as the main source of fat. Several epidemiological studies have evidenced an inverse correlation between a greater adherence to the Mediterranean diet and the incidence of major chronic diseases, such as atherosclerosis, cardiovascular disease and certain types of cancer [1]. The beneficial effects of the Mediterranean diet on human health have been partly attributed to virgin olive oil consumption [2]. It has been proven that olive oil intake is associated with a reduced risk of cancer in different sites such as breast, endometrium, prostate and the gastrointestinal tract [3].

Traditionally, many of the beneficial properties associated with virgin olive oil have been ascribed to its high oleic acid content. However, it is becoming increasingly evident that the benefits associated with the consumption of virgin olive oil may be due to its minor compounds such as polyphenols which are present in small amounts in all the other plant oils [3].

Hydroxytyrosol (HT), tyrosol, secoiridoids and lignans are the most important phenols found in olive oil. Among the olive oil phenols, HT has been carefully studied because of its interesting biological activities [4] which have received growing interest in the last few years. Indeed, 943 papers have been published on this topic (Figure 1 A and B) since 1988 when it appeared for the first time in a scientific publication.

As expected, most of the studies were carried out in Spain and Italy, since these countries have a long tradition of olive cultivation and olive oil production and therefore consume large quantities of olive oil (Table1).

Unlike other phenols, which can be obtained exclusively from natural sources, HT can also be produced by means of synthetic procedures with at a lower production cost. This fact has lead to extensive research aimed at investigating its biological effects and extending the use of HT as a nutraceutical and in the cosmetic and food industry [5].

Since cardiovascular diseases are the first cause of death in the western world, most of the studies have focused on the effects of HT on cardiovascular system. However, over the last few decades more and more attention has been paid to the role played by HT in cancer prevention. It is believed that six hallmarks are essential for cancer development: sustained proliferative signalling, insensitivity to antigrowth signals, limitless replicative potential, evasion of apoptosis, sustained angiogenesis, and tissue invasion and metastasization [6]. Furthermore, immune inflammatory cells and inflammation have recently been added as the 7th cancer hallmark: indeed, it has been demonstrated that inflammatory cells may be involved in tumour promotion [7].

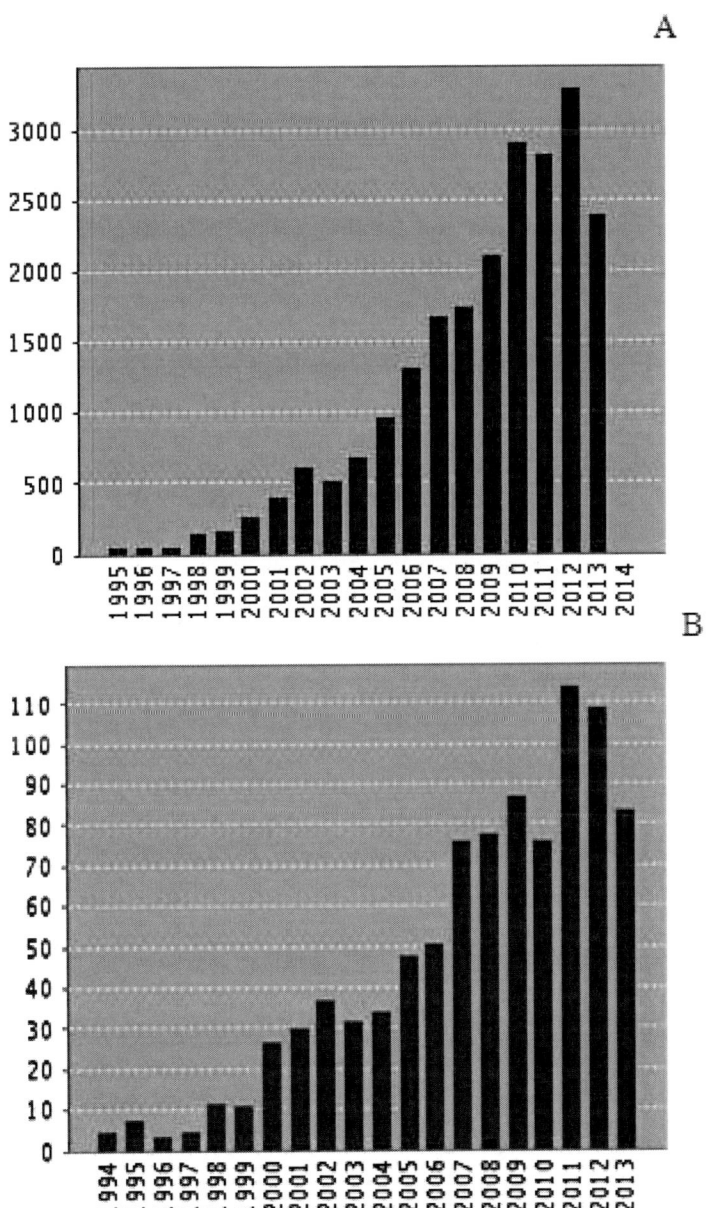

Data were obtained from Web of Science site: Science citation index 1988-2013 (Thomson Reuters).

Figure 1. Scientific interest on Hydroxytyrosol over time. Number of times the word "Hydroxytyrosol" is cited in scientific publications in each year (Figure 1A). Number of publications on "Hydroxytyrosol" in each year (Figure 1B).

Table 1. Number of studies on Hydroxytyrosol carried out in various nations

Field: Countries/Territories	Record Count	% Of total (943 record)
SPAIN	287	30.435
ITALY	285	30.223
TUNISIA	80	8.484
GREECE	60	6.363
FRANCE	46	4.878
USA	45	4.772
PEOPLES R CHINA	35	3.712
PORTUGAL	32	3.393
ENGLAND	29	3.075
JAPAN	29	3.075
GERMANY	27	2.863
AUSTRALIA	25	2.651
TURKEY	22	2.333
NETHERLANDS	20	2.121
SWITZERLAND	16	1.697
MOROCCO	14	1.485
BELGIUM	9	0.954
IRAN	9	0.954
ISRAEL	9	0.954
BRAZIL	7	0.742
FINLAND	7	0.742
SCOTLAND	7	0.742
CANADA	6	0.636
IRELAND	6	0.636
SOUTH KOREA	6	0.636
SINGAPORE	5	0.530
INDIA	4	0.424
SAUDI ARABIA	4	0.424
ARGENTINA	3	0.318
AUSTRIA	3	0.318
CROATIA	3	0.318
DENMARK	3	0.318
CYPRUS*	2	0.212

*CZECH REPUBLIC, HUNGARY, JORDAN, MALAYSIA, NEW ZEALAND, NORTH IRELAND, ROMANIA, SLOVAKIA, SLOVENIA, SWEDEN, TAIWAN, VENEZUELA: Record Count =2; % of Total (943 records) 0.212.

Data were obtained from Web of Science site: Science citation index 1988-2013 (Thomson Reuters).

HT is capable of counteracting most of the above mentioned cancer hallmarks as shown by various authors. In fact, it has been proven that HT has anti-proliferative and pro-apoptotic activities as well as its anti-inflammatory effects and that it is also a potent antioxidant [8, 9]. Moreover, a number of studies have been carried out in order to comprehend the mechanisms through which HT exerts its chemo-preventive activities [10]. Most of the studies were carried out in vitro using breast, colon and, to a lesser extent, lung and prostate tumor cell lines, while less is known regarding its in vivo chemo-preventive effects. Furthermore, there is a lack of data regarding the anticancer effects of HT on humans since all the studies concerned the effects of polyphenol-rich olive oil or HT on cardiovascular diseases [11, 12].

In Vitro Studies

The chemo-preventive effects of HT can be summarized as follows:

1) Antioxidant activities;
2) Prevention of DNA damage;
3) Anti-proliferative and pro-apoptotic effect;
4) Inhibition of angiogenesis;
5) Regulation of inflammation.

1) Antioxidant activities: since the first paper on this subject was published in 1988 [13] more and more publications have aimed at investigating the HT antioxidant potential [14].

HT may act both as a direct antioxidant and as an indirect antioxidant. Due to its o-dihydroxy structure, HT acts directly as an efficient chelator of transition metals inhibiting oxygen-free radical production; in addition it is able to scavenge both superoxide anions generated by either human polymorphonuclear cells or the xanthine/xanthine oxidase system and hypochlorous acid, a potent oxidant produced in vivo at the inflammation site [15,16].

Regarding the protective effect of HT toward H_2O_2-induced damage, several studies have demonstrated that:

- HT (50-100 µM) in red blood cells was able to reduce haemolysis and membrane lipo-peroxidation [17].
- HT (10-100 µM) in human activated neutrophils possessed a clear scavenging activity directed towards H_2O_2 [18].
- HT (6.25-100 µM) in the immortalized non-tumorigenic MCF10A human breast epithelial cell line reduced both the basal level of ROS and H_2O_2-induced ROS levels [19].
- HT interacting with important intracellular signalling pathways, as MAP kinase and PI3 kinase, and reducing the ERK1/2, JNK1/2/3 and Akt phosphorylation, was capable of protecting renal cells against oxidative injury induced by H_2O_2 [20].

In addition, in M14 UVA- irradiated melanoma cell line, HT (0-400 µM) prevented the increase of some typical oxidative stress markers, such as lipid peroxidation products (TBARS), 2'7'-dichlorofluorescein (DCF) fluorescence intensity and altered L-isoAsp residues [21]. The antioxidant potential of HT concerning the generation of reactive oxygen species by the faecal matrix has been also investigated. The data show that HT is a potent inhibitor of free radical generated in abundance by the faecal matrix [22].

The indirect antioxidant effect of HT regards its ability to regulate enzymes and signalling molecules implicated in the control of oxidative stress. It has been demonstrated that:

- In HepG2 cells HT induced antioxidant/detoxificant enzymes and Nrf2 translocation via extracellular regulated kinases and phosphatidylinositol-3-kinase/protein kinase B pathways [23]. Goya and coll. also found that pre-treatment of HepG2 cells with 10-40 μM of HT for 2 or 20 h completely prevented cell damage, the decrease of reduced glutathione and the increase of malondialdehyde evoked by tertbutylhydroperoxide (t-BOOH) [24].
- In retinal pigment epithelial cells HT (100 μM) activated the nuclear factor-E2-related factor-2/antioxidant responsive element (Nrf2/ ARE) pathway, resulting in the increased transcription of both antioxidant enzymes (e.g. Cu/ZnSOD, MnSOD, peroxiredoxin 3 and 5, thioredoxin-2) and phase II detoxifying enzymes (e.g. GSH S-transferase, NAD(P)H:quinone oxidoreductase 1 or GSH reductase) [25].

2) Prevention of DNA damage: DNA damage represents one of the main events leading to cancerogenesis and the role played by HT in its prevention has been well documented. In fact, it was found that:

- HT (<50 μM) prevented ONOO-induced DNA damage and tyrosine nitration by scavenging peroxynitrite [26].
- In the human skin keratinocyte HaCaT cell line exposed to UVB, HT significantly reduced DNA strand breaks by decreasing intracellular ROS formation and 8-hydroxydeoxyguanosine levels [27].
- In various t cancer cell types such as those derived from gut [16], blood [28], prostate [29], and breast [19], HT, at potential physiological concentrations (1-100 μM), protected against oxidative damage induced by H_2O_2.

3) Anti-proliferative and pro-apoptotic effects: The inhibition of cancer-cell proliferation is a common goal of chemotherapies. HT is able to inhibit proliferation and promote apoptosis in several tumour-cell lines such as those derived from the colon, breast, liver and prostate. These effects are mediated by various mechnisms. For example, in HT29 human colon adenocarcinoma cell line, HT induced significant changes in cell cycle distribution also causing an increase in the number of apoptotic cells. The possible mechanisms behind these effects could be the up-regulation of both p53 and the peroxisome proliferator-activated receptor gamma (PPARγ) expression, and the decrease of the HIF-1α protein expression [30]. It has also been demonstrated that the arrest of growth and the induction of apoptosis in HT29 cells can be mediated by a prolonged stress of the endoplasmic reticulum (ER), which activates the unfolded protein response (UPR), including the Ire1/XBP-1/GRP78/Bip and PERK/eIF2alpha arms [31]. In addition, treatment of HT29 cells with HT led to: over-expression of the pro-apoptotic factor CHOP/GADD153 and persistent activation of the Jun-NH2-terminal kinase/activator protein-1 signalling pathway; activation of the extracellular signal-regulated kinase 1/2 and Akt/PKB pro-survival factors by altering their phosphorylation status; inactivation of the phosphorylation of the nuclear factor inhibitor-kappaB kinase which in turn inhibits tumour necrosis factor-alpha-induced nuclear factor-kappaB activation. Finally, HT specifically activates PP2A, which plays a key-initiating role in various pathways that lead to the apoptosis of HT29 colon cancer cells [31].

HT showed anti-proliferative effects and apoptosis inducing activity in another human colorectal cancer cell line, SW620, by suppressing the fatty acid synthase activity [32]. Corona and coll. also demonstrated that in Caco2 colon adenocarcinoma cells, HT induced a block in the G2/M phase of the cell cycle. These anti-proliferative effects were preceded by a strong inhibition of extracellular signal-regulated kinase (ERK) 1/2 phosphorylation and by a downstream reduction of cyclin D1 expression, rather than inhibition of p38 activity and cyclooxygenase-2 (COX-2) expression [33].

It has been proved that by using Affymetrix microarrays, HT inhibited the SW620 cell proliferation and arrested the cell cycle by up-regulating p21 and CCNG2, and by down-regulating the CCNB1 protein expression. HT also up-regulated BNIP3, BNIP3L, PDCD4 and ATF3 which are thespecific genes involved in apoptosis and activated caspase-3 [34]. Moreover, this olive phenol enhances carcinogen detoxification by increasing the activity of xenobiotic metabolizing enzymes UGT1A10 and CYP1A1, [34].

In epato-carcinoma cell lines HepG2 and Hep 3B following treatment with increasing doses of HT (0-30-80-100-200 μM), cell proliferation was inhibited in a dose-dependent manner as a consequence of a down regulation of both FAS and FPPS (farnesyl diphosphate synthase), the key lipogenic enzymes which are necessary for tumour growth and survival [35, 36].

In the breast cancer cell line MCF-7, Sirianni and coll. showed that HT was able to arrest proliferation by inhibiting estrogen-dependent receptor kinase (ERK1/2) involved in uncontrolled tumour cell growth [37]. The same authors underlined that this inhibition could be mediated by a G-protein-coupled receptor named GPER/GPR30 [38]. The cytotoxic effect of HT on MCF-7 was also confirmed by Bouallagui and coll. who showed that the growth inhibition of MCF-7 cells was due to the cell cycle arrest in the G0/G1 phase and it was mediated by a down-expression of the peptidyl-prolyl cis-trans isomerase Pin1 which in turn decreased the level of a G1 key protein Cyclin D1 [39].

In the same MCF-7 breast cancer cell line Han and coll. recently found that HT (324 μM) inhibited cell proliferation through the arrest of cell cycle inn G0/G1 phase and the induction of apoptosis [40].

In human prostate cancer cells PC-3, after treatment with HT (80 μM) the reduction of cells viability was coupled with a significant increase in superoxide anion production, apoptosis activation, mitochondrial dysfunction, defects in autophagy, and activation of MAP kinases. Moreover, the addition of ROS scavengers (i.e. NAC, catalase, pyruvate, SOD) to the growth media prevented the HT induced cell viability loss, suggesting that extracellular ROS could be involved in the anti-proliferative effect of HT in prostate cancer cells [41].

In the human promyelocytic leukemia HL60 cell line, as well as in resting and activated peripheral blood lymphocytes, Della Ragione and coll. demonstrated that HT (50 and 100 μM) reduced proliferation by arresting cell cycle and inducing apoptosis. These effects were mediated by both the release of cytochrome c, which in turn activated the effector caspase-3, and the activation of the c-jun NH2-terminal kinase (JNK) transduction pathway which, through phosphorylation inactivated the anti-apoptotic protein Bcl-2 [42].

4) Inhibition of angiogenesis: the altered balance between stimulators and inhibitors of angiogenesis plays an important role in the development of several pathologies, including inflammatory diseases and cancer. In this regard it has been observed that HT is able to reduce both endothelial cell tube formation on matrigel and migration in wound healing assays. The reduced angiogenesis was coupled with the inhibition of the PMA-induced COX-2 expression, prostanoid production, MMP-9 release and gelatinolytic activity. These effects were also associated to a significant reduction of intracellular reactive oxygen species levels and to the activation of the redox-sensitive transcription factor nuclear factor NF-kB [43].

5) Regulation of inflammation: although most studies suggest that the anti-inflammatory effects of HT are mostly mediated by its antioxidant activity, other mechanisms may also be involved. In fact, Maiuri and coll. demonstrated that in a dose-dependent manner HT (100-200 μM) down-regulated the production of PGE2 in LPS-stimulated J774 murine macrophages and this effect was coupled with an inhibition of COX-2 gene expression [44].

Richard and coll. showed that HT (25 μM) inhibited PGE2 as well as cytokines (IL-12, IL-6, IL-1, and TNFα), chemokines (CXCL10/IP-10, CCL2/MCP1, and MIP-1β) and enzymes (MMP-9, PGE2 synthase, and inducible nitric oxide synthase, iNOS) production in LPS-stimulated RAW264.7 murine macrophages [45]. Similarly Giner and coll. reported that in LPS-stimulated peritoneal macrophages, treatment with HT (100 μM) had an inhibiting effect on cytokine (IL-1β, IL-6, and TNFα) production [46].

A similar anti-inflammatory activity was also seen in LPS-stimulated THP-1 monocytes, where HT (50-100 μM) inhibited TNFα production and secretion [47]. In the same cell line Zhang and coll. showed that HT (50-100 μM) suppressed COX-2 expression [48]. The HT-mediated inhibitory production of pro-inflammatory factors in stimulated macrophages could be mediated by the inhibition of some transcriptional factors, including NF-kB [44, 45, 48]. In macrophages, NF-kB regulates the expression of a large variety of genes encoding for immuno-regulatory mediators (e.g. inflammatory cytokines and COX2), which may be implicated in carcinogenesis [49].

Finally, the anti-inflammatory effect of HT was also observed in the epatocarcinoma cell line HepG2 where it induced a significant decrease of IL-6 production [32,35]. Although there is a low concentration of HT in olive oil, a daily intake of olive oil could lead to levels producing similar effects to NSAIDs.

In Vivo Studies

Animal Studies

Several animal studies have proved that HT protects against chemical-induced carcinogenesis. Granados-Principal and coll. demonstrated that HT was able to exert an anti-proliferative and pro-apoptotic effects on mammary tumours in Sprague–Dawley rats caused by Dimethylbenz[α]anthracene. These effects were coupled with an inhibition of cell growth by altering several genes associated with cell proliferation, apoptosis and the Wnt signalling pathway [50].

Moreover, the chemo-preventive effect of HT has been also investigated on cholangiocarcinoma (CCA) xenografts in mice. It was shown that this phenol significantly inhibited the growth of CCA through the G2/M phase cell cycle arrest and apoptosis induction. Moreover, the inhibition of phosphor-ERK was also observed [51].

Regarding the in vivo anti-inflammatory effects of HT, Bitler and coll. found that in LPS-treated mice, olive vegetation water highly enriched in HT (125 mg/mouse; 500 mg/kg) reduced scrum TNFα levels [52].

Another study on anti-inflammatory effects of HT was conducted by Gong and coll. using rats with acute inflammation induced by intravenous injections of carrageenan. The rats received different dosages (100, 250, and 500 mg/kg of body weight) of a preparation called HT-20 in which the main ingredient was HT (22%). In rats treated with HT20, inflammation was significantly reduced as demonstrated by the reduction of both the IL-1β and TNFα levels [53].

More recently, Sánchez-Fidalgo and coll. reported that HT-enriched extra-VOO attenuated dextran sulfate sodium (DSS)-induced chronic colitis in mice and decreased the COX2 and iNOS expression down regulating the p38MAPK [54].

Studies on animal models have demonstrated that HT is able to exert an antioxidant activity also in vivo. Indeed, following ingestion HT increased the antioxidant capacity of plasma [55] and prevented passive smoking-induced oxidative stress [56]. Moreover, the effectiveness of a HT-rich extract for attenuating $Fe2^{+}$ and nitric oxide (NO)-induced cytotoxicity in murine-dissociated brain cells was demonstrated [57].

It should be noted that most of the experiments described above were performed with mixture of olive phenols among which HT proved to be the most active ingredient. Therefore the synergy of HT with other olive phenols cannot be excluded at present.

DISCUSSION

Our research team has been studying the chemo-preventive activities of extra-virgin olive oil polyphenols for a long time. In particular, several in vitro studies have been carried out in order to highlight HT's anticancer potential and the results obtained are summarized below.

1) Anti-Proliferative and Pro-Apoptotic Effects

Data from literature shows that plant phenols such as resveratrol and epigallocathechin-3-gallate, found respectively in grapes and green tea, possess anticancer properties. Indeed, it has been proved that treatment with these compounds inhibited cell growth due to cell cycle arrest and apoptosis induction in several tumor cell lines [58, 59].

It is therefore interesting to investigate whether HT could have the same chemo-preventive effects on tumor cell lines from various organs, such as colon (HCT116, SW480), prostate (PC3 and LnCap), breast (MDA and MCF-7) and an human promielocytic cell line (HL60).

In order to evaluate cell viability after treatment over time (24-48-72-96 h) with increasing doses of HT (0-25-50-100 μM), trypan blue exclusion dye and MTT assays were used for non-adherent HL60 cells and for adherent cell lines, respectively. The results obtained have been reported in several publications [60-62]. The anti-proliferative effects of HT on different cell lines are summarized in Table 2, which indicates the percentage of proliferating cells after 72 h of treatment with different doses of HT. HT was able to inhibit the proliferation of all the cell lines tested, even though the effect was stronger in HL60 cells (83% of growth inhibition at 100 μM) whereas the prostate derived cells (LnCap and PC3) proved to be the least sensitive (about 33% of growth inhibition at 100 μM). Furthermore, breast cell lines proved to be the most sensitive to HT treatment among the solid tumor-derived cell lines.

The effect of HT on the cell cycle distribution was studied on HL60 in order to fully comprehend the mechanisms behind cell-growth inhibition t. The inhibition of HL60 proliferation proved to be associated to a block of the cell cycle in the G1 phase, to a decrease of Cdk6 and to an increase of Cdk inhibitors p21WAF/Cip1 and p27Kip1 [63]. Since ROS are implicated in the regulation of proliferation, differentiation and apoptosis [64], we hypothesised that HT, acting as pro-oxidant, could enhance ROS generation. Indeed, as demonstrated by other authors, molecules such as β-carotene [65], ascorbate [66] and quercetin [67], which at low concentrations act as antioxidants, exert a pro-oxidant activity inhibiting proliferation in high doses. In order to test this hypothesis, HL60 cells were treated with HT 100 μM in presence of N-acetylcysteine (NAC), which is a widely used scavenger of free radicals for demonstrating the involvement of ROS in various cell systems and experimental conditions [68]. The inclusion of NAC in the culture medium partially reduced the anti-proliferative activity of HT (15% with HT vs. 35% with HT+NAC) after 72 h of incubation (Table 3) [69].

The inhibition of cell growth was associated to the induction of apoptosis in HL60 exposed to increasing concentration of HT (0-25-50-100 μM) for 24 h and was assessed using various methods: fluorescence microscopy, subG1 peak detection and annexin V staining. NAC was added to culture medium in order to determine whether ROS were also involved in the pro-apoptotic activity of HT, the results of which are summarized in Table 4. The pro-apoptotic effect of HT on HL60 was dose-dependent and a maximum effect was reached at 100μM as proved by all three methods used, although the highest values were obtained with the fluorescence microscopy method. As expected, the presence of NAC in the culture medium effectively inhibited apoptosis induced by HT in HL60 cells after 24 h of treatment (Table 4) [69]. These results suggest that HT at 100 μM is able to evoke an oxidative stress leading to a growth arrest as well as apoptosis induction. It is important to note that HT in non-tumorigenic human cells failed to inhibit proliferation and induce apoptosis [60].

The effect of HT on HL60 cell differentiation was also examined.

Table 2. The antiproliferative effect of HT on various tumour cell lines after 72 h of treatment

HT, μM	Proliferating cells, %						
	HL60	MDA	MCF-7	PC3	LnCap	SW480	HCT116
0	100	100	100	100	100	100	100
25	67	97	90	84	96	100	90
50	42	55	65	93	82	83	62
75	29	33	38	72	76	75	43
100	17	13	20	63	67	48	21

Table 3. Effect of NAC on the HT anti-proliferative ability toward HL60 cells after 72 h of treatment

	Proliferating cells, %	
HT, µM	- NAC	+ NAC
0	100	85
100	15	35

Table 4. Effects of HT and NAC on apoptosis of HL60 cells after 24 h of treatment

	Apoptotic cells, %					
	Fluorescence microscopy		SubG1 pick		Annexin V	
HT, µM	- NAC	+ NAC	- NAC	+ NAC	- NAC	+ NAC
0	5	8 [a]	13	10	5	8
25	25	n.d.	n.d.	n.d.	n.d.	n.d.
50	62	n.d.	n.d.	n.d.	n.d.	n.d.
100	73	20	49	11	48	14

n.d. not detected

Table 5. Effect of catalase (CAT, 100U/ml) on HT induced apoptosis in HL60 and on H_2O_2 accumulation in the culture medium

	Apoptotic cells, % (HL60)		H_2O_2 in culture medium (µM)	
HT, µM	- CAT	+ CAT	- CAT	+ CAT
0	8	12	1	1.5
100	75	20	5	0.8

The cells were treated with HT 50 and 100 µM, harvested after 24, 48 and 72 h, and then subjected to the NBT reduction assay. The ability of cells in reducing NBT, a functional marker of granulocyte/monocyte differentiation of HL60, increased significantly after 48 and 72 h of treatment with HT at both concentration levels tested. The NBT assay measures the PMA-induced superoxide ion production as a functional marker for granulocyte/monocyte differentiation of HL60 cells. This assay cannot distinguish whether the cells acquire a granulocyte or monocyte phenotype. However, microscopic observation showed that cells did not form aggregates and did not adhere to the flask. These results suggest that HT treatment induced HL60 cells to differentiate along the granulocytic lineage [63].

For the first time we have demonstrated that at high doses HT exerted (100 µM) its chemo preventive effects by means of a pro-oxidant action consisting in the generation of hydrogen peroxide (H_2O_2) in the culture medium.

As shown in Table 5, adding the H2O2-scavenging enzyme catalase to the culture medium efficiently prevented this effect. By treating the HL60 cells with HT 100 μM in the presence of CAT (100 U/mL) the percentage of apoptotic cells was reduced, suggesting that H_2O_2 was the main compound responsible for the HT-induced apoptosis in HL60 cells. In addition, since CAT is not cell permeable, the results suggest that the H_2O_2 is produced in the extracellular spaces. In order to test this hypothesis we measured the concentration of H_2O_2 in the HL60 culture medium after 24 h of incubation with HT. As shown in Table 5, HT caused an evident accumulation of H_2O_2 (5 μM) in the HL60 culture medium which was totally inhibited by the addition of CAT [69].

Further experiments were carried out by incubating HT with various cellular densities in order to determine whether the HL60 cells were involved in the HT-mediated H_2O_2 generation.

The results reported in Table 6 show that the HL60 cells drastically reduced the accumulation of H_2O_2 in the medium after 24 h of incubation. The H_2O_2 accumulation in the culture medium and the percentage of apoptotic cells were inversely correlated to the cell number (Table 6). Conversely, the percentage of apoptotic cells and the concentration of H_2O_2 in the medium were directly and significantly correlated (Table 6) [69].

In presence of exogenously added H_2O_2 (50 μM) HL60 cells efficiently eliminated this compound from the medium with a kinetic highly dependent on cell density (Table 7) [69].

Table 6. Effect of various cellular densities on H_2O_2 accumulation and apoptosis of HL60 cells

	HT 100μM	
Cells/well	**H_2O_2 in culture medium (μM)**	**% of apoptotic cells (HL60)**
100000	8	100
200000	3.5	75
400000	2	40
800000	1	35

Table 7. Effect of various cellular densities on the kinetics of H_2O_2 elimination from culture medium

	Time, minutes				
	0	**5**	**15**	**30**	**60**
Cells/well	H_2O_2, μM				
0	50	48	45	42	37
100000	50	37	25	18	5
200000	50	32	18	12	8
400000	50	18	8	5	2
800000	50	5	2	2	0

In order to identify whether the culture medium components influenced H_2O_2 accumulation, the time-dependent H_2O_2-producing activity of HT (100 μM) in various culture media (RPMI, MEM, DMEM, McCOY'S) was detected in the absence of cells. The H_2O_2 production rate was similar in RPMI, MEM and McCOY media while the formation of H_2O_2 was completely prevented in D-MEM. As the analysis of the D-MEM components revealed that it is the only medium containing pyruvate (110 mg/L, 1 mM), we hypothesized that this compound could be responsible for the reduction in H_2O_2 concentration. Indeed, RPMI medium enriched with 1 mM pyruvate significantly prevented H_2O_2 accumulation.

Consequently, when HL60 was exposed to HT 100μM in RPMI containing pyruvate apoptosis was reduced but not completely inhibited suggesting that other mechanisms could be involved in the pro-apoptotic activity of HT [61] in addition to the H_2O_2-releasing activity.

2) HT-Preventive Activity on Oxidative DNA Damage

In order to analyse the effect of HT on oxidative DNA damage, human cells were exposed to oxidative stress induced by both H_2O_2 and phorbole-myristate-acetate (PMA)-activated monocytes. While in the first case the cells were treated with extreme non-physiological levels of a single oxidant (H_2O_2) in the second case an ex vivo model was used which had recently been developed in our laboratory [70], and is capable of mimicking the physiological condition in which a wide range of oxidative species are continuously produced at low concentrations. The oxidative DNA damage was detected by the highly sensitive comet assay [71]. Briefly: the microscope images obtained at the end of the comet assay and analysed at 400×magnification using a fluorescence microscope (Zeiss, R.G.) equipped with a 50-W mercury lamp, revealed circular shapes (undamaged DNA) and "comet-like" shapes in which the DNA had migrated out from the head to form a tail (damaged DNA).

The extension of each comet was analysed with a computerised image-analysis system (Comet assay II, Perceptive Instruments, UK) which among several other parameters provided the "tail moment", which is considered to be the most directly related parameter to DNA damage. In fact, the tail moment is defined as the product of DNA in the tail and the mean distance of its migration in the tail. The calculation of the extent of DNA damage, which was not homogeneous, was based on the analysis of 100 randomly-selected comets from each slide, divided into 5 classes according to the tail moment (t.m.) values as follows: class 0 (t.m. <1; no damage), class 1 (t.m. 1–5; slightly damaged), class 2 (t.m. 5–10; medium damage), class 3 (t.m. 10–20; highly damaged) class 4 (t.m. >20; completely damaged).

Table 8. Effect of HT on DNA damage of HL60 cells and PBMC induced by H_2O_2 40 μM

HT, μM	DNA damage, AU	
	HL60	PBMC
0	125	150
1	48	90
3	23	47
5	12	23
10	12	9

The overall score expressed in arbitrary units (AU) for each slide ranged from 0 (100% of the comets in class 0) to 400 (100% of the comets in class 4) [71].

HT-Preventive Activity on Oxidative DNA Damage Induced by H₂O₂

HL60 cells and PBMC were treated with H_2O_2 40 μM for 30 min at 37°C in order to induce an appreciable amount of DNA damage without citotoxicity. The data obtained (Table 8) showed that HT in a dose-dependent manner reduced DNA damage in both cell types.

It is important to note that this effect was already significant at the lowest HT dose tested (1μM) [72].

HT-Preventive Activity on Oxidative DNA Damage Induced by PMA-Activated Monocytes

Freshly isolated lymphocytes were co-incubated with monocytes (attached to the bottom of the 96-well plate) both stimulated and not stimulated with PMA 2 μM, a protein kinase C activator. After 1 h of incubation at 37°C, the lymphocytes were removed and assayed for the DNA damage. PMA increased the DNA damage in the lymphocytes and this effect was significantly reduced in the presence of HT 10 μM (Table 9) [72].

3) Modulation of Inflammatory Pathways by HT

Our attention has recently been focused on the study of the HT effect on inflammatory pathways. In particular, we studied the ability of HT in influencing the release of superoxide anions (O2⁻), prostaglandin E2 (PGE2), tumour necrosis factor α (TNFα) and the expression of cyclooxygenase2 (COX2) in freshly-isolated human monocytes from healthy donors. O2 was measured by superoxide dismutase-inhibitable cytochrome c reduction and PGE2 and TNFα production was determined in culture medium with appropriate enzyme immunoassay kits. COX2 expression at the level of mRNA and protein was evaluated by means of quantitative reverse transcription-polymerase chain reaction and Western immunoblotting, respectively.

Superoxide Anion Production in PMA-Stimulated Human Monocytes

In this ex vivo model of human inflammation, HT caused a dose-dependent reduction of the O_2 which became statistically significant ($P<0.05$) at the dose of 100 μM after 24 h of incubation (Table 10). In all experimental conditions shown in Table 10, HT neither reduced cell viability below 80% nor influenced the basal level of O2⁻ production.

Table 9. Effect of HT 10 μM on lymphocyte DNA damage induced by PMA-activated monocytes

HT, μM	DNA damage, AU	
	- PMA	+ PMA
0	12	90
10	14	50

These results suggest that HT could inhibit NADPH oxidase activity, which is the enzyme that catalyses $O2^-$ production. As mentioned above, HT in the RPMI medium generates H_2O_2, therefore it was important to investigate whether H_2O_2 was responsible for the inhibition of NADPH oxidase. For this purpose the production of $O2^-$ by monocytes was measured in the presence of catalase (CAT 100 U/ml). However, CAT did not reduce the O^- production suggesting that this phenomenon was not H_2O_2 mediated [73].

Cyclooxygenase2 (COX2) Expression and PGE2 Production in LPS-Stimulated Human Monocytes

HT acts as a powerful inhibitor of COX2 expression. In fact, in freshly-isolated human monocytes activated with LPS (5 μg/ml) for 24 h, the level of both COX2 mRNA and protein drastically decreased in the presence of HT 50 and 100 μM (Table 11).

Moreover, in the same experimental model, the increase of PGE2 in the culture medium was inhibited in presence of HT, at both 50 and 100 μM (Table 11). Catalase (CAT 100 U/ml) was added to the culture medium in order to verify whether the effect of HT on PGE2 production was mediated by H_2O_2. Since the CAT did not have any effect on PGE2 production we concluded that H_2O_2 was not involved in HT-induced PGE2 reduction [73].

Table 10. Effect of HT on superoxide anion ($O2^-$) production in freshly-isolated human monocytes stimulated with PMA (2 μM)

HT, μM	% $O2^-$ (nmol/h) produced	
	2h	24h
0	100	100
10	95	82
50	82	74
100	80	60

Table 11. Effect of HT on COX2 expression and PGE2 production in LPS-stimulated human monocytes

HT, μM	% COX2 mRNA (qPCR)	% COX2 protein (WB)	% PGE2 Production (ELISA assay)
0	100	100	100
50	45	81	55
100	33	54	29

Table 12. Effect of HT on TNFα production in LPS-stimulated human monocytes

HT, μM	% TNFα production (ELISA assay)
0	100
50	182
100	209

TNFα Production in LPS-Stimulated Human Monocytes

In LPS-activated human monocytes the release of TNFα in culture medium significantly increased. Intriguingly, this effect was enhanced when the cells were treated with increasing doses of HT (Table 12) and was not mediated by H_2O_2 accumulation in the cell culture medium since the addition of CAT did not modify the TNFα levels [73].

CONCLUSION

The studies reported in this chapter indicate that HT possesses several chemo-preventive properties. Indeed, it has been proven that this olive oil phenol counteracts the main hallmarks of cancerogenesis acting as an antioxidant, anti-proliferative and anti-inflammatory agent. It is important to note that this knowledge comes from in vitro studies which did not reproduce the in vivo behaviour, especially concerning the possible level of concentration, but they are important for obtaining a better understanding of the potential biological activities of HT. Hence, more clinical research is required in order to confirm or deny the beneficial effects of olive oil phenols such as HT, on humans making them useful tools for cancer prevention and treatment.

However, due to the absence of toxic effects exerted by HT and its safety profile [74], several industries are already using this compound as an additive in cosmetology and food science.

Finally, it is wrong to believe that one type of food can cure or prevent disease. The secret is to follow a well-balanced and variegated diet whose components have beneficial biological activities.

REFERENCES

[1]   de Lorgeril M., Salen P. (2006) The Mediterranean-style diet for the prevention of cardiovascular diseases. *Public Health Nutr.,* 9(1A):118–23.

[2]   Harwood J. L., Yaqoob P. (2002) Nutritional and health aspects of olive oil. *Eur. J. Lipid Sci. Technol.,* 104:685–97.

[3]   Tutino V., Caruso M. G., Messa C., Perri E., Notarnicola M. (2012) Antiproliferative, antioxidant and anti-inflammatory effects of hydroxytyrosol on human hepatoma HepG2 and Hep3B cell lines. *Anticancer Res.,* 32(12):5371-7.

[4]   Granados-Principal S., Quiles J. L., Ramirez-Tortosa C. L., Sanchez-Rovira P., Ramirez-Tortosa M. C. (2010) Hydroxytyrosol: from laboratory investigations to future clinical trials. *Nutr Rev.,* 68(4):191-206. Review.

[5]   Visioli F., Bernardini E. (2011). Extra virgin olive oil's polyphenols: biological activities. *Curr. Pharm. Des.,* 17(8):786-804. Review.

[6]   Hanahan D., Weinberg R. A. (2000) The Hallmarks of cancer. *Cell,*100:57-70.

[7]   Hanahan D., Weinberg R. A. (2011) Hallmarks of cancer: the next generation. *Cell,* 144(5):646-74. Review.

[8]   Lee K. W., Lee H. J. (2006) The roles of polyphenols in cancer chemoprevention. *Biofactors,* 26(2):105-21. Review.

[9] Cornwell D. G., Ma J. (2008) Nutritional benefit of olive oil: the biological effects of hydroxytyrosol and its arylating quinone adducts. *J. Agric. Food Chem.*, 56(19): 8774-86.

[10] Bernini R., Merendino N., Romani A., Velotti F. (2013) Naturally occurring hydroxytyrosol: synthesis and anticancer potential. *Curr. Med. Chem.*, 20(5):655-70.

[11] Salvini S., Sera F., Caruso D., Giovannelli L., Visioli F., Saieva C., Masala G., Ceroti M., Giovacchini V., Pitozzi V., Galli C., Romani A., Mulinacci N., Bortolomeazzi R., Dolara P., Palli D. (2006) Daily consumption of a high-phenol extra-virgin olive oil reduces oxidative DNA damage in postmenopausal women. *Br. J. Nutr.*, 95(4):742-51.

[12] Covas M. I., Nyyssönen K., Poulsen H. E., Kaikkonen J., Zunft H. J., Kiesewetter H., Gaddi A., de la Torre R., Mursu J., Bäumler H., Nascetti S., Salonen J. T., Fitó M., Virtanen J., Marrugat J. (2006) The effect of polyphenols in olive oil on heart disease risk factors: a randomized trial. EUROLIVE Study Group. *Ann. Intern. Med.*, 145(5):333-41.

[13] Chimi H., Sadik A., Letutour B. (1988) Antioxidant activity of tyrosol, hydroxytyrosol, caffeic acid, oleuropein and BHT in olive oil. *Revue Francaise Des Corps Gras,*. 35 (8-9): 339-344.

[14] Perez-Jimenez F., Alvarez de Cienfuegos G., Badimon L., Barja G., Battino M., Blanco A., Bonanome A., Colomer R., Corella-Piquer D., Covas I., Chamorro-Quiros J., Escrich E., Gaforio J. J., Garcia Luna P. P., Hidalgo L., Kafatos A., Kris-Etherton P. M., Lairon D., Lamuela-Raventos R., Lopez-Miranda J., Lopez-Segura F., Martinez-Gonzalez M. A., Mata P., Mataix J., Ordovas J., Osada J., Pacheco-Reyes R., Perucho M., Pineda-Priego M., Quiles J. L., Ramirez-Tortosa M. C., Ruiz-Gutierrez V., Sanchez-Rovira P., Solfrizzi V., Soriguer-Escofet F., de la Torre-Fornell R., Trichopoulos A., Villalba-Montoro J. M., Villar-Ortiz J. R., Visioli F. (2005) International conference on the healthy effect of virgin olive oil. *Eur. J. Clin. Invest.*, 35(7):421-4. Review.

[15] Visioli F., Galli C. (1998) The effect of minor constituents of olive oil on cardiovascular disease: new findings. *Nutr. Rev.*, 56(5 Pt 1):142-7. Review.

[16] Manna C., Galletti P., Cucciolla V., Moltedo O., Leone A., Zappia V. (1997) The protective effect of the olive oil polyphenol (3,4-dihydroxyphenyl)-ethanol counteracts reactive oxygen metabolite-induced cytotoxicity in Caco-2 cells. *J. Nutr.*, 127(2): 286-92.

[17] Manna C., Galletti P., Cucciolla V., Montedoro G., Zappia V. (1999) Olive oil hydroxytyrosol protects human erythrocytes against oxidative damages. *J. Nutr. Biochem.*, 10(3):159-65.

[18] O'Dowd Y., Driss F., Dang P. M., Elbim C., Gougerot-Pocidalo M. A., Pasquier C., El-Benna J. (2004) Antioxidant effect of hydroxytyrosol, a polyphenol from olive oil: scavenging of hydrogen peroxide but not superoxide anion produced by human neutrophils. *Biochem. Pharmacol.*, 68(10):2003-8.

[19] Warleta F., Quesada C. S., Campos M., Allouche Y., Beltrán G., Gaforio J. J. (2011) Hydroxytyrosol protects against oxidative DNA damage in human breast cells. *Nutrients*, 3 (10):839-57.

[20] Incani A., Deiana M., Corona G., Vafeiadou K., Vauzour D., Dessì M. A., Spencer J. P. (2010) Involvement of ERK, Akt and JNK signalling in H2O2-induced cell injury and

protection by hydroxytyrosol and its metabolite homovanillic alcohol. *Mol. Nutr. Food Res.*, 54(6):788-96.

[21] D'Angelo S., Ingrosso D., Migliardi V., Sorrentino A., Donnarumma G., Baroni A., Masella L., Tufano M. A., Zappia M., Galletti P. (2005) Hydroxytyrosol, a natural antioxidant from olive oil, prevents protein damage induced by long-wave ultraviolet radiation in melanoma cells. *Free Radic. Biol. Med.*, 38(7):908-19.

[22] Owen R. W., Giacosa A., Hull W. E., Haubner R., Spiegelhalder B., Bartsch H. (2000) The antioxidant/anticancer potential of phenolic compounds isolated from olive oil. *Eur. J. Cancer*, 36(10):1235-47.

[23] Martín M. A., Ramos S., Granado-Serrano A. B., Rodríguez-Ramiro I., Trujillo M., Bravo L., Goya L. (2010) Hydroxytyrosol induces antioxidant/detoxificant enzymes and Nrf2 translocation via extracellular regulated kinases and phosphatidylinositol-3-kinase/protein kinase B pathways in HepG2 cells. *Mol. Nutr. Food Res.*, 54(7):956-66.

[24] Goya L., Mateos R., Bravo L. (2007) Effect of the olive oil phenol hydroxytyrosol on human hepatoma HepG2 cells. Protection against oxidative stress induced by tert-butylhydroperoxide. *Eur. J. Nutr.*, 46(2):70-8.

[25] Zhu L., Liu Z., Feng Z., Hao J., Shen W., Li X., Sun L., Sharman E., Wang Y., Wertz K., Weber P., Shi X., Liu J. (2010) Hydroxytyrosol protects against oxidative damage by simultaneous activation of mitochondrial biogenesis and phase II detoxifying enzyme systems in retinal pigment epithelial cells. *J. Nutr. Biochem.*, 21(11):1089-98.

[26] Deiana M., Aruoma O. I., Bianchi M. L., Spencer J. P., Kaur H., Halliwell B., Aeschbach R., Banni S., Dessi M. A., Corongiu F. P. (1999) Inhibition of peroxynitrite dependent DNA base modification and tyrosine nitration by the extra virgin olive oil-derived antioxidant hydroxytyrosol. *Free Radic. Biol. Med.*, 26(5-6):762-9.

[27] Guo W., An Y., Jiang L., Geng C., Zhong L. (2010) The protective effects of hydroxytyrosol against UVB-induced DNA damage in HaCaT cells. *Phytother. Res.*, 24(3):352-9.

[28] Nousis L., Doulias P. T., Aligiannis N., Bazios D., Agalias A., Galaris D., Mitakou S. (2005) DNA protecting and genotoxic effects of olive oil related components in cells exposed to hydrogen peroxide. *Free Radic. Res.*, 39(7):787-95.

[29] Quiles J. L., Farquharson A. J., Simpson D. K., Grant I., Wahle K. W. (2002) Olive oil phenolics: effects on DNA oxidation and redox enzyme mRNA in prostate cells. *Br. J. Nutr.*, 88(3):225-34; discussion 223-4.

[30] Cárdeno A., Sánchez-Hidalgo M., Rosillo M. A., Alarcón de la Lastra C. (2013) Oleuropein, a secoiridoid derived from olive tree, inhibits the proliferation of human colorectal cancer cell through downregulation of HIF-1α. *Nutr. Cancer*, 65(1):147-56.

[31] Guichard C., Pedruzzi E., Fay M., Marie J. C., Braut-Boucher F., Daniel F., Grodet A., Gougerot-Pocidalo M. A., Chastre E., Kotelevets L., Lizard G., Vandewalle A., Driss F., Ogier-Denis E. (2006) Dihydroxyphenylethanol induces apoptosis by activating serine/threonine protein phosphatase PP2A and promotes the endoplasmic reticulum stress response in human colon carcinoma cells. *Carcinogenesis*, 27(9):1812-27.

[32] Notarnicola M., Pisanti S., Tutino V., Bocale D., Rotelli M. T., Gentile A., Memeo V., Bifulco M., Perri E., Caruso M. G. (2011) Effects of olive oil polyphenols on fatty acid synthase gene expression and activity in human colorectal cancer cells. *Genes. Nutr.*, 6(1):63-9.

[33] Corona G., Deiana M., Incani A., Vauzour D., Dessì M. A., Spencer J. P. (2009) Hydroxytyrosol inhibits the proliferation of human colon adenocarcinoma cells through inhibition of ERK1/2 and cyclin D1. *Mol. Nutr. Food Res.*, 53(7):897-903.

[34] Pereira-Caro G., Mateos R., Traka M. H., Bacon J. R., Bongaerts R., Sarriá B., Bravo L., Kroon P. A. (2013) Hydroxytyrosyl ethyl ether exhibits stronger intestinal anticarcinogenic potency and effects on transcript profiles compared to hydroxytyrosol. *Food Chem.*, 138(2-3):1172-82.

[35] Siperstein M. D. (1984) Role of cholesterogenesis and isoprenoid synthesis in DNA replication and cell growth. *J. Lipid Res.*, 25(13):1462-8. Review.

[36] Holstein S. A., Wohlford-Lenane C. L., Hohl R. J. (2002) Isoprenoids influence expression of Ras and Ras-related proteins. *Biochemistry*, 41(46):13698-704.

[37] Sirianni R., Chimento A., De Luca A., Casaburi I., Rizza P., Onofrio A., Iacopetta D., Puoci F., Andò S., Maggiolini M., Pezzi V. (2010) Oleuropein and hydroxytyrosol inhibit MCF-7 breast cancer cell proliferation interfering with ERK1/2 activation. *Mol. Nutr. Food Res.*, 54(6):833-40.

[38] Chimento A., Casaburi I., Rosano C., Avena P., De Luca A., Campana C., Martire E., Santolla M. F., Maggiolini M., Pezzi V., Sirianni R. (2013) Oleuropein and hydroxytyrosol activate GPER/ GPR30-dependent pathways leading to apoptosis of ER-negative SKBR3 breast cancer cells. *Mol. Nutr. Food Res.*, 2013 Sep 9. doi: 10.1002/mnfr.201300323. [Epub ahead of print]

[39] Bouallagui Z., Han J., Isoda H., Sayadi S. (2011) Hydroxytyrosol rich extract from olive leaves modulates cell cycle progression in MCF-7 human breast cancer cells. *Food Chem. Toxicol.*, 49(1):179-84.

[40] Han J., Talorete T. P., Yamada P., Isoda H. (2009) Anti-proliferative and apoptotic effects of oleuropein and hydroxytyrosol on human breast cancer MCF-7 cells. *Cytotechnology*, 59(1):45-53.

[41] Luo C., Li Y., Wang H., Cui Y., Feng Z., Li H., Li Y., Wang Y., Wurtz K., Weber P., Long J., Liu J. (2013) Hydroxytyrosol promotes superoxide production and defects in autophagy leading to anti-proliferation and apoptosis on human prostate cancer cells. *Curr. Cancer Drug Targets*, 13(6):625-39.

[42] Della Ragione F., Cucciolla V., Borriello A., Della Pietra V., Manna C., Galletti P., Zappia V. (2000) Pyrrolidine dithiocarbamate induces apoptosis by a cytochrome c-dependent mechanism. *Biochem. Biophys. Res. Commun.*, 268(3):942-6.

[43] Scoditti E., Calabriso N., Massaro M., Pellegrino M., Storelli C., Martines G., De Caterina R., Carluccio M. A. (2012) Mediterranean diet polyphenols reduce inflammatory angiogenesis through MMP-9 and COX-2 inhibition in human vascular endothelial cells: a potentially protective mechanism in atherosclerotic vascular disease and cancer. *Arch. Biochem. Biophys.*, 527(2):81-9.

[44] Maiuri M. C., De Stefano D., Di Meglio P., Irace C., Savarese M., Sacchi R., Cinelli M. P., Carnuccio R. (2005) Hydroxytyrosol, a phenolic compound from virgin olive oil, prevents macrophage activation. *Naunyn Schmiedebergs Arch. Pharmacol.*, 371(6): 457-65.

[45] Richard N., Arnold S., Hoeller U., Kilpert C., Wertz K., Schwager J. Hydroxytyrosol is the major anti-inflammatory compound in aqueous olive extracts and impairs cytokine and chemokine production in macrophages. *Planta Med.*, 77(17):1890-7.

[46] Giner E., Andújar I., Recio M. C., Ríos J. L., Cerdá-Nicolás J. M., Giner R. M. (2011) Oleuropein ameliorates acute colitis in mice. *J. Agric. Food Chem.*, 59(24):12882-92.

[47] Zhang X., Cao J., Zhong L. (2009) Hydroxytyrosol inhibits pro-inflammatory cytokines, iNOS, and COX-2 expression in human monocytic cells. *Naunyn Schmiedebergs Arch. Pharmacol.*, 379(6):581-6.

[48] Zhang X., Cao J., Jiang L., Zhong L. (2009) Suppressive effects of hydroxytyrosol on oxidative stress and nuclear Factor-kappaB activation in THP-1 cells. *Biol. Pharm. Bull.*, 32(4):578-82.

[49] Karin M. (2006) Nuclear factor-kappaB in cancer development and progression. *Nature*, 441(7092):431-6. Review.

[50] Granados-Principal S., Quiles J. L., Ramirez-Tortosa C., Camacho-Corencia P., Sanchez-Rovira P., Vera-Ramirez L., Ramirez-Tortosa M. C. (2011) Hydroxytyrosol inhibits growth and cell proliferation and promotes high expression of sfrp4 in rat mammary tumours. *Mol. Nutr. Food Res.*, 55 Suppl 1:S117-26.

[51] Li S., Han Z., Ma Y., Song R., Pei T., Zheng T., Wang J., Xu D., Fang X., Jiang H., Liu L. (2014) Hydroxytyrosol inhibits cholangiocarcinoma tumor growth: An in vivo and in vitro study. *Oncol. Rep.*, 31(1):145-52.

[52] Bitler C. M., Viale T. M., Damaj B., Crea R. (2005) Hydrolyzed olive vegetation water in mice has anti-inflammatory activity. *J. Nutr.*, 135(6):1475-9.

[53] Gong D., Geng C., Jiang L., Cao J., Yoshimura H., Zhong L. Effects of hydroxytyrosol-20 on carrageenan-induced acute inflammation and hyperalgesia in rats. *Phytother. Res.*, 23(5):646-50.

[54] Sánchez-Fidalgo S., Sánchez de Ibargüen L., Cárdeno A., Alarcón de la Lastra C. (2012) Influence of extra virgin olive oil diet enriched with hydroxytyrosol in a chronic DSS colitis model. *Eur. J. Nutr.*, 51(4):497-506.

[55] Visioli F., Caruso D., Plasmati E., Patelli R., Mulinacci N., Romani A., Galli G., Galli C. (2001) Hydroxytyrosol, as a component of olive mill waste water, is dose-dependently absorbed and increases the antioxidant capacity of rat plasma. *Free Radic. Res.*, 34(3):301-5.

[56] Visioli F., Galli C., Plasmati E., Viappiani S., Hernandez A., Colombo C., Sala A. (2000) Olive phenol hydroxytyrosol prevents passive smoking-induced oxidative stress. *Circulation*, 102(18):2169-71.

[57] Schaffer S., Podstawa M., Visioli F., Bogani P., Müller W. E., Eckert G. P. Hydroxytyrosol-rich olive mill wastewater extract protects brain cells in vitro and ex vivo. *J. Agric. Food Chem.*, 55(13):5043-9.

[58] Whitlock N. C., Baek S. J. (2012) The anticancer effects of resveratrol: modulation of transcription factors. *Nutr. Cancer*, 64(4):493-502. Review.

[59] Mak J. C. (2012) Potential role of green tea catechins in various disease therapies: progress and promise. *Clin. Exp. Pharmacol. Physiol.*, 39(3):265-73. Review.

[60] Fabiani R., De Bartolomeo A., Rosignoli P., Servili M., Montedoro G. F., Morozzi G. (2002) Cancer chemoprevention by hydroxytyrosol isolated from virgin olive oil through G1 cell cycle arrest and apoptosis. *Eur. J. Cancer Prev.*, 11(4):351-8.

[61] Fabiani R., De Bartolomeo A., Rosignoli P., Servili M., Selvaggini R., Montedoro G. F., Di Saverio C., Morozzi G. (2006) Virgin olive oil phenols inhibit proliferation of human promyelocytic leukemia cells (HL60) by inducing apoptosis and differentiation. *J. Nutr.*, 136(3):614-9.

[62] Fabiani R., Sepporta M. V., Rosignoli P., De Bartolomeo A., Crescimanno M., Morozzi G. (2012) Anti-proliferative and pro-apoptotic activities of hydroxytyrosol on different tumour cells: the role of extracellular production of hydrogen peroxide. *Eur. J. Nutr.,* 51(4):455-64.

[63] Fabiani R., Rosignoli P., De Bartolomeo A., Fuccelli R., Morozzi G. (2008) Inhibition of cell cycle progression by hydroxytyrosol is associated with upregulation of cyclin-dependent protein kinase inhibitors p21(WAF1/Cip1) and p27(Kip1) and with induction of differentiation in HL60 cells. *J. Nutr.,* 138(1):42-8.

[64] Boonstra, J., Post, J. A. (2004) Molecular events associated with reactive oxygen species and cell cycle progression in mammalian cells. *Gene,* 337, 1–13.

[65] Palozza, P., Calviello, G., Serini, S., Maggiano, N., (2001) ⊔-carotene at high concentrations induces apoptosis by enhancing oxy-radical production in human adenocarcinoma cells. *Free Radic. Biol. Med.,* 30, 1000 –100.

[66] Park, S., Han, S. S., Park, C. H., Hahm, E. R., (2004) L-Ascorbic acid induces apoptosis in acute myeloid leukemia cells via hydrogen peroxide-mediated mechanisms. *Int. J. Biochem. Cell Biol.,* 36, 2180 –2195

[67] Robaszkiewicz, A., Balcerczyk, A., Bartosz, G. (2007) Antioxidative and prooxidative effects of quercetin on A549 cells. *Cell Biol. Int.,* 31, 1245 –1250.

[68] Zafarullah, M., Li,W. Q., Sylvester, J., Ahmad, M. (2003) Molecular mechanisms of N-acetylcysteine actions. *Cell Mol. Life Sci.,* 60, 6–20.

[69] Fabiani R., Fuccelli R., Pieravanti F., De Bartolomeo A., Morozzi G. (2009) Production of hydrogen peroxide is responsible for the induction of apoptosis by hydroxytyrosol on HL60 cells. *Mol. Nutr. Food Res.,* 53(7):887-96.

[70] Fabiani R., De Bartolomeo A., Rosignoli P., Morozzi G. (2001) Antioxidants prevent the lymphocyte DNA damage induced by PMA-stimulated monocytes. *Nutr. Cancer,* 39:284–91.

[71] Collins A. R. (2004) The comet assay for DNA damage and repair. *Mol. Biotechnol.,* 26:249–61.

[72] Fabiani R., Rosignoli P., De Bartolomeo A., Fuccelli R., Servili M., Montedoro G. F., Morozzi G. (2008) Oxidative DNA damage is prevented by extracts of olive oil, hydroxytyrosol, and other olive phenolic compounds in human blood mononuclear cells and HL60 cells. *J. Nutr.,* 138(8):1411-6.

[73] Rosignoli P., Fuccelli R., Fabiani R., Servili M., Morozzi G. (2013) Effect of olive oil phenols on the production of inflammatory mediators in freshly isolated human monocytes. *J. Nutr. Biochem.,* 24(8):1513-9.

[74] Babich H., Visioli F. (2003) In vitro cytotoxicity to human cells in culture of some phenolics from olive oil. *Farmaco.,* 58(5):403-7.

In: Virgin Olive Oil
Editor: Antonella De Leonardis

ISBN: 978-1-63117-656-2
© 2014 Nova Science Publishers, Inc.

Chapter 11

VIRGIN OLIVE OIL AS A SOURCE OF ANTI-INFLAMMATORY AGENTS

Susana M. Cardoso[1,2]*, *Marcelo D. Catarino*[1], *Marta S. Semião*[1]
and Olívia R. Pereira[1,2,3]

[1]CERNAS, School of Agriculture, Polytechnic Institute of Coimbra, Bencanta,
Coimbra, Portugal
[2]CIMO, School of Agriculture, Polytechnic Institute of Bragança, Bragança, Portugal
[3]DTDT, School of Health Sciences, Polytechnic Institute of Bragança,
Bragança, Portugal

ABSTRACT

Virgin olive oil (VOO) has many potential health benefits, including the amelioration of inflammatory processes. In part, this is known to occur through the modification of the endothelial function, leading to a decrease of the levels of cell-adhesion molecules (CAMs), including the inter-cellular adhesion molecule 1 (ICAM-1) and the vascular cell adhesion molecule 1 (VCAM-1). Importantly, virgin olive oil is able to inhibit the tumor necrosis factor-alpha (TNF-α), that is a key cytokine in controlling distinct types of cell functions and a particular therapeutic target for inflammatory diseases. Moreover, *in vitro* and *in vivo* assays with virgin olive oil or its main components clearly indicate a marked modulation of signaling pathways regulating the activation of pro-inflammatory mediators, including the nuclear transcriptional factor NF-κβ, the cytokines interleukin-1 (IL-1), and interleukin-6 (IL-6), and the enzymes cyclooxygenase-2 (COX-2), 5-lipoxygenase (5-LOX) and inducible nitric oxide synthase (iNOS). So far, the cellular and molecular anti-inflammatory mechanisms of virgin olive oil have been particular associated with its high amounts of phenolic compounds, as well as to its composition in mono and polyunsaturated fatty acids. Still, the available data is disperse and needs consolidation, in order to allow solid conclusions on this issue. The present chapter summarizes the epidemiological data and intervention trials focusing the effects of virgin olive oil in inflammatory processes and/or inflammatory related-

---

* Corresponding author. Email: scardoso@esac.pt.

diseases, as well as the main virgin olive oil constituents associated to the protection process and their underlying mechanisms of action.

Keywords: Virgin olive oil; inflammation; atherosclerosis; bowel diseases; cancer; hydroxytyrosol; oleic acid; oleuropein; unsaponifiable fraction

INTRODUCTION

Mediterranean diet has been associated to the prevention of distinct diseases such as coronary diseases, bowel ailments, cancer and aging [1-3]. Olive oil is one of the prime ingredients of the Mediterranean diet, thus playing an important contribution for its beneficial properties. In particular, the latter have been closely associated to VOO, i.e., the oil resultant only from the pressing of olives [4].

The main beneficial property attributed to VOO is undoubtedly its capacity to prevent cardiovascular diseases. The U.S. Food and Drug Administration (FDA) recognized that, due to its high content in monounsaturated fatty acids (MUFAs, in particular oleic acid C18:1), the daily ingestion of approximately two tablespoons (23 g) of VOO exerts benefic effects on the risk of coronary heart disease [5]. However, supposing that these beneficial effects would only be due to the oil´s MUFAs contents, then any type of oleic acid-enriched oils (such as rapeseed oil) or any MUFAs-rich food would have similar health benefits. As this is not observed, it was assumed that other VOO's components must also contribute to its beneficial effects [6].

It has been suggested that the VOO´s minor components, in particular the phenolic compounds, are as well key agents contributing for its health benefits. These are also believed to exert synergistic effects with MUFAs [7]. Besides those, triterpenic dialcohols, phytosterols, tocopherols, hydrocarbons and volatile and aromatic compounds are equally important in establishing the VOO´s bioactive properties [6]. The present chapter summarizes the main claimed anti-inflammatory properties of VOO´s, as well as the individual constituents associated to this property.

1. Inflammatory Process

The inflammation is a body defense mechanism whose goal is to eliminate the injury caused by pathogens or by the action of physical agents. Overall, the inflammatory process comprises two interconnected defense mechanisms, i.e., an unspecific response (innate immunity) and a highly specific one (adaptive immunity) [8]. The cells of the innate system residing in tissues (e.g. macrophages, fibroblasts, mast cells and dendritic cells) as well as circulating cells (e.g. monocytes and neutrophils) express pattern recognition receptors (PRRs), i.e. proteins able to directly or indirectly recognize pathogen-associated molecular patterns (PAMPs) or damage-associated molecular patterns (DAMPs) that are released by injured cells [9]. When activated, PRRs form multi-subunit oligomeric complexes which then trigger signaling cascades that help to contain the infection and the activation of the adaptive immune response [10]. In turn, the activation of this immune system causes the increment of microbial-specific leukocytes (e.g. T and B cells), a process that is highly effective and specific but that takes days to fully develop [10].

Hence, overall, the common hallmarks of inflammation, i.e. swelling, redness, pain and heat are primarily initiated by the innate response. Notably, the cells of the adaptive immune response can contribute to and exacerbate these effects, but those of the innate immune system are also strictly associated to the inflammation terminus process [11].

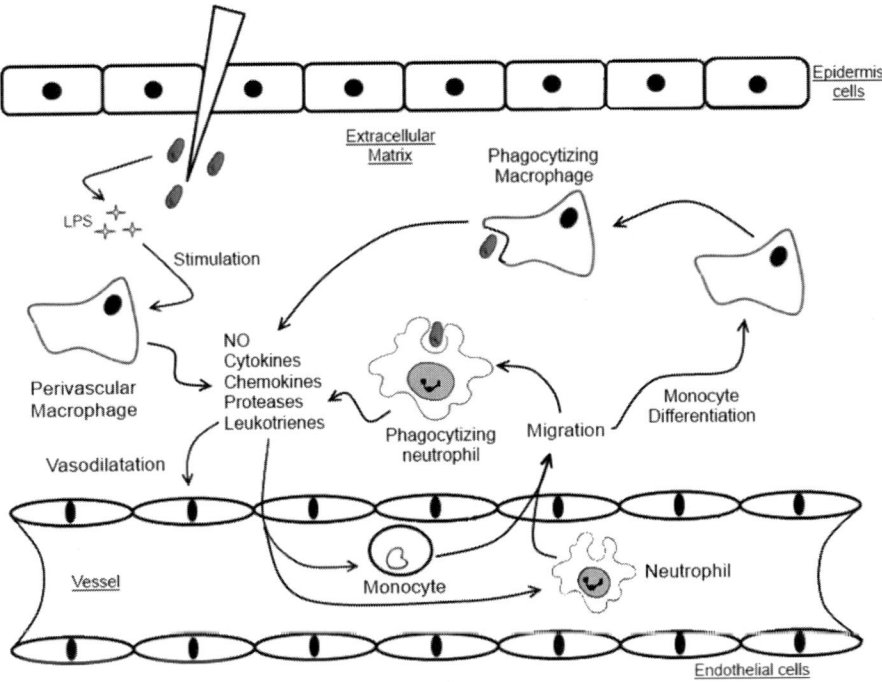

Figure 1. Cells and mediators of the innate inflammatory response. The injury or the pathogenic stimuli activate the perivascular macrophages around the injured area. These begin to phagocyte the infectious agents and release inflammatory mediators to the blood that will increase the permeability of the vessel and recruit more neutrophils and monocytes. The leukocytes on the blood cross the vessel and migrate to the injured area, where the monocytes differentiate into macrophages and, together with neutrophils, they will phagocyte the pathogenic agents and produce more inflammatory mediators until the clearance of the pathogenic.

Inflammation usually begins in a localized area, although depending on the injury´s severity, it can quickly become systemic. The acute phase of inflammation starts immediately upon injury (see Figure 1) and rapidly turns severe but notably, it persists only for a short period and is usually beneficial to the host [12]. This is mainly characterized by the rapid increase in blood flow to the affected region, which is supplied by the dilated arterioles. The capillaries also become more permeable, so that fluid and blood proteins move into the interstitial spaces, together with neutrophils and possibly some macrophages. Note that neutrophils are the most abundant leukocytes in the blood and they represent a first line of attack of the immune system. Likewise macrophages, they are capable of emitting phagocytes cytoplasmic processes involving foreign particles through the digesting enzymes present in cells (Figure 1) [13]. Overall, these events allow the removal of the noxious stimulus via phagocytosis, which is then followed by the resolution of the inflammation. Sore throat from a cold or flu, a cut on the skin or a blow are typical conditions that can result in acute inflammation [14].

On the other hand, if the inflammation lasts for a long period of time, a second stage (or chronic inflammation) is settled and may predispose the host to various chronic inflammatory illnesses. The beginning of this type of inflammation is characterized by the replacement of neutrophils by macrophages and other cells, including the T cells (also known as T lymphocytes i.e., a sub-class of leukocytes that plays a key role in the regulation of the immune system, in particular the adaptive) [15]. The evolution of this condition is also histologically associated with the proliferation of blood vessels, fibrosis and necrosis. Typical examples of chronic inflammation conditions include, in between many, asthma, tuberculosis, rheumatoid arthritis and lupus [12].

Notably, vascular and cellular reactions of acute and chronic inflammation are mediated by many chemical mediators (proteins or plasma cells) which are produced and/or released by the cells of the immune system. Once activated and/or released by the cells, most of these mediators have a short life and exert their biological activity through binding to specific receptors on target cells, although others possess enzymatic activity (lysosomal proteases). Note that a mediator may itself stimulate the release of other mediators by target cells. The most relevant inflammatory mediators are further detailed in below [16].

1.1. Chemical Mediators

Inflammatory mediators can be generally grouped with respect to their origin: plasma or cellular. While the former are present in plasma in the form of precursors and suffer activation through a number of proteolytic cleavages, the cellular mediators are usually preformed and further stored in intracellular granules (e.g. histamine) or alternatively, synthesized de novo (e.g. prostaglandins and cytokines) [17].

The two vasoactive amines histamine and serotonin, together with lysosomal enzymes, comprise the most relevant preformed cellular mediators. These mediators are stored in secretory granules and are involved in the initial stage of the inflammatory process [18]. In particular, histamine is synthesized by basophils (a subtype of granulocytes leukocytes), platelets, and especially by mast cells (resident cells of several types of tissues) and this mediator acts by interacting with three distinct receptors present on target cells, known as H_1, H_2 and H_3 [19]. From those, the H_1 receptor causes contraction of bronchial smooth muscle, intestine and uterus and increases the permeability of venous capillaries. In turn, serotonin is mainly found in the lining of the gut, in platelets and in central nervous system. Such as histamine, serotonin is a vasodilator and increases vascular permeability [19]. Its release from the platelets is triggered by platelet aggregation, upon contact with collagen, thrombin, adenosine diphosphate and antigen-antibody complex [14]. In turn, as previously mentioned, lysosomal enzymes are released by macrophages and neutrophils with the direct purpose of destroying the pathogenic agents [14].

From the newly synthesized inflammatory mediators, arachidonic acid (AA) and other polyunsaturated fatty acids (PUFAs) plus their products deserve special mention in view of their role in inflammation, resolution of inflammation, and inhibition of production of proinflammatory cytokines [20]. Arachidonic acid is a polyunsaturated acid present in phospholipids of the plasma membrane (particularly in phosphatidylcholine, phosphatidylethanolamine and phosphatidylinositides) that can be enzymatically released by the action of enzyme such as phospholipase A2 (PLA2) [21].

The AA is subsequently transformed through cyclooxygenase (COX) and lipoxygenase (LOX) pathways to various biologically active metabolites (e.g. prostaglandins, thromboxane, leukotrienes and lipoxins) collectively termed as eicosanoids [22]. Note that eicosanoid production is considerably increased during inflammation and both COX and LOX pathways are of clinical relevance. Particular interest has been given to the isoenzyme COX-2 [23] that is primarily expressed at sites of inflammation and produces pro-inflammatory eicosanoids and to 5-LOX, which is the key enzyme in leukotriene biosynthesis (considered as potent mediators locally released at the inflammation site by leukocytes and other 5-LOX expressing cells) [24].

Cytokines, a diverse group of proteic mediators (anti-inflammatory and pro-inflammatory), are also determinant in the inflammatory response [25]. Any disorder in the regulation of cytokines can lead to the development of inflammatory diseases [26]. Tumor necrosis factor-α (TNF-α) is one of the most important inflammatory cytokines that controls different types of cell functions [27]. This is produced in its active form mainly by macrophages, but also by other immune cells including mast cells, neutrophyls, T cells and natural killer cells (NK, i.e, effector lymphocytes of the innate immune system that limit the spread of tumors or microbial infections). At the cellular level, TNF-α is a potent activator of neutrophils, mediating adhesion, cellular degranulation and chemotaxis, as well as interacting with the endothelial cells to induce the appearance of adhesion molecules such as the intercellular adhesion molecule-1 (ICAM-1), the vascular cell adhesions molecule-1 (VCAM-1) and E-selectin [28].

The pro-inflammatory effects of TNF-α are mainly due of its ability to activate nuclear factor kappa B (NF-κB) [29], through a series of complex signaling cascades leading to the degradation of the IκBκ (an inhibitor of NF-κB activation). When activated, the NF-κB translocates to the nucleus, binds to the promoter or enhancer regions of target genes to enhance transcription [26]. There are many genes known to be regulated by NF-κB, including TNF-α itself and others such as COX-2, 5-LOX, cell-adhesion molecules (CAMs), inflammatory cytokines and inducible nitric oxide synthase (iNOS). Due to its central role in inflammation process, this pathway is a particular therapeutic target [26].

Besides the activation of NF-κB, TNF-α is also capable of activating the mitogen-activated protein kinases (MAPKs) pathway, a class of proteins that are involved in the regulation and/or activation of a series of transcription factors. In particular the P38 MAPKs are involved in cell differentiation, apoptosis and autophagy [30].

Interleukins (IL) are also a central group of cytokines. These are primarily produced by T helper cells (T$_H$ or CD4 lymphocytes i.e, a specific population of T cells), but also by monocytes, macrophages and endothelial cells. They mainly induce the development and differentiation of T cells as well as of B cells (a lymphocyte population whose main function is the production of antibodies), fibroblasts and endothelial cells [9]. Each IL acts on a limited set of cells which express the appropriate receptors [31]. The best characterized pro-inflammatory interleukins are the IL-1, IL-2, IL-6 and IL-8, while IL-4 and IL-10 are two major anti-inflammatory ILs.

Interferon gama (IFN-γ) is another key cytokine, that is mainly produced by T cells, B and NK cells. Its main function is the macrophage activation, rendering them able to exert its microbicidal functions [32]. In addition, it also promotes the differentiation of T helper lymphocytes to the Th1 subpopulation (the host immunity effectors against intracellular bacteria and protozoa). IFN-γ induces the transcription of many genes in macrophages, including those for the production of antimicrobial molecules such as oxygen free radicals and nitric oxide, which represent one of the best effector mechanisms for elimination [32].

Note that superoxide anion (O_2^-), hydrogen peroxide (H_2O_2), singlet oxygen ($^1O_2$), inducible nitric oxide (iNO), and other reactive oxygen species (ROS) are indeed key mediators with an important role in vascular and cellular components of inflammatory reactions [14]. These are produced not only by macrophages, but also by other immune cells including activated neutrophils and monocytes, T-cells, Kupffer cells and glial cells [33].

2. Targets of inflammation by VOO

2.1. Anti-Inflammatory General Mechanisms

Several studies report the fact that VOO and/or its components hamper numerous inflammatory processes. In fact, Eisner et al. [34] observed that leukocytes collected from VOO-intestinally treated rats were less susceptible to lipopolysaccharide (LPS).

From all the VOO´s components, the phenolics, and in particular hydroxytyrosol, are the main compounds associated to the oil´s anti-inflammatory properties. Recently, Pontoniere [35] referred that the VOO´s phenolic fraction was able to block the activation of NF-κB and consequently, the expression of cytokines, of LOX and COX, also affecting the production of adhesion molecules and eicosanoids derived from arachidonic acid. In turn, studies performed in LPS-stimulated murine macrophages (RAW 264.7 cells) treated with hydroxytyrosol have demonstrated that this phenolic is able to attenuate iNOS and COX-2, and to decrease the secretion of prostaglandin E_2, plus of several pro-inflammatory citokynes (TNF-α, IL-1 α and β, IL-6 and IL-12) and chemokines (a family of small cytokines), as well as to reduce the gene expression of the metalloproteinase-9 (MMP-9, a protease of the MMP´s family, which are closely associated to the inflammatory process) [36-38]. Moreover, it was also demonstrated that this phenolic compound exerted a negative effect on NF-κB pathway [38, 39]. Similar results were obtained in phorbol 12-myristate 13-acetate (PMA)-stimulated human monocytes. Indeed, in this cellular model, hydroxytyrosol was able to decrease the transcription of iNOS and TNF-α, the COX-2 expression and the production of superoxide ion and of prostaglandin PEG_2 [40, 41].

Besides hydroxytyrosol, oleuropein and/or oleuropein glucoside plus other VOO´s phenolics were also reported to counteract inflammatory events. In this sense, Visioli et al. [36] demonstrated that hydroxytyrosol and oleuropein showed great biological activities, including the ability to inhibit platelet aggregation, as well as to scavenge hypochlorous acid, superoxide ion and other ROS, overall resulting in an increased plasma antioxidant capacity. Notably, the authors concluded that these compounds were more effective than butylated hydroxytoluene (BHT), which is a potent synthetic antioxidant. Also, De la Puerta et al. [42] reported that hydroxytyrosol, oleuropein, caffeic acid, and tyrosol could inhibit 5-LOX activity, thus diminishing the leukotriene B_4 production on rat peritoneal leukocytes, while Miles et al. [43] reported that oleuropein glycoside, caffeic acid and kaempferol caused a high inhibition on the IL-1β and PEG_2 production levels, on human blood cultures. Furthermore, as demonstrated by Dell´Agli et al. [44], oleuropein is also an intervening phenolic on the reduced expression and secretion of MMP-9, as demonstrated in TNF-α-stimulated monocyte

(THP-1) cells. The authors showed that this effect was due to the impairing effect of oleuropein on the NF-κB signaling pathway [44].

Additionally, oleocanthal (another VOO's potent antioxidant) is being a target of interest due to its strong anti-inflammatory properties. This phenol has been proved to have strong inhibitory effects on the prostaglandin biosynthesis by hampering the activity of COX-1 and COX-2, mimicking the anti-inflammatory effects of the well-known drug ibuprofen. Even more, for equimolar concentrations, oleocanthal exhibited greater inhibitory effects on COX than ibuprofen [9, 45].

Recently, the work of Cardeno et al. [46] has demonstrated, for the first time, the anti-inflammatory and antioxidant properties of the VOO's unsaponifiable fraction, on a LPS-stimulated murine macrophages model. In particular, the authors showed that this fraction exerted a strong inhibitory ability on intracellular ROS and NO production. This also decreased the expression of COX-2 and iNOS and, more importantly, it induced down-regulation of the NF-κB signaling and MAPK phosphorylation pathways.

Additional studies also reported the anti-inflammatory effects of some individual unsaponifiable components. One of those components is α-tocopherol, which has been shown to inhibit the expression of 5-LOX, COX-2 and IL-1β [37, 47, 48]. Furthermore, β-sitosterol, the main VOO's phytosterol, was shown to influence the reduction of ROS production and arachidonic acid release, as well as COX-2 activity and PGE2 production in phorbol ester-induced macrophages [49]. These results are in agreement to those of Moreno et al. [50], whom verified that ROS and NO production, arachidonic acid release and arachidonic acid metabolites synthesis (through the COX and LOX pathways) were impaired in PMA-stimulated macrophages RAW 264.7, when exposed to β-sitosterol and two other minor components of VOO, i.e, squalene and tyrosol.

2.2. Atherosclerosis

Atherosclerosis is the main inflammatory disease in which VOO is claimed to exert beneficial effects. Inflammation in atherosclerosis is due to injury or change of endothelium function that might be triggered by several causes (step 1 in Figure 2), including an excess of reactive oxygen species, or the exposure to toxic agents (e.g. oxidized low density lipoprotein cholesterol, oxLDL), to infectious agents or advanced glycosylated end products (the result of an oxidation reaction with glucose that results in a type of oxidant commonly found in the blood of diabetics) [51, 52].

The inflamed endothelium then expresses selective adhesion molecules, namely the VCAM-1 and ICAM-1 respectively (step 2 in Figure 2). As a result, and in opposition to the normal epithelium, the inflamed one is able to bind various classes of leukocytes (step 3 in Figure 2), which in turn are exponentially recruited by the increased expression of specific cytokines and of their pro-inflammatory downstream events, e.g. NF-κB, COX and LOX (steps 4 and 5 in Figure 2). Increased endothelial stress also stimulates the arterial smooth muscle cells to produce proteoglycans that are able to bind and retain lipoprotein particles and increase their oxidative modification, also contributing for the reinforcement of the inflammatory process [53].

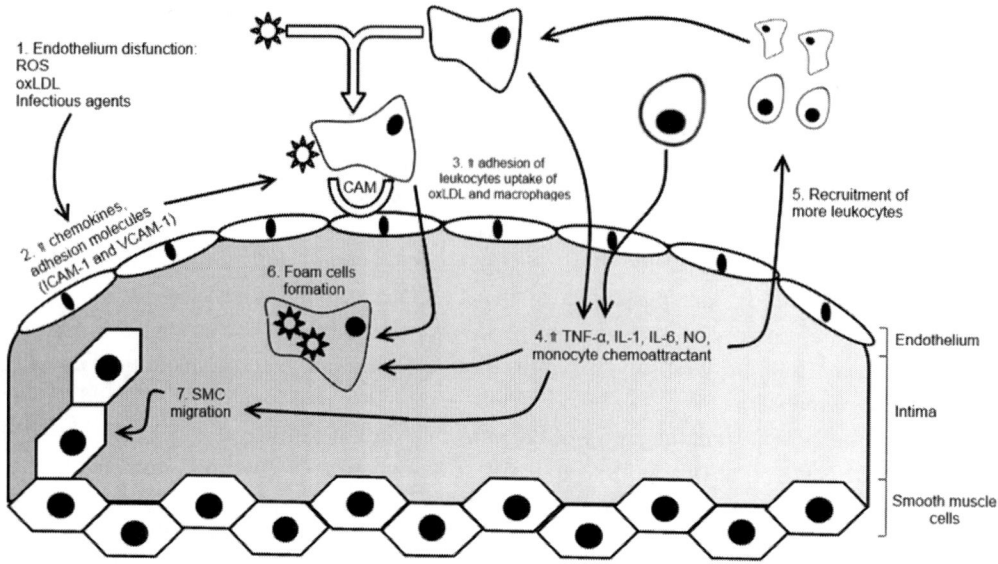

Figure 2. Endothelium dysfunction and atheroma formation. Factors as reactive oxygen species (ROS), oxidized LDL cholesterol (oxLDL) or other infectious agents can promote endothelium dysfunction. The production of cellular adhesion molecules (CAM) allow the binding of blood circulating leukocytes, which in turn will migrate into the intima and further release inflammatory mediators. The macrophages start to phagocyte the oxLDL particles, thus forming foam cells and, at the same time, more leukocytes are recruited to the intima, thus amplifying the inflammatory mediators. As the inflammation and the deposition of foam cells grow, the smooth muscle cells start to migrate to form a coat around the injured area.

In particular, monocytes differentiate into macrophages and express the scavenging receptors, allowing them to bind and engulf these modified lipoproteins, leading to the formation of lipid-laden macrophages i.e. foam cells (step 6 in Figure 2). Macrophages and other leukocytes also release cytokines and grown factors that are central for stimulating the migration and proliferation of smooth muscular cells (step 7 in Figure 2), ending up with the formation of a typical dense extracellular matrix [54].

Besides its crucial role in initiation and establishment of atheroma, one should highlight that inflammatory processes are also decisive in the acute thrombotic complications of atheroma, as the combined action of macrophages and lymphocytes contribute for the production of proteinases such as MMP-9, which degrade extracellular matrix proteins including collagen and elastin and hence cause the narrowing of the fibrous cap, rendering it susceptible to rupture [53].

On the other hand, literature data support the fact that VOO and/or its components hamper numerous atherosclerosis-related inflammatory processes. In particular, Camargo et al. [55] reported that a Mediterranean diet enriched in VOO can indirectly inhibit the expression of NF-κB and that of metalloproteinase MMP-9. Moreover, in high/medium-risk cardiovascular disease patients, the administration of VOO has resulted in a reduction of white blood cells, as well as of the endothelium ICAM expression, pointing for an amelioration of the endothelium function [56]. Additionally, Urpi-Sarda et al. [57] reported that a Mediterranean diet enriched in VOO promoted the reduction of IL-6 and ICAM molecules on plasma of high-risk cardiovascular disease patients. The IL-6 reduction effect of VOO has also been shown by Fitó et al. [58], in a similar test model.

According to literature data, health-promoting VOO´s components in atherosclerosis-related inflammatory processes mainly enclose oleic acid, as well as some phenolics and terpenes. Oleic acid is very well known for its ability to reduce the blood levels of ROS, LDL and its oxidative modification into oxLDL cholesterol (Figure 3) [59-61]. Moreover, the consumption of oleic acid by Man has been described to increase the levels of HDL cholesterol and to protect it against oxidation [62]. Note that while ROS and oxLDL cholesterol are known to increase the risk of atherosclerosis, elevated HDL cholesterol levels are believed to ameliorate the lipid efflux from the foam cells to the HDL and hence, to counteract the inflammatory atherosclerotic process [4] (Figure 3). HDL particles can also transport antioxidant enzymes that can break down oxidized lipids and neutralize their pro-inflammatory effects [53].

Furthermore, incorporation of oleic acid in total cell lipids of an *in vitro* stearic acid-induced model of early atherogenesis caused the decrease on the incorporation of this fat acid in the phospholipids, as well as a decrement of NF-κB activation, thereby down-regulating the expression of several endothelial and leukocyte adhesion molecules, among which the VCAM-1 and ICAM-1 (represented in Figure 3) [63, 64]. This was reinforced by the work of Sanadgol et al. [65] which reported the suppressed expression of the two CAMs on LPS-stimulated human bone marrow endothelial cells treated with oleic acid.

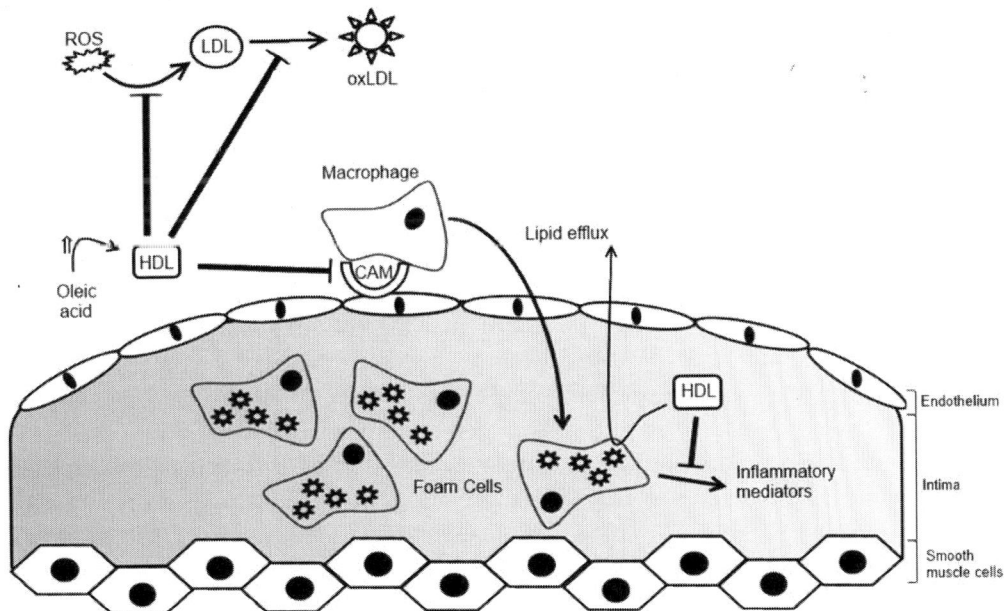

Figure 3. Influence of oleic acid and HDL cholesterol on the atheroma formation. Oleic acid promotes the increase of HDL cholesterol levels, which in turn interfere with reactive oxygen species (ROS) and block the oxidation of the LDL into oxLDL. HDL cholesterol also impairs the adhesion of the leukocytes to the endothelium, preventing their accumulation in the atheroma. Moreover, it can inhibit the release of inflammatory mediators and interact with the foam cells, by promoting the lipid efflux from the cells to the bloodstream and hence ameliorate the inflammatory condition.

Regarding the protective effects of polyphenols, Covas et al. [66] have shown that the intake of polyphenols-enriched VOO decreased the levels of LDL cholesterol and it simultaneously increased those of HDL cholesterol. Moreover, a similar study reported an increased up-regulation of several lipid efflux related genes including peroxisome proliferator activated receptors (PPARs)(α, γ and δ), ATP-binding cassette transporter-A1 (ABCA1) and COX-1 [62]. As a consequence of the activation of PPARs, the authors have described an increase on the HDL cholesterol levels and a down-regulation of CD40-ligand expression and of its related products, such as vascular endothelial growth factor, ICAM-1 and IFN-γ [67, 68].

Amongst the VOO´s phenolic constituents, hydroxytyrosol is by far the most associated to its claimed protective effects on atherosclerosis. In more detail, Zrelli et al. [69] described that hydroxytyrosol was able to induce an indirect dose-dependent inhibition of the NF-κB on vascular endothelium cells. Furthermore, this polyphenol was shown to inhibit MMP-9 release, an effect that was potentiated in the presence of oleic acid [7].

Notwithstanding, hydroxytyrosol is not the only polyphenol capable of inhibiting MMP-9, as oleuropein was also observed to have similar preventive effects on this metalloproteinase [3]. Besides, it was also reported that hydroxytyrosol has the ability to restrain platelet aggregation and the synthesis of eicosanoids in human blood, as well as the reduction of the expression of VCAM-1 in endothelial cells [70, 71].

The potent antioxidant activity of hydroxytyrosol was also shown to prevent increasing levels of ROS formation in pulmonary artery endothelial cells in the presence of H_2O_2 [72]. This ROS preventing ability was reinforced by Scoditti et al. [3] in a study showing that hydroxytyrosol and oleuropein caused the decrease of the intracellular levels of ROS formation, as well as a reduction of the activation of NF-κB, in PMA-stimulated human umbilical vein endothelial cells (HUVEC). The authors also tested the ability of these polyphenols to counteract the expression of COX-2 in the same HUVEC model, concluding that both hydroxytyrosol and oleuropein were able to cause a decrease of 50% on the expression of this enzyme, without affecting the constitutive expression of COX-1.

Another study focusing the effects of oleuropein, hydroxytyrosol and homovanillyl alcohol on HUVEC surface and on the mRNA expression of three adhesion molecules (ICAM-1, VCAM-1 and E-selectin) concluded that the two first polyphenols induced a decrement on the ICAM-1 and VCAM-1 expression (surface and mRNA), while homovanillyl alcohol caused a reduction on the cell surface expression of the three adhesion molecules, although no effect was observed for their corresponding mRNA levels [73].

Moreover, expression of surface scavenger receptors on LDL-stimulated macrophages was shown to be reduced in the presence of the phenols tyrosol and hydroxytyrosol and of the hydrocarbon squalene, thus lowering the macrophages lipid intake [74]. Moreover, experimental tests performed on rabbits submitted to an atherosclerotic diet against diets with squalene or hydroxytyrosol were compared, revealing that the polyphenol reduced the endothelium inflammatory activation while the hydrocarbon decreased the fibrosis [75].

The unsaponifiable fraction of the VOO is also rich in antioxidant and anti-inflammatory compounds. α-Tocopherol has a great potential as it can influence many pro-inflammatory molecules. An inhibitory effect on LDL-induced cytokines and on the expression of adhesion molecules was attributed to α-tocopherol. In fact, this compound also reduced the expression of ICAM-1, VCAM-1 and E-selectin in IL-1β-stimulated endothelial cells [76].

Moreover, a great number of experiments have been carried out to elucidate the effect of triterpenes on atherothrombotic risk factors such as lipid profile, oxidative stress, hyperglycemia, endothelial dysfunction, hypertension and inflammation [54]. In this context, the consumption of triterpene-enriched VOO (particularly in oleanolic acid) improved endothelial function in spontaneous hypertensive rats, by increasing the endothelial NO-mediated relaxation of aortic rings through an enhanced eNOS expression [77]. Moreover, a decrease of COX-2 gene expression was noted in human activated monocytes of subjects submitted to a diet rich in oleanolic acid [78]. Additionally, maslinic acid has shown a potent dose-dependent antioxidant effect on a chemical model of LDL peroxidation, whereas uvaol and erythrodiol acids exhibited both antioxidant and antithrombotic activities [79].

2.3. Ulcerative Colitis (UC) and Crohn´s Disease (CD)

Ulcerative colitis (UC) and Crohn's disease (CD) are the two main types of inflammatory bowel diseases. While UC is characterised by diffuse mucosal inflammation limited to the colon, the CD is characterised by patchy, transmural inflammation, which may affect any part of the gastrointestinal tract [80].

Evidences point that genetic factors, enteric microflora and host response are the causes of inflammation in the two diseases [81]. LPS stimuli of the bacteria lead to the cleavage of IκBα and to the concomitant activation of NF-κB in the epithelial cells (Figure 4). This in turn triggers the inflammatory cell signaling pathways (COX-2 and LOX expression), the release of chemokines and pro-inflammatory cytokines TNF-α, IL-1β, IL-6 and IL-8, thus promoting the recruitment of other leucocytes [82]. At this point, the characteristics of tissue damage will define one of two inflammation pathways: an excessive T helper 1 (T_H1, as stated before, are immunity effectors against intracellular bacteria and protozoa) response, which is associated with CD or alternatively, an excessive T_H2 phenotype that is linked to the development of UC. Secretion of cytokines influences T lymphocytes maturation. In particular, the overproduction of IL-12 shifts the immune response in a T_H1 direction, and the concomitant increased production of IFN-γ, TNF-α, IL-1β, IL-2, and IL-6, resulting in a self-sustaining cycle of activation. In turn, an excessive T_H2 cell response is associated with increased secretion of IL-4, IL-5, IL-10 and IL-13.

It is also believed that a deficient production of transforming growth factor (TGF)-β, IL-10 and other immune-inhibitory cytokines by T_H3 and T regulatory 1 cells (T_R1) precipitates the loss of tolerance of the mucosa, turning it sensitive to ordinary antigens of the microflora [82, 83].

Some studies have shown that a dietary rich in VOO can attenuate bowel inflammation. Hegazi et al. [84] demonstrated that the introduction of VOO in the diet of IL-10 knock-out mices caused modulation of chronic colitis, inhibiting the expression of COX-2 and decreased dysplasia, an early stage of pre-cancer lesion. In dextran sodium sulphate (DSS)-induced colitis rats, the intake of VOO resulted in a reduction of ROS and NO levels, due to the inhibition of iNOS. Moreover, in the same model, rats fed with (n-3) PUFA-enriched VOO exhibited decreased expression of iNOS and of LPS-mediated TNF-α, together with the inhibition of the NF-κB signaling, which is a fundamental piece in the gene activation of the previous mediators [85].

Phenolic compounds have been closely associated to impairment of inflammatory processes in bowel diseases. Recent studies demonstrated the ability of an VOO extract enriched in polyphenols (e.g. oleuropein aglycone, hydroxytyrosol, tyrosol, etc.) to significantly reduce the expression of TNF-α and COX-2, iNOS and a chemokine that promotes the migration and infiltration of monocytes (the chemotactic protein (MCP)-1), in DSS colitis-induced mice. The authors also referred that during the same treatment, the activation of MAPKs (particularly p38 and JNK MAPKs) and the degradation of IκBα were both diminished, while the activity of the peroxisome proliferator activated receptor γ (PPARγ, a key transcription factor in maintenance of gut homeostasis) was increased [86, 87].

The previous results are in agreement with earlier studies that attested a significant improvement on the production of IL-10 and an inhibitory effect on iNOS and COX-2 expression, and p38 MAPK activation in colitis-induced mice fed with hydroxytyrosol-enriched VOO, thus suggesting that hydroxytyrosol is one of the VOO compounds responsible for modulating these proteins [1]. Giner et al. [88] recently proved that oleuropein interacts in the inflammatory pathways associated to bowel diseases. Their study demonstrated that the production of the pro-inflammatory mediators IL-6 and IL-1β was significantly decreased in oleuropein-treated DSS chronic colitis-induced mice, while the IL-10 levels were increased [88].

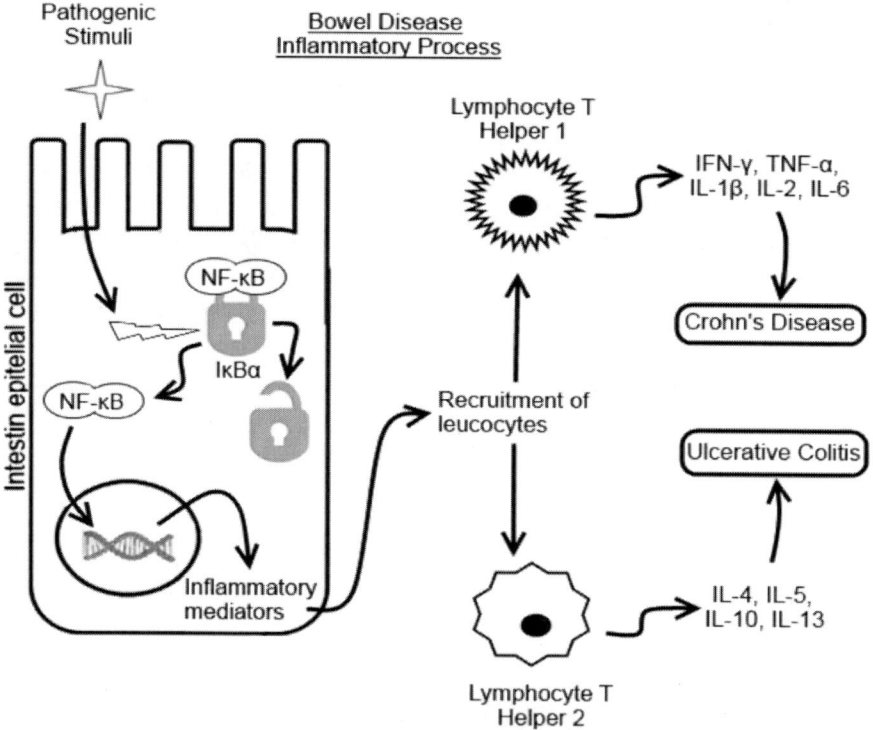

Figure 4. Developing of bowel inflammation. The pathogenic stimulus promotes the degradation of the IκBα, thus releasing the NF-κB. When activated, this nuclear transcription factor migrates to the nucleus where it is responsible for the transcription of several inflammatory mediators genes. Depending on the mediators produced, one of two pathways can be triggered: 1- activation of the lymphocytes T helper 1 resulting on a Crohn's Disease; 2- activation of the lymphocytes T helper 2 leading to an Ulcerative Colitis.

Additionally, the treatment of chronic colitis mices with oleuropein lowered the expressions of COX-2 and iNOS, the decrease phosphorylation of p38 sub-unity of MAPK pathway and the reduction of NF-κB activation. It was also concluded that oleuropein remarkably stimulated the production of Annexin A1, a potent endogenous anti-inflammatory agent capable of inhibiting COX-2 and iNOS, in injured colon tissue [88].

A recent study demonstrated, for the first time, the protective effects of the VOO unsaponifiable fraction on bowel diseases. Sánchez-Fidalgo et al. [89] demonstrated that unsaponifiable enriched diets were able to block the activation of TNF-α and MCP-1 and also mediate COX-2 and iNOS downregulation, thus ameliorating the first stage of ulcerative colitis. In addition, the authors referred a blockage of the MAPKs and of NF-κB pathways, through decrement of the p38 activation and IκBα degradation, respectively. The consequent reduction of the expression of other mediators (e.g. COX-2, TNF-α, iNOS) was equally shown.

Besides that study, other authors have reported the anti-inflammatory properties of some minor components of the VOO´s unsaponifiable fraction in colitis models. For instance, α-tocopherol effects were assessed in ulcerative colitis mice models demonstrating suppression of serum IL-6 and serum kinase C, as well as promotion of serum IL-10, overall resulting in a decreased colonic damage [90]. Also, β-sitosterol has been demonstrated to inhibit the expression of TNF-α, IL-1β IL-6, COX-2, as well as the activation of NF-κB, in 2,4,6-trinitrobenzenesulfonic acid (TNBS)-induced colitic mice colons [45].

2.4. Other Effects

Cancer

Chronic inflammation also plays an important role in the development and progression of phenomena related with cancer and was estimated to contribute to 15–20% of all cancers [91, 92]. The relevance of inflammation in carcinogenesis was firstly proposed in the 19th century by Rudolf Virchow after verifying the presence of leukocytes within tumors [93]. Presently, it is generally accepted that inflammation is involved at distinct stages of carcinogenesis, particularly in tumor initiation (mutations, genomic instability, and epigenetic modifications) and along progression, by activating tissue repair responses, promoting cells surveillance and proliferation of premalignant cells, by enhancing their survival and causing localized immunosuppression and accumulation of additional mutations, in between others [94]. Moreover, the microenvironment of tumor is characterized by the presence of several inflammatory cells such as neutrophils, eosinophils, lymphocytes, mast cells, dendritic cells, tumor-associated macrophages (TAM) and NKcells. These cells produce PLA2, COX-2, 5-LOX, MMPs and serine and cysteine proteases, as well as membrane perforating agents, reactive oxygen and nitrogen species and also several cytokines (TNF-α, IL-1, IL-6, IL-8, interferons, chemokines) that are associated to DNA damage, increase of mutation rates, development of tumor growth and promotion of metastasis genomic instability. The activation of transcription factors such as NF-κB, nuclear factor erythroid 2 related factor 2 (Nrf2, which regulates environmental stress response by activating the expression of genes for antioxidants and detoxification enzymes) or signal transducers and activators of transcription-3 (STAT3, a protein of the STAT family that mediates the expression of a variety of genes in response to cell stimuli, and thus plays a key role in many cellular processes such as cell growth and apoptosis) are involved in these effects [95].

Epidemiological studies using VOOs have evidenced their preventive effect on some types of cancer, including the breast cancer [96-98]. In particular, VOO was able to improve the efficacy treatment of letrozole and anastrozole, two commercially available therapies for early and late stages of breast cancer, as shown in the human breast cancer MCF-7 cells model. In the same model, a similar treatment promoted the mitochondrial release of cytochrome c and apoptosis, which were mediated by a depletion of reduced glutathione levels, disruption of membrane potential and the increase of lipid peroxidation [99].

The antitumor activity of VOO was also observed in SKBr3, MCF-7 and JIMT-1 breast cancer cell lines, with IC_{50} values ranging from 200 to 300 μg/mL, as measured by MTT assay [100]. Moreover, VOO has a chemopreventive effect in the ulcerative colitis-associated colorectal cancer, as demostrated in a C57BL/6 mice model. Animals feed with VOO presented a minor number of dysplastic macroscopic lesions, a reduction of TNF-α, IL-6 and IFN-γ and COX-2 levels and a decreased expression of iNOS in the colonic tissue [101].

Amongst VOO´s polyphenols, hydroxytyrosol and oleuropein have been recognized as being anti-inflammatory and also modulators of several signal transduction pathways associated with cell survival, proliferation and apoptosis. The mechanisms of hydroxytyrosol include the decreased expression of 5-LOX, reduced synthesis of prostaglandin E2 and alteration of tumour eicosanoid biosynthesis. Apoptosis promotion and prevention of oxidative DNA damage are other described mechanisms for hydroxytyrosol and oleuropein, as concluded from studies performed in human tumour-cell lines, blood mononuclear cells as HL60 cells (the latter a human leukemia cells) [102-104]. Moreover, in cultured endothelial cells, hydroxytyrosol and oleuropein have promoted a suppression of the inflammatory angiogenesis through MMP-9 downregulation, together with an inhibitory effect on COX-2 expression and on the ROS levels [105].

At last, the anticancer effects demonstrated for an unsaponifiable fraction of VOO (in HT-29 human colon adenocarcinoma cells) have been associated to its ability in downregulating COX-2 through PPARγ and NF-kB signaling pathways, apoptosis promotion and modulation levels of p53 suppressor protein (a nuclear protein that functions as a regulator of transcription) [106].

Neurodegenerative Diseases

Mediterranean-style diet has also been related to decreased risks for neurodegenerative diseases [107, 108] and in particular, the dietary component VOO, included in Mediterranean diet, has been reported to promote beneficial neuroprotective effects in Alzheimer's and Parkinson's diseases, stroke, traumatic brain injury and multiple sclerosis [102, 109]. Note also that the protective role of Mediterranean diet against cognitive decline has been related to the attenuation of inflammation [110]. A described mechanism includes the decrease of C-reactive protein levels, which is an inflammatory marker upregulated in Alzheimer's disease associated to the presence of neuritic plaques and neurofibrillary tangles [111].

Despite the protective effects of VOO in neurodegenerative-inflammatory related processes are still scarcely studied, in vivo experiments in the mouse model SAMP8 showed that VOO significantly improved learning and memory [110]. Moreover, oleocanthal, which as stated before exerts anti-inflammatory properties through COX-1 and COX-2 inhibition [112], is also an inhibitor of Aβ-amyloid oligomers and of fibrillization of tau-protein, counteracting functional damage in synapses and the formation of neurofibrillary tangles, respectively [113-116]. Besides that, oleuropein is known to form a non-covalent complex with Aβ-amyloid peptide, thus decreasing or preventing the Aβ-aggregation, and consequently, it counteracts the deposition of these neurotoxic structures [117].

CONCLUSION

VOO has become recognized as an important ingredient of Mediterranean diet due to its claimed health benefits. This chapter summarizes existing information on the beneficial properties of VOO, or of its components, in inflammatory-related diseases such as atherosclerosis, bowel diseases, cancer and neurodegenerative disorders.

The anti-inflammatory activities of VOO includes the inhibition of NF-κB activation, together with the expression decrement of cytokines and chemokines, LOX and COX-2 enzymes, with concomitant lowering levels of prostaglandins, leukotrienes and thromboxanes. Other mechanism includes the decrease of intracellular oxidative stress. Overall, the main VOO's components that have been associated to its anti-inflammatory effects enclose the oleic acid, phenolic compounds (particularly hydroxytyrosol, oleuropein and oleocanthal) and some unsaponifiable components, namely the α-tocopherol and β-sitosterol.

REFERENCES

[1]   Sánchez-Fidalgo S, Sánchez de Ibargüe L, Cárdeno A, & Alarcón de la Lastra C (2012) Influence of extra virgin olive oil diet enriched with hydroxytyrosol in a chronic DSS colitis model. *European Journal of Nutrition*, 51, 497–506.

[2]   Santos-González M, López-Miranda J, Pérez-Jiménez F, Navas P, & Villalba J M (2012) Dietary oil modifies the plasma proteome during aging in the rat. *Age*, 34, 341–358.

[3]   Scoditti E, Calabriso N, Massaro M, Pellegrino M, Storelli C, Martines G, De Caterina R, & Carluccio M A (2012) Mediterranean diet polyphenols reduce inflammatory angiogenesis through MMP-9 and COX-2 inhibition in human vascular endothelial cells: a potentially protective mechanism in atherosclerotic vascular disease and cancer. *Archives Biochemistry and Biophysics*, 527, 81–89.

[4]   Cicerale S, Lucas L, & Keast R (2010) Biological activities of phenolic compounds present in virgin olive oil. *Internacional Journal of Molecular Sciences*, 11, 458–479.

[5]   Covas M-I (2008) Bioactive effects of olive oil phenolic compounds in humans: reduction of heart disease factors and oxidative damage. *Inflammopharmacology*, 16, 216–218.

[6]   Ghanbari R, Anwar F, Alkharfy K M, Gilani A-H, & Saari N (2012) Valuable nutrients and functional bioactives in different parts of olive (*Olea europaea* L.)-A Review. *International Journal of Molecular Sciences*, 13, 3291–3340.

[7]   Massaro M, Scoditti E, Carluccio M A, & De Caterina R (2010) Nutraceuticals and prevention of atherosclerosis: focus on omega-3 polyunsaturated fatty acids and Mediterranean diet polyphenols. *Cardiovascular Therapeutics*, 28, 13–19.

[8]   Alberts B, Johnson A, Lewis J, Raff M, Roberts K, & Walter P (2002) Molecular Biology of the Cell. 4th edition. New York: Garland Science, pp. 605.

[9]   Newton K, Dixit V M (2012) Signaling in innate immunity and inflammation cold. *Spring Harbor Perspectives in Biology*, 4, 1–19.

[10] Barton G M (2008) A calculated response : control of inflammation by the innate immune system. *The Journal of Clinical Investigation*, 118, 413–420.

[11] Ashley N T, Weil Z M, & Nelson R J (2012) Inflammation: Mechanisms, costs, and natural variation. *Annual Review of Ecology, Evolution and Systematics*, 43, 385–406.

[12] Costa G, Francisco V, Lopes M C, Cruz M T, & Batista MT (2012) Intracellular signaling pathways modulated by phenolic compounds: Application for new anti-inflammatory drugs discovery. *Current Medicinal Chemistry*, 19, 2876–2900

[13] Moira K (1993) Inhibition functional of apoptosis and prolongation of neutrophil longevity by inflammatory mediators. *Infection and Immunity*, 54, 283–288.

[14] Enoch S, & Leaper D J (2008) Basic science of wound healing. *Surgery (Oxford)*, 26, 31–37.

[15] Medzhitov R (2010) Inflammation 2010: new adventures of an old flame. *Cell*, 140, 771–776.

[16] Engeli S, Feldpausch M, Gorzelniak K, Hartwig F, Heintze U, Janke J, Möhlig M, Pfeiffer A F H, Luft F C, & Sharma A M (2003) Association between adiponectin and mediators of inflammation in obese women. *Diabetes*, 52, 942–947.

[17] Souza J R M, Oliveira R T, Blotta M H S L, & Coelho O R (2008) Serum levels of interleukin-6 (IL-6), interleukin-18 (IL-18) and C-reactive protein (CRP) in patients with type-2 diabetes and acute coronary syndrome without ST-segment elevation. *Arquivos Brasileiros de Cardiologia*, 90, 86–90.

[18] Kumar V, Abbas A K & Aster J C, Robbins Basic Pathology, *Chemical Mediators and Regulators of Inflammation*, Elsevier, 9th edition.

[19] Macglashan D (2003) Histamine: A mediator of inflammation. *The Journal of Allergy and Clinical Immunology*, 112, 53–59.

[20] Das U N (2010) Current and emerging strategies for the treatment and management of systemic lupus erythematosus based on molecular signatures of acute and chronic inflammation. *Journal of Inflammation Research*, 3, 143–170.

[21] Serhan C N (2007) Resolution phase of inflammation: novel endogenous anti-inflammatory and proresolving lipid mediators and pathways. *Annual Review of Immunology*, 25, 101–137.

[22] Harizi H, Corcuff J-B, & Gualde N (2008) Arachidonic-acid-derived eicosanoids: roles in biology and immunopathology. *Trends in Molecular Medicine*, 14, 461–469.

[23] Turini M E, & DuBois R N (2002) Cyclooxygenase-2: A Therapeutic Target. *Annual Review of Medicine*, 53, 35–57.

[24] Funk C D (2006) Arteriosclerosis, Thrombosis, and Vascular Biology. *American Heart Association*, 26, 1204–1206.

[25] Dinarello CA (1997) Role of pro- and anti-inflammatory cytokines during inflammation: experimental and clinical findings. *Journal of Biological Regulators and Homeostatic Agents*, 11, 91–103.

[26] Iqbal M, Verpoorte R, Korthout J, & Mustafa N R (2012) Phytochemicals as a potential source for TNF-α inhibitors. *Phytochemistry Reviews*, 12, 65–93.

[27] Scallon B J, Moore M A, Trinh H, Knight D M, & Ghrayeb J (1995) Chimeric anti-TNF-α monoclonal antibody cA2 binds recombinant transmembrane TNF-α and activates immune effector functions. *Cytokine*, 7, 251–259.

[28] Barone F C, & Feuerstein G Z (1999) Inflammatory mediators and stroke: new opportunities for novel therapeutics. *Journal of Cerebral Blood Flow and Metabolism*, 19, 819–834.

[29] Numcrof R P, Dinarello C & Asadullah K *Cytokines as potential therapeutic targets for inflammatory skin diseases.* Springer Berlin Heidelberg, 2006, vol 56.

[30] Romagnani S (1991) Type 1 T helper and type 2 T helper cells: functions, regulation and role in protection and disease. *International Journal of Clinical and Laboratory.* 21,152–158.

[31] Janeway C A, Travers P, Walport M, & Shlomchik M J Immunobiology, *The Immune System in Health and Disease*, New York: Garland Science, 2001, 5th edition.

[32] Cavalcanti Y V N, Brelaz M C A, Neves J K D A L, Ferraz J C, & Pereira V R A (2012) Role of TNF-Alpha, IFN-Gamma, and IL-10 in the development of pulmonary tuberculosis. *Pulmonary Medicine*, 1–10.

[33] Liu B, Gao H-M, Wang J-Y, Jeohn G-H, Cooper C L, & Hong J-S (2002) Role of nitric oxide in inflammation-mediated neurodegeneration. *Annals of the New York Academy of Sciences*, 962, 318–331.

[34] Eisner F, Jacob P, Frick J-S, Feilitzsch M, Geisel J, Mueller M H, Küper M, Raybould H E, Königsrainer I, & Glatzle J (2011) Immunonutrition with long-chain fatty acids prevents activation of macrophages in the gut wall. *Journal of Gastrointestinal Surgery*, 15, 853–859.

[35] Pontoniere P (2013) Inflammation and olive polyphenols-A perspective review of supporting literature. *Agro Food Industries*, 23, 69–71.

[36] Visioli F, Poli A, & Galli C (2002) Antioxidant and other biological activities of phenols from olives and olive oil. *Medicinal Research Reviews*, 22, 65–75.

[37] Perona J S, Moruno R C, & Gutierrez V R (2006) The role of virgin olive oil components in the modulation of endothelial function. *Journal of Nutritional Biochemistry*, 17, 429–445.

[38] Maiuri MC, Stefano D De, Meglio P Di, Irace C, Savarese M, Sacchi R, Cinelli MP, & Carnuccio R (2005) Hydroxytyrosol, a phenolic compound from virgin olive oil, prevents macrophage activation. *Naunyn-Shmiedebergs Archives of Pharmacology*, 371, 457–465.

[39] Richard N, Arnold S, Hoeller U, Kilpert C, Wertz K, & Schwager J (2011) Hydroxytyrosol is the major anti-inflammatory compound in aqueous olive extracts and impairs cytokine and chemokine production in macrophages. *Planta Medica*, 77, 1890–1897.

[40] Rosignoli P, Fuccelli R, Fabiani R, Servili M &, Morozzi G (2013) Effect of oil phenols on the production of inflammatory mediators in freshly isolated human monocytes. *The Journal of Nutritional Biochemistry*, 24, 1513–1519.

[41] Zhang X, Cao J &, Zhong L (2009) Hydroxytyrosol inhibits pro-inflammatory cytokines, iNOS, and COX-2 expression in human monocytic cells. *Naunyn Schmiedeberg's Archives of Pharmacology*, 379, 581–586.

[42] Puerta R de la, Gutierrez V Ruiz, & Hoult J R (1999) Inhibition of leukocyte 5-lipoxygenase by phenolics from virgin olive oil. *Biochemical Pharmacology*, 57, 445–449.

[43] Miles E, Zoubouli P, & Calder P C (2005) Differential anti-inflammatory effects of phenolic compounds from extra virgin olive oil identified in human whole blood cultures. *Nutrition*, 21, 389–394.

[44] Dell'Agli M, Fagnani R, Galli G V, Maschi O, Gilardi F, Bellosta S, Crestani M, Bosisio E, Fabiani E, & Caruso D (2010) Olive oil phenols modulate the expression of metalloproteinase 9 in THP-1 cells by acting on nuclear factor-kappaB signaling. *Journal of Agricultural and Food Chemistry*, 58, 2246–2252.

[45] Beauchamp G K, Keast R S J, Morel D, Lin J, Pika J, Han Q, Lee C-H, Smith A B, & Breslin P S (2005) Phytochemistry: ibuprofen-like activity in extra-virgin olive oil. *Nature*, 437, 45–46.

[46] Cardeno A, Sanchez-Hidalgo M, Aparicio-Soto M, & Alarcón-de-la-Lastra C (2014) Unsaponifiable fraction from extra virgin olive oil inhibits the inflammatory response in LPS-activated murine macrophages. *Food Chemistry*, 147, 117–123.

[47] Jialal I, Devaraj S, & Kaul N (2001) Symposium : Molecular Mechanisms of Protective Effects of Vitamin E in Atherosclerosis The effect of α -tocopherol on monocyte proatherogenic activity 1. *The Journal of Nutrition*, 389–394.

[48] Wu D, Hayek M G & Meydani S N (2001) Symposium : Molecular mechanisms of protective effects of vitamin e in atherosclerosis vitamin e and macrophage cyclooxygenase regulation in the aged 1, 2. *The Journal of Nutrition*, 382–388.

[49] Vivancos M, Moreno JJ (2008) Effect of resveratrol, tyrosol and beta-sitosterol on oxidised low-density lipoprotein-stimulated oxidative stress, arachidonic acid release and prostaglandin E2 synthesis by RAW 264.7 macrophages. *British Journal of Nutrition*, 99, 1199–1207.

[50] Moreno J J (2003) Effect of olive oil minor components on oxidative stress and arachidonic acid mobilization and metabolism by macrophages RAW 264.7. *Free Radical Biology & Medicine*, 35, 1073–1081.

[51] Inhibitors H R, Rice R Y, Polyphenols O O, Patrick L, & Uzick M (2001) Cardiovascular Disease : C-Reactive Protein and the Inflammatory Disease Paradigm : A Review of the Literature. *Alternative Medicine Review*, 6, 248–271.

[52] Vila L (2004) Cyclooxygenase and 5-lipoxygenase pathways in the vessel wall: role in atherosclerosis. *Medicinal Research Reviews*, 24, 399–424.

[53] Haddy N (2003) IL-6, TNF-α and atherosclerosis risk indicators in a healthy family population: the STANISLAS cohort. *Atherosclerosis*, 170, 277–283.

[54] Lou-Bonafonte J M, Arnal C, Navarro M, & Osad a J (2012) Efficacy of bioactive compounds from extra virgin olive oil to modulate atherosclerosis development. *Molecular Nutrition & Food Research*, 56, 1043–1057.

[55] Camargo A, Delgado J, Garcia-Rios A, Cruz-Teno C, Yubero-Serrano E M, Perez-Martinez P, Gutierrez-Mariscal F M, Lora-Aguilar P, Rodriguez-Cantalejo F, Fuentes-Jimenez F, Tinahones F J, Malagon M M, Perez-Jimenez F, & Lopez-Miranda J (2012) Expression of proinflammatory, proatherogenic genes is reduced by the Mediterranean diet in elderly people. *British Journal of Nutrition*, 108, 500–508.

[56] Widmer R J, Freund M, Flamme J, Sexton J, Lennon R, Romani A, Mulinacci N, Vinceri F F, Lerman L O, & Lerman A (2013) Beneficial effects of polyphenol-rich

olive oil in patients with early atherosclerosis. *European Journal of Nutrition,* 52, 1223–1231.

[57] Urpi-Sarda M, Casas R, Chiva-Blanch G, Sau E, Andres-Lacueva C, Llorach R, Ruiz-gutierrez V, Lamuela-Raventos R M, & Estruch R (2012) The Mediterranean diet pattern and its main components are associated with lower plasma concentrations of tumor necrosis factor receptor 60 in patients at high risk for cardiovascular disease 1 – 4. *The Journal of Nutrition and Disease,* 1019–1025.

[58] Fitó M, Cladellas M, Torre R de la, Martí J, Muñoz D, Schröder H, Alcántara M, Pujadas-Bastardes M, Marrugat J, López-Sabater M C, Bruguera J, & Covas M I (2008) Anti-inflammatory effect of virgin olive oil in stable coronary disease patients: a randomized, crossover, controlled trial. *European Journal of Clinical Nutrition,* 62, 570–574.

[59] Allman-Farinelli M, Gomes K, Favaloro E J & Petocz P (2005) A diet rich in high-oleic-acid sunflower oil favorably alters low-density lipoprotein cholesterol, triglycerides, and factor VII coagulant activity. *Journal of the American Dietetic Association,* 105, 1071–1079.

[60] Kris-Etherton P M, Pearson T A, Wan Y, Hargrove R L, Moriarty K, Fishell V, & Etherton T D (1999) High – monounsaturated fatty acid diets lower both plasma cholesterol and triacylglycerol concentrations 1–3 Experimental design. *The American Journal of Clinical Nutrition,* 70, 1009–1015.

[61] Parthasarathy S, Khoo J C, Miller E, Barnett J, Witztum J L, & Steinberg D (1990) Low density lipoprotein rich in oleic acid is protected against oxidative modification: implications for dietary prevention of atherosclerosis. *Proceedings of the National Academy of Sciences of the United States of America,* 87, 3894–3898.

[62] Farràs M, Valls R M, Fernández-Castillejo S, Giralt M, Solà R, Subirana I, Motilva M-J, Konstantinidou V, Covas M-I, & Fitó M (2013) Olive oil polyphenols enhance the expression of cholesterol efflux related genes in vivo in humans. A randomized controlled trial. *Journal of Nutritional Biochemistry,* 24, 1334–1339.

[63] Massaro M, Carluccio M A, & Caterina R De (1999) Direct vascular antiatherogenic effects of oleic acid: a clue to the cardioprotective effects of the Mediterranean diet. *Cardiologia,* 44, 507–513.

[64] Harvev K A, Walker C L, Xu Z, Whitlev P, Pavlina T M, Hise M, Zaloga G P, & Siddigui R A (2010) Oleic acid innibits stearic acid-induced inhibition of cell growth and pro-inflammatory responses in human aortic endothelial cells. *The Journal of Lipid Research,* 51, 3470–3480.

[65] Sanadgol N, Mostafaie A, Bahrami G, Mansouri K, Ghanbari F, & Bidmeshkipour A (2010) Elaidic acid sustains LPS and TNF-α induced ICAM-1 and VCAM-1 expression on human bone marrow endothelial cells (HBMEC). *Clinical Biochemistry,* 43, 968–972.

[66] Covas M-I, Nyyssönen K, Poulsen H E, Kaikkonen J, Zunft H-J F, Kiesewetter H, Gaddi A, Torre R, Mursu J, Bäumler H, Nascetti S, Salonen J T, Fitó M, Virtanen J, & Marrugatv J (2006) The effect of polyphenols in olive oil on heart disease risk factors: a randomized trial. *Annals of Internal Medicine,* 145, 333–341.

[67] Covas M, Khymenets O, Nyyssonen K, Konstantinidou V, Zunft H, Castan O, Vila J, Torre R D, & Mun D (2012) Protection of LDL from oxidation by olive oil polyphenols

is associated with a downregulation of CD40-ligand expression and its downstream. *The American Journal of Clinical Nutrition*, 95, 1238–1244.

[68] Sprecher D L, Massien C, Pearce G, Billin A N, Perlstein I, Willson T M, Hassall D G, Ancellin N, Patterson S D, Lobe D C, & Johnson T G, (2007) Triglyceride: high-density lipoprotein cholesterol effects in healthy subjects administered a peroxisome proliferator activated receptor delta agonist. *Arteriosclerosis, Thrombosis, and Vascular Biology*, 27, 359–365.

[69] Zrelli H, Wei Wu C, Zghonda N, Shimizu H, & Miyazaki H (2013) Combined Treatment of Hydroxytyrosol with Carbon Monoxide-Releasing Molecule-2 Prevents TNF α -Induced Vascular Endothelial Cell Dysfunction through NO Production with Subsequent NF κ B Inactivation. *BioMed Research International*, 1–10.

[70] Carluccio M A, Siculella L, Ancora M A, Massaro M, Scoditti E, Storelli C, Visioli F, Distante A, & Caterina R D (2003) Olive oil and red wine antioxidant polyphenols inhibit endothelial activation: antiatherogenic properties of Mediterranean diet phytochemicals. *Arteriosclerosis, Thrombosis and Vascular Biology*, 23, 622–629.

[71] Correa J A G, López-Villodres J A, Asensi R, Espartero J L, Rodríguez-Gutiérez G, & De La Cruz J P (2009) Virgin olive oil polyphenol hydroxytyrosol acetate inhibits in vitro platelet aggregation in human whole blood: comparison with hydroxytyrosol and acetylsalicylic acid. *British Journal of Nutrition*, 101, 1157–1164.

[72] Zrelli H, Matsuoka M, Kitazaki S, Zarrouk M, & Miyazaki H (2011) Hydroxytyrosol reduces intracellular reactive oxygen species levels in vascular endothelial cells by upregulating catalase expression through the AMPK-FOXO3a pathway. *European Journal of Pharmacology*, 660, 275–282.

[73] Dell'Agli M, Fagnani R, Mitro N, Scurati S, Masciadri M, Mussoni L, Galli G V, Bosisio E, Crestani M, Fabiani E De, Tremoli E, & Caruso D (2006) Minor components of olive oil modulate proatherogenic adhesion molecules involved in endothelial activation. *Journal of Agricultural and Food Chemistry*, 54, 3259–3264.

[74] Granados-Principal S, Quiles J L, Ramirez-Tortosa C L, Ochoa-Herrera J, Perez-Lopez P, Pulido-Moran M, & Ramirez-Tortosa M (2012) Squalene ameliorates atherosclerotic lesions through the reduction of CD36 scavenger receptor expression in macrophages. *Molecular Nutrition & Food Research*, 56, 733–740.

[75] Bullon P, Quiles J L, Morillo J M, Rubini C, Goteri G, Granados-Principal S, Battino M, & Ramirez-Tortosa M (2009) Gingival vascular damage in atherosclerotic rabbits: hydroxytyrosol and squalene benefits. *Food and Chemical Toxicology*, 47, 2327–2331.

[76] Perona J S, Cabello-Moruno R, & Ruiz-Gutierrez V (2006) The role of virgin olive oil components in the modulation of endothelial function. *Journal of Nutritional Biochemistry*, 17, 429–445.

[77] Rodriguez-Rodriguez R, Herrera M D, Sotomayor M A, & Ruiz-Gutierrez V (2007) Pomace olive oil improves endothelial function in spontaneous hypertensive rats by increasing endothelial nitric oxide synthase expression. *American Journal of Hypertension*, 20, 728–734.

[78] Graham V S, Lawson C, Wheeler-Jones C P D, Perona J S, Ruiz-Gutierrez V, & Botham K M (2012) Triacylglycerol-rich lipoproteins derived from healthy donors fed different olive oils modulate cytokine secretion and cyclooxygenase-2 expression in macrophages: the potential role of oleanolic acid. *European Journal of Nutrition*, 51 301–309.

[79] Allouche Y, Beltrán G, Gaforio J J, Uceda M, & Mesa M D (2010) Antioxidant and antiatherogenic activities of pentacyclic triterpenic diols and acids. *Food and Chemical Toxicology*, 48, 2885–2890.

[80] Carter M J, Lobo J, & Travis S P (2004) Guidelines for the management of inflammatory bowel disease in adults. *Gut*, 53, 1–16.

[81] Shanahan F (2000) Probiotics and inflammatory bowel disease: is there a scientific rationale. *Inflammatory Bowel Disease*, 6, 107–115.

[82] Hanauer S B (2006) Inflammatory bowel disease: epidemiology, pathogenesis, and therapeutic opportunities. *Inflammatory Bowel Disease*, 12, 3–9.

[83] Head K, & Jurenka J S (2003) Inflammatory bowel disease Part 1: ulcerative colitis-pathophysiology and conventional and alternative treatment options. *Alternative Medicine Review*, 8, 247–283.

[84] Hegazi R F, Saad R S, Mady H, Matarese L E, O'Keefe S & Kandil H M (2006) Dietary fatty acids modulate chronic colitis, colitis-associated colon neoplasia and COX-2 expression in IL-10 knockout mice. *Nutrition*, 22, 275–282.

[85] Camuesco D, Galvez J, Nieto A, Comalada M, Rodriguez-Cabezas M E, Concha A, Xaus J, & Zarzuelo A (2005) Biochemical and molecular actions of nutrients dietary olive oil supplemented with fish oil, rich in EPA and DHA (n-3) polyunsaturated fatty acids, attenuates colonic inflammation in rats with DSS-induced colitis. *The Journal of Nutrition*, 135, 687–694.

[86] Adachi M, Kurotani R, Morimura K, Shah Y, Sanford M, Madison B B, Gumucio D L, Marin H E, Peters J M, Young H, & Gonzalez F J (2006) Peroxisome proliferator activated receptor gamma in colonic epithelial cells protects against experimental inflammatory bowel disease. *Gut*, 55, 1104–1113.

[87] Sánchez-Fidalgo S, Cárdeno A, Sánchez-Hidalgo M, Aparicio-Soto M, & Lastra C A de la (2013) Dietary extra virgin olive oil polyphenols supplementation modulates DSS-induced chronic colitis in mice. *The Journal of Nutritional Biochemistry*, 24, 1401–1413.

[88] Giner E, Recio M-C, Ríos J-L, & Giner R M (2013) Oleuropein protects against dextran sodium sulfate-induced chronic colitis in mice. *Journal of Natural Products*, 76, 1113–1120.

[89] Sánchez-Fidalgo S, Cárdeno A, Sánchez-Hidalgo M, Aparicio-Soto M, Villegas I, Rosillo M, & Lastra C A de la (2012) Dietary unsaponifiable fraction from extra virgin olive oil supplementation attenuates acute ulcerative colitis in mice. *European Journal of Pharmaceutical Science*, 48, 572–581.

[90] Hiratsuka T, Inomata M, Hagiwara S, Kono Y, Shiraishi N, Noguchi T, & Kitano S (2013) Bolus injection of newly synthesized vitamin E derivative ETS-GS for the treatment of acute severe ulcerative colitis in a mouse model. New vitamin E derivative for acute severe UC. *International Journal of Colorectal Disease*, 28, 305–311.

[91] Marx J (2004) Inflammation and cancer: The link grows stronger. *Cancer research*, 306, 966–968.

[92] Balkwill F, & Coussens LM (2004) Cancer: An inflammatory link. *Nature*, 431, 405–406.

[93] Grivennikov, S I, Greten, F R, & Karin, M (2010). Immunity, Inflammation, and Cancer. *Cell*, 140, 883–899.

[94] Coussens LM, & Werb Z (2002) Inflammation and cancer. *Nature*, 420, 860-867.

[95] Lu HT, Ouyang W M, & Huang C S (2006) Inflammation, a key event in cancer development. *Molecular Cancer Research*, 4, 221–233.

[96] Binukumar B, & Mathew A (2005) Dietary fat and risk of breast cancer. *World Journal of Surgical Oncology*, 3, 45–45.

[97] La Vecchia C, & Bosetti C (2006) Diet and cancer risk in Mediterranean countries: Open issues. *Public Health Nutrition*, 9, 1077–1082.

[98] Panagiotakos D B, Dimakopoulou K, Katsouyanni K, Bellander T, Grau M, Koenig W, Lanki T, Pistelli R, Schneider A, Peters A, & Grp A S (2009) Mediterranean diet and inflammatory response in myocardial infarction survivors. *International Journal of Epidemiology*, 38, 856–866.

[99] Ismail A M, In L L A, Tasyriq M, Syamsir D R, Awang K, Omer Mustafa A H, Idris O F, Fadl-Elmula I, & Hasima N (2013) Extra virgin olive oil potentiates the effects of aromatase inhibitors via glutathione depletion in estrogen receptor-positive human breast cancer (MCF-7) cells. *Food and Chemical Toxicology*, 62, 817–824.

[100] Fu S P, Arraez-Roman D, Segura-Carretero A, Menendez J A, Menendez-Gutierrez M P, Micol V, & Fernandez-Gutierrez A (2010) Qualitative screening of phenolic compounds in olive leaf extracts by hyphenated liquid chromatography and preliminary evaluation of cytotoxic activity against human breast cancer cells. *Analytical and Bioanalytical Chemistry*, 397, 643–654.

[101] Sanchez-Fidalgo S, Villegas I, Cardeno A, Talero E, Sanchez-Hidalgo M, Motilva V, & Alarcon de la Lastra C (2010) Extra-virgin olive oil-enriched diet modulates DSS-colitis-associated colon carcinogenesis in mice. *Clinical Nutrition*, 29, 663–673.

[102] Fernandez-Mar M I, Mateos R, Garcia-Parrilla M C, Puertas B, & Cantos-Villar E (2012) Bioactive compounds in wine: Resveratrol, hydroxytyrosol and melatonin: A review. *Food Chemistry*, 130, 797–813.

[103] Bernini R, Merendino N, Romani A, & Velotti F (2013) Naturally Occurring Hydroxytyrosol: Synthesis and Anticancer Potential. *Current Medicinal Chemistry*, 20, 655–670.

[104] Cornwell D G, & Ma J (2008) Nutritional benefit of olive oil: The biological effects of hydroxytyrosol and its arylating quinone adducts. *Journal of Agricultural and Food Chemistry*, 56, 8774–8786.

[105] Scoditti E, Calabriso N, Massaro M, Pellegrino M, Storelli C, Martines G, De Caterina R, & Carluccio MA (2012) Mediterranean diet polyphenols reduce inflammatory angiogenesis through MMP-9 and COX-2 inhibition in human vascular endothelial cells: A potentially protective mechanism in atherosclerotic vascular disease and cancer. *Archives of Biochemistry and Biophysics*, 527, 81–89.

[106] Cardeno A, Sanchez-Hidalgo M, Cortes-Delgado A, & Alarcon de la Lastra C (2013) Mechanisms involved in the antiproliferative and proapoptotic effects of unsaponifiable fraction of extra virgin olive oil on HT-29 cancer cells. *Nutrition and Cancer*, 65, 908–918.

[107] Opie R S, Ralston R A, & Walker K Z (2013) Adherence to a Mediterranean-style diet can slow the rate of cognitive decline and decrease the risk of dementia: a systematic review. *Nutrition & Dietetics*, 70, 206–217.

[108] Sofi F, Cesari F, Abbate R, Gensini G F, & Casini A (2008) Adherence to Mediterranean diet and health status: meta-analysis. *British Medical Journal*, 337, 1–7.

[109] Khalatbary A R (2013) Olive oil phenols and neuroprotection. *Nutritional Neuroscience*, 16, 243–249.

[110] Pérez-López F R, Chedraui P, Haya J, & Cuadros J L (2009) Effects of the Mediterranean diet on longevity and age-related morbid conditions. *Maturitas*, 64, 67–79.

[111] Farr S A, Price T O, Dominguez L J, Motisi A, Saiano F, Niehoff M L, Morley J E, Banks W A, Ercal N, & Barbagallo M (2012) Extra virgin olive oil improves learning and memory in SAMP8 mice. *Journal of Alzheimers Disease*, 28, 81–92.

[112] Beauchamp G K, Keast R S J, Morel D, Lin J, Pika J, Han Q, Lee C-H, Smith A B, & Breslin P A S (2005) Phytochemistry: Ibuprofen-like activity in extra-virgin olive oil. *Nature*, 437, 45–46.

[113] Pitt J, Roth W, Lacor P, Smith A B, Blankenship M, Velasco P, De Felice F, Breslin P, & Klein W L (2009) Alzheimer's-associated Abeta oligomers show altered structure, immunoreactivity and synaptotoxicity with low doses of oleocanthal. *Toxicology and Applied Pharmacology*, 240, 189–197.

[114] Li W, Sperry J B, Crowe A, Trojanowski J Q, Smith AB, & Lee V M Y (2009) Inhibition of tau fibrillization by oleocanthal via reaction with the amino groups of tau. *Journal of Neurochemistry*, 110, 1339–1351.

[115] Monti M C, Margarucci L, Riccio R & Casapullo A (2012) Modulation of tau protein fibrillization by oleocanthal. *Journal of Natural Products*, 75, 1584–1588.

[116] Abuznait A H, Qosa H, Busnena B A, El Sayed K A, & Kaddoumi A (2013) Olive-Oil-derived oleocanthal enhances beta-amyloid clearance as a potential neuroprotective mechanism against Alzheimer's Disease: In Vitro and in Vivo Studies. *ACS Chemical Neuroscience*, 4, 973–982.

[117] Omar S H (2010) Cardioprotective and neuroprotective roles of oleuropein in olive. *Saudi Pharmaceutical Journal*, 18, 111–121.

In: Virgin Olive Oil
Editor: Antonella De Leonardis

ISBN: 978-1-63117-656-2
© 2014 Nova Science Publishers, Inc.

Chapter 12

OLIVE OIL: MOLECULAR MECHANISMS AND CARDIOVASCULAR PROTECTIVE ROLE

Sherif Y. Shalaby, Brandon J. Sumpio and Bauer E. Sumpio[*]
Yale School of Medicine, New Haven, CT, US

ABSTRACT

The Mediterranean diet, with its high consumption of olive oil, has gained worldwide recognition in the possible prevention of cardiovascular related diseases through epidemiological and clinical studies that illustrate the possible long term health benefits. This chapter will review the evidence based studies related to the wide scale benefits of olive oil. We will also present the diverse effects of olive oil and its chemical constituents on various molecular mechanisms that affect vascular cell function.

Keywords: Atherosclerosis, cardiovascular disease, olive oil, Mediterranean diet

INTRODUCTION

Although, the benefits of the Mediterranean diet has been recognized by modern scientific approaches, based on archaeo-botany and written records, this diet existed for centuries in Ancient Egyptian, Greek, biblical, and Roman times [1]. The Mediterranean diet is characterized by a high intake of monounsaturated fatty acids (MUFA) because of olive oil's role as the primary source of fat intake, specifically oleic acid, and low in saturated fatty acids (SFA) [2]. It is also rich in plant proteins, whole grains, and fish, with a moderate intake of alcohol, and low consumption of red meats, refined grains, and sweets [3]. The Mediterranean diet has been gaining attention in the science of nutrition and has increased popularity worldwide. The Mediterranean diet continues to withstand the dietary modification of the industrial age; inhabitants of Southern European and North African regions

[*] Yale School of Medicine, 333 Cedar Street, BB 204 New Haven, CT 06510 USAEmail: bauer.sumpio@yale.edu.

surrounding the Mediterranean Sea have a longer life expectancy and lower risk of chronic diseases than any other regions of the world [4].

A large study analyzed the regional variation in cardiovascular morality within Europe (Figure 1). The report confirmed the lower cardiovascular mortality in Mediterranean countries compared with northern European countries or the United States [5] and the inverse association between the Mediterranean diet [6, 7] or olive-oil consumption [8] on the incidence of stroke.

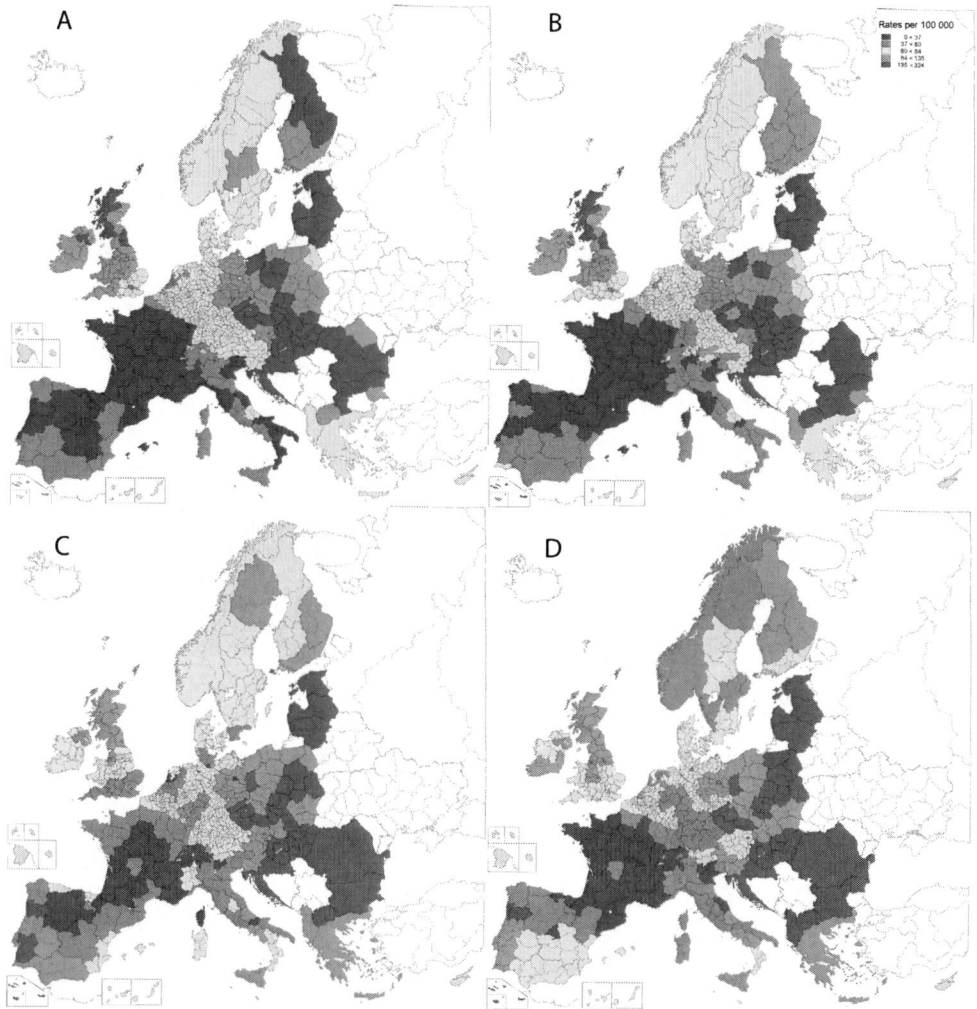

Figure 1. Age-standardized mortality of cardiovascular diseases in European regions (age group 45–74 years; year 2000). Ischemic heart disease (IHD) in men (A) and women (B); Cerebrovascular disease (CVD) in men (C) and women (D). From Reference [5].

OLIVE OIL AND GLOBAL HEALTH IMPACT

Olive oil is a hallmark of the Mediterranean diet and dates back to some 2000 years B.C. where trade between Palestine and Egypt of olive oil and wine has been documented on hieroglyphs. During ancient times, olives were consumed by farmers, accompanied travelers, and nomads. The trade route expanded to make the Mediterranean countries the largest consumers, producers, and exporters of both olives and olive oil, accounting for approximately 90% of the total olive oil produced.

Epidemiological studies comparing high olive oil consuming countries with other neighboring countries demonstrated that the Mediterranean countries have a reduced cardiovascular risk compared to their northern counterparts. The 1968 "Seven Countries" study evidenced low cardiovascular mortality in the Mediterranean region after a 25 year follow-up of 11 579 men and proposed that diet could be one of the reasons. For instance, it was observed that coronary heart disease was low in men who inhabited the Greek island of Crete [9]. In Crete, olive oil provides approximately 29% of the total dietary energy. The occurrence of myocardial infarction was 26 in 10,000 Cretans despite their high fat diet (33% to 40% of caloric intake) [10]. This was in contrast to the Finland cohort, where the rate was 1,074 in 10,000 [9]. This study also demonstrated that the high intake of fat predominantly from olive oil was not associated with increased plasma cholesterol levels. Although a variety of factors play a role in heart disease mortality rates, including differing regional diets, health care quality, and socioeconomic status, it is generally accepted that the Finnish diet of fatty red meats, butter, and bread is highly conducive to heart disease when compared with the Mediterranean diet. Fat and calorie dense foods support the physically demanding Finnish lifestyle of living and laboring under cold and wet conditions. Other factors, including genetic differences, stress levels, and Finland's arduous work environments could also contribute to these observed health disparities. In contrast, whole grains, fresh vegetables, and olive oil provide sufficient calories in the temperate Mediterranean basin. It should be noted, that other confounding variables, such as lifestyle, exercise, and environment could play a contributory role in the health of people in the Mediterranean region [11].

In a recent large multicenter randomized trial, 7447 patients aged from 55 – 80 years, half of them with chronic diseases or with significant dyslipidemia, received free provision of extra-virgin olive oil (50 g/ day) or mixed nuts (30 g of walnut, almonds, and hazelnuts), or small non food gifts [12]. The patients were followed up for a median of 4.8 years and the olive oil or mixed nuts intervention resulted in statistically significant reduction of the rate of major cardiovascular events (myocardial infarction, stroke, or death from cardiovascular causes) (Figure 2). However, only reduction of stroke reached statistical significance. The reason for this finding is unclear but could be due to stronger effects of the olive oil and nuts on specific risk factors for stroke or to a beta error of low numbers to identify effects on myocardial infarction. Although there were some weaknesses of this study including lack of genetic variations (mostly European white), a change in the protocol halfway through the trial, and a dropout in the control group, the results are compelling. Further trials that answers the benefits of the Mediterranean diet supplemented with olive oil and nuts for individuals with low or no risk of cardiovascular events needs to be investigated.

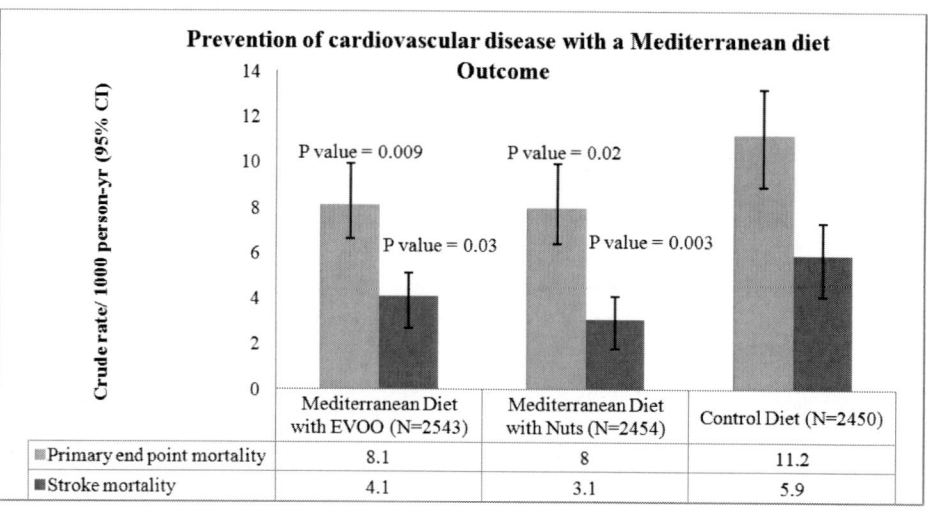

Figure 2. Statically significant clinical trial outcome of prevention of cardiovascular disease with a Mediterranean diet. The primary end point was a composite of myocardial infarction, stroke, and death from cardiovascular causes. Mediterranean Diet with extra virgin olive oil (EVOO) or nuts showed protection from cardiovascular related mortality in general and stroke specifically compared to a controlled diet. Adapted from reference [12].

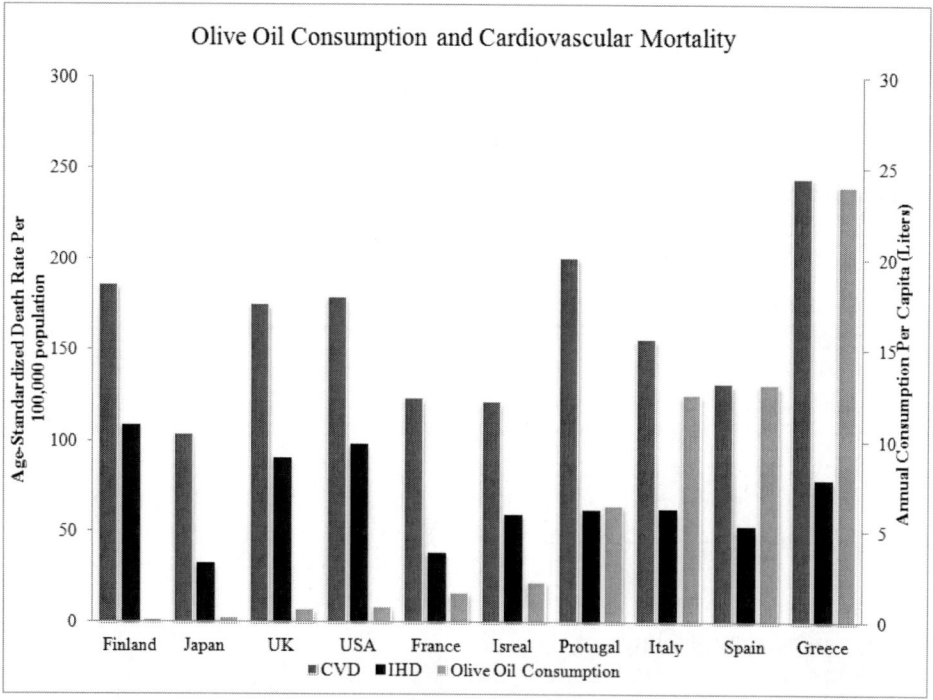

Figure 3. Cardiovascular disease (CVD) and ischemic heart disease (IHD) mortality (left axis) in selected countries in order of increasing olive oil consumption (right axis). CVD and IHD from World Health Organization (WHO, 2004) [74]. Olive oil consumption data from Olive oil council statistics [75].

The World Health Organization MONICA (monitoring trends and determinants in cardiovascular disease) project reported on the 10-year cardiovascular morbidity from 37 populations in 21 countries across four continents [13]. High consuming olive oil countries such as Spain and Italy had the lowest incidence of coronary events in men and women while, low consuming olive oil countries such as Finland and United Kingdom had the highest incidence. Interestingly, the highest ischemic heart disease mortality was observed (Figure. 3) in some countries that consume modest amounts of olive oil. Greece had the highest consumption of olive oil but also highest mortality of ischemic heart disease as well as cerebrovascular disease (Figure 3). This appears contradictory to the hypothesis of olive oil having a cardio-protective role. However, a subsequent case control study on a Greek population with exclusive use of olive oil was associated with a 47% lower chance of having acute coronary syndrome comparedto nonusers, after adjusting for physical and lifestyle characteristics [14]. These studies illustrate that there may not be an absolute relationship between olive oil consumption, cardiovascular, and cerebrovascular mortality. For example, Finland has the highest incidence of coronary events rate but the lowest rate for coronary mortality in the MONICA study. This emphasizes the need for further well controlled studies to determine whether olive oil can effectively reduce the incidence of cardiovascular diseases in the population after carefully accounting for heterogeneous epidemiological factors.

OLIVE OIL COMPOSITION

The Mediterranean climate consists of dry summers and mild winters, which provide a favorable extended growing season for vegetables and fruits [15]. The region's substantial sun exposure has been correlated to a high antioxidant content in plants. It has been postulated that vegetation native to the area has augmented its production of antioxidants to defend against reactive oxygen species produced during photosynthesis. The most widespread species of olive is the Oleaeuropaea and its genus includes 35 species of evergreen shrubs and trees [16]. Olive trees have an unusual ability to develop roots from temporary buds at the lower end of their trunks, are resistant to severe weather conditions, and are able to grow in infertile soil. Olive fruit maturation spans several months and its taste and chemical composition is dependent on growing conditions, including latitude, water availability, and temperature. The maturation, harvesting, and developing process of olives is heavily dependent on regional techniques.

Olive oil extraction is conducted through pressure, centrifugation, and percolation. Nonedible olive oil undergoes a refining process and is blended with edible oils to produce regular olive oil [16]. Virgin olive oil is obtained under mechanical conditions that do not alter its composition and it is not mixed with other oils. Extra virgin olive oil (EVOO) is the highest quality olive oil and accounts for only 10% of oil produced. It has a free acidity, primarily oleic acid, of 0.8%. Experts judge it for taste, mouth feel, and aroma; the oil tends to be most delicate in flavor. Refined olive oil has a free acidity of 0.3% while regular olive oil, a blend of refined and virgin olive oils has a free acidity of 0.1% [16].

Table 1. Composition of olive oil

	Extra virgin	Virgin	Refined*[†]
Glyceride fraction			
Fatty acids (g/100 g)			
16:0 Palmitic	6.6	8.6	9.1
16:1n-7 Palmitoleic	0.4	1.1	0.6
18:0 Stearic	2.8	1.9	3.4
18:1n-9 Oleic	83.1	78.7	78.6
18:2n-6 Linoleic	5.1	8.3	6.2
18:3n-3 α-Linoleic	0.6	—	0.4
18:3n- γ-Linoleic	0.4	—	0.5
Nonglyceride fraction*			
Component (mg/kg)			
Aliphatic alcohols			
C_{18}–C_{30} alcohols	≤200	≤200	≤200
Triterpene alcohols	500–3,000	500–3,000	500–3,000
Total sterols[‡]	1,260.8	687.4	1,366.6
Cholesterol	1.9	2.8	2.0
Δ^5-Avenasterol	91.5	35.1	82.7
β-Sitosterol	1,124.4	640.9	1,268.8
Sitostanol	7.3	2.3	1.1
Stigmasterol	8.2	6.4	12.0
Proteins (μg/kg)	1.76	1.76	1.26
Nonglyceride esters	100–250	100–250	100–250
Waxes	≤250	≤250	≤350
Hydrocarbons			
Squalene[§]	4,277	ND	2,598
β-carotene	0.33–4.0	ND	ND
Polyphenols			
Lipophilic			
α-tocopherols[¶]	300	ND	200
Tocotrienols	ND	ND	None
Hydrophilic	40–1,000	40–1,000	None
Hydroxytyrosol p-Tyrosol	Considered to be the major constituents of polyphenols, although data is inconsistent in quantifying the total concentrations, which depend on irrigation and harvesting techniques.		

†Phenols are removed completely during the refining process and so are not present in refined oil. Currently there is no standard measurement for quantification of phenols in olive oil and present values cannot be compared with total polyphenol content or that of individual compounds. § Squalene can make up to 40% of the weight of minor components. ¶ α-tocopherols make up to 95% of total tocopherols. ND, No data. From reference [76].

The major components of olive oil are divided into glyceride fractions. Glycerols represent 98% of total oil weight and are composed mainly of triacylglycerols (Table 1). Oleic acid makes up 70% to 80% of the fatty acids in olive oil. Minor components are present in about 2% of oil weight and include 230 chemical compounds. These minor components are present almost exclusively in virgin olive oil because the refining process expunges these compounds. Considerable research has centered on EVOO and virgin olive oil with the belief that these minor components may contain important cardiovascular protective effects [11]. Several components of olive oil have beneficial health effects on the atherosclerotic and thrombotic pathways, which include lipid oxidation, hemostasis, platelet aggregation, coagulation, and fibrinolysis. Oleic acid, a major component, and the polyphenols (tocopherol, hydroxyty-rosol, and oleuropein) have been the most studied for their antiatherosclerotic effects [11].

OLEOCANTHAL CARDIOVASCULAR BENEFITS

Oleocanthal is a compound present in newly pressed EVOO that can be identified by the stinging of the throat; the degree of irritation correlates with its concentration [17]. Oleocanthal has natural anti-inflammatory properties that has a potency profile strikingly similar to that of ibuprofen and the two molecules share irritant properties. Ibuprofen is a potent modulator of inflammation and analgesia [18], associated with a reduction in the risk of developing some cancers [19] and of platelet aggregation in the blood [19]. Oleocanthal causes dose-dependent inhibition of COX-1 and COX-2 activities but has no effect on lipoxygenase in vitro [20]. These findings raise the possibility that long term consumption of oleocanthal may help to protect against some diseases by virtue of its ibuprofen-like COX-inhibiting activity [19], [21]. In addition, this compound may contribute to the lower incidence of heart disease on the Mediterranean population.

OLEIC ACID CARDIOVASCULAR BENEFITS

1. Low Density Lipoproteins (LDL), High Density Lipoproteins (HDL), Oxidation and Chylomicrons (CM)

Oleic acid is preventive in the development of atheromas and subsequent thrombi through their establishment of larger MUFA -to- PUFA and MUFA -to- SFA ratios, increased resistance to oxidation, and induction of larger hydrolysable CM. Increased levels of LDL are important factors in arteriosclerosis, as they facilitate transport of cholesterol to arteries. LDL, which carries about two-thirds of plasma cholesterol, can infiltrate the arterial wall and attract macrophages, smooth muscle cells, and endothelial cells. Once embedded in the intima, LDL undergoes oxidation to oxidized LDL (oxLDL), contributing to foam-cell formation. Circulating oxLDL induces transcription of adhesion factors and is chemotactic for monocytes and leukocytes, thereby inhibiting egression of macrophages from plaques [22]. Conversely, HDL are antiatherogenic. Unlike their larger counterpart, HDL primarily deliver cholesterol to the liver to be metabolized and excreted or reused. It is also hypothesized that HDL are able to dislodge cholesterol molecules from atheromas in arterial walls. LDL are less susceptible to free radical oxidation in a diet enriched by MUFA. MUFA are more stable than PUFA and more resistant to oxidation [23]. A supplemental diet of EVOO was reported to decrease LDL oxidation in rabbits with experimentally induced arteriosclerosis [24]. It also resulted in lower atherosclerotic lesions in all aortic fragments isolated from the rabbits [25]. In addition, MUFA consumption, specifically that of oleic and linoleic acids, has been linked to a decrease in human plasma levels of LDL and an increase in serum HDL [25]. In one trial, 24 human subjects diagnosed with peripheral vascular disease were fed EVOO or refined olive oil for 3 months. In the EVOO group, LDL susceptibility to oxidation was considerably lower [26].

Additional studies suggest that in a meal enriched with MUFA, the larger, more beneficial CM are secreted and rapidly cleared. CM transport dietary cholesterol and fats to the liver and periphery and are strongly atherogenic. They can penetrate the artery wall and facilitate foam-cell formation. One trial reported statistically significant differences between olive oil's ability to increase the entry of CM after ingestion of a meal compared with fish, safflower, and palm oils [27]. Patients that ingested oleic acid maintained high CM levels for the longest postprandial period of time in comparison with the group that were treated with sunflower oil, mixed oil, and beef tallow [28].

2. Hemostasis, Platelet Aggregation and Fibrinolysis

Oleic acid consumption decreases platelet sensitivity and aggregation, lowers levels of the coagulation factor VII (FVII), and increases fibrinolysis [29]. An appropriate mechanistic explanation of the ability of dietary MUFA to decrease platelet aggregation has yet to be determined. Studies suggest that changes in membrane lipid fluidity and long-chain n-3 fatty acid from oleic acid can reduce platelet sensitivity to collagen and other coagulation factors, induce a hypersensitivity to the aggregation antagonist, ADP, and inhibit thromboxane and prostaglandin synthesis [29], [30]. Fifty-one healthy adults participated in a 4-month trial with diets of high and moderate MUFA intake (18% and 15% of caloric intake, respectively) and a diet high in saturated fats (16% intake). At 8 weeks, those on the high and moderate MUFA diets demonstrated a decrease in platelet aggregation when exposed to platelet agonists, such as ADP, arachidonic acid, and collagen. The reduction in aggregatory response to ADP and arachidonic acid was sustained in the high-MUFA group for the entire 16-week trial [31].

Several animal studies have confirmed olive oil's ability to reduce thromboses. Rats fed an EVOO-enriched diet had a lower rate of thrombotic occlusion in an "aortic loop" model, a lower incidence of venous thrombosis, and an extended bleeding time relative to a control group fed a normal diet [32]. Hypercholesterolemic rabbits fed a virgin olive oil diet compared with those on a SFA diet had a substantial decrease in platelet hyperactivity, sub-endothelial thrombogenicity, and platelet lipid peroxide production. In addition to marked changes in cholesterol, triglycerides, and HDL, the olive oil supplement stimulated endothelial synthesis of prostacyclin, and lowered thromboxane B2 plasma levels [33]. Von Willebrand Factor (vWF) induces irreversible binding of platelets to the subendothelial collagen layer [34]. In human trials, 3 weeks of a MUFA-enriched diet resulted in a substantial reduction in vWF levels. In another study of 25 people on low-fat, high-MUFA, or high- SFA diets, the MUFA diet induced a statistically significant decrease in VWF activity, 71.8% compared with 78.6% for the SFA diet [2].

During plaque rupture, tissue factor and FVII proteins are released as a component in the coagulation cascade. High FVII levels increase the risk of fatal coronary heart disease because of coronary thrombosis [35]. Tissue factor complexes with FVII to activate the fibrin cascade. There are varying results on the effects of oleic acid on tissue factor and its inhibitor, tissue factor pathway inhibitor (TFPI). A high MUFA diet has been linked to decreased levels of FVII [36]. A recent study with an isocaloric replacement of a MUFA-enriched

Mediterranean diet found a reduction in TFPI. Although this can be seen as detrimental, it has been suggested that low TFPI levels indicate the presence of the protease in the endothelium, which has a regulatory effect on thrombogenesis [2]. In human trials, the effect of MUFA, oleic acid, and other vegetable oils on tissue plasminogen activator (tPA) and plasminogen activator inhibitor-1 (PAI-1) have been assessed. When 21 young healthy males were given two low-fat diets and two oleic acid enriched diets from virgin olive oil with the same dietary cholesterol as the low-fat diets, there was a decrease in PAI-1 plasma levels with both oleic acid enriched diets. Substantial decreases in insulin levels and PAI-1 activity were observed, suggesting an improvement in insulin sensitivity during high-MUFA olive oil diets [37]. This was confirmed in another intervention which compared the isocaloric substitution of a palmitic acid diet for a low-fat or a MUFA diet in 25 healthy male subjects. Both diets decreased PAI-1 plasma levels with a higher reduction with the Mediterranean diet [2].

A study involving urban and rural populations in western Sicily reported that conversion from an urban diet to a Mediterranean diet for 8 weeks substantially reduced FVIIc and PAI-1 activity. Conversely, a rural Mediterranean diet population that switched to an urban diet developed substantial increases in FVIIc, t-PA antigen, PAI-1 activity, and fibrinogen [38]. In a study with 15 volunteers ingesting either an SFA-rich diet or MUFA-rich diet, the diet with MUFA from high oleic acid sunflower oil resulted in a lower concentration of FVIIc, LDL cholesterol, and triglycerides [39]. Another study comprised of 69 students in a controlled feeding environment, demonstrated that sunflower oil decreased FXIIa, FXIIc, and FIXc after 4 weeks. Rapeseed oil induced no change while the olive oil diet induced a decrease in FVIIc, FXIIc, FXIIa, and FXc [40].

POLYPHENOLS CARDIOVASCULAR BENEFITS

The minor constituents of olive oil also have substantial vascular and cardioprotective effects. The nonglyceride fraction of EVOO is rich in hydrocarbons, nonglyceride esters, tocopherols, flavonoids, sterols, and phenolic constituents (Table 2). The proportions of these minor compounds depend on the manufacturing processes of oil. Because these processes vary by oil mill, it is difficult to quantify the precise dietary intake of these components. In addition, Mediterranean countries tend to consume EVOO, which is much richer in phenolic compounds than refined oils. The main antioxidants in olives are carotenoids and polyphenols. The primary polyphenols are oleuroepein, hydroxytyrosol, and -tocopherol. Oleuropein, the major polyphenol, consists of up to 14% of the total net weight [41]. Hydroxytyrosol is a byproduct of oleuropein [42]. α-tocopherol, also known as an active form of vitamin E, is highly resistant to oxidative degradation [43]. Although it exists in relatively low concentrations in olive oil, its daily consumption augments the overall antioxidant content in the human body and protects against free radicals and lipid peroxidation in humans [44].

Table 2. Results of recent research on oleic acid and polyphenol effects on cardiovascular function

	Oleic acid effects	Polyphenol effects
Lipoproteins	↑ Serum HDL[23] ↓ Plasma LDL ↓ Oxidation of LDL [24], [77]	↑ Lag phase before oxidation [51] Maintain vitamin E basal levels [45] ↓ Oxidation of LDL[45] ↑ Inhibition of thiobarbituric acid reactive substances[54] ↑ Phenolic concentration
Chylomicrons	↑ CM secretion ↑ CM size[28]	
Platelets	↓ Collagen sensitivity ↓ Platelet aggregation High-MUFA diet ↓ vWF[30] ↓ Venous thrombosis ↓ Thrombotic occlusion ↑ Bleeding time[32] ↓ Platelet hyperactivity[33] ↓ Platelet aggregation ↑ Prostacyclin ↓ Sensitivity to ADP and arachidonic acid ↓ FVIIc response[31]	↓ Collagen-induced thromboxane production by 94%[61] ↓ C-reactive protein and interleukins, markers of inflammation[64] ↓ Inhibited platelet aggregation and camp-PDE[62] ↓ Aggregation when exposed to APD and collagen agonists[61]
Coagulation	↓ FVII[36] ↓ FVIIc[39] ↓ LDL cholesterol, triglycerides ↓ FXIIc, FXIIa, FXc[40]	↓ Postprandial increase in FVIIa[58]
Fibrinolysis	↓ Plasma PAI-1 ↑ t-PA[2] ↑ Insulin sensitivity[37]	↓ PAI-1 plasma concentration and activity[58]
Vasodilation		↑ Postprandial vasodilation ↑ Final products of NO Reversal of cholesterol-induced vasoconstriction[58] ↑ LDL oxidation resistance[55] ↑ Nitric oxide ↓ Lipoperoxides[58]
Adhesion molecules	↓ VCAM-1 and E-selection expression[2] ↓ Monocyte chemotaxis and cell adhesion	↓ Monocyte cell adhesion to endothelium ↓ VCAM-1 levels[66] ↓ Cell surface expression of ICAM-1 and VCAM-2[65]

CM, chylomicrons; ICAM, intercellular adhesion molecule; MUFA, monounsaturated fatty acids; PAI-1, plasminogen activator inhibitor-1; PDE, phosphodiesterase; t-PA, tissue plasminogen activator; VCAM, vascular cell-adhesion molecule; vWF, vonWillebrand Factor.

Polyphenols interfere with the chain reactions initiated and supported by free radicals [45]. This prevents DNA damage, lipid hydroperoxide formation, and lipid peroxidation [46]. In addition, exogenous antioxidants increase the concentration of antioxidants present in the body and protect against degenerative diseases [47]. Flavonoids contribute by sparing the basal levels of β-carotene, urate, and vitamins C and E activity [22]. The phenolic compounds have also been reported to decrease the presence of cell-adhesion molecules, increase nitric oxide (NO) disposability, suppress platelet aggregation, and boost total phenolic content of LDL to delay arteriosclerosis, reduce inflammation, and inhibit oxygen use in neutrophils [48].

1. LDL and Oxidation

Free radicals oxidize plasma LDL into atherogenic ox-LDL. The latter have deleterious effects on cell membranes and cytoplasmic structures, which can initiate cardiovascular disease [22]. The ortho-diphenolic structure of hydroxytyrosol and oleuropein confers an especially strong antioxidant property [49]. It is believed that these polyphenols exert their antioxidant activity by chelating free metal ions, such as copper and iron, and by scavenging free radicals [50]. The effects of polyphenols on LDL susceptibility to copper-mediated oxidation in rabbits with normalized levels of vitamin E have been reported. Rabbits fed EVOO demonstrated no change in cholesterol or vitamin E levels after 6 weeks. They had a 30% longer lag phase before oxidation compared with rabbits fed the refined olive oil or sunflower oil [51]. Rats fed diets of olive oil also had a decreased concentration of lipoproteins and thiobarbituric acid reactive substances, that are end products of lipid peroxidation [52].

An epidemiologic study conducted in the primarily urban Greek province of Attica examined the correlation between Mediterranean diet adherence and levels of antioxidants and oxLDL. Those with the highest Mediterranean diet score, on average, had 11% higher levels of total antioxidant capacity than those in the lowest percentile. In addition, the study found that those with the strongest adherence to the Mediterranean diet, on average, had 19% lower oxLDL-cholesterol levels than those with the lowest dietary score [53]. The Prevención con DietaMediterránea Studies (PREDIMED), was a multicenter study randomized controlled trial of asymptomatic subjects at high risk for cardiovascular disease. Seven-hundred and seventy two participants from 10 Spanish primary care centers were placed on a low-fat diet or a traditional Mediterranean diet supplemented with either nuts or virgin olive oil. A 3-month followup found substantial decreases in oxLDL levels in both Mediterranean diet groups [54].

2. Nitric Oxide (NO) Activity and Endothelial Dysfunction

During hypercholesterolemia, the production of superoxide anions and other free radical species is increased in endothelial cells, smooth muscle cells, and monocytes when compared with that of normocholesterolemic controls [55]. These species degrade NO and a damaged endothelium cannot produce sufficient NO, which can lead to monocyte recruitment, platelet aggregation, and thrombosis. Diminished NO bioactivity can cause constriction of coronary arteries during exercise and vascular inflammation leading to lipoprotein oxidation and foam-cell formation [56]. On the other hand, some studies have found that ox- LDL stimulates NO synthase transcription and synthesis in bovine aortic cells [57]. Studies also demonstrated increased expression of NO and NO synthase in atherosclerotic rabbit aorta tissue and human atherosclerotic plaques. It is postulated that this overproduction accompanies rapid oxidative inactivation or conversion of NO to toxic nitrogen oxides because of accumulation of superoxide anions and free radicals [56]. Studies suggest that vascular dysfunction can be reversed through intake of agents able to scavenge these radicals [44]. For example, consumption of a meal with high phenolic EVOO improved endothelium-dependent microvascular vasodilation during the first 4 hours of the postprandial period in hypercholesterolemic volunteers. Subjects fed the phenol rich meal displayed a higher concentration of NO and lower lipo-peroxide levels than those fed a low-phenol meal. This improvement is linked with a decrease in oxidative stress and increase in the final products of NO [58]. In addition, oleuropein stimulated NO production in mouse macrophages and activated the inducible form of NO synthase [58]. Rabbits with cholesterol-induced impaired endothelial dilation were given a supplement of α-tocopherol or B-carotene. α-tocopherol led to complete reversal of vasoconstriction. Additionally, increased concentrations of α-tocopherol were detected in the plasma, aorta, and LDL. α-tocopherol was able to increase the resistance to LDL ex vivo oxidation after exposure to copper [55].

A survey of 20,343 subjects, as part of the Greek cohort of the EPIC (European Prospective Investigation into Cancer and Nutrition) study, found that adherence to the Mediterranean diet was inversely related to systolic and diastolic blood pressure levels. Olive oil, vegetables, and fruits were the principal factors responsible for the overall effect of the Mediterranean diet on arterial blood pressure. Interestingly, consumption of cereals, generally considered to be beneficial to health, correlated with increases in arterial blood pressure [59]. The PREDIMED study investigated the effect of the low fat and two Mediterranean diet diets on 772 adults with high risk for cardiovascular disease. The Mediterranean diet participants with hypertension showed statistically significant reductions from baseline values in systolic blood pressure, and the low-fat group showed a mean increase in blood pressure. The Mediterranean diet participants demonstrated improved lipid profiles, decreased insulin resistance, and reduced concentrations of inflammatory molecules compared with the low-fat group [60].

3. Platelet Aggregation

An in vitro study examined the effects of hydroxytyrosol and oleuropein on platelet aggregation. Hydroxytyrosol completely inhibited ADP and collagen induced platelet aggregation in platelet-rich plasma [61]. The aggregation inhibition potency of hydroxytyrosol was found to be equivalent to that of aspirin. In the same study, hydroxytyrosol was also found to inhibit collagen and thrombin-induced thromboxane B2 production.

In human volunteers, it has been reported that the pure olive leaf extract with 5.40 mg/mL polyphenol oleuropein, was capableof inhibiting in vitro platelet activation in healthy, nonsmoking male individuals [62]. Dell'Agli and colleagues [63] examined the effects of olive oil extracts of high and low phenol content and single phenols on platelet aggregation to prove that cAMP and cGMP phosphodiesterases might be a biologic target of platelet aggregation inhibition. Of the polyphenols examined—oleuropein, hydroxytyrosol, tyrosol, and the flavonoids quercetin, luteolin, and apigenin—oleuropein had the most substantial effect on platelet aggregation inhibition. The olive oil extracts and the single phenols exhibited concentration-dependent inhibition on aggregation and on cAMP-phosphodiesterases.

In a randomized controlled trial in Naples, Italy, 180 subjects with confirmed metabolic syndrome were instructed to follow a Mediterranean-style diet supplemented with olive oil. After 2 years, there was marked improvement in endothelial function, with statistically significant decrease in blood pressure, cholesterol, insulin and glucose levels and platelet aggregation response to L-arginine. Also, there were substantial reduction in markers of systemic vascular inflammation, including C-reactive protein and interleukins. Sixty of ninety participants in the intervention group experienced reductions in risk factors, such that they were no longer classified as having metabolic syndrome [64]. It is still unclear through what exact pathway polyphenols inhibit platelet aggregation. It has been hypothesized to be mediated by either a phenol-induced decrease in eicosanoid production or through the degradation of cAMP in conjunction with inhibition of the arachidonic acid cascade [63].

4. Endothelial Adhesion Molecules

The inflammatory response during atherogenesis includes adhesion of leukocytes, monocytes, and lymphocytes to the endothelium. Adhesion of these molecules is facilitated by intercellular adhesion molecule-1 (ICAM-1), vascular cell adhesion molecule-1 (VCAM-1) and E-selection [65]. One study reported that physiologically relevant dosages of phenolic extract from EVOO reduced cell surface expression of ICAM-1 and VCAM-1 [66]. A mixture of olive oil polyphenols, including oleuropein, hydroxytyrosol, and tyrosol, also induced a decrease in VCAM-1 mRNA levels and promoter activity and ICAM-1 and E-selection expression [66]. The PREDIMED study on human subjects found that the Mediterranean diet supplemented with olive oil resulted in statistically significant reductions in inflammatory markers including C-reactive protein, interleukin-6, ICAM-1, and VCAM-1 when compared with a low-fat diet [60].

MOLECULAR MECHANISMS OF OLIVE OIL

There is a paucity of research on the effect of the Mediterranean diet on the vascular wall. The PEDIMED study examined the effect of olive oil on morphologic changes in the vascular wall, using carotid intima-media thickness (IMT) as surrogate of severity of atherosclerotic disease. An increase of 0.2 mm in IMT has been reported to imply a 28% and 31% increase in risk for stroke and myocardial infarction, respectively [67]. Data collected on 190 participants found that those who consumed the least amount of olive oil had the highest IMT. The study found a statistically significant difference in IMT between those who consumed 6 to 34 g and 35 to 74 g of olive oil per day [68].

Although there is extensive research on the role of olive oil on primary prevention of cardiovasculardisease, there is scarce studies on utilization of olive oil as a therapeutic treatment for cardiovascular related diseases. In a double blinded randomized dietary intervention study examined the influence of a polyphenol-rich olive oil on women with high-normal BP (a systolic pressure of 120–139 mm Hg or diastolic pressure of 80–89 mm Hg) or stage 1 essential hypertension (a systolic pressure of 140–159 mm Hg and diastolic pressure of 90–99 mm Hg) [69]. The polyphenol rich olive oil diet was capable of reducing blood pressure in newly diagnosed women where 22 of 24 participants had a diastolic BP of 90 mm Hg or less (Table 3). Also, there was decrease in serum asymmetric dimethylarginine, ox-LDL, and plasma C-reactive protein; increase in plasma nitrites/nitrates and hyperemic area post cuff inflation. These results were not related to age or body mass index. This study gives an optimistic anticipation of olive oil utilization in treatment of hypertension and endothelial dysfunction and larger studies need to be carried out to further justify such utilization.

Table 3. Endothelial function, oxidative stress, and inflammation biomarkers in young women with high-normal BP or stage 1 essential hypertension after 4 months on a Mediterranean-style diet and changes after 2 months on the polyphenolrich or the polyphenol-free olive oil diets

| Biomarker | Baseline | Changes from baseline | | P value* |
		Polyphenol-rich olive oil	Polyphenol-free olive oil	
Nitrites/nitrates (μmol/l)	19.7 ± 2.6	+4.7 ± 6.6	+0.8 ± 4.1	<0.001
AdMA (μmol/l)	0.82 ± 0.04	−0.09 ± 0.01	−0.04 ± 0.03	<0.01
Ox-LdL (μg/l)	153.0 ± 51.0	−28.2 ± 28.5	−6.9 ± 22.2	<0.01
cRP (mg/l)	1.6 ± 0.9	−1.9 ± 1.3	−0.6 ± 0.9	<0.001
Blood pressure (mm Hg)				
Systolic	134.14 ± 9.32	−7.91 ± 9.51	−1.65 ± 8.22	<0.001
diastolic	84.64 ± 8.52	−6.65 ± 6.63	−2.17 ± 7.24	<0.001
IRH measurement (PU)				
HA	1,084 ± 266	+345 ± 386	+36 ± 367	<0.001

Table values are mean ± SD, n = 24. ADMA, asymmetric dimethylarginine; BP, blood pressure; CRP, C-reactive protein; HA, hyperemic area; IRH, ischemia-reactive hyperemia; ox-LDL, oxidized low-density lipoprotein; PU, perfusion units. *P value for the comparison across the intervention groups by ANOVA. From reference [69]

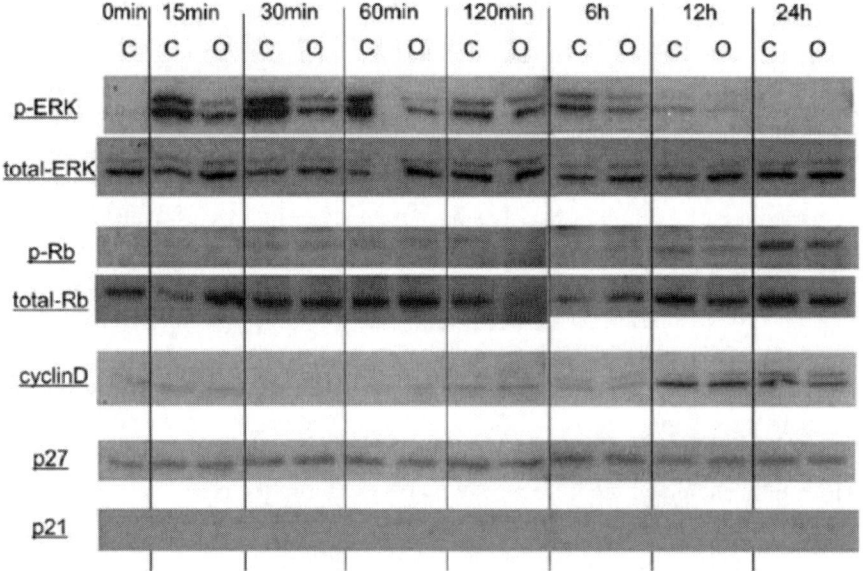

Figure 4. Immunoblot for phospho/total-ERK1/2, phospho/total-Rb, cyclinD, p27, and p21 of SMC in the absence or presence of 100 mMoleuropein. C: control, O: oleuropein-treated group. The data shows inhibition of ERK1/2, a key regulator of cell cycle, while showing no effect on other cell cycle regulators. From reference [70].

The interaction of olive oil components on a cellular level are not well established but it is important in the causality of olive oil contributing to an atheroprotective role.

Our laboratory has shown that the oleuropein inhibits vascular smooth muscle cells proliferation in bovine aortic cells [70] by primarily blocking the G1 phase of the cell cycle. The known major regulator of cell cycle progression at the G1 -S phase such as p21, p27, and cyclin D were not affected by oleuropein. However, Erk1/2 which is a major regulator of G1 phase also inhibited in a dose dependant and reversible manner (Figure 4) [71]. Hydroxytyrosol also inhibited SMC proliferation via ERK1/2 pathway (Figure 5).

This dose dependent inhibition byphenolic extracts has also been observed in human cell monocytes, which play a key role in the pathogenesis of atherosclerosis, inhibiting metalloproteinase (MMP) 9, a member of the MMPs family acting on the extracellular matrix and inflammatory cytokines [72]. The most effective inhibitors were flavonoids (with 80% reduction in secretion of MMP9) while tyrosol and hydroxytyrosol had no effect. Phenolic extracts did not have a direct inhibitory effect on enzymatic activity but interfered with the nuclear factor-kB transcription factor. Apigenin, a minor phenol component, has been reported to abolish the tissue necrosis factor alpha stimulatory effect (figure 6). Olive oil's capability to reduce MMP9 secretion and protein expression may elucidate olive oil capability to decline invasiveness of tumor cells [73].

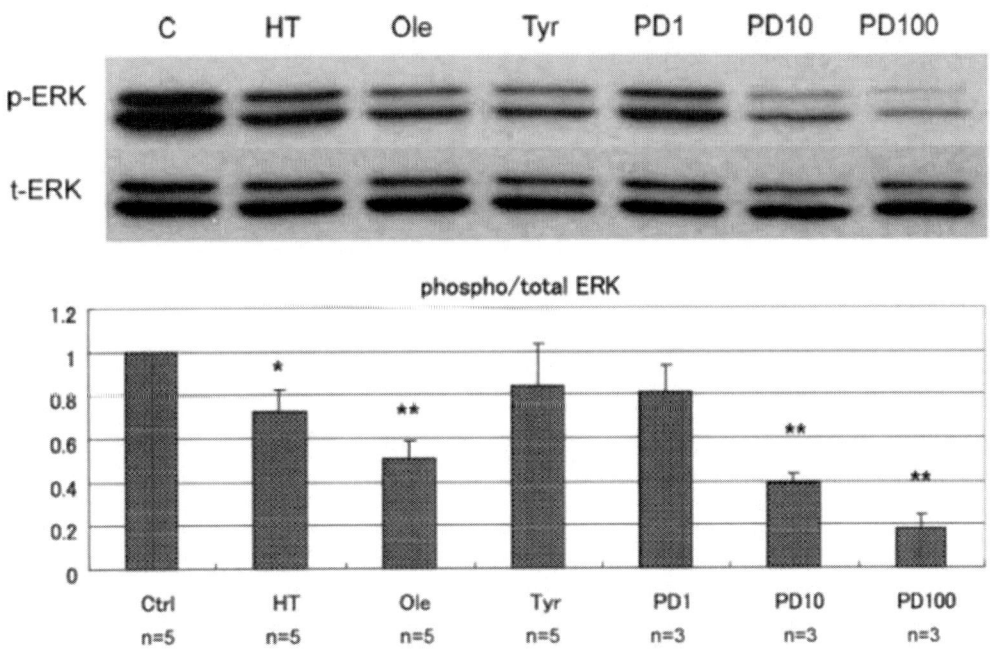

Figure 5. ERK1/2 expression of smooth muscle cells (SMC) exposed to 100 μMhydroxytyrosol (HT), 100 μMoleuropein (Ole), 100 μMtyrosol (Tyr), or 1, 10, or 100 μM PD98059 for 60 minutes. Data are presented as mean ± standard error. *, p < 0.05; **, p < 0.01 compared with control; n, number for performed. From reference [71].

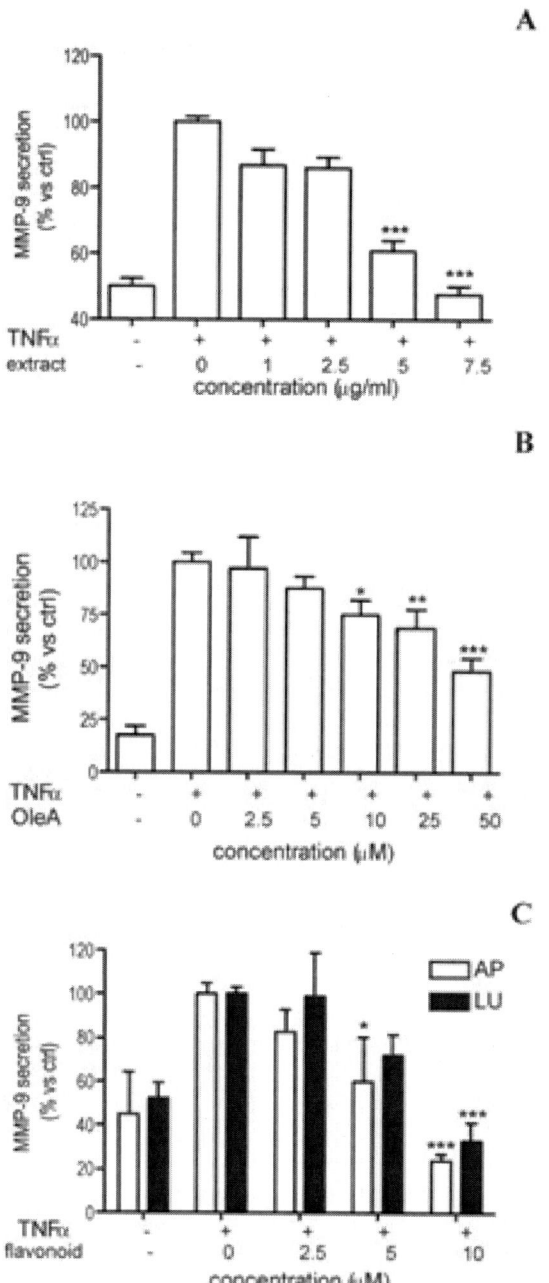

Figure 6. Effect of olive oil phenolic extract and pure polyphenols onMMP-9 secretion. THP-1, a human monocyte-like cell line, cells were incubated for 6 h in the presence of 10 ng/mL TNF-α and increasing concentrations of olive oil phenolic extract (A), oleuropeinaglycone (B), apigenin (AP), and luteolin (LU), (C). The amount of MMP-9 secreted into the media was evaluated by zymography. Results are expressed as arbitrary optical density units normalized against the protein content and calculated as percentages versus the normalized values measured in stimulated cells (+TNF-α). Each bar represents the mean ± SD of three independent experiments performed in duplicate (**p < 0.01). From reference [72].

CONCLUSION

Recent studies consistently support the concept that the olive oil enrich Mediterranean diet is compatible with healthier aging and increased longevity. In countries where the population adheres to the Mediterranean diet, such as Spain, Greece and Italy, and olive oil is the principal source of fat, rates of coronary events are lower than in northern European countries. Experimental and human cellular studies have provided new evidence on the potential protective effect of olive oil on cardiovascular related diseases. Phenolic compounds in olive oil have shown antioxidant and anti-inflammatory properties, prevent lipoperoxidation, induce favorable changes of lipid profile, improve endothelial function, and disclose antithrombotic properties. Much more research needs to be conducted especially at the cellular level, to more fully understand the pathways by which oleic acid and the polyphenols in olive oil help to reduce cardiovascular disease risk factors. Nevertheless, current epidemiological and clinical studies show evidence that suggests the components within EVOO exert a beneficial effect on cardiovascular health. Olive oil's biological properties that contribute to dampen the inflammatory and the proliferative stages of atherosclerosis gives a promising outlook on continuing utilization of olive oil as a preventative and potentially a curative natural compound to cardiovascular and chronic diseases in general.

REFERENCES

[1]   E. M. Berry, Y. Arnoni, and M. Aviram, 'The Middle Eastern and Biblical Origins of the Mediterranean Diet', *Public Health Nutr,* 14 (2011), 2288-95.

[2]   F. Perez-Jimenez, P. Castro, J. Lopez-Miranda, E. Paz-Rojas, A. Blanco, F. Lopez-Segura, F. Velasco, C. Marin, F. Fuentes, and J. M. Ordovas, 'Circulating Levels of Endothelial Function Are Modulated by Dietary Monounsaturated Fat', *Atherosclerosis,* 145 (1999), 351-8.

[3]   W. C. Willett, F. Sacks, A. Trichopoulou, G. Drescher, A. Ferro-Luzzi, E. Helsing, and D. Trichopoulos, 'Mediterranean Diet Pyramid: A Cultural Model for Healthy Eating', *Am J Clin Nutr,* 61 (1995), 1402S-1406S.

[4]   R. Ortega, 'Importance of Functional Foods in the Mediterranean Diet', *Public Health Nutr,* 9 (2006), 1136-40.

[5]   J. Muller-Nordhorn, S. Binting, S. Roll, and S. N. Willich, 'An Update on Regional Variation in Cardiovascular Mortality within Europe', *Eur Heart J,* 29 (2008), 1316-26.

[6]   C. M. Kastorini, H. J. Milionis, A. Ioannidi, K. Kalantzi, V. Nikolaou, K. N. Vemmos, J. A. Goudevenos, and D. B. Panagiotakos, 'Adherence to the Mediterranean Diet in Relation to Acute Coronary Syndrome or Stroke Nonfatal Events: A Comparative Analysis of a Case/Case-Control Study', *Am Heart J,* 162 (2011), 717-24.

[7]   F. Sofi, R. Abbate, G. F. Gensini, and A. Casini, 'Accruing Evidence on Benefits of Adherence to the Mediterranean Diet on Health: An Updated Systematic Review and Meta-Analysis', *Am J Clin Nutr,* 92 (2010), 1189-96.

[8]   C. Samieri, C. Feart, C. Proust-Lima, E. Peuchant, C. Tzourio, C. Stapf, C. Berr, and P. Barberger-Gateau, 'Olive Oil Consumption, Plasma Oleic Acid, and Stroke Incidence: The Three-City Study', *Neurology,* 77 (2011), 418-25.

[9]   A. Keys, 'Coronary Heart Disease in Seven Countries. 1970', *Nutrition,* 13 (1997), 250-2; discussion 249, 253.

[10] P. M. Kris-Etherton, 'Aha Science Advisory. Monounsaturated Fatty Acids and Risk of Cardiovascular Disease. American Heart Association. Nutrition Committee', *Circulation,* 100 (1999), 1253-8.

[11] C. L. Huang, and B. E. Sumpio, 'Olive Oil, the Mediterranean Diet, and Cardiovascular Health', *J Am Coll Surg,* 207 (2008), 407-16.

[12] R. Estruch, E. Ros, J. Salas-Salvado, M. I. Covas, D. Corella, F. Aros, E. Gomez-Gracia, V. Ruiz-Gutierrez, M. Fiol, J. Lapetra, R. M. Lamuela-Raventos, L. Serra-Majem, X. Pinto, J. Basora, M. A. Munoz, J. V. Sorli, J. A. Martinez, M. A. Martinez-Gonzalez, and Predimed Study Investigators, 'Primary Prevention of Cardiovascular Disease with a Mediterranean Diet', *N Engl J Med,* 368 (2013), 1279-90.

[13] H. Tunstall-Pedoe, K. Kuulasmaa, M. Mahonen, H. Tolonen, E. Ruokokoski, and P. Amouyel, 'Contribution of Trends in Survival and Coronary-Event Rates to Changes in Coronary Heart Disease Mortality: 10-Year Results from 37 Who Monica Project Populations. Monitoring Trends and Determinants in Cardiovascular Disease', *Lancet,* 353 (1999), 1547-57.

[14] M. D. Kontogianni, D. B. Panagiotakos, C. Chrysohoou, C. Pitsavos, A. Zampelas, and C. Stefanadis, 'The Impact of Olive Oil Consumption Pattern on the Risk of Acute Coronary Syndromes: The Cardio2000 Case-Control Study', *Clin Cardiol,* 30 (2007), 125-9.

[15] J. P. Mackenbach, 'The Mediterranean Diet Story Illustrates That "Why" Questions Are as Important as "How" Questions in Disease Explanation', *J Clin Epidemiol,* 60 (2007), 105-9.

[16] Boskou D, 'Olive Oil: Chemistry and Technology', *AOCS Press* (1996).

[17] P. A. S. Breslin, Gingerich, 'T. N. & Green', *B. G. Chem. Sens* (2001), 55-66.

[18] J. R. Vane, and R. M. Botting, 'New Insights into the Mode of Action of Anti-Inflammatory Drugs', *Inflamm Res,* 44 (1995), 1-10.

[19] R. E. Harris, J. Beebe-Donk, H. Doss, and D. Burr Doss, 'Aspirin, Ibuprofen, and Other Non-Steroidal Anti-Inflammatory Drugs in Cancer Prevention: A Critical Review of Non-Selective Cox-2 Blockade (Review)', *Oncol Rep,* 13 (2005), 559-83.

[20] G. K. Beauchamp, R. S. Keast, D. Morel, J. Lin, J. Pika, Q. Han, C. H. Lee, A. B. Smith, and P. A. Breslin, 'Phytochemistry: Ibuprofen-Like Activity in Extra-Virgin Olive Oil', *Nature,* 437 (2005), 45-6.

[21] Y. Zhou, Y. Su, B. Li, F. Liu, J. W. Ryder, X. Wu, P. A. Gonzalez-DeWhitt, V. Gelfanova, J. E. Hale, P. C. May, S. M. Paul, and B. Ni, 'Nonsteroidal Anti-Inflammatory Drugs Can Lower Amyloidogenic Abeta42 by Inhibiting Rho', *Science,* 302 (2003), 1215-7.

[22] B. E. Sumpio, A. C. Cordova, D. W. Berke-Schlessel, F. Qin, and Q. H. Chen, 'Green Tea, the "Asian Paradox," and Cardiovascular Disease', *J Am Coll Surg,* 202 (2006), 813-25.

[23] M. Kratz, P. Cullen, F. Kannenberg, A. Kassner, M. Fobker, P. M. Abuja, G. Assmann, and U. Wahrburg, 'Effects of Dietary Fatty Acids on the Composition and Oxidizability of Low-Density Lipoprotein', *Eur J Clin Nutr*, 56 (2002), 72-81.

[24] M. C. Ramirez-Tortosa, C. M. Aguilera, J. L. Quiles, and A. Gil, 'Influence of Dietary Lipids on Lipoprotein Composition and Ldl Cu(2+)-Induced Oxidation in Rabbits with Experimental Atherosclerosis', *Biofactors*, 8 (1998), 79-85.

[25] C. M. Aguilera, M. C. Ramirez-Tortosa, M. D. Mesa, C. L. Ramirez-Tortosa, and A. Gil, 'Sunflower, Virgin-Olive and Fish Oils Differentially Affect the Progression of Aortic Lesions in Rabbits with Experimental Atherosclerosis', *Atherosclerosis*, 162 (2002), 335-44.

[26] M. C. Ramirez-Tortosa, G. Urbano, M. Lopez-Jurado, T. Nestares, M. C. Gomez, A. Mir, E. Ros, J. Mataix, and A. Gil, 'Extra-Virgin Olive Oil Increases the Resistance of Ldl to Oxidation More Than Refined Olive Oil in Free-Living Men with Peripheral Vascular Disease', *J Nutr*, 129 (1999), 2177-83.

[27] K. G. Jackson, M. D. Robertson, B. A. Fielding, K. N. Frayn, and C. M. Williams, 'Olive Oil Increases the Number of Triacylglycerol-Rich Chylomicron Particles Compared with Other Oils: An Effect Retained When a Second Standard Meal Is Fed', *Am J Clin Nutr*, 76 (2002), 942-9.

[28] S. W. Sakr, N. Attia, M. Haourigui, J. L. Paul, T. Soni, D. Vacher, and A. Girard-Globa, 'Fatty Acid Composition of an Oral Load Affects Chylomicron Size in Human Subjects', *Br J Nutr*, 77 (1997), 19-31.

[29] L. F. Larsen, J. Jespersen, and P. Marckmann, 'Are Olive Oil Diets Antithrombotic? Diets Enriched with Olive, Rapeseed, or Sunflower Oil Affect Postprandial Factor Vii Differently', *Am J Clin Nutr*, 70 (1999), 976-82.

[30] F. Perez-Jimenez, J. Lopez-Miranda, and P. Mata, 'Protective Effect of Dietary Monounsaturated Fat on Arteriosclerosis: Beyond Cholesterol', *Atherosclerosis*, 163 (2002), 385-98.

[31] R. D. Smith, C. N. Kelly, B. A. Fielding, D. Hauton, K. D. Silva, M. C. Nydahl, G. J. Miller, and C. M. Williams, 'Long-Term Monounsaturated Fatty Acid Diets Reduce Platelet Aggregation in Healthy Young Subjects', *Br J Nutr*, 90 (2003), 597-606.

[32] S. Brzosko, A. De Curtis, S. Murzilli, G. de Gaetano, M. B. Donati, and L. Iacoviello, 'Effect of Extra Virgin Olive Oil on Experimental Thrombosis and Primary Hemostasis in Rats', *Nutr Metab Cardiovasc Dis*, 12 (2002), 337-42.

[33] J. P. De La Cruz, M. A. Villalobos, J. A. Carmona, M. Martin-Romero, J. M. Smith-Agreda, and F. S. de la Cuesta, 'Antithrombotic Potential of Olive Oil Administration in Rabbits with Elevated Cholesterol', *Thromb Res*, 100 (2000), 305-15.

[34] D. D. Wagner, and P. C. Burger, 'Platelets in Inflammation and Thrombosis', *Arterioscler Thromb Vasc Biol*, 23 (2003), 2131-7.

[35] L. Iacoviello, and M. B. Donati, 'Interpretation of Thrombosis Prevention Trial', *Lancet*, 351 (1998), 1205; author reply 1206-7.

[36] T. W. Meade, V. Ruddock, Y. Stirling, R. Chakrabarti, and G. J. Miller, 'Fibrinolytic Activity, Clotting Factors, and Long-Term Incidence of Ischaemic Heart Disease in the Northwick Park Heart Study', *Lancet*, 342 (1993), 1076-9.

[37] F. Lopez-Segura, F. Velasco, J. Lopez-Miranda, P. Castro, R. Lopez-Pedrera, A. Blanco, J. Jimenez-Pereperez, A. Torres, J. Trujillo, J. M. Ordovas, and F. Perez-

Jimenez, 'Monounsaturated Fatty Acid-Enriched Diet Decreases Plasma Plasminogen Activator Inhibitor Type 1', *Arterioscler Thromb Vasc Biol,* 16 (1996), 82-8.

[38]   Cordova R Avellone G, Scalffidi L, Bompiani G., 'Effects of Mediterranean Diet on Lipid, Coagulative, and Fibrinolytic Parameters
in Two Randomly Selected Population Samples Inwestern Sicily.', *Nutr Metab Cardiovasc Dis* (1998), 287–296.

[39]   M. A. Allman-Farinelli, K. Gomes, E. J. Favaloro, and P. Petocz, 'A Diet Rich in High-Oleic-Acid Sunflower Oil Favorably Alters Low-Density Lipoprotein Cholesterol, Triglycerides, and Factor Vii Coagulant Activity', *J Am Diet Assoc,* 105 (2005), 1071-9.

[40]   R. Junker, M. Kratz, M. Neufeld, M. Erren, J. R. Nofer, H. Schulte, U. Nowak-Gottl, G. Assmann, and U. Wahrburg, 'Effects of Diets Containing Olive Oil, Sunflower Oil, or Rapeseed Oil on the Hemostatic System', *Thromb Haemost,* 85 (2001), 280-6.

[41]   Fleuriet A Amiot M, Macheix J, 'Importance and Evolution of Phenolic Compounds in Olive During Growth and Maturation', *J Agric Food Chem* (1986), 823–826.

[42]   Esti M Cinquanta L, La Notte E, 'Evolution of Phenolic Compounds in Virgin Olive Oil During Storage', *J Am Oil Chem Soc* (1997), 1259-1264.

[43]   E. Tripoli, M. Giammanco, G. Tabacchi, D. Di Majo, S. Giammanco, and M. La Guardia, 'The Phenolic Compounds of Olive Oil: Structure, Biological Activity and Beneficial Effects on Human Health', *Nutr Res Rev,* 18 (2005), 98-112.

[44]   J. S. Perona, R. Cabello-Moruno, and V. Ruiz-Gutierrez, 'The Role of Virgin Olive Oil Components in the Modulation of Endothelial Function', *J Nutr Biochem,* 17 (2006), 429-45.

[45]   F. Visioli, and C. Galli, 'The Effect of Minor Constituents of Olive Oil on Cardiovascular Disease: New Findings', *Nutr Rev,* 56 (1998), 142-7.

[46]   S. Katiyar, and H. Mukhtar, 'Tea in Chemoprevention of Cancer', *Int J Oncol,* 8 (1996), 221-38.

[47]   F. Visioli, G. Bellomo, G. Montedoro, and C. Galli, 'Low Density Lipoprotein Oxidation Is Inhibited in Vitro by Olive Oil Constituents', *Atherosclerosis,* 117 (1995), 25-32.

[48]   F. Perez-Jimenez, G. Alvarez de Cienfuegos, L. Badimon, G. Barja, M. Battino, A. Blanco, A. Bonanome, R. Colomer, D. Corella-Piquer, I. Covas, J. Chamorro-Quiros, E. Escrich, J. J. Gaforio, P. P. Garcia Luna, L. Hidalgo, A. Kafatos, P. M. Kris-Etherton, D. Lairon, R. Lamuela-Raventos, J. Lopez-Miranda, F. Lopez-Segura, M. A. Martinez-Gonzalez, P. Mata, J. Mataix, J. Ordovas, J. Osada, R. Pacheco-Reyes, M. Perucho, M. Pineda-Priego, J. L. Quiles, M. C. Ramirez-Tortosa, V. Ruiz-Gutierrez, P. Sanchez-Rovira, V. Solfrizzi, F. Soriguer-Escofet, R. de la Torre-Fornell, A. Trichopoulos, J. M. Villalba-Montoro, J. R. Villar-Ortiz, and F. Visioli, 'International Conference on the Healthy Effect of Virgin Olive Oil', *Eur J Clin Invest,* 35 (2005), 421-4.

[49]   C. Galli, and F. Visioli, 'Antioxidant and Other Activities of Phenolics in Olives/Olive Oil, Typical Components of the Mediterranean Diet', *Lipids,* 34 Suppl (1999), S23-6.

[50]   Miller N Rice-Evans C, Paganga G, 'Antioxidant Properties of Phenolic Compounds',(1997), 159-160.

[51]   S. A. Wiseman, J. N. Mathot, N. J. de Fouw, and L. B. Tijburg, 'Dietary Non-Tocopherol Antioxidants Present in Extra Virgin Olive Oil Increase the Resistance of Low Density Lipoproteins to Oxidation in Rabbits', *Atherosclerosis,* 120 (1996), 15-23.

[52] C. Scaccini, M. Nardini, M. D'Aquino, V. Gentili, M. Di Felice, and G. Tomassi, 'Effect of Dietary Oils on Lipid Peroxidation and on Antioxidant Parameters of Rat Plasma and Lipoprotein Fractions', *J Lipid Res,* 33 (1992), 627-33

[53] C. Pitsavos, D. B. Panagiotakos, N. Tzima, C. Chrysohoou, M. Economou, A. Zampelas, and C. Stefanadis, 'Adherence to the Mediterranean Diet Is Associated with Total Antioxidant Capacity in Healthy Adults: The Attica Study', *Am J Clin Nutr,* 82 (2005), 694-9.

[54] M. Fito, M. Guxens, D. Corella, G. Saez, R. Estruch, R. de la Torre, F. Frances, C. Cabezas, C. Lopez-Sabater Mdel, J. Marrugat, A. Garcia-Arellano, F. Aros, V. Ruiz-Gutierrez, E. Ros, J. Salas-Salvado, M. Fiol, R. Sola, M. I. Covas, and Predimed Study Investigators, 'Effect of a Traditional Mediterranean Diet on Lipoprotein Oxidation: A Randomized Controlled Trial', *Arch Intern Med,* 167 (2007), 1195-203.

[55] J. F. Keaney, Jr., and J. A. Vita, 'Atherosclerosis, Oxidative Stress, and Antioxidant Protection in Endothelium-Derived Relaxing Factor Action', *Prog Cardiovasc Dis,* 38 (1995), 129-54.

[56] R. O. Cannon, 3rd, 'Role of Nitric Oxide in Cardiovascular Disease: Focus on the Endothelium', *Clin Chem,* 44 (1998), 1809-19.

[57] K. Hirata, N. Miki, Y. Kuroda, T. Sakoda, S. Kawashima, and M. Yokoyama, 'Low Concentration of Oxidized Low-Density Lipoprotein and Lysophosphatidylcholine Upregulate Constitutive Nitric Oxide Synthase Mrna Expression in Bovine Aortic Endothelial Cells', *Circ Res,* 76 (1995), 958-62.

[58] J. Ruano, J. Lopez-Miranda, F. Fuentes, J. A. Moreno, C. Bellido, P. Perez-Martinez, A. Lozano, P. Gomez, Y. Jimenez, and F. Perez Jimenez, 'Phenolic Content of Virgin Olive Oil Improves Ischemic Reactive Hyperemia in Hypercholesterolemic Patients', *J Am Coll Cardiol,* 46 (2005), 1864-8.

[59] T. Psaltopoulou, A. Naska, P. Orfanos, D. Trichopoulos, T. Mountokalakis, and A. Trichopoulou, 'Olive Oil, the Mediterranean Diet, and Arterial Blood Pressure: The Greek European Prospective Investigation into Cancer and Nutrition (Epic) Study', *Am J Clin Nutr,* 80 (2004), 1012-8.

[60] R. Estruch, M. A. Martinez-Gonzalez, D. Corella, J. Salas-Salvado, V. Ruiz-Gutierrez, M. I. Covas, M. Fiol, E. Gomez-Gracia, M. C. Lopez-Sabater, E. Vinyoles, F. Aros, M. Conde, C. Lahoz, J. Lapetra, G. Saez, E. Ros, and Predimed Study Investigators, 'Effects of a Mediterranean-Style Diet on Cardiovascular Risk Factors: A Randomized Trial', *Ann Intern Med,* 145 (2006), 1-11.

[61] A. Petroni, M. Blasevich, M. Salami, N. Papini, G. F. Montedoro, and C. Galli, 'Inhibition of Platelet Aggregation and Eicosanoid Production by Phenolic Components of Olive Oil', *Thromb Res,* 78 (1995), 151-60.

[62] I. Singh, M. Mok, A. M. Christensen, A. H. Turner, and J. A. Hawley, 'The Effects of Polyphenols in Olive Leaves on Platelet Function', *Nutr Metab Cardiovasc Dis,* 18 (2008), 127-32.

[63] M. Dell'Agli, O. Maschi, G. V. Galli, R. Fagnani, E. Dal Cero, D. Caruso, and E. Bosisio, 'Inhibition of Platelet Aggregation by Olive Oil Phenols Via Camp-Phosphodiesterase', *Br J Nutr,* 99 (2008), 945-51.

[64] K. Esposito, R. Marfella, M. Ciotola, C. Di Palo, F. Giugliano, G. Giugliano, M. D'Armiento, F. D'Andrea, and D. Giugliano, 'Effect of a Mediterranean-Style Diet on

Endothelial Dysfunction and Markers of Vascular Inflammation in the Metabolic Syndrome: A Randomized Trial', *JAMA,* 292 (2004), 1440-6.

[65]   M. Dell'Agli, R. Fagnani, N. Mitro, S. Scurati, M. Masciadri, L. Mussoni, G. V. Galli, E. Bosisio, M. Crestani, E. De Fabiani, E. Tremoli, and D. Caruso, 'Minor Components of Olive Oil Modulate Proatherogenic Adhesion Molecules Involved in Endothelial Activation', *J Agric Food Chem,* 54 (2006), 3259-64.

[66]   M. A. Carluccio, L. Siculella, M. A. Ancora, M. Massaro, E. Scoditti, C. Storelli, F. Visioli, A. Distante, and R. De Caterina, 'Olive Oil and Red Wine Antioxidant Polyphenols Inhibit Endothelial Activation: Antiatherogenic Properties of Mediterranean Diet Phytochemicals', *Arterioscler Thromb Vasc Biol,* 23 (2003), 622-9.

[67]   E. de Groot, G. K. Hovingh, A. Wiegman, P. Duriez, A. J. Smit, J. C. Fruchart, and J. J. Kastelein, 'Measurement of Arterial Wall Thickness as a Surrogate Marker for Atherosclerosis', *Circulation,* 109 (2004), III33-8.

[68]   P. Buil-Cosiales, P. Irimia, N. Berrade, A. Garcia-Arellano, M. Riverol, M. Murie-Fernandez, E. Martinez-Vila, M. A. Martinez-Gonzalez, and M. Serrano-Martinez, 'Carotid Intima-Media Thickness Is Inversely Associated with Olive Oil Consumption', *Atherosclerosis,* 196 (2008), 742-8.

[69]   R. Moreno-Luna, R. Munoz-Hernandez, M. L. Miranda, A. F. Costa, L. Jimenez-Jimenez, A. J. Vallejo-Vaz, F. J. Muriana, J. Villar, and P. Stiefel, 'Olive Oil Polyphenols Decrease Blood Pressure and Improve Endothelial Function in Young Women with Mild Hypertension', *Am J Hypertens,* 25 (2012), 1299-304.

[70]   R. Abe, J. Beckett, R. Abe, A. Nixon, A. Rochier, N. Yamashita, and B. Sumpio, 'Olive Oil Polyphenol Oleuropein Inhibits Smooth Muscle Cell Proliferation', *Eur J Vasc Endovasc Surg,* 41 (2011), 814-20.

[71]   R. Abe, J. Beckett, R. Abe, A. Nixon, A. Rochier, N. Yamashita, and B. Sumpio, 'Olive Oil Polyphenols Differentially Inhibit Smooth Muscle Cell Proliferation through a G1/S Cell Cycle Block Regulated by Erk1/2', *Int J Angiol,* 21 (2012), 69-76.

[72]   M. Dell'Agli, R. Fagnani, G. V. Galli, O. Maschi, F. Gilardi, S. Bellosta, M. Crestani, E. Bosisio, E. De Fabiani, and D. Caruso, 'Olive Oil Phenols Modulate the Expression of Metalloproteinase 9 in Thp-1 Cells by Acting on Nuclear Factor-Kappab Signaling', *J Agric Food Chem,* 58 (2010), 2246-52.

[73]   Y. Z. Hashim, I. R. Rowland, H. McGlynn, M. Servili, R. Selvaggini, A. Taticchi, S. Esposto, G. Montedoro, L. Kaisalo, K. Wahala, and C. I. Gill, 'Inhibitory Effects of Olive Oil Phenolics on Invasion in Human Colon Adenocarcinoma Cells in Vitro', *Int J Cancer,* 122 (2008), 495-500.

[74]   'Available At: Https://Apps.Who.Int/Infobase/. Accessed: October, 2013'.

[75]   'Available At: Http://Www.Internationaloliveoil.Org/. Accessed: October, 2013'.

[76]   Qiles J., 'Olive Oil and Health', *CABI Publishing* (2006).

[77]   M. B. Katan, P. L. Zock, and R. P. Mensink, 'Effects of Fats and Fatty Acids on Blood Lipids in Humans: An Overview', *Am J Clin Nutr,* 60 (1994), 1017S-1022S.

In: Virgin Olive Oil
Editor: Antonella De Leonardis

ISBN: 978-1-63117-656-2
© 2014 Nova Science Publishers, Inc.

Chapter 13

MODERN PSYCHOPHYSICS AT WORK: DISSOCIATION BETWEEN SENSORY AND DECISION PROCESSES WITHIN THE QUALITY ASSESSMENT OF OLIVE OIL

Teresa L. Martín-Guerrero, Concepción Paredes-Olay,
Juan M. Rosas, and Manuel M. Ramos-Álvarez[*]

Department of Psychology, University of Jaén, Spain

ABSTRACT

The analysis of the quality of olive oils is a complex task but highly relevant to achieve the transfer of the benefits associated to its consumption to the general population. Highly specialized tasting panels usually conduct measurement of its properties. However, traditional methods confound sensory precision with other aspects related to the taster's decision process, such as his/her beliefs, motivations, and preferences. We present and evaluate a new method of research in olive oil tasting based on Signal Detection Theory (SDT) that allows separating the evaluation of sensory and decision processes in a tasting situation. The proposed procedure, based on a dissociative model, allows for obtaining independent measures of sensory and cognitive factors, and also the calculation of the Receiver Operating Curves. The results of studies where this method was applied to olive oil tasting are presented and discussed here. This methodological proposal has also been proven to be useful to study psychological processes involved in tasting situations, including the evaluation of training programs conducted to improve tasters' discriminative abilities.

From a practical point of view, this dissociation between sensory and decision performance may contribute to an optimization of the evaluation of the quality of the olive oil, aiding its potential acceptance as a functional food, and facilitating the comparison between experts' and regular consumers' evaluations.

---

[*] Corresponding author: Manuel M. Ramos-Álvarez. Department of Psychology, University of Jaén, 23071, Jaén, Spain. E-mail address: mramos@ujaen.es (M.M. Ramos-Álvarez).

Keywords: Olive oil tasting, sensory process, decision criterion, Signal Detection Theory, Receiver Operating Curves, training programs, functional foods, consumers' evaluations

1. INTRODUCTION

Some of the prevalent pathologies in modern societies are related to environmental factors including inappropriate food habits. The better known pathologies related to inadequate nutrition are cardiovascular diseases, diabetes, cancer, or the cognitive impairment associated to premature ageing. All of them are apparently related to the greater cell oxidization promoted by a diet high in saturated fat, e.g. [1-8]. In order to confront this problem, a new biopsychosocial approach to the health concept has changed the deterministic focus on the process of illness, giving the individual an active role in preventing and promoting his/her own health by adopting a healthy lifestyle. Contemporary research associates the prevention of psychological and physiological impairments, and fosters a good quality of life through a balanced diet. In that sense, the World Health Organization (WHO) published a document with the nutritional parameters recommended to prevent six food-related pathologies, highlighting the importance of the monounsaturated fat on health maintenance [9].

The Mediterranean diet has been proposed as one of the most suitable models to keep a correct nutritional status. Some of the foods used in this diet are currently aspiring to being classified as functional foods, see for instance [10]. Functional foods are those that have a beneficial effect for one or several functions in the organism, so that they provide a better state of health and sense of wellbeing. These foods may exert a preventive role, reducing the risk factors that lead to the occurrence of different diseases, or they may even act therapeutically on certain pathologies. Organisms such as the Food Information Council (FIC), the International Life Sciences Institute (ILSI), or its European section, the Functional Food Science in Europe (FUFOSE), define this type of foods as foods that provide benefits for health beyond their basic nutritional value [10].

The Mediterranean diet is defined with a pyramidal structure. On the base of this structure we find vegetables. Within those foods of vegetal origin, olive oil is probably the most common link across Mediterranean countries, being on the basis of every Mediterranean culture. Mediterranean people have used olive oil as food for several millennia. However, the scientific knowledge about its nutritional properties and health benefits is relatively new. Among those properties, we can enumerate benefits such as longevity, cardiovascular disease prevention, and protection against cancer, as qualities that are currently associated to regular consumption of olive oil as a source of monounsaturated fat, for a review, see [11]. The therapeutic value of olive oil is related to its chemical composition, formed by triglycerides in a 99%. The most representative component of this fraction is the oleic acid; an element that is considered to have protective properties against oxidative stress [12]. The 1% that is left is formed by components of high biological and nutritional value associated to virgin olive oils. The refining processes may alter these components, so that the refined oil usually loses vitamins, color, aroma, general flavor and nutritive quality.

These biologically and nutritionally relevant minor components add some supplemental positive effects in health, and provide qualitative and quantitative differences to diverse virgin varieties, e.g. [13], in press. Due to its high specificity, these minor components are used as quality and authenticity criteria when evaluating olive oils, e.g. [14].

In the last few decades it has been concluded that consumer's acceptance should be one of the key factors for the definition of the new categories of functional foods. Once that the features of functional foods have been established, and being olive oil a firm candidate to become one of them, an important issue is how to transmit the notion of functional food to the general population, given the influence of the diet for public health, and the essential role that consumer's acceptance has on the consumption of functional foods. As a matter of fact, this acceptance has been considered as the key factor for the definition of the new categories of functional foods since there has been verified that consumers do not accept easily a new food for the mere fact of which it has effects stated for the health until such food turns out to be agreeable to the palate [15]. This issue raises the importance of knowing the impact that different types of information may exert on the perception and acceptance of functional foods.

Not surprisingly, some researchers have identified food flavor as the key factor on consumers' choice, even in the case of functional foods. According to these researchers, taste beliefs and experiences have been reported as extremely critical factors when selecting this type of food, e.g. [16].

Thus, acceptance of healthy foods is not unconditional. Findings seem to relate consumption loyalty to consumers' hedonic evaluation, though consumption is also related to the specific benefits that provides for health, and to how reliable is the information about these benefits for the consumer, e.g. [15]. As a summary of these ideas, Hilliam [17] suggests that the food industry should focus on improving the flavor of functional foods so that they end reaching the consumer.

Some researchers have gone even further, suggesting that consumers are hardly ready to renounce to the appetitive quality of the flavor to increase the benefits of their diet, see [18, 19]. Twisting the issue a little more, recent studies that have examined the beliefs and attitudes individuals have towards food have found that they are strongly dependent on cultural traditions, education, cooking habits, hedonic appreciation, etc., though these attitudes are also susceptible of change [20]. Results in this area suggest that factors such as nutritional qualities, health benefits, ingredients and uses of foods, its price, brand or other purely contextual or social issues influence behavior, and they may also generate hedonic expectancies, e.g. [21-26]. Focusing on the case of olive oil, the study conducted by Caporale et al. [21] found that the information about the origin of the olive oil affects the perception of the sensory profile of this olive oil when provided by regular consumers, see also, [27-29]. Köster [24] goes even further, analyzing the different fallacies that are often accepted as true by professionals and consumers, such as the assumed uniformity and consistency in consumers' behavior, the idea that choice is based on conscious, rational thoughts, or the perceptual fallacy that sustains that only what is perceived may be remembered. Those fallacies have been dismounted by the scientific literature, drawing researchers' attention to the difficulty of the analyses of consumers' behavior, and emphasizing the need of restructuring the methods used in both, sensory and consumer's behavior research. The previous paragraphs raise the importance of considering multiple factors when promoting functional foods and its introduction as food habits for the population. Most of the research studies discussed in this section coincide on assuming the multiple determinants of choice behavior when it comes to select a food, including sensory qualities, and also some extrinsic and intrinsic factors that are interconnected in the consumers' evaluation of a product.

A precise methodology that would allow for isolating the role of the different psychological variables on choice would help the analysts in sorting out the role of the different factors within this complex situation. Advancing towards the development of such methodology has been the goal of our research efforts in the last few years. Our advances will be summarized in the following sections of the chapter.

1.1. Organoleptic Evaluation of Virgin Olive Oil: Quality Criteria

The classification of different oil types and the evaluation of their quality are complex and relevant processes, given that they determine commercialization of the product, consumers' preference and the eventual transfer of the benefits associated with its consumption to the general population. Oil quality and purity are established by evaluating positive and negative features of each variety. The evaluation process essentially involves the use of chemical analyses. Mainly attending to chemical parameters, olive oils are classified into three different categories: extra-virgin olive oil, virgin olive oil, and "lampante" olive oil. However, the complexity of olive oil and the variety of its components forces an organoleptic evaluation, or normalized analysis through the human senses. Sensory analysis is the mechanism used to evaluate the differences among varieties and to determine whether consumers like them, or even to establish consumers' preferences across varieties [30].

Given the importance of these sensory aspects, food industry in general, and olive oil industry in particular, has devoted a great effort to develop analytic methods in order to establish a correspondence between chemical components, such as volatile or phenolic compounds, and sensory analysis. In spite of these efforts, sensory analysis is still the most effective method to evaluate qualitative and quantitative differences in the sensations caused by the food.

The only way to know the final result of the product at the sensory level is by the evaluation of the food in highly specialized tasting panels [31]. A tasting panel consists of a group of judges trained to measure and describe the noticeable attributes of a given food. Their goal is to evaluate foods in an objective way, minimizing possible mistakes and estimation-measurement biases. The reliance on their decisions and thus, in the quality of the food depends on the accuracy of this process.

Evaluation methods are regulated by international organisms that unify and supervise the properness of the sensory practice. Specifically, the International Olive Council (IOC, formerly the International Olive Oil Council, IOOC), as the organization that protects developing and growing of the olive grove and its products, has established the general method for the organoleptic evaluation of oils, see [32].

Tasters' selection is conducted by using a highly structured process that determines potential tasters' sensibility thresholds to basic attributes of the oil (fusty, winey, burnt, and bitter), as well as the tasters' group mean threshold. The best candidates go through a training phase in which the coherence and reliability of their evaluations is tested.

In order to attain normalization in the evaluation process, the identity and singularity of olive oils has been acknowledged and measured. This has allowed professionals to find ways to improve their quality, and has also provided consumers reference patterns that help them to learn to distinguish this quality. The absence of homogeneity in panels' evaluations encouraged successive revisions of the sensory method so as to improve its efficacy.

Some of the control measures adopted by the revised norms were: the high level of training required for professionals, essay replications, parameters normalization (temperature, light, quantity, and sample presentation), adoption of a mean criterion, and the statistical treatment of the results [33]. However, these cautions have not quite completely solved the problem of the variability of the judgments. Sensory analysis of food is not an exact science: it presents a number of limitations that steam from the human nature of the measurement instruments, from the classical psychophysics methodology used, and from other factors related to each individual panel.

At the commercial level, sensory analysis may offer a way to discover whether consumers enjoy food sensory stimuli, as well as to identify the consumers' preferences. In this sense, the evolution of the consumers' preferences and their causes can be analyzed, explaining possible changes in the selection of oils. Organoleptic exploration becomes more complex, evaluating new and subtler attributes with the goal of improving the excellence and exclusivity of the olive oils. At the same time, the advances derived from the research about the connection between food and health may promote the consumption of olive oils of greater quality, with added health benefits, although the general population may not easily accept some attributes, such as high acidity. In this sense, a rigorous methodology that allows isolating the factors that influence experts' and consumers' judgments and that allows for a comparison between their estimations is required.

1.2. Classical Psychophysical Theories of Threshold

As mentioned above, the technical documents of the IOC show that the method for tasters' selection and training is conducted by a process that determines the potential tasters' sensitivity thresholds to basic attributes of the oil. For instance, determining the mean sensitivity threshold of aspirants for characteristic attributes will allow to eliminate less sensitive tasters, see [32]. This approach assumes that sensations directly depend on the intensity of the attributes of the olive oil, so that the stimulus is perceived only when its intensity is above some specific absolute level or threshold. Thus, as the IOC method is based on the tasters' sensibility thresholds, it shows its reliance on the principles of classical psychophysical theory.

Psychophysics has the merit of introducing the measurement in psychology, substituting the mere metaphysic speculation with the search of functional relationships between mind and body. The measurement of the sensory experience allowed establishing psychophysical laws. Weber and Fechner developed the classic psychophysics based on the concept of sensory threshold [34]. In their two-state theory (sensation versus no-sensation), the relevant concept was the absolute threshold. In the second half of the nineteenth century, Weber and Fechner accepted the idea that for any given attribute (e.g., sweetness) there is a stimulus value such that only greater intensities should produce the state of sensation (e.g., the degree of sugar concentration needed to detect sweetness). Fechner [35] devised three methods to measure the absolute threshold, all of them based in statistical-mathematical techniques: The limits, the adjustment, and the constant stimuli methods. Fechner also considered Weber's work, which some years earlier had measured a different threshold: the differential threshold, or the amount of stimulus increase that is needed to establish the difference between two stimuli of the same modality [36].

Given the sensory variation, both thresholds were established in probabilistic terms. Absolute threshold was defined as the stimulus that produces a 50% of "yes" responses in detection psychophysics tasks, and the differential threshold was defined as the stimulus increase that produces a 75% of responses "greater than" in discrimination tests.

Classical psychophysics opened the possibility of studying human perception through scientific methods, providing sensitivity indexes valid for the different sensory channels. However, classic psychophysics found a serious number of experimental limitations, such as the lack of universality of its principles for all the stimulus ranges or sensory modalities, see [34, 36, 37]. In the next section of the chapter we will focus on the limitations of classical psychophysical theory, and how these limitations affect olive oil tasting methods.

1.3. Limitations of Classical Psychophysics and Traditional Methods

Food industry has evolved quickly in the last fifty years. This development has been promoted by the research and innovation in food biotechnology, the engineering of the industrial processes and quality management, and also by people's focus on preventive health. Markets and consumers are nowadays establishing new requirements for the industry that should be met.

However, sensory analysis as the compendium of techniques for the evaluation of olive oil has changed insufficiently, carrying out the measurement of properties and attributes through traditional methods based on the threshold concept [32]. Classical psychophysics regarded the sensations experimented in front of a stimulus as directly reflecting the intensity of the properties of such a stimulus, so that those stimulus properties that are under a specific threshold are not perceived [36]. Evaluation in tasting panels has been developed following the classic idea that sensations are discrete elements, assigning panelists the function of evaluating the intensity of desirable and undesirable attributes, but without considering the subjective factors in the perceptual process that might affect the taster's criterion. These factors would go unnoticed in traditional methods of tasters' selection, and also on their evaluations.

Additionally, discrimination may be required in ambiguous situations, with concurrent or previous stimuli that show slight differences with the stimulus currently tested. Ambiguity favors the presence of biases and mistakes of different nature, such as biased evaluation responses. In these situations of maximum uncertainty or extreme similarity among attributes, the taster's internal aspects may become as relevant for the task as the physical features of the stimulus. Threshold psychophysics models cannot separate the measurement of attributes or properties from other effects that are not directly related with the stimuli, or with the sensory capacity for the evaluation.

However, modern psychophysics has adopted a different approach, changing the discrete conception of the sensations on which threshold theories were sustained. Sensations are currently understood as a continuum that depends on the stimulus properties and the perceiver's sensitivity, but also on other individual factors such as the taster's motivation, beliefs or interests.

Decision-making processes take an essential role in sensory evaluation, and thus in the tasting process. The new methodology based in contemporary psychophysical models allows estimating the contribution of cognitive aspects to psychophysical performance.

In the middle of the twentieth century, research in human sensory discrimination focused its attention in a new model that allowed estimating the influence of non-sensory processes in the observers' response, the Signal Detection Theory, and this theory might be of straight-forward application to the field of olive oil tasting, see [34, 36, 37].

1.4. Modern Psychophysics: Signal Detection Theory

Signal detection theory (SDT) provided a general framework to describe the decisions adopted in situations of uncertainty or ambiguity [38]. The development of SDT as a direct translation of the theory of statistical decision to the explanation of the mental processes that underlie subjective judgments has produced a number of applications in experimental psychology, especially in areas in which discrimination is central, such as in research about perceptive processes [39] or in the area of recognition memory [40]. In applied sciences, a variety of diagnosis systems used this paradigm to measure its inherent precision [41]. The versatility of this model has favored its application in fields as different as weather forecast, clinical medicine, and survey research, staff selection or in sensory analysis of drinks and foods, e.g. [42-45].

Any situation of discrimination between two stimuli may be analyzed in SDT terms. In a typical experiment of signal detection, the observer should decide in each trial whether the experimented sensory effects are due to a background irrelevant activity, or to the superimposed signal. Through the analysis of the performance of the observer we may know the type of response given by him or her (correct/incorrect). If we consider a Yes/No classic task in which the participant should judge the presence or the absence of a small amount of olive oil in a stimulus compound, the taster may be understood as a detector of the presence of the component in the solution. Let us imagine that it is exposed to 10 samples randomly arranged, five with a small amount of olive oil diluted in other oil varieties (Signal+Noise stimulus for SDT model), and five without olive oil (Noise on SDT). This experimental situation allows for four different responses: Hits (the judge identifies the sample that contains olive oil, saying yes to a Signal+Noise trial); Misses (the judge does not identify the olive oil in a sample that contains olive oil, saying no to a Signal+Noise trial); False Alarms (the judge identifies olive oil in a sample that does not have it, saying yes to a Noise trial); and Correct Rejections (the judge does not identify the presence of olive oil in a sample that does not contain it, saying no to a Noise trial). Table 1 presents the confusion matrix with the four possible responses adapted to a classical experiment of flavor detection.

Following with our example, if our taster says "YES" to 4 of the 5 samples that actually contain olive oil and to 1 of the 5 Noise trials (saying "NO" to the other trials), taster's performance may be estimated by considering his or her Hits (4/5 or 0.8) and False Alarms (1/5 or 0.2) rates. These two results define the space of response, given that performance is redundant two by two once the amount of trials with Signal and Noise is fixed (the same result is obtained calculating the Rate of Hits together with the Rate of False Alarms, and the Rate of Correct Rejections together with that of Misses). Table 1 shows the confusion matrix with the four possible responses adapted to a classical experiment of flavor detection.

The most interesting feature of SDT's application to the field of olive oil tasting is that it allows for separating the sensory and decision processes. With respect to the sensory process, the threshold concept is discarded.

Table 1. Confusion matrix in a Tasting situation

Stimulus

	Flavor + Noise (FN)	Noise (N)	
"Yes: Signal was present"	Hits (H) (4)	False Alarms (FA) (1)	Responses YES (5)
"No: Signal was not present"	Misses (M) (1)	Correct Rejections (CR) (4)	Responses NO (5)
	Signal Trials (5)	Noise Trials (5)	

Response (row label for the two response categories)

$$P(H) = \frac{H}{H + M} \qquad P(FA) = \frac{FA}{FA + CR}$$

Notes: The four possible responses are adapted to a classical experiment of flavor detection. The numbers within brackets represent the responses given by participants in the hypothetical experiment.

The result of the sensory process is a group of sensation values with a specific probability of appearing when the signal is presented with the noise, and with a different one when only the noise is presented.

Initial SDT approaches understood that the distribution of the density of probability associated to the Signal and the Noise may be adjusted to normal curves of equivalent variance [46].

This allows for a characterization of the sensitivity or detectability of a given observer in a test of sensory discrimination by the separation of Signal and Noise distributions.

If those stimuli are experienced as very different, sensory performance will be better and then there will be little overlapping between the functions, as they are more separated than when the stimuli are similar. The contrary is true when the stimuli are very alike, given that the noise and signal distributions will overlap significantly. This process depends on objective or external factors such as the degree of contrast between the signal and the noise, or the physiological state of the organism, being fairly stable across equivalent stimulus conditions.

Now let us imagine that the participant in our experiment is more willing to inform of the presence of olive oil throughout the session, either because of a bias that comes from his or her own experience, or because of a manipulation in the conditions introduced by the experimenter.

The tendency of response is likely to shift towards a more lenient criterion, saying yes more often, so that the obtained rates will change even though the presented samples are the same. From this perspective, performance in a perceptual learning task is partially determined by the criterion adopted to make decisions. Criterion is established in terms of the "perceptual evidence" needed by the observer to give a positive response to the signal presentation. So it may change or it might be manipulated by conditions that are not necessarily associated to the intensity of the stimulus or to the discriminative abilities of the observer. SDT allows separating this decision process from the purely sensory process.

SDT suggests that sensory processes are related to the stimulus physical properties (i.e., its intensity), and to the physiological status of the individual. Decision processes are assumed to be related to the knowledge about the situation that the observer has. This information is a source of theories and hypotheses about the nature of the sensory signals that the participant expects will occur. Thus, sensory and decision processes should be affected by the manipulation of different factors. The sensory process should be modified by changes in the contrast between the signal and the noise, or by the duration of the task. However, the decision process should be modified by instructions, payoff matrix (the relative cost of making the two types of errors, FA versus M on Table 1, and the relative benefit of the two types of correct responses, H versus CR on Table 1), or by changes in the probability of signal and noise (see [36], for an empirical review; and [38], for theoretical implications).

According to SDT, the proper combination of hits and false alarms rates allow for computing different indexes for sensory and decision processes. Receiver Operating Characteristic (ROC) curves represent the relationship between hits and false alarms rates when a stimulus is shown at a constant intensity. ROC curves are graphical representations of pairs of hit and false alarm rates when the factors affecting sensitivity are kept constant. They are also called isosensitivity curves, as all the dots in the curve represent the same level of sensitivity, or signal-noise discrimination. To build a ROC curve the same Yes/No task needs to be applied a number of times while keeping constant all the factors that are supposed to affect sensitivity (i.e., signal/noise contrast), changing only those factors that supposedly affect decision criterion. Each of the dots in the ROC curve represents participants' performance for a different decision criterion. This decision criterion may be manipulated through at least the three factors newly mentioned: signal-noise probability, payoff matrix, and instructions. Figure 1 presents an example of ROC curves.

The normal distributions with equal variance model predict a symmetrical ROC with respect to the diagonal that connect the coordinates (0, 1) and (1,0). As the curve gets closer to the superior left vortex from the ROC space (0, 1) the most sensitivity it reflects, and the closest it is to the diagonal, the lowest the sensitivity. Within each isosensitivity curve, the closer the dot from the upper right vortex (coordinates 1, 1), the more lenient/relaxed is the observer's criterion, and vice versa, the closer the dot to the lower left vortex (0,0) the stricter/conservative the observer's criterion (see L and C points on Figure 1, respectively for lenient and stricter cases).

Using the Hits and False Alarms rates a measure of the capacity of the individual to detect the signal may be estimated, as well as another measure of the decision criterion (or the sensation value that the taster needs to respond when he or she has tasted olive oil). The theoretical model allows computing two different indexes to dissociate those two processes. About 30 varieties have been proposed to quantify the sensory-decision performance, e.g. [47-49].

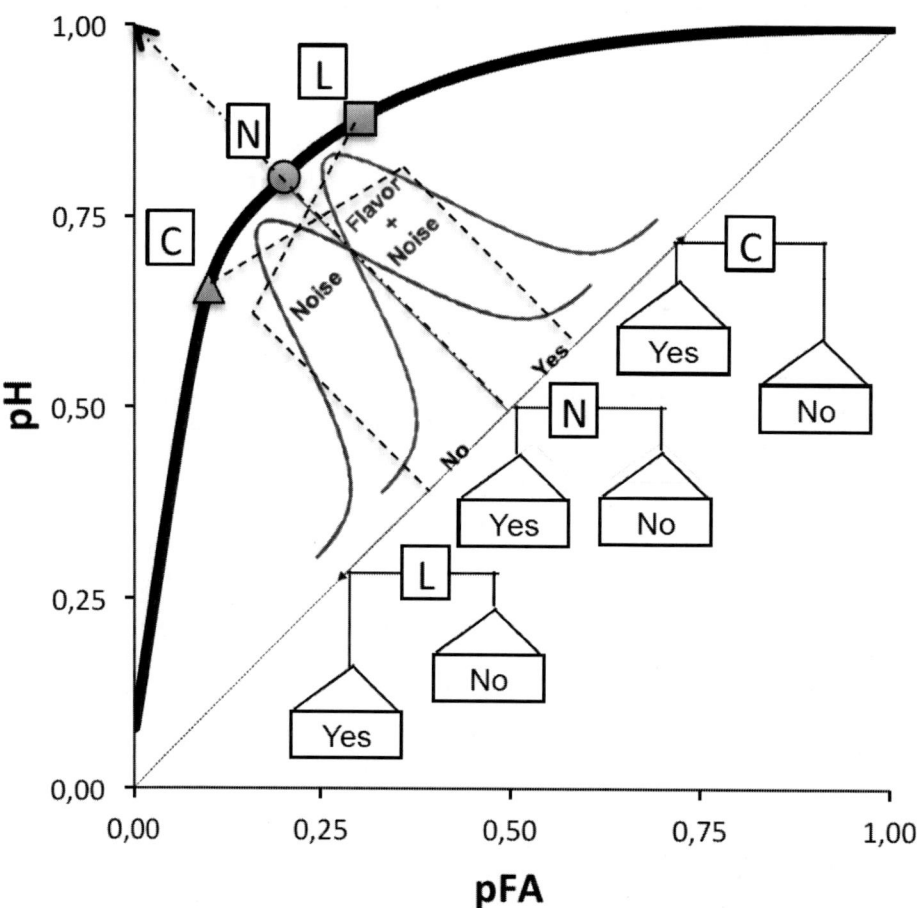

Notes: False Alarm probabilities (pFA) are presented as a function of the Hits probabilities (pH). The dashed line represents the isosensitivity curve. The three points in the line correspond to the variations in the response criteria: Neutral (N), Lenient (L), and Conservative (C). Within the chart the Normal SDT model, a conventional one, has been superimposed. According to this model the signal and the noise lead to distribution models based on the Gaussian distribution. The middle rectangles represent each of the decision criteria: The one that favors Yes answers (L), the second one favoring No answers (C), and the middle one favoring neither of the two answers (N).

Figure 1. ROC graph.

Classical indexes d' and Beta have been the most commonly used in psychophysical tasks. Both are based on the assumption of normal distributions of equal variance associated to signal and noise [46].

Sensory index d' represents the distance between the distributions associated to signal and noise, while the index associated to the decision process (Beta) represents the relative distance between the heights of the two distributions, e.g. [46]; but see [38].

The use of d' and β is only justified when the mathematical assumptions of normality and homogeneity of the variances about the density and probability underlying functions have been confirmed, limiting its general use in situations in which the information is limited [36, 41].

Aside of the parametric approach associated to the normal distribution, different families of theoretical models that do not need the normality assumptions, such as models of thresholds, or nonparametric models based on geometric measures, have been proposed, see [36]. These models might be more appropriate for the limitations imposed by the tasting situation [50].

Once established an alternative method to classical psychophysics that may be appropriate in the field of olive oil tasting, the next step is to test the bonhomie of this method on tasting situations. In the following sections we will briefly present the results obtained in our research when using SDT in the field of olive oil tasting. The first section will be devoted to summarizing those studies that have addressed the improvement of the methods used to select tasters and test their performance. Subsequently we will summarize the research conducted to explore the effects of different training methods on olive oil discrimination.

2. ADAPTING SDT METHODOLOGY
TO THE STUDY OF OLIVE OIL TASTING

Our research aims at developing an experimental set up that would allow applying the SDT to the study of olive oil tasting under controlled conditions. This methodological approach fulfills the requirements of Signal Detection Theory, allowing for quantifying the sensory precision but dissociating it at the same time from other non-sensory aspects of performance. Below, we will describe the general features of the method and procedure used in all the studies, and then describe the results obtained [50-53].

Participants. Participants were undergraduate students of the University of Jaén. They were regular consumers of olive oil, with no explicit training in sensory evaluation.

Their participation was voluntary.

To avoid distortion of their discriminative ability, they were requested not to smoke, or ingest any food or beverages for 30 minutes before the experimental sessions. With the same goal, participation of individuals that were affected by cold, allergies or any other seasonal alteration of taste was declined. Assignation of participants to the different experimental conditions was done randomly, where needed.

Materials and Apparatus. Preliminary studies allowed to standardize both the facilities to implement sensory panels, and the protocol of the tasting process, such as reflected in Figure 2, see [53]. The tasting laboratory consisted of 8 individual cabins following the IOC guide for the installation of a test room [54]. Cabins were separated by mobile panels that precluded participants from seeing each other while performing the task. Each cubicle was equipped with a desk and a chair (Workstation), a Water bath, and a computer (PC- control). A numbered plastic sheet was placed at the table so that participants could leave the samples after tasting them (Sheet Coaster). A coaster to place the trial sample, a place mat in which a glass of water and a plate full of green apple chunks were located to wash the palate between tastings (Taste Wiping), a 17" TFT computer screen, a rubber keyboard and a mouse were also placed at the desk in front of the participant to interact with the PC Control.

The final setting allowed presenting up to 24 samples of 5 ml each, and two different types of stimuli arrangements: Signal and noise.

Signal was arranged by including an amount of extra virgin olive oil diluted on a mixture of sunflower oil and food paraffin. The amount of olive oil depended on the design. *Noise* was identical to signal, with the exception that no olive oil was included in the mixture. In both cases, proportion of sunflower oil was kept constant (85%).

Samples were presented in blue tasting glasses covered with Watch-Glasses following the IOC guidelines for the standardized presentation of oil samples [55]. Samples were maintained between 27 and 29 degrees Celsius, IOC [54], by using industrial water-bath heaters that allowed keeping up to 28 samples at the same time for any given participant (see Water Bath on Figure 2). Water and apple chunks were provided to the participants between sample tastings, instructing them on how to cleanse their palate between trials.

Experiments were developed in individual cabins provided by a computer with *LearnOlive* software. This software was explicitly developed within our research team to conduct flavor experiments, and allows participants to easily follow the complete sequence of the task without intervention of the experimenter.

General Procedure. Experiments followed a 5-phase protocol: (1) Participants' recruiting and debriefing about the requirements to participate in the experiment, (2) Stimulus and taste sample preparation an hour before the experiment, (3) Cubicle preparation: Sample distribution according to the experimental setting and the randomizing provided by the software, and arranging of the setting for the participants including an arrangement of water, apple chunks, and cleansing elements, (4) Participants' reception, random assignation of the experimental location for each participant, and general instructions about how to perform the task, and (5) Performance of the experimental task.

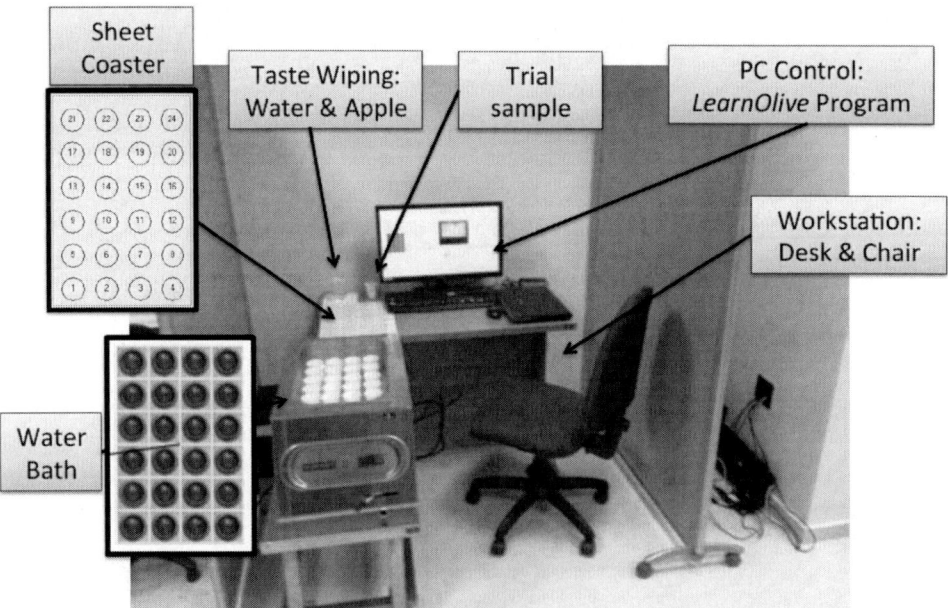

Notes: The standard Tasting Cabin was equipped with a desk and a chair (Workstation), a Water Bath, a numbered plastic sheet (Sheet Coaster), a coaster to place the Trial Sample, a place mat to place the Taste Wiping, and a computer running LearnOlive software (PC- control).

Figure 2. The Tasting Cabin.

After signing the written consent, and filling a socio-demographic questionnaire, participants received general instructions about how to taste the olive oil samples (Turn-Smell-Taste-Clean procedure, see [51] for details).

The experimental session started immediately afterwards. Participants confronted a simple-detection task in which they should indicate whether each presented sample contained olive oil. The protocol followed by each participant was: (1) Picking the sample according to the established sequence, (2) tasting the sample, (3) giving the response about its content (Yes/No), (4) leaving the glass aside, and (5) cleansing her or his palate. Participants either received or did not receive feedback with respect to whether their response had been correct, depending on the design.

Dependent Variables and Statistical Analysis. Yes/No responses to each sample were recorded to compute Hits and False Alarms rates with the goal of computing the sensitivity and decision criterion indexes. The choice of the proper indexes should depend on the specific field of application. Sensory evaluation tests, and especially those using flavors, require the use of a limited number of trials so that saturation of the senses is avoided. This requirement does not favor the fulfillment of the mathematical assumptions in which classic measures are based: normality of the data and homogeneity of the variances. That is the reason why the choice of a nonparametric alternative is more appropriate within this field. In olive oil tasting tasks Non parametric Grier's A' [56], and Donaldson's B'$_D$ [57] have been proposed as sensory and decision indexes, respectively [51]. Ramos-Álvarez et al. [53] performed a comparison between different indexes (d' and A' for sensitivity, and c, Logβ, B', and B'$_D$ for decision criterion) with respect to the effect size of the factors manipulated in their study, finding that indexes A' and B'$_D$ are at least as appropriate as classical parametric indexes to characterize performance in tasting situations. Figure 3 summarizes these results.

As shown, sensory indexes have a higher value when the sensory factor appears, whereas the same phenomenon happens with decision indexes. Moreover, sensory indexes d' and A' are very similar, and optimal values for decision indexes are C and B'$_D$ Thus, the studies presented here have used those nonparametric indexes A' and B'$_D$, see also [48].

Under the same reasoning, parametric analyses of variance may not be most appropriate ones to be used in tasting situations, given the difficulties to verify Normality and Homo-scedasticity assumptions because of the small number of data. Similarly, outliers are likely to appear in tasting tasks, according to our experience, invalidating the use of conventional statistics.

To explore these assumptions while summarizing the data, variability and central tendency indexes are presented in Box-Plot graphs according to robust statistics based on median [58]. Although this is not a requirement if the parametric assumptions are fulfilled by the data, our experience has inclined us to favor the use of robust statistics as a general analytic approach to tasting research, see [53].

2.1. Optimization of Olive Oil Sensory Analysis: Selection and Performance

Tasting situations are extraordinarily complex due to the confluence of physiological, psychological, social and even cultural factors and processes. These features make it difficult to analyze the tasting situation even under strictly controlled laboratory conditions.

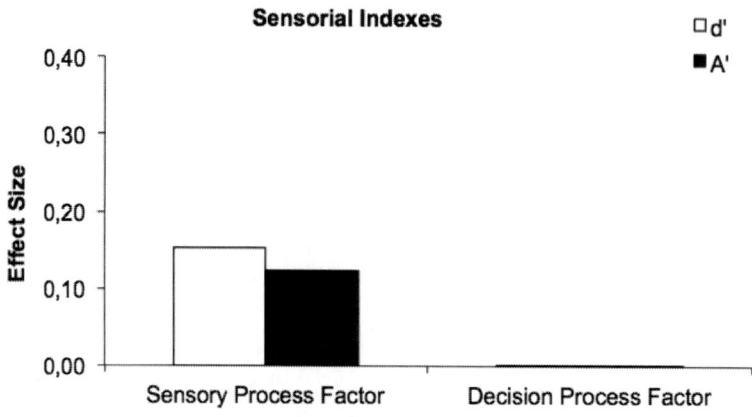

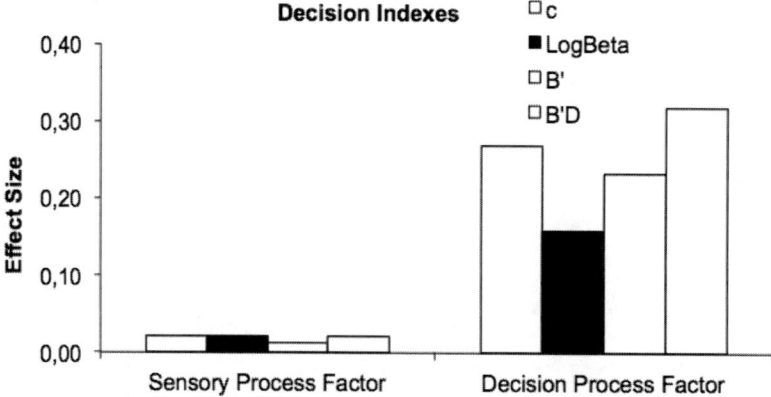

Notes: Effect size was computed for an ANOVA where manipulated factors were the two variables affecting the Sensory or Decision process. On the top part computations for sensory indexes put forward for the SDT (d') are shown, together with its non-parametric alternative (A'). In the lower part the table shows the estimations for decision indexes, either using the SDT (c and LogBeta), or using the non-parametric alternative (Gief B' and Donaldson B'$_D$). Based on Ramos-Álvarez et al. [53].

Figure 3. Comparison of the different psychophysical indexes with respect to the Effect Size.

In a recent study by Jiménez and Carpio [59], they point out two different categories of variables that may affect the evaluation of the attributes in olive oil tasting: Factors that affect the sensory acuity of the taster (genetic factors, concentration ability, adaptation, saturation, sensory fatigue, physiology of the taster, and sample features among others), and other factors that are related to tendencies, response biases or task specific conditions that may affect judgments without affecting sensory acuity (motivation, test type, or quantity and quality of the information provided, among others). This idea is in agreement with the modern psychophysics that sustains performance is based on two different processes, the purely sensory one that derives from the combination of the physical qualities of the stimuli and the discriminative ability of the taster, and the decision process related to the cognitive aspects and the decision strategy of the observer, with his motivation, preferences, expectancies, and interests, see [41].

Considering a psychological perspective on the evaluation of olive oil may be of use to improve the internal processes of the panels, such as the selection of the components. Methods that would allow for quantifying separately the psychological processes involved in the tasting situation would imply an essential development on the optimization of the professional tasters' selection.

After all, the goal of such a selection process is to identify the individuals with a greater capacity to detect and discriminate among olive oil attributes, aside of the internal aspects that may bias their responses.

Human sensory mechanisms have a limited capacity and they are fairly imprecise. This imprecision of the mechanisms, as well as our own psychological nature makes us vulnerable to internal and external factors when living a sensory experience. It is in this limitation of the human mechanisms, in the adaptation capacity of psychophysical judgments, where our approach to sensory discrimination of olive oil as a decision-making process lays, a process in which the taster's response tendency may be guided to a goal, and changes in his or her strategy of behavior may be induced by experimental manipulations: Using rewards, instructions, or favoring expectancies about the presence of attributes and properties.

Paredes-Olay et al. [51] began the exploration of applying modern psychophysics to the study of olive oil tasting, searching for a separation of purely sensory processes from other factors that may affect performance, such as tasters' cognitive biases. These authors conducted a calibration experiment, based in classical psychophysics, using a task akin to the constant stimuli method adapted from the IOC normative [32], with the goal of selecting the concentrations of olive oil that would be used in subsequent experiments. The concentrations selected were 1.6%, 3.2% and 12.8%. The first two were detected by the 50-60% of participants, while the last one was detected by over 70% of participants. Thus, in the subsequent experiment there were two concentrations of some difficulty, and one that was easy to detect by most participants.

In their second experiment, Paredes-Olay et al. [51] implemented a detection task adapted to the specific methodology of Signal Detection Theory, allowing for a calculation of ROC curves and the separate calculation of sensory and decision indexes. The three levels of olive oil concentration were presented to three different groups. Each participant was exposed to 10 samples (5 Signal+Noise and 5 Noise). Participants did not receive feedback about their performance.

The results of this experiment are presented in Figures 4 and 5. Figure 4 presents a ROC graph of participants' mean performance. False Alarm rates and Hits rates are presented as a function of the three levels of olive oil concentration manipulated in this experiment: 1.6%, 3.2%, and 12.8%. The dashed lines represent the estimated isosensibility curve for each of the three points from SDT Model. The curves showed an increment in participant's sensitivity as the salience of the signal increased. With respect to the decision criterion, participants kept it basically neutral since the three points fall down closely together of the secondary diagonal (from the center towards the left vortex). Results obtained with indexes A' and B', which are presented in the left top and bottom charts of Figure 5, respectively, confirmed the impressions based on the ROC graph. Sensitivity criterion A' medians increased as olive oil concentration increased. Decision criterion B' was kept around zero (left bottom chart), an expected result given that no factors affecting decision criterion were manipulated in this experiment. These results are in agreement with dissociative predictions derived from SDT [36].

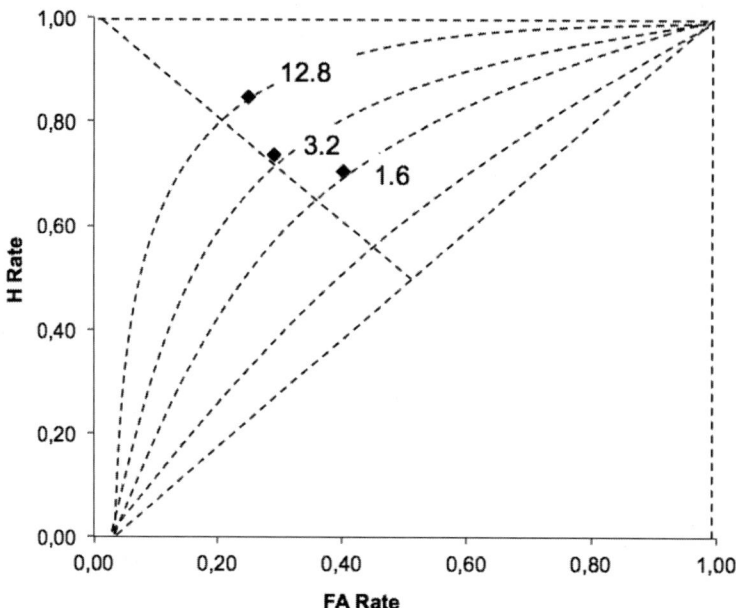

Notes: In this experiment, three olive oil concentrations (1.6%, 3.2% and 12.8%) were manipulated
between groups, with a neutral situation regarding the sensory process. In the ROC graph, dashed
lines represent isosensitivity curves for different sensitivity changes in the sensory process.
Besides, points for the three manipulated conditions have been signaled. Based on Paredes-Olay et
al. [51].

Figure 4. Olive Oil Tasting ROC.

This experimental series allowed for comparing two procedures based on two different
psychophysics approaches.

The first one, regularly used in tasters' selection, allowed characterizing the
discriminative capacity of participants with different olive oil concentrations. However, that
methodology did not allow estimating whether tasters' performance was affected by any
response bias that may or may not be playing a role in that situation. As it happens in other
discriminative tasks, sensory evaluation of olive oil may be affected by cognitive biases that
would go undetected when using these procedures.

In fact, undesirable deviations from a neutral criterion have been already found with other
foods [44]. The second experiment applied a dissociative approach that allowed computing a
pure measure of sensitivity, finding at the same time the lack of response biases in this
situation by using an independent index. This methodological approach presents important
advantages with respect to the traditional ones, allowing for a separate test of tasters'
sensitivity and tasters' criterion biases.

After showing that Signal Detection Theory methods may be applied to olive oil tasting,
Ramos-Álvarez et al. [53] conducted an experiment in which they applied the logic of the
double dissociation additive test based on Sternberg additive factors method [60] that allows
for assessing the interaction or independence of the sensory and criterion indexes within a
single experimental situation.

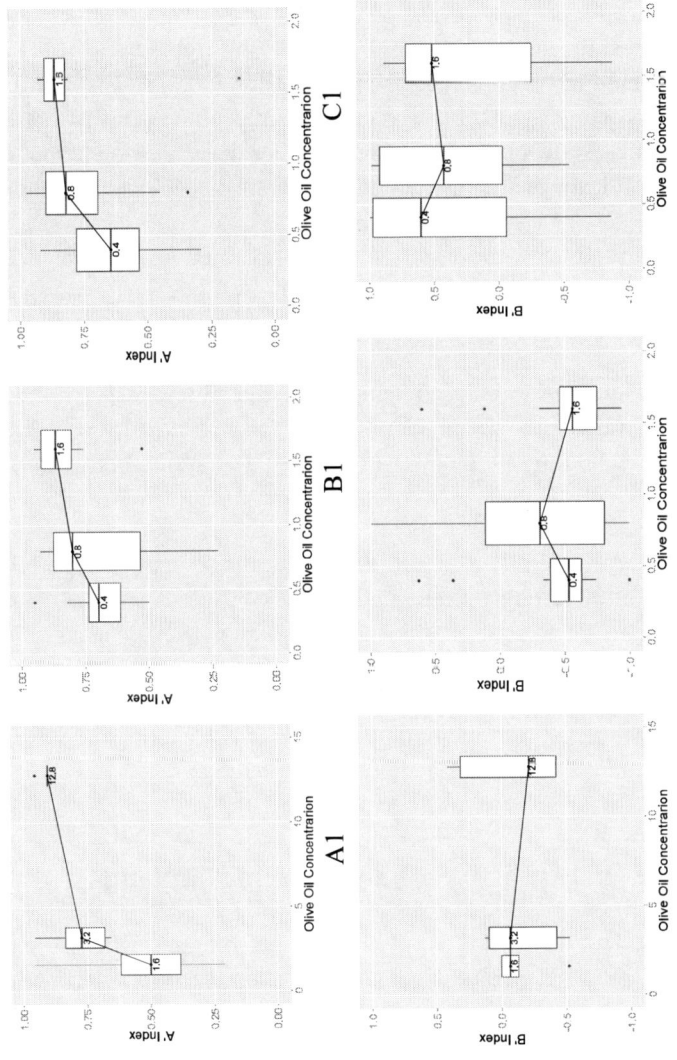

A1) Selection Experiment. A'Index. A2) Selection Experiment. A'Index. B1) Dissociation Experiment. Lenient Condition. B'Index. C1) Dissociation Experiment. Conservative condition. B'Index. C2) Dissociation Experiment. A'Index. Notes: Panel A (left) presents the charts corresponding to the Selection Experiment, based on Paredes-Olay et al. [51]; and Panels B and C, charts for Dissociative experiment adapted from Ramos-Álvarez et al. [53]. Panel B (middle) corresponds to the condition induced bias type "lenient", and Panel C (right) corresponds to the "conservative" condition. Top charts (A1, A2, and A3) show the nonparametric sensitivity index A', while bottom charts (B1, B2, and B3) present the nonparametric bias index B'. In all cases, three olive oil concentrations were manipulated between groups: 1.6%, 3.2%, and 12.8% on the Selection Experiment (left); and 0.4%, 0.8%, and 1.6% on the Dissociative Experiment (middle and right). In each of the box-plot graphs, straight lines represent the median, boxes represent variability, and outliers are represented by dots outside the boxes.

Figure 5. Olive Oil Tasting Statistical Analysis.

Two independent factors were simultaneously manipulated in a complete factorial design. Olive oil concentration was manipulated with three levels (0.4%, 0.8% and 1.6%).

Instructions were manipulated so that participants were induced to be either lenient or strict. Olive oil concentration was expected to affect only the detection process (A' type indexes), while instructions should affect only the decision processes (B'$_D$ type indexes).

In this double dissociation test, the absence of interaction would point to the independence of the psychological processes associated to each factor.

The results of this experiment are summarized in the middle and right charts of Figure 5. Sensitivity index A' is represented in the top charts, while criterion index B'$_D$ is represented in the bottom charts. Sensitivity increased with the concentration level, and this increase was not affected by the instructions, as shown by the two parallel lines that connect the median of each concentration level at each level of the instructions (compare top-middle and top-right charts).

The opposite result was found with the criterion index. Instructions that intended to induce a conservative bias did so regardless of the concentration level (see positive values for the median of the criterion index for every concentration level presented at the bottom-right chart of Figure 5).

Similarly, instructions that intended to induce a lenient bias did so regardless of the olive oil concentration (see negative values for the median of the criterion index at every level of oil concentration presented at the bottom-middle graph of Figure 5). Parallel lines connecting boxes at each concentration level suggest a lack of interaction that was confirmed by statistical analyses.

Manipulation of factors that theoretically should affect specifically either sensitivity or criterion decision did so without affecting the alternative index. These results confirm the usefulness of the SDT based methodology on the tasting situations, even using a small number of samples and a small number of participants.

In fact, these conditions were selected on purpose, with the idea of making the experimental situation as akin to the natural situation of a tasting panel as possible, so that the results of these experiments could be easily translated to applied settings.

2.2. Optimization of Olive Oil Sensory Analysis: Tasters' Training

Once the panelists are selected, the main goal of any training program within a sensory panel is to improve tasters' ability to detect and judge with precision the sensory properties of the food. In the context of olive oil tasting, achieving this goal involves an improvement of professionals' abilities to discriminate among the different olive oil varieties. Research in the field of perceptual learning may help to understand how those abilities are developed, contributing to the design of more efficient training programs, see for instance, [61, 62].

The concept of perceptual learning involves the effects of an increase in discriminability of the stimuli because of the experience with them [63]. One of those perceptual effects with potential implications for tasters' training is the transfer of discrimination along a continuum, e.g. [64, 65], or the easy-to-hard effect, e.g. [66]. It turns out that experiencing an easy discrimination makes it easier for subsequent learning of a difficult discrimination about stimuli that vary across the same dimension as compared to a group that is only trained with the difficult discrimination, e.g. [67].

This phenomenon has been studied in humans, and also with animals under different sensory modalities, including flavor discrimination, e.g. [66]. However, traditional approaches to the easy-to-hard effect have focused on the improvement of perceptual abilities. The use of the dissociative approach of SDT applied to the study of perceptual learning would allow for testing the potential participation of other strategic factor within the discriminative training procedure. This issue was approached by Moreno-Fernández et al. [52] in an experiment conducted within a situation akin to the ones described above.

Two groups of participants were exposed to a tasting situation in which they received one 45-minute session in each of four days in a row. In each session participants received 20 samples, 10 with the signal, and 10 with only the noise. Participants received feedback about the presence or the absence of olive oil in each sample. In the testing session participants were trained with a 0.5% olive oil concentration, a concentration that is difficult to discriminate by non-trained participants. Groups differed in the treatment received before the test. The Hard-to-hard group was trained with the same concentration throughout the 4 sessions. However, for the Easy-to-hard group, the difficulty of the discrimination increased from session to session, starting with a concentration easy to detect (2%) and decreasing progressively the concentration level of olive oil across sessions (1.5% and 1%) before being trained with the same concentration at testing.

Performance at testing was better in the Easy-to-Hard group (A' index of 0.76) than in the Hard-to-hard group (A' index of 0.62). Both groups of participants showed learning, but performance was clearly better for the group that received progressive training than for the group that was directly trained with the difficult discrimination. These results extend to the field of olive oil tasting results previously obtained with other discriminations in human and non-human animals, e.g. [66, 67], suggesting that the easy-to-hard training may be an useful training schedule for tasters. However, the use of the SDT methodology, aside of providing a sensitivity bias-free index, also allowed for detecting potential response biases that in any other way would have gone undetected.

In fact, this was the case with progressively trained participants. They became conservative as the training progressed, showing a significant tendency to categorize incorrectly signal samples as noise (B'$_D$ index was 0.2 at the test session).

Thus, the easy-to-hard training schedule improves the ability for detecting an attribute, but it also produces a bias towards a more conservative non neutral strategy of response, suggesting that, when applied to tasters' training, this program should be combined with a corrective strategy, such as the use of specific instructions that may prevent trainees to be overly conservative, e.g. [53]. It should be as well obvious that the results of this experiment point towards the need to consider both, sensitivity improvements and the strategy used by participants, that is, the presence of responding biases on performance.

CONCLUSION

This chapter presents a new method of research and training for the sensory analysis of olive oil based in modern psychophysics theory that focuses on perceptive processes from the decision-making perspective. It is assumed that stimuli cannot be perceived in isolation from other psychological cognitive factors that are part of our human nature.

The proposed methodology based on a dissociative model, allows obtaining independent measures of the effects generated by purely sensitivity factors from purely decision factors that are present in all discriminations. The experiments used above to explain the dissociative SDT based dissociative model have been able to reveal the advantages of this methodology with respect to the one traditionally used in tasting panels, based on the threshold concept. Traditional methods do not allow separating the influence of factors related with tendencies, response biases, or task conditions that may affect tasters' judgments and are thus confounded with sensory acuity.

At face value, the use of an SDT based methodology applied to the olive oil tasting process will improve the quality of the sensory evaluation.

In addition, improving the quality of sensory evaluation including the evaluation of the cognitive aspects of the tasting process has a number of important implications. First, it will favor the approximation of experts' and consumers' criteria. This approach may be achieved by favoring the compatibility between their evaluations, something that can be accomplished by developing new advanced methods of analysis facilitating discrimination of variables in determination of quality criteria, acceptance and appetency.

Second, this approach may be beneficial for the research conducted in consumption and health. Given the complexity of organoleptic exploration, the quality of the evaluation will be improved by a precise measure of the sensory sensations that are part of the identity and singularity of the olive oils. These new method allows for the estimation of factors that are not necessarily attached to the perceptive experience, but that influence decisions of both, experts and consumers.

In the case of olive oil as a possible functional food, it is necessary to know what determines consumers' choice behavior, on which factors they base their choices, how they change their criterion, etc. so that processes to stimulate acceptance of the new functional foods may be developed.

Third, the SDT based method as has been exposed here has a direct translation to selection and training of tasters. The use of the dissociative method allows for an improvement of the selection of potential candidates for tasters, as measures of their acuity may be obtained free of biases.

As for its influence on tasters' training, the last experiment we summarized here not only shows that the gradual increase on the difficulty of the discrimination is a useful strategy to produce better training results; it also shows the importance of measuring the presence of strategic factors on the evaluation of any training program. And we should remember that the influence of those factors would go undetected in traditional selection and training methods based on classic psychophysics as in the case of olive oil tasting.

Finally, we should acknowledge that this line of research is just being born. The potential applications of dissociation methods based on Signal Detection Theory to olive oil tasting are yet to be fully determined.

Experimental psychology and olive oil tasting would benefit each other from an association that would not only allow for an improvement of the selection and formation of tasters, or for a detection of critical sensorial and criterion differences on consumers preferences, or on the differences between regular and potential consumers. It will also allow for a better understanding of the psychological processes involved in olive oil tasting and consumers' behavior, opening interesting, exciting and socially relevant perspectives on these fields.

ACKNOWLEDGMENTS

The research reported in this chapter was funded by research grants HUM642, and P06-HUM-01391, from the Consejería de Innovación, Ciencia y Empresa of Junta de Andalucía; and by ACS7PP2005, and UJA2010/12/51 of the University of Jaén, Spain. T. L. Martín-Guerrero participation was funded by fellowships of the University of Jaén's Plan for Supporting Research (Action 16, fellowships for Research Fellows' training, 2010), and AP2012-2934 from the FPU program of the Spanish Ministry of Science and Innovation.

REFERENCES

[1]   Colditz, G. A., Samplin-Salgado, M., Ryan, C. T., Dart, H., Fisher, L., Tokuda, A., Rockhill, B., and Harvard Center for Cancer Prevention (2002) Harvard report on cancer prevention, volume 5: fulfilling the potential for cancer prevention: policy approaches. *Cancer Causes and Control,* 13, 199-212.

[2]   Feskens, E. J., Stengard, J., Virtanen, S. M., Pekkanen, J., Räsänen, L., Nissinen, A., Tuomilehto, J., and Kromhout, D. (1995) Dietary factors determining diabetes and impaired glucose tolerance. A 20-year follow-up of the Finnish and Dutch cohorts of the Seven Countries Study. *Diabetes Care,* 18, 1104-1112.

[3]   Keys, A. (1980) *Seven Countries: A Multivariate Analysis of Death and Coronary Heart Disease.* Harvard University Press, Cambridge.

[4]   Kushi, J. K. and Giovannucci, E. (2002) Dietary fat and cancer. *The American Journal of Medicine,* 113(9B), 63-70S.

[5]   Kushi, L. H., Lew, R. A., Stare, F. J., Ellison, C. R., el Lozy, M., Bourke, G., Daly, L., Graham, I., Hickey, N., and Mulcahy, R. (1985) Diet and 20-year mortality from coronary heart disease. The Ireland-Boston Diet-Heart Study. *New England Journal of Medicine* 312, 811-818.

[6]   Posner, B. M., Cobb, J. L., Belanger, A. J., Cupples, L. A., D'Agostino, R. B., and Stokes, J. (1991) Dietary lipid predictors of coronary heart disease in men. The Framingham Study. *Archives of Internal Medicine* 151, 1181-1187.

[7]   Tannenbaum, A. (1942) The genesis and growth of tumors. III Effects of a high fat diet. *Cancer Research* 2, 468-475.

[8]   Van Dam, R. M., Stampfer, M. J., Willett, W. C., Hu, F. B., and Rimm, E. B. (2002) Dietary fat and meat intake in relation to risk of type 2 diabetes in men. *Diabetes Care* 25, 417-424.

[9]   WHO/FAO (2003) *Diet, nutrition and the prevention of chronic diseases.* Who Technical Report Series; 916. Retrieved from: http://whqlibdoc.who.int/trs/who_trs_916.pdf 26.11.2013.

[10]   European Commission (2010) Functional Foods. Studies and Reports. European Research Areas. Food, Agriculture and Fisheries and Biotechnology. Brussels, Belgium. Retrieved from: ftp://ftp.cordis.europa.eu/pub/fp7/kbbe/docs/functional-foods_en.pdf 26.11.2013.

[11]   Quiles, J. L., Ramírez-Tortosa, M. C. and Yaqoob, P. (2006) *Olive Oil and Health.* Wallingford, Oxfordshire: CABI Publishing.

[12] Pellegrini, N. and Battino, M. (2006) Total antioxidant capacity of olive oils. In: J. L. Quiles, M. C. Ramírez-Tortosa and P. Yaqoob (Eds.) *Olive Oil and Health* (pp. 63-73). Wallingford, Oxfordshire: CABI Publishing.

[13] Sánchez-Quesada, C., López-Biezma, A., Warleta, F., Campos, M., Beltrán, G., and Gaforio, J. J. (in press). Bioactive Properties of the Main Triterpenes Found in Olives, Virgin Olive Oil, and Leaves of *Olea europaea*. *Journal of Agricultural and Food Chemistry, Article ASAP,* doi: 10.1021/jf403154e. Retrieved from: http://pubs.acs.org/doi/pdf/10.1021/jf403154e.

[14] Serrano, L. and Lezcano, C. (2005) El aceite: valor Nutritivo. In: J. A. Pinto y J. R. Martínez (Ed.), *El aceite de oliva y la dieta mediterránea.* (pp. 25-50). Madrid: Nueva Imprenta, S. A.

[15] Verbeke, W. (2006) Functional foods: Consumer willingness to compromise on taste for health? *Food Quality and Preference,* 17, 126–131. doi:10.1016/j.foodqual.2005.03.003.

[16] Tuorila, H. and Cardello, A. V. (2002) Consumer response to an off-flavour in juice in the presence of specific health claims. *Food Quality and Preference,* 13, 561–569.

[17] Hilliam, M. (2003) Future for dairy products and ingredients in the functional foods market. *Australian Journal of Dairy Technology,* 58, 98–103.

[18] Augustin, M. (2001) Functional foods: an adventure in food formulation. *Food Australia,* 53, 428–432.

[19] Cox, D. N., Koster, A. and Russell, C. G. (2004) Predicting intentions to consume functional foods and supplements to offset memory loss using an adaptation of protection motivation theory. *Appetite,* 33, 55–64.

[20] Issanchou, S. (1996) Consumer expectations and perceptions of meat and meat product quality. *Meat Science,* 43, S5–S19.

[21] Caporale, G., Policastro, S., Carlucci, A., and Monteleone, E. (2006) Consumer expectations for sensory properties in virgin olive oils. *Food Quality and Preference,* 17, 116–125.

[22] Jaeger, S. R. (2006) Non-sensory factors in sensory science research. *Food Quality and Preference,* 17, 132–144.

[23] Köster, E. P. (2003) The psychology of food choice: some often encountered fallacies. *Food Quality and Preference,* 14, 359–373.

[24] Köster, E. P. (2009) Diversity in the determinants of food choice: A psychological perspective. *Food Quality and Preference,* 20, 70–82.

[25] Meiselman, H. L. (1996) The contextual basis for food acceptance, food choice and food intake: The food, the situation and the individual. In: H. L. Meiselman and H. J. H. MacFie (Eds.), *Food choice, acceptance and consumption* (pp. 239–263). London: Blackie Academic 7 Professional.

[26] Stefani, G., Romano, D. and Cavicchi, A. (2006) Consumer expectations, liking and willingness to pay for specialty foods: Do sensory characteristics tell the whole story? *Food Quality and Preference,* 17, 53–62.

[27] Caporale, G. and Monteleone, E. (2001) Effect of expectations induced by information on origin and its guarantee on the acceptability of a traditional food: olive oil. *Sciences des Aliments,* 21, 243–254.

[28] Siret, F. and Issanchou, S. (2000) Traditional process: influence on sensory properties and on consumers expectation and liking. Application to "pâte de campagne". *Food Quality and Preference*, 11, 217–228.

[29] Tuorila, H., Meiselman, H. L., Cardello, A. V., and Lesher, L. L. (1998) Effect of expectations and the definition of product category on the acceptance of unfamiliar foods. *Food Quality and Preference*, 9, 421–430.

[30] Carrasco, A., García, R., Zarrouk, W., and Fernández, A. (2009) Calidad sensorial del aceite de oliva. In: A. Fernández y A. Segura (Ed.), *El Aceite de Oliva Virgen: Tesoro de Andalucía*. (pp. 223-241). Málaga: Fundación Unicaja Editorial.

[31] Aparicio, R. and Harwood, J. (2003) *Manual del aceite de oliva*. Madrid: AMV y Mundi-Prensa.

[32] International Olive Oil Council (2013) *Guide for selection, training and monitoring of skilled virgin olive oil tasters* (COI/T.20/Doc. 14/Rev.4). Retrieved from: http://www. internationaloliveoil.org/documents/viewfile/3681-orga5eng 26.11.2013.

[33] International Olive Oil Council (2012) Standardising olive products. *Olivae*, 117, 35-38. Retrieved from: http://www.internationaloliveoil.org/store/download/8822 26.11. 2013.

[34] McNicol, D. (2005, 2nd Ed.) *A Primer of Signal Detection Theory*. Mahwah, NJ: Lawrence Erlbaum Associates.

[35] Fechner, G. T. (1860/1966) *Elements of Psychophysics. Volume I* (HE Adler, Trans.). *New York: Holt, Rinehart and Winston. (Original* work published 1860: Elemente der Psychophysik. Leipzig: Breitkopf und Härtel. Retrieved from: http://ia700404.us. archive.org/8/items/elementederpsych01fech/elementederpsych01fech.pdf 26.11.2013).

[36] Macmillan, N. A. and Creelman, C. D. (2005) *Detection Theory: A User's Guide*. Mahwah, NJ: Lawrence Erlbaum Associates.

[37] Green, D. M. and Swets, J. A. (1988) *Signal detection theory and psychophysics*. Los Altos, California: Peninsula Publishing.

[38] Wickens, T. D. (2002) *Elementary signal detection theory*. Oxford; New York: Oxford University Press.

[39] Stillman, J. A., Brown, G. M. and Troscianko, T. (2000) Influence of sensitivity on response bias in taste and audition. *Perception and Psychophysics*, 62, 1645-1654.

[40] Snodgrass, J. C. and Corwin, J. (1998) Pragmatics of Measuring Recognition Memory: Applications to Dementia and Amnesia. *Journal of Experimental Psychology*, 117, 34-50.

[41] Swets, J. A. (1996) *Signal detection theory and ROC analysis in psychology and diagnostics: Collected papers* (Scientific Psychology Series). Mahwah, NJ: Lawrence Erlbaum Associates.

[42] Lawless, H. (1985) Psychological perspectives of wine tasting and recognition of volatile flavours. In: G. Birch and M. Lindley (Eds.), *Alcoholic beverages*. London: Elsevier Applied Science.

[43] Moskowitz, H. (1988) *Applied sensory analysis of foods*. Boca Raton, Florida: CRC Press.

[44] O'Mahony, M. and Hautus, M. J. (2008) The Signal Detection Theory Roc Curve: Some Applications in Food Sensory Science. *Journal of Sensory Studies*, 23, 186-204.

[45] Parr, W. V., Heatherbell, D. and White, K. G. (2002) Demystifying wine expertise: Olfactory threshold, perceptual skill and semantic memory in expert and novice wine judges. *Chemical Senses,* 27, 747-755.

[46] Tanner, W. P. and Swets, J. A. (1954) A decision-making theory of visual detection. *Psychological Review,* 61, 401-409.

[47] Balakrishnan, J. D. (1998) Some More Sensitive Measures of Sensitivity and Response Bias. *Psychological Methods,* 3, 68-90.

[48] Macmillan, N. A. and Creelman, C. D. (1990) Response bias: Characteristics of Detection theory, Threshold theory, and "nonparametric" indexes. *Psychological Bulletin,* 107, 401-413.

[49] Stanislaw, H. and Todorov, N. (1999) Calculation of signal detection theory measures. *Behavior Research Methods, Instruments and Computers,* 31, 137-149.

[50] Ramos-Álvarez, M. M., Paredes-Olay, C., Moreno-Fernández, M. M., Callejas-Aguilera, J. E., Abad, M. J., and Rosas, J. M. (2008) Psychophysics of the taste process of virgin olive oil. *International Journal of Psychology,* 43 (3/4) 650. doi: 10.1080/00 207594.2008.10108486.

[51] Paredes-Olay, C., Moreno-Fernández, M. M., Rosas, J. M., and Ramos-Álvarez, M. M. (2010) ROC analysis in olive oil tasting: A signal detection theory approach to tasting tasks. *Food Quality and Preference,* 21, 562-568. doi:10.1016/j.foodqual.2010.03.003.

[52] Moreno-Fernández, M. M., Ramos-Álvarez, M. M., Paredes-Olay, C., and Rosas, J. M (2012) Effects of progressively increasing the difficulty of training on sensitivity and strategic factors in olive oil tasting. *Food Quality and Preference,* 24, 225-229. doi:10. 1016/j.foodqual.2011.12.001.

[53] Ramos-Álvarez, M. M., Moreno-Fernández, M. M., Paredes-Olay, C., and Rosas, J. M. (2013) A methodological proposal based on signal detection theory for the study of dissociation between sensory and decision processes in the context of olive oil. *Food Quality and Preference,* 28, 71-76. doi: 0.1016/j.foodqual.2012.09.004.

[54] International Olive Oil Council (2007b) Guide for the installation of a test room (COI/ T.20/Doc. 6/Rev.1). Retrieved from: http://www.internationaloliveoil.org/documents/ viewfile/3673-orga3 26.11.2013.

[55] International Olive Oil Council (2007a) *Glass for oil tasting* (COI/T.20/Doc. 5/Rev.1). Retrieved from: http://www.internationaloliveoil.org/documents/viewfile/3669-orga2 26.11.2013.

[56] Grier, J. B. (1971) Nonparametric indexes for sensitivity and bias: Computing formulas. *Psychological Bulletin,* 75, 424-429.

[57] Donaldson, W. (1992) Measuring recognition memory. *Journal of Experimental Psychology: General,* 121, 275-277.

[58] Wilcox, R. R. (2005) *Introduction to robust estimation and hypothesis testing.* San Diego, California: Academic Press.

[59] Jiménez, B. and Carpio, A. (2008) *La cata de aceites: aceite de oliva virgen. Características organolépticas y análisis sensorial.* Junta de Andalucía: Consejería de Agricultura y Pesca. Sevilla: Servicio de Publicaciones y Divulgación. Retrieved from: http://www.besana.es/sites/default/files/la_cata_de_aceites_baja_0.pdf 26.11.2013.

[60] Sternberg, S. (1998) Discovering mental processing stages: The method of additive factors. In: D Scarborough and S Sternberg (Eds.), *An Invitation to Cognitive Science, Volume 4: Methods, Models, and Conceptual Issues* (pp. 635-702). Cambridge: MIT.

[61] Gibson, J. J. and Gibson, E. J. (1955) Perceptual learning: Differentiation or enrichment? *Psychological Review,* 62, 32-41.

[62] Hall, G. (1991) *Perceptual and associative learning.* Oxford, England: Claredon Press.

[63] Gibson, E. J. (1963) Perceptual learning. *Annual Review of Psychology,* 14, 29-56.

[64] Lawrence, D. H. (1952) The transfer of a discrimination along a continuum. *Journal of Comparative and Physiological Psychology,* 45, 511-516.

[65] Walker, M. M., Lee, Y. and Bitterman, M. E. (1990) Transfer along a continuum in the discriminative learning of honeybees (apis mellifera). *Journal of Comparative Psychology,* 104, 66-70.

[66] Scahill, V. L. and Mackintosh, N. J. (2004) The easy to hard effect and perceptual learning in flavor aversion conditioning. *Journal of Experimental Psychology: Animal Behavior Processes,* 30, 96-103.

[67] Liu, E. H., Mercado, E., Church, B. A., and Orduña, I. (2008) The easy-to-hard effect in human (homo sapiens) and rat (rattus norvegicus) auditory identification. *Journal of Comparative Psychology,* 122, 132-145.

In: Virgin Olive Oil
Editor: Antonella De Leonardis

ISBN: 978-1-63117-656-2
© 2014 Nova Science Publishers, Inc.

Chapter 14

THE ROLE OF VIRGIN OLIVE OIL IN THE TRADITIONAL MEDITERRANEAN CUISINE

Antonella De Leonardis, Vincenzo Macciola and Francesco Lopez*

Department of Agriculture, Environmental and Food Sciences (DiAAA),
University of Molise, Via De Sanctis, Campobasso, Italy

ABSTRACT

In several Mediterranean countries, especially in Greece, Italy, Portugal and Spain, virgin olive oil (VOO) is typically preferred not only in salad dressing, but also as the oil of choice in food preparation. Specifically, both in domestic and industrial use, VOO is the main fat used in cooking, in baked goods, as a preserving agent for vegetable and fish preserves and finally, as an ingredient in various sauces.

It is a well known fact that VOO distinguishes itself from the other vegetable oils due to the high content of monounsaturated fatty acids (oleic acid) and the presence of various minor components, deriving from the olives during oil production. This composition give to VOO well-known healthy, organoleptic and antioxidant properties. However, nutritional and organoleptic properties can be significantly affected when VOO is used in the food preparation, due to the conditions of food processing and/or the interaction with other food components.

The role of VOO in food preparation is presented in this Chapter by reviewing the relevant scientific literature.

Keywords: Virgin olive oil, cooking, bakery, in-oil-preserves, sauces

ABBREVIATIONS

VOO	virgin olive oil
SFA	saturated fatty acids
UFA	unsaturated fatty acids

---

* Corresponding author: Email: antomac@unimol.it.

MUFA	mono-unsaturated fatty acids
PUFA	poly-saturated fatty acids
SCFA	short chain fatty acids
FFA	free fatty acids
PV	peroxide values
TPC	total polar compounds
MR	Maillard reaction
PPO	polyphenoloxidases

INTRODUCTION

A well-known combination of wheat, olives and grapes and their derivates (such as bread, pasta, pizza, couscous, olive oil, wine), together with fruits and vegetables, fish, meat in reduced quantities, and dairy products, are the basis of the diet of the Mediterranean populations.

Recently, these dietary habits, associated with a correct lifestyle, have been linked with a reduced incidence of coronary heart disease and low blood pressure levels. Thus, the so-called 'Mediterranean diet' is recognized worldwide as a healthy dietary model [1].

Undoubtedly, a key component of the Mediterranean diet is the extra virgin olive oil, hereon in called virgin olive oil (VOO).

In a few countries of the Mediterranean area, VOO is used traditionally not only crude in salad dressing, but also in combination with other ingredients in food preparation. In home cooking especially, VOO is used almost exclusively as cooking fat, as an ingredient in various sauces and baked goods and as covering oil in the preparation of vegetables and fish preserves. These many uses of VOO explain the high consumption per capita by some populations, in particular those living in Greece, Italy, Portugal and Spain (Figure 1) where olive oil is traditionally the basis for their local cuisine.

Recently, the use of VOO is increasing also in the industrial preparation of specific foods, e.g. baked goods, sauces and canned foods. In these cases, the replacement of refined oils with VOO is generally emphasized on labels in order to address the consumer demand of choice toward healthy, natural, and traditional foods.

As is well-known, VOO is produced by using only mechanical extraction process and may be directly consumed without any further treatment, as opposed to other vegetable oils that are usually refined before being eaten. Consequently, VOO preserves a particular aroma and a significant content of specific polyphenol compounds, that distinguish it markedly it from all other edible oils and fats.

Another remarkable distinctive element for olive oil is its high level of the monounsaturated oleic acid, in a range between 65 to 85% of fatty acid content.

The distinctive elements mentioned above affect the organoleptic and health properties of VOO, especially when used crude in salad dressing. Nevertheless, these same elements could have a different impact when VOO is used in food preparation. In fact, both high cooking temperatures and the interaction between oil and food may significantly modify the initial VOO composition with consequent loss of beneficial properties.

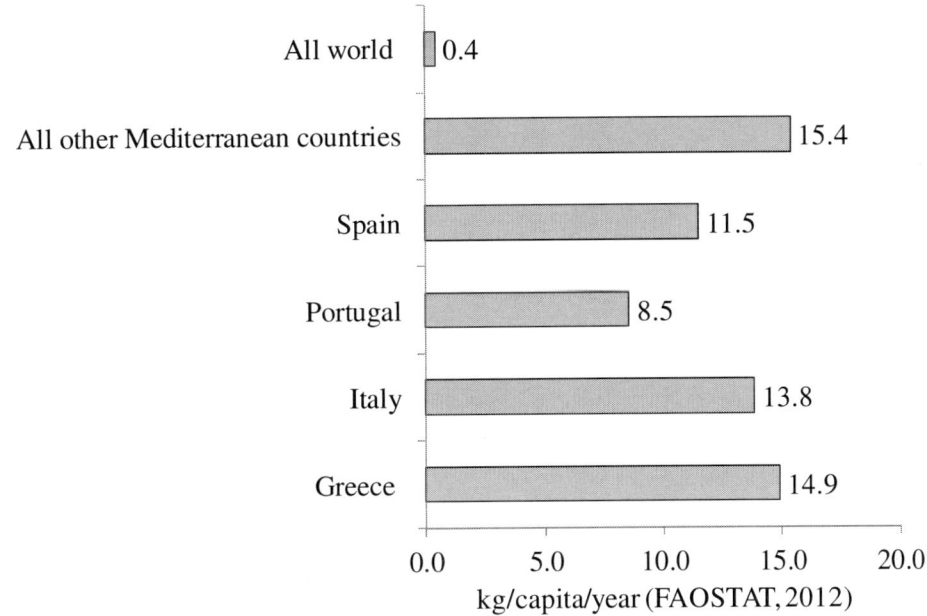

Source: Faostat.org.

Figure 1. Consumption per capita (kg/per capita/year) of olive oil by people living in the Mediterranean countries. Data relative to the year 2009.

In this Chapter the role of VOO in the preparation of typical Mediterranean foods is reported by reviewing the data of the scientific literature on the matter.

1. COOKING OIL

Cooking in oil or fat is an old and popular culinary method worldwide, both in commercial and home food preparation. Oils/fats represent the most efficient modes of heat transfer by leaving a good taste to the cooked food.

Frying, cooking in the oven (microwave or conventional heating) and boiling in water are the main cooking-in-oil practices.

During the cooking process, oils/fats are exposed to high temperatures, ranging from 100 to 240°C, for an extended period of time. A high temperature, coupled with the presence of oxygen and water from the food, produces important chemical changes in oils/fats by leaving oxidized and polymerized products that are known to be undesirable from a health point of view [2].

Thus, a good cooking oil/fat has to satisfy the sensorial needs of consumers while, at the same time, ensure the maxima health safety.

Fried foods are certainly the most popular foods throughout the world [3]. Frying may be considered technically as a combination of rapid cooking and drying, due to the volume of fried food decreasing as a result of water loss.

Pan-frying and deep-frying are two possible technical variants. Deep-frying is the process in which food is totally immersed in the hot oil, so that all the flavors and juices are retained within the crispy crust. Commonly, in deep-frying the oil is continuously and repeatedly used, often regularly replenished with fresh oil. Conversely, in the pan-frying minimal amount of cooking oil/fat is used.

In household cooking, pan-frying and deep-frying are usually performed in uncovered open steel pans and in covered casseroles, respectively. In this condition, fried oil receives maximum oxidative and thermal abuse due to the presence of atmospheric oxygen. One alternative is the use of a closed electric frying machine.

Surface-to-volume ratio plays an important role in the stability of oil/fat during cooking. Generally, in pan-frying surface-to-volume ratio is significantly higher than in deep-frying and this difference seems to cause more marked changes on the principal oxidation parameters [4].

In oven cooking conventional or microwave heating systems may be used. The use of the microwave oven for cooking has increased considerably during the past few decades [5]. Microwave ovens change regular electricity into high frequency microwaves that water, fat and sugar can absorb causing food particle vibration, and thus the heating of the foodstuff. In microwave ovens the transference of thermal energy is 10–20 times higher than in a conventional oven.

Prolonged exposure of oil/fat at high temperatures induces several reactions, such as hydrolysis, oxidation, polymerization and cyclization [6]. The grade of oil/fat thermal alteration is due to the combination of intrinsic and extrinsic factors, as such oil/fat composition, food moisture, atmospheric oxygen adsorption, and both the temperature and time at which the cooking operation takes place [3].

Among intrinsic factors fatty acids composition and the content of antioxidants of the cooking oil/fat are those most affecting the oxidative stability.

Generally, oils/fats with a high content of saturated fatty acids (SFA) like tallow, coconut, butter, lard or palm oils are considered more adequate for cooking due to their higher resistance to form free radicals. Conversely, oils with a high content of polyunsaturated fatty acids (PUFA), like sunflower, corn, soy or safflower are most likely to become oxidized. VOO is in an intermediate position being rich of monounsaturated fatty acids (MUFA).

Susceptibility to oxidation of fatty acids follows this order: linolenate> linoleate> oleate> stearate; however, at very high temperature the differences are minimized and the linoleate and linolenate radicals act as initiators for the oxidation of oleate. Moreover, the percentage of palmitic and linoleic acids influenced oxidation stability more than that of oleic acid [7].

Actually, both the palmitic/linoleic and SFA/UFA (UFA: unsaturated fatty acid) ratios appear to be the principal parameters affecting oxidative stability of the oil [8]. Indicatively, a SFA/UFA quotient near 1 (one) seems to be an optimal stable combination [9].

Thus, fatty acid composition of oils/fats affects their oxidative stability. However, unlike from all other oils/fats, the fatty acid composition of VOO is not stable and varies significantly with drupe cultivar, geographical origin, the season, agronomical conditions and oil extraction process method [10]. Consequently, oxidative stability of one VOO is not easily predictable because it depends on its actual fatty acid profile.

Performance of the VOO under high temperature in comparison to other edible oils has been more and more investigated. Results are not always directly comparable because the experimental designs are very different in regards to the heating process (i.e. continuous or discontinuous), surface/volume oil ratio, the temperature and length of time of heating, the presence and kind of cooked food, and eventual adding of fresh oil.

Rancimat apparatus allows researchers to compare the oxidation resistance of different oils/fats stressed under controlled conditions of high temperature and aeration, but in the absence of food [11].

For example, Figure 2 shows the oxidative resistance (induction time in hours measured by rancimat at a temperature of 130°C and 20 L/h air flow) of one VOO sample in comparison to itself after being chemically-dephenolized, one refined olive oil and finally, one palm oil.

Oil obtained from palm fruits (*Elaies guineensis*) has grown to be one of the most important vegetable oils due to its advantageous properties such as high productivity, low price, high oxidation stability, fatty acid composition and, finally, good plasticity at room temperature [12]. Nowadays, palm oil and palm-based fractions are widely used in various food products, such as margarines, shortenings and fractioned oils used as cooking oils in restaurants and industrial process.

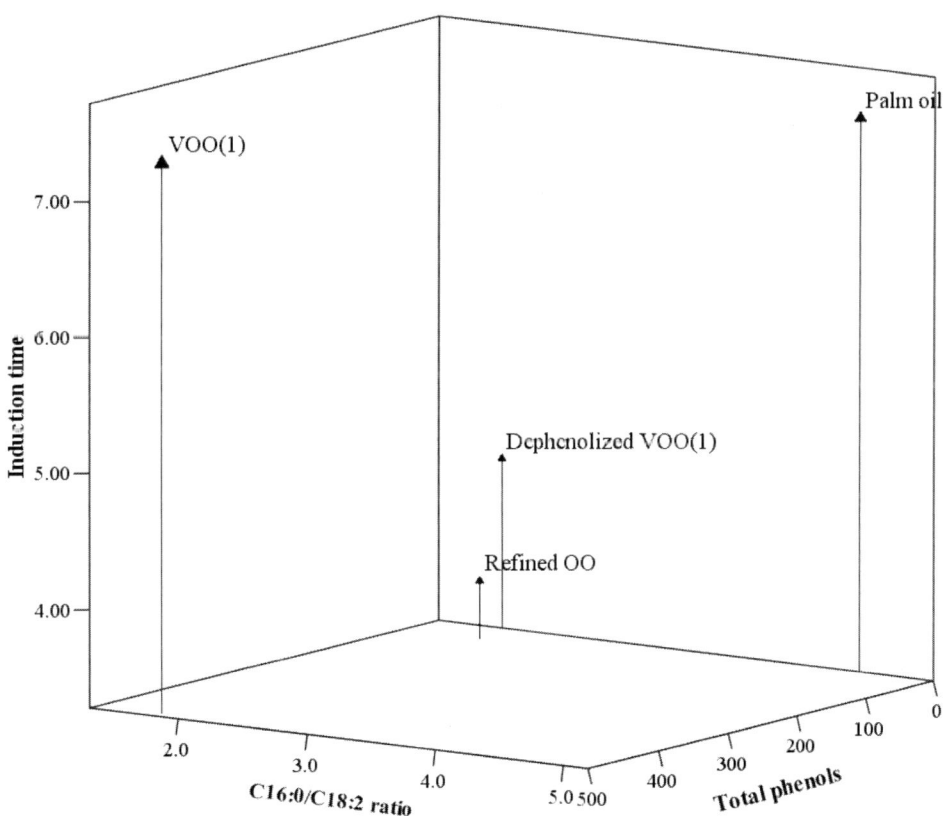

Figure 2. Oxidative resistance (induction time in hours) measured by rancimat method (130°C temperature and 20 L/h air flow) of one virgin olive oil sample and again after it has been chemically-dephenolized, one refined olive oil and one palm oil. Oxidative resistance is related with C16:0/C18:2 ratio and total phenol content of each oil sample.

The oxidation stability of palm oil and that of VOO is very different [9]. The stability of VOO is due primarily to its high natural antioxidant content, while the stability of palm oil is due primarily to its high saturation level (Figure 2). In the case of VOO, it appear very significant the role of polyphenol content; indeed, both the refining (refined OO sample) and chemical dephenolization (dephenolized VOO(1) sample) have resulted in a drastic reduction of oxidative stability in comparison to VOO(1) sample.

Finally, it has observed that also the position of fatty acids into the triacylglycerol may affect the oxidizability of an oil [13,14].

1.1. Effects of Thermal Oxidation on the Glycerides

As far as the mechanism for the thermal oxidation of glycerides, the behavior of VOO is substantially similar to that of other edible oils/fats.

In Figure 3 the most relevant alteration mechanisms that occur during the thermal oxidation of oil/fat are given.

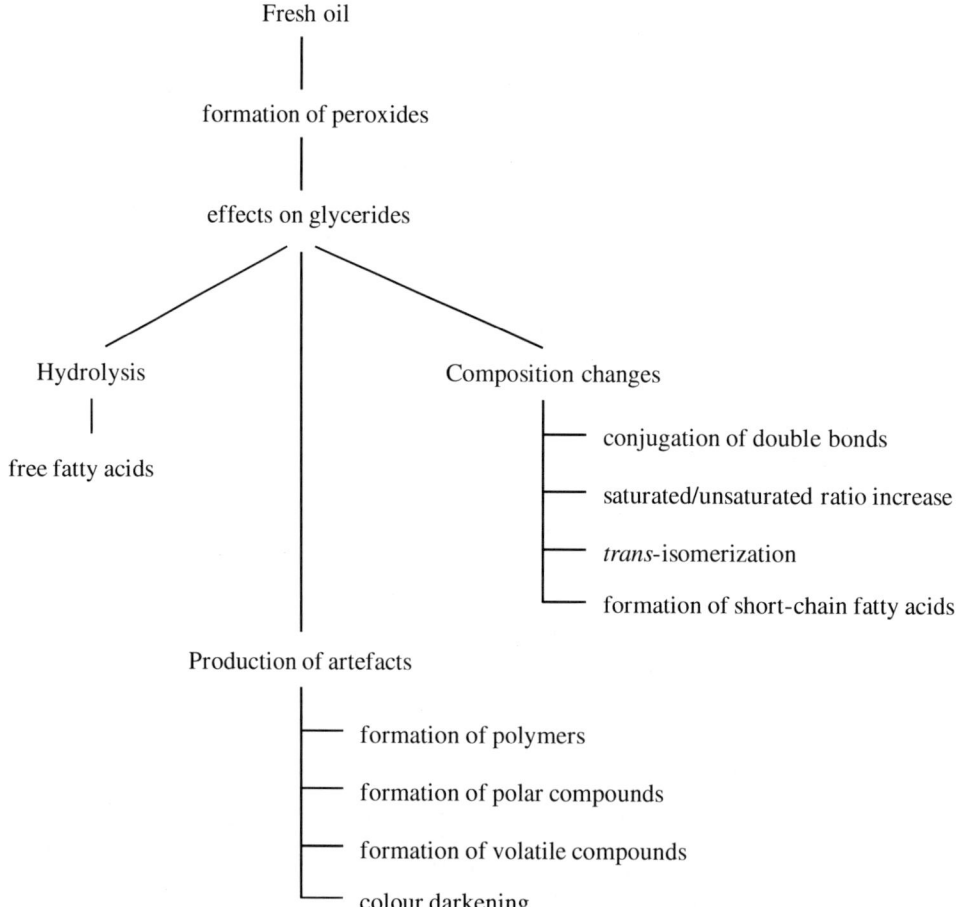

Figure 3. Most relevant alteration mechanisms that occur during the thermal oxidation of oils/fats.

Generally, thermo-oxidative reactions induced under the heating conditions seem to be faster than those under the frying conditions due to the protective role of the water released by food during the frying process [15].

Initially, an increase of peroxide values is observed in the heated oils/fats. Formation of peroxides starts by the abstraction of a hydrogen atom from the carbon atom in the allylic position of an unsaturated fatty acid (RH) to generate an alkyl radical (R•). This combines with molecular oxygen to produce a peroxide radical (ROO•). The ROO• propagate the oxidation reaction by selectively abstracting hydrogen from another RH to form lipid hydroperoxides (ROOH) and another alkyl radical (R•).

Peroxides, being transient chemical compounds, may not be directly correlated with the thermal oil/fat oxidation. Indeed, these are unstable at high temperatures and decompose rapidly in a wide range of volatiles and nonvolatile compounds [15].

However, the formation grade of peroxide values is similar for VOO and other refined oils with similar fatty acid compositions [4,15,16].

High temperatures cause a hydrolytic alteration which results in an increase of free fatty acids (FFA); the co-presence of food moisture may strongly contribute to this result.

Similar to peroxide values, FFA content is also a dynamic parameter in measuring the heated-stressed oil, because part of the produced FFA have sufficient vapor pressure at high temperatures to evaporate from the surface [17].

From another point of view, it is known that the content of both FFA and PV before heating strongly influences the triggering of the radical stages of the heating.

Usually, FFA and PV in refined vegetable oil are less (generally less than 0.1% and 10 meqO$_2$/kg, respectively) than those of VOO, due to the refining process. Actually, in accordance with its commercial category, an extra virgin olive oil would have up to 0.8% of FFA and 20 meqO$_2$/kg of PV, respectively. Thus, the presence of a significant pre-heating level of FFA and PV in VOO could sensibly reduced its oxidative stability [16]. Consequently, when cooking, it is good practice to choose only the VOO that exhibits very low initial values of FFA and PV.

Fatty acid composition of heated oil/fat is transformed as a result of the thermal oxidation. Specifically, the SFA/UFA ratio changes, conjugation of double bonds occurs, trans-isomers increase, and short-chain fatty acids are formed. These changes happen similarly on the fatty acids in both free and glycerol-bounded forms.

Similarly, for both VOO and other oils/fats a significant loss of PUFA occurs during heating. Generally, PUFA degradation is fastest at higher temperatures and is associated with an increase of SFA, while the oleic acid percentage remains substantially unchanged [4, 9].

In the oxidized oils a significant increase of UV extinctions is registered, thought the measurement of K232 and K270 indexes. The K232 index is correlated with the formation of both peroxides and conjugated dienes of PUFA, whereas the K270 is indicative of the presence of primary and secondary oxidation products, including conjugated trienes and carbonyl compounds.

An increase of trans-isomers is found to be negligible even after repeated frying in all kinds of fried oils.

Conversely, the formation of short-chain fatty acids (SCFA), especially octanoic acid (8:0) and heptanoic (7:0), is a very real phenomena to this issue [18].

Octanoic acid (8:0) and heptanoic (7:0) are produced by the degradation of oleic and linoleic acids, respectively. SCFA are generally volatile, but actually they remain bound to the parent TG in the thermal-oxidized oils, while the other volatile newly-formed compounds evaporate from the oil surface. For this behavior, SCFA, especially the octanoic acid, has been proposed as the markers for oxidative deterioration in fats and oils that do not originally contain them, like the VOO [19].

Heat oxidation of VOO leads to faster production of SCFA when compared to palm oil [9].

By prolonging the high temperature treatment, the fatty acid peroxides degrade by yielding multiple secondary products with different molecular weight (MW) with respect to starting triglycerides, which include volatile compounds (lower MW), oxidized triglycerides (medium MW) and triacylglycerols polymers (higher MW). The greater part of these compounds are very harmful to humans [20].

Polymerization and formation of high MW compounds determine as secondary effects an increase of viscosity of the thermal-oxidized oil [21]. Therefore, the measurements of oil viscosity have been proposed as a index of oil degradation [22]. Actually, it is observed that the oil viscosity increases faster in the first step than in the second step of thermal oxidation, following a polynomial curve.

However, the principal evidence of heated-stressed oil degradation is the increase of the namely 'total polar compounds' (TPC). Hydrolysis products (diglycerides, monoglycerides and free fatty acids), oxidized and dimerized triglycerides, oxidized derivates of minor oil components and all the other compounds with polar behavior are included in the TPC.

TPC partially remain in the oil and partially are absorbed by food.

Several European countries have passed specific laws and regulations concerning the TPC levels for the frying oils/fats, setting a cutoff point between 20% and 27% [23].

Formation of TPC during repeated frying operations has been shown to increase mainly with the degree of oil unsaturation degree and the surface-to-oil-volume ratio, while differences in temperature do not cause significant changes [24].

Usually, in commercial frying services, mostly restaurants and industries, as well as in research laboratories, TPC are evaluated by commercial rapid tests, mostly based on colorimetric readings, which have proven to be well correlated with the values obtained by the laborious official standard method in which the sample is eluted through a silica column.

However, it has been demonstrated that under typical household practice of deep-frying and pan-frying, the VOO, similar to the other tested vegetable oils, rarely reaches the TPA rejection limits [4,25].

Heated oils/fats generate volatile compounds that have a high evaporation rate into the atmosphere through smoke, whereas the remaining compounds in the oil undergo either further chemical reactions or become absorbed by the cooked food [15].

Alkanes and alkanals are the main classes of identified volatile compounds; alcohols, ketones, acids, and aromatic heterocyclic compounds are also generated in minor amounts.

Alkenes detected in the cooking oil fumes are mainly acetaldehyde, acrolein, propanal, butanal, 2-butenal, pentanal, 2-pentenal, hexanal, 2-hexenal, heptanal, 2-heptenal, 2,4-hetadienal, 2-octenal, nonanal, 2-nonenal, and 2-decenal.

The main alkanals are octane, heptane, hexane, pentane, butane, 2-propanone, and 1-pentanol.

Volatile aldehydes represent the 60-70% of the total volatile compounds and their formation is related to the temperature. Moreover, aldehydes derived from linoleic acid (acetaldehyde, pentanal, hexanal, 2-heptenal, 2-octenal, and 2-nonenal) are generally higher than those derived from oleic acid (nonanal and 2-decenal). Indeed, the n-nonanal, followed by n-pentanal, n-hexanal, n-heptanal, is the major volatile compounds revealed in heated VOO [26].

In the last stages of thermal oxidation, decomposition of glycerol yield acrolein at a level that increases significantly when the cooking temperature is raised from 180 to 240 °C. Acrolein which is among the newly-formed volatile compounds is one of the most monitored because it is hazardous [27].

Generally, the 'smoke point' of an oil/fat is considered to be the temperature at which smoke containing acrolein and other toxins, mutagenic and carcinogenic compounds begins to be emitted. The smoke point of VOO oil is around 190-200°C. Some refined oils, such as palm, peanut, safflower and soybean oils can have smoke points at around 230°C to 260°C, but their corresponding unrefined oils can have smoke points in the low hundreds [27].

Finally, heated oil/fat shows a darker colour than that of its raw state [6]. The increase in color is attributed to the α- and β-unsaturated carbonyl compounds, which are intermediates of nonvolatile decomposition products with the ability to absorb energy of visible light wavelengths [17].

Colour darkening is faster and more intense with the presence of food as a result of the amino-carbonyl reaction between thermally oxidized oil and amino acids exuded by cooked food [21].

1.2. The Role of VOO Minor Compounds

It is well known that, different from the refined vegetable oil, VOO maintains considerable amounts of several minor components which occur naturally in the olive fruits. These compounds are mainly tocopherols, sterols, hydrocarbons (squalene), carotenoids, polyphenols, and chlorophyll.

Fresh VOO delaines a typical spectrum into visible region exhibiting points of maxima absorbance typical of carotenoid (415, 455 and 485 nm), flavonoid (535 nm) and chlorophyll (610 and 670 nm) pigments.

This is well visible in Figure 4 where the 400-700 nm absorption spectra of one VOO, before and after heating-stress (rancimat treatment at 130°C temperature and 20 L/h air flow for about 7 hours), is reported.

After prolonged heating, the VOO spectrum changed significantly and there is evidence of a lower level of chlorophylls, carotenoids and flavonoids.

Actually, minor compounds may induce opposite effects during the thermal oxidation of the VOO.

The presence of chlorophylls and their derivatives greatly affect the oxidative stability of VOO, especially in the presence of light [28]. As a result of heating, chlorophyll pigment degrades in pheophytin due to a loss of the magnesium atom; pheophytin provides a brown colour that gives at the darker colour of VOO after prolonged heating [21].

Conversely, carotenoids are pigments possessing conjugated hydrocarbons and have been found to be potent protectors against photosensitized oxidation, acting as singlet oxygen quenchers [16].

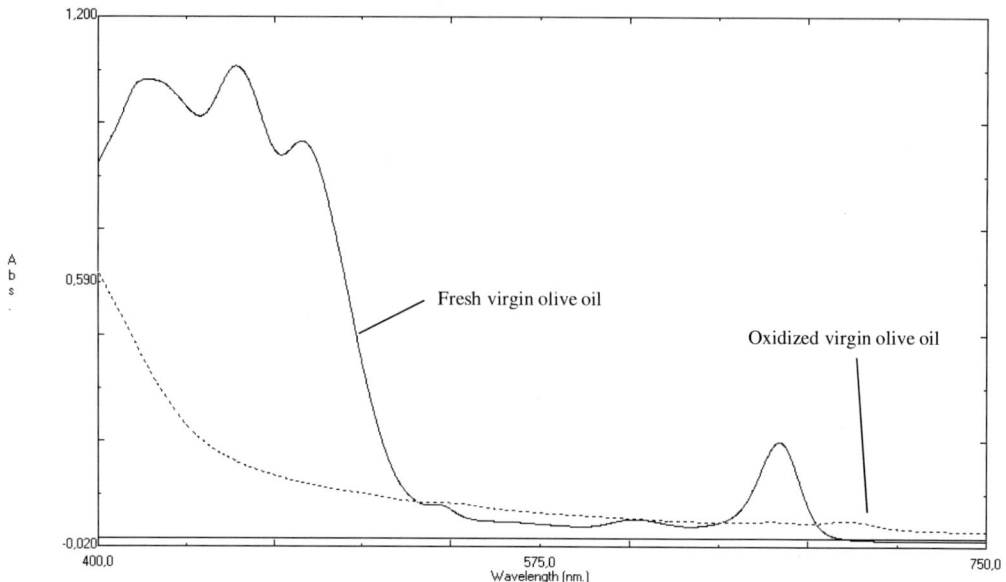

Figure 4. Absorption spectra between 400 and 700 nm (light visible region) of a virgin olive oil, before (fresh oil) and after applying heat (rancimat at 130°C and 20 L/h air flow for about 7 hours).

However, crucial factors in the VOO oxidative resistance are certainly tocopherols and mainly polyphenols [11, 29, 30].

Tocopherols seem to play a minor role in thermal oxidation being destroyed or inactivated at 180°C [4,16,31]. Moreover, α-tocopherol, the predominant tocopherol in VOO, is less stable than δ-tocopherol, while β- and γ- tocopherol degrade at an intermediate rate [32, 33].

In regards to polyphenols, it is well-known fact that the higher performing antioxidants are those belonging to the *o*-diphenol family, mainly hydroxytyrosol and the compounds that contain it, such as verbascoside and oleuropein. Conversely, the monophenol tyrosol and its secoiridoid derivatives show a negligible antioxidant activity.

VOO's polyphenols suffered important changes during the heating process. All components decreased in concentration, although the rate of loss depended on the chemical structure, the antioxidant activity and on both the heating time and temperature [29,34,35].

Thermal loss grade of each phenolic compound seems to be correlated with its antioxidant ability; indeed, hydroxytyrosol and its related compounds decrease more rapidly than the compounds of the tyrosol-family. E.g., Brenes et al. (2002) [32] has found that after twelve frying operations the hydroxytyrosol-family was completely degraded (over 95%), while the tyrosol-family compounds decreased by only about 20% of the initial value.

Only a slight loss of the lignans 1-acetoxy-pinoresinol and pinoresinol has been registered during prolonged oil heating [32]. These phenolic compounds have aroused great interest recently because they seem to possess *in vitro* antioxidant activity and *in vivo* ability to inhibit lipid peroxidation when they are absorbed [36].

1.3. The Impact of VOO on Cooked Food

Cooked food absorb about 15-25% of the heated oil/fat while at the same time, it leaches out simultaneously its fat soluble molecules and water. Both the weight and length of the food affect the amount of water loss and oil absorption [34].

Moreover, the amount of oil/fat adsorbed into the cooked-food is inversely correlated with the food's fat content. E.g. lean fish tends to absorb the cooking oils, while fatty fish shows a decrease in fat content [37].

Prolonged heating influences food/oil swapping due to the high oil/fat viscosity that increases the amount of oil on the food surface and reduces the interfacial tension of the oil/food [22].

The fatty acid composition of fried-food trends to be similar to that of the cooking fat [37,38]. Specifically, in fish fried in VOO a decreasing of SFA and PUFA (such as eicosapentaenoic and docosahexaenoic acids) and conversely, an increasing of MUFA and the n-6/n-3 fatty acid ratio occur [39].

Conversely, conventional and microwave oven-baking affect only slightly the loss of fat and the changes of fatty acid composition in the cooked fish [39].

Thus, an increase of MUFA in food fried with VOO is a realistic expectation.

Moreover, VOO absorbed during frying seems to leaves a significant enrichment of α-tocopherol and phenols in the fried foods [34].

Polarity of the VOO antioxidants play a significant role in their distribution between the cooking oil and the water-containing food. Specifically, the higher polar polyphenols tend to migrate in the cooked-food [40], the less polar terpenic acids (oleanolic, maslinic, and ursolic) remain mainly in the oil phase [34], while the intermediate polar α-tocopherol tends to distribute more uniformly between oil and fried foods [32].

However, it has been observed that VOO polyphenols migrated in the water-containing foods survive better from the oxidation degradation than those remaining in the hot VOO. Moreover, it seems that VOO could be re-used for frying more times than other vegetable oils before leaving a significant fried-food quality deterioration [17].

During the cooking of meat, the high temperature may increase at ppb levels the formation of mutagenic/carcinogenic heterocyclic amines (HAs) [41]. Cooking without oil yields lower amounts of HAs because of the oil is a more efficient heat transfer. However, when meat is cooked in VOO, the polyphenols seem to prevent the formation of HAs [42].

2. BAKERY AND PASTRY PRODUCTS

Bakery and pastry products constitute one of the most consumed foods in the world.

Fat, sugar, egg, milk and flour are the main ingredients used in making sweet baked goods, while only fat, flour, salt and water are generally used to prepare savory products. Pastry products, such as cookies and cakes, may contain up to 20-30% fat, while in savory baked goods, such as rusks, crackers, breadsticks, the fat content generally ranges between 5 to 15%.

Oils/fats play a crucial functional role in the structure and quality of the final product, performing three basic functions: (a) entrapment of air bubbles into the dough during mixing by easing the product leavening; (b) coating the protein and starch particles in order to tenderize and lubricate better the mouthfeel; (c) emulsifying large amounts of liquid thereby contributing to increased crumb moisture and softness in the product [43].

However, oil/fat is chosen on the basis of dough malleability, rheology and sensory properties, shelf-life of the products and finally, the specific needs of the consumers.

Shortening, margarines, hydrogenated oils and fats, refined vegetable oils, as well as animal fats are the oils/fats commonly used in the bakery and pastry applications.

Actually, plastic shortenings have begun to be the major fat ingredient. Plastic shortenings are semi-fluid fats in which the high presence of β' crystals provides a greater plasticity. Shortenings are usually formulated from blends of processed oils/fats, which may be derived from vegetable, marine or animal sources; among them, partially hydrogenated soybean oil is certainly the most used [44].

The crystal polymorph structure of VOO's triacylglycerols has been little studied although it is of remarkable importance with respect to functional, sensorial and nutritional properties of a large number of formulated foods [45]. During cooling and subsequent heating in VOO two crystalline β1 phases have been found; the so-called β1a, which emerged at higher temperatures and largely constituted by TAG rich in palmitic and/or stearic acid and a β1b phase, probably related to crystals formed by TAG composed mainly by oleic and/or linoleic acids [46].

It is known that oils/fats deriving from a refining or hydrogenation process may contain high amounts of both triacylglycerol degradation compounds (triacylglycerol oligopolymers, oxidized triacylglycerols and diacylglycerols) and *trans*-isomers of unsaturated fatty acids. The latter increases the risk of coronary heart disease by raising the levels of LDL cholesterol and lowering the levels of HDL cholesterol [47].

Recently, many companies began using the relatively inexpensive palm oil instead of hydrogenated soybean oil to make zero-*trans* fat shortenings [48]. Nevertheless, palm oil is rich in saturated fats that are associated with various health disorders, such as obesity, cancer, high blood cholesterol and coronary disease [49].

In a few Mediterranean countries VOO is largely used as an ingredient in the production of baked products, especially in domestic preparations; however, in commercial manufacturing the trend to replace, partly or wholly, the refined oils with VOO is also spreading.

Typical Italian baked foods made with VOO are 'pizza', 'focaccia' and 'taralli'. In the commercial preparation of these products, olive oil, olive–pomace oil and palm oil generally are used alternatively.

'Pizza' is a known worldwide. 'Focaccia' is similar to 'pizza'; it is a very popular baked bread product, made from regular bread dough rolled out into varying thicknesses, seasoned with simple or richer toppings, salt, oil and baked in a low pan.

An increase of both free fatty acids and peroxide value has been observed during the cooking process of the 'focaccia', especially that with tomato topping [50].

'Taralli' made with VOO presents similar friability and taste when compared with other refined vegetable oils, but appear to have a longer shelf-life due to a delay of the lipid oxidation product. Specifically, during the storing of 'taralli' the VOO leaves lower values of hydroperoxides, K232 and K270 indexes, triacylglycerol oligopolymers and oxidized triacylglycerols [51].

During the baking process oil/fat blends are often used. It has been observed that heat-oxidation stability of a palm oil/VOO binary blend is strongly influenced by the relative percentages of the oil partners [9]. Specifically, a negative impact on the stability of blends occurs when olive oil is added in a quantity greater than 20%.

Literature on the use of VOO in the preparation of pastry products is very rare. When compared to the control group, 'Madeira cakes' made with VOO or VOO/margarine blend as a substitute to margarine, exhibit a higher volume and a lower baking loss, higher values of hardness and cohesiveness, and finally, more complex volatile profile [52].

In the preparation of pastry the development of the typical flavor is fundamental. Flavor compounds are mainly the result of the Maillard reaction (MR), which occurs between reducing sugars and the -NH$_2$ function of amino acids, peptides and proteins. The selection of raw materials can modify the aroma composition.

MR and caramelization are the most relevant chemical reactions during baking process, affecting the textural, physicochemical and organoleptic changes which take place during cooking. It is known that oxidized lipids compete very efficiently with carbohydrates for carbonyl-amine reactions [53].

Moreover, under elevated temperatures and medium-low moisture contents, acrylamide may be produced as a consequence of the reaction between sugar and asparagine; in secondary pathway, the acrolein formed during the thermal oil/fat oxidation may enhance the formation of acrylamide [54].

Acrylamide is formed mainly in French fries, potato crisps, coffee, biscuits and bread and its concentration significantly increased only at the final cooking stage. It is strongly suspected of carcinogenic, neurological and reproductive diseases [55].

Recent studies suggest a positive role of VOO phenolic compounds to limit the formation of acrylamide [56]. The reductive response for o-diphenolic compounds is double the total phenols. Protection offered by VOO polyphenols might be related to the blocking of the radical-mediated degradation of sugars from caramelisation and/or to delay the formation of lipid oxidation products that will increase acrylamide formation [57].

Finally, application of microwave heating may induce higher oxidative alterations than conventional oven [58,59]. Specifically, the effect of microwave heating on the visible spectrum, K232 and K270, density, viscosity and trans-isomer contents of fats and oils appears to be worse than that produced by heating the same fats in a conventional oven at the same temperature, time, surface/volume ratio, and light conditions.

On the other hand, o-diphenols and lignans have demonstrated a high microwave heating resistance and this is of great importance from a nutritional standpoint [60].

3. VEGETABLE AND FISH PRESERVES

Vegetable and fish packaging in oils (in-oil-preserves) are typical foods of the traditional Mediterranean gastronomy.

The role of the covering oil is primarily to isolate the food from air by reducing both the growth of aerobic micro-organisms and oxidative processes. Secondly, the covering oil plays a part in influencing the organoleptic quality and taste of the final product.

In the commercial preparation of in-oil preserves, sunflower seed oil is commonly used, although most consumers of the Mediterranean area prefer olive oil. Actually, in a few Mediterranean countries, VOO is used exclusively in the domestic preparation of the preserves, while refined olive oil is generally used by the industry. Recently, numerous industries have also begun to pack the preserves in VOO with the aim of satisfying consumer demand for natural foods.

The original composition and quality of the VOO may change during preparation and/or storage as a result of the specific operation of the canning process or through the interaction with food.

Foods are appropriately blanched, acidified, salted, fermented, and cooked before being packed with oil. Pasteurization of closed jars of artichokes, tomatoes, little onions, eggplants, mushrooms, tomatoes, table olives and fish preserves packaged in VOO, induces only a small increase of hydrolytic and oxidative degradation in the covering oil [61-63].

However, a few chemical parameters, as a percentage of free fatty acids, peroxide value and spectrophotometric indexes, during storage may exceed the European legal limits relative to the VOO (2568/91 EEC Regulation); as a consequence, it could be difficult to establish the genuineness and the market category of the oil used at the time of packaging (e.g. refined olive oil or VOO) [64,65].

Moreover, preserved food may release fatty acids with subsequent modification of the original VOO fatty acid profile. E.g. an absorption of highly unsaturated fatty acids (C20:5 and C22:6), typical of fish lipids, may occur in the oil used in the preservation of fish [66-68].

It is well known that most vegetables synthesize specific copper enzymes, generically referred to as polyphenol oxidases (PPO), which catalyze in the presence of molecular oxygen the oxidation of phenols at reactive oxygen species and quinones [69]. The last two, undergo spontaneous non-enzymatic polymerization, rise to the formation of dark brown and stable polymers named melanin. The whole process is known as 'enzymatic browning' and leaves, a rapid depletion of phenols in addition to a color-change in food.

Enzymatic browning is documented abundantly for olive leaves and fruits [70,71]. In the olives, endogenous PPO catalyze the progressive browning of the fruit during the ripening process on the tree. Successively, during olive milling and kneading, the olive PPO still active are involved in the oxidization of polyphenols and subsequent reddish-brown coloration of the olive pastes and wastewaters; at the same time, olive PPO are progressively deactivated.

Nevertheless, it is possible that a minimal amounts of oxidative enzymes deriving from the olives (e.g. lipoxygenase and PPO) remain active into the VOO [72].

Pre-treatment of vegetables are generally sufficient to inactivate the natural-occurring PPO; nevertheless, a residual enzyme activity, left by an inadequate treatment, may promote oxidation of VOO's phenols with serious consequences on the quality of the food products. Indeed, it has been shown that immediately after the jar pasteurization, PPO deriving from vegetables may cause the depletion of intrinsic VOO polyphenols. As a result, the VOO depleted of antioxidants may alters itself at a higher rate than that of the refined oils, with a negative consequence on the quality of the final products [61, 73-75].

Effectively, it has been proven that the endogenous phenolic profile of VOO is modified, both qualitatively and quantitatively, when adding PPO to the oil [76,77]. Specifically, a significant reductions of the o-diphenols occurs and, subsequently that of the total phenols as well.

The end of oxidized phenols in the VOO is complex because the newly-formed compounds are involved in simultaneously coupling non-enzymatic reactions through different physical-chemical pathways; specifically, these may form self-coupling polymers, cross-link the enzyme, and/or remain in the oil matrix in an undetermined transitional oxidized form.

Actually, the formation of insoluble polymers may also be macroscopically visible also for VOO as result of the enzyme activity. As example, the precipitate formed in VOO treated with laccase from *Trametes versicolor* is shown in Figure 5; conversely, in the same experimental condition, precipitate is totally absent in the control oil (oil without laccase).

In olive fruits and olive pomace the disappearance of oleuropein is related to the formation of the nearly stable phenolic oligomers [78,79]. Moreover, although the polymerization of the oxidized phenols proceeds rapidly in an aqueous solution, it may be inhibited in the oil [80]. Consequently, a fraction of partial-oxidized phenols remains in the oil leaving controversial effects on VOO oxidative stability.

It has been observed that, at an initial state, partial-oxidized phenols exhibit a higher radical scavenging activity than the non-oxidized ones; whereas, at the advanced oxidation stage, phenol antioxidant properties may decrease [77].

Moreover, partially-oxidized phenols, at a low temperature and under a limited oxygen supply, show an antioxidant power higher than that of the phenols naturally occurring in the fresh VOO. Conversely, the same partially-oxidized phenols show an apparent pro-oxidant effect at high temperatures and a high oxygen supply.

VOO-PBS   **VOO-LAC**

Figure 5. The formation of sediments in virgin olive oil treated with laccase from *Trametes versicolor*. (VOO-PBS: control sample with phosphate buffer solution; VOO-LAC: oil treated with laccase).

In conclusion, phenol-oxidase enzymes deriving from vegetables can oxidize *in situ* polyphenols of VOO, leaving in it part of enzymatic-oxidation by-products that may affect positively or negatively on the oxidative stabilities of the oil itself.

4. SAUCES

A number of various sauces are used throughout the world as a condiment; oils/fats are fundamental ingredients in both domestic and commercial preparation of all sauces.

A typical habit of a few Mediterranean countries, especially Italy, is to prepare 'red', 'white' and 'green' sauces with VOO. The role of VOO in these sauces has been little studied.

'Red' sauces are those with a tomato base. Tomato-based products (sauce, paste, puree) are regularly consumed in many parts of the world. The protective effects of a regular consumption of raw tomatoes and tomato sauce is due to high levels of bioactive compounds, such as lycopene and other carotenoids, but also to the minerals, flavonoids and vitamins E and C [81]. Specifically, epidemiological and clinical studies have confirmed the beneficial effects of carotenoid intake on chronic and degenerative diseases. Tomato processing, as well as the simultaneous intake of fats and triglycerides, demonstrate an increase of carotenoid bioavailability [82].

Processing induced the thermal cell wall rupture and the disruption of lycopene-protein complexes, as well as heat-improved carotenoid extraction into the lipophilic fraction. Moreover, the addition of oil to tomato puree may favor carotenoids isomerization and oxidation, as a consequence of their protective action towards oil oxidation [83].

Recent researches have also shown that oil added to tomato sauce may influence the bioavailability of phenolic compounds as well by modifying their bioaccessibility from the food matrix, modulating the gastric emptying and/or the intestinal and hepatic metabolism of the absorbed phenolic compounds. These have been observed for the flavonol quercetin, the flavanone naringenin and some hydroxycinnamates, that are among the most abundant phenolics in tomato sauces [84, 85].

No change, neither in lycopene concentration nor in the chain-breaking activity of the lipophilic fractions, have been observed when both laboratory and industrial scale heating treatments are performed on peeled-tomato puree containing 5% virgin olive oil [86].

VOO is also widely used in the preparation of 'white' sauces, such as mayonnaise, which are emulsions with pH values generally in the acid range.

The antioxidant behavior of VOO polyphenols is more complex in emulsions than in oil alone because more variables influence lipid oxidation, including emulsifiers and pH [87]. Presence of the aqueous phase often decreases the activity of antioxidants because hydrogen bonded complexes formed with water are less effective in scavenging lipid radicals by hydrogen donation.

In emulsions, oxidation is a reaction which is initiated at the interface between the two phases and is influenced by both the presence of pro-oxidant and antioxidant compounds and the interactions between the different components of the system.

Polarity of the antioxidant is not the only parameter to determine antioxidant efficiency in emulsions; there are other variables as well, such as antioxidant activity in terms of capacity and rapidity in donating an hydrogen atom and localization among the three phases (lipid, water and interface), can affect the protective role towards lipid auto-oxidation in emulsions [88 90].

The most common problem with sauces is their destabilization after preparation and/or during the storage. Proteins (such as milk and derivates, whey, eggs and derivates) are very often used both as emulsifiers and stabilizers. Some polysaccharides (such as starch and gum) may be also used for giving additional stabilization. A sample white sauce containing corn starch, casein and olive oil form an pseudoplastic emulsion, shown to be unstable during storage; however, sauce stability improves significantly by the addition of hydrocolloids, such as xanthan and locust bean gum [91].

Finally, a typical 'green' sauce is the 'pesto', an Italian basil-based pasta sauce which is becoming well known all over the world for its enjoyable taste and its very pleasant aroma. Traditional pesto originates in Liguria, a North Italian region that stretches along a wide tract of the Mediterranean coast. Pesto is a very complex sauce, and the other traditional ingredients are cheese, extra-virgin olive oil, pine nuts and/or walnuts and garlic [92, 93].

CONCLUSION

In general, though sometimes contrasting results have been found, VOO seems to be a valid partner in the preparation of food, especially when the cooking time is not extensive and the temperature is not very high.

Olive oil has proven to be resistant to degradation under domestic cooking conditions independently of its category label; although, extra-virgin olive oil seems to be more resistant. Moreover, the monovarietal olive oils may exhibit a different oxidative stability due to the variable content of oleic acid, vitamin E and phenol amounts.

In conclusion, virgin olive oil has been demonstrated a crucial element of the Mediterranean Diet, not only for its inherent nutritional values but also for the cumulative benefits of the foods that are typically prepared with it.

However, the VOO chosen for cooking have to be preferably of excellent quality, containing very low content of acidity and peroxide levels.

ACKNOWLEDGMENTS

The authors are very grateful to Toni Caza for the help given in the editing the manuscript.

REFERENCES

[1]   Serra-Majema, L., Ngo de la Cruz, J., Ribas, L. & Salleras, L. (2004). Mediterranean diet and health: is all the secret in olive oil? *Pathophysiology of Haemostasis and Thrombosis, 33,* 461-465.

[2]   Fritsch, C. W., Egberg, D. C. & Magnuson, J. S. (1979). Changes in dielectric constant as a measure of frying oil deterioration. *Journal of the American Oil Chemists' Society*, 56, 746-750.

[3]   Saguy, I. S. & Dana, D. (2003). Integrated approach to deep fat frying: engineering, nutrition, health and consumer aspects. *Journal of Food Engineering*, 56(2), 143-152.

[4]   Andrikopoulos, N. K., Tzamtzis, V. A., Giannopoulos, G. A., Kalantzopoulos, G. K. & Demopoulos, C. A. (1989). Deterioration of some vegetable oils. I. During heating or frying of several foods. *Revue Française Des Corps Gras*, 36, 127 - 129.

[5]   Sumnu, G. (2001). A review on microwave baking of foods. *International Journal of Food Science and Technology*, 6, 117–127.

[6]   Paul, S. & Mittal, G. S. (1997). Regulating the use of degraded oil/fat in deep-fat/oil food frying. *Critical Review of Food Science and Nutrition*, 37, 636-662.

[7]   Kamal-Eldin, A., Velasco, J. & Dobarganes, C. (2003). Oxidation of mixtures of triolein and trilinolein at elevated temperatures. *European Journal of Lipid Science and Technology*, 105, 165–170.

[8]   Kamal-Eldin, A. (2006). Effect of fatty acids and tocopherols on the oxidative stability of vegetable oils. *European Journal of Lipid Science and Technology*, 58, 1051-1061.

[9]   De Leonardis, A. & Macciola, V. (2012). Heat-oxidation stability of palm oil blended with extra virgin olive oil. *Food Chemistry*, 135, 1769–1776.

[10]   Chiavaro, E., Vittadini, E., Rodriguez-Estrada, M. T., Cerretani, L., Bonoli, M., Bendini, A. & Lercker, G. (2007). Monovarietal extra virgin olive oils: correlation between thermal properties and chemical composition. *Journal of Agricultural and Food Chemistry*, 55(26), 10779-10786.

[11]   Aparicio, R., Roda, L., Albi, M. A. & Gutiérrez, F. (1999). Effect of various compounds on virgin olive oil stability measured by Rancimat. *Journal of Agriculture and Food Chemistry*, 47(10), 4150-4155.

[12]   Nor Aini, I. & Miskandar, M. S. (2007). Utilization of palm oil and palm products in shortenings and margarines. *European Journal of Lipid Science and Technology*, 109, 422–432.

[13]   Frankel, E. N., Selke, E., Neff, W. E. & Miyashita, K. (1992). Autoxidation of polyunsaturated triacylglycerols. Volatile decomposition products from triacylglycerols containing linoleate and linolenate. *Lipids*, 27(6), 442-446.

[14]   Neff, W. E. & El-Agaimy, M. (1996). Effect of linoleic acid position in triacylglycerols on their oxidative stability. *LWT-Food Science and Technology*, 29(8), 772-775.

[15]   Romano, R., Giordano, A., Vitiello, S., Le Grottaglie, L. & Spagna Musso, S. (2012). Comparison of the frying performance of olive oil and palm superolein. *Journal of Food Science*, 77, 519-531.

[16]   Casal, S., Ricardo Malheiro, R., Sendas, A., Oliveira, B. P. P. & Pereira, J. A. (2010). Olive oil stability under deep-frying conditions. *Food and Chemical Toxicology*, 48, 2972–2979.

[17]   Chatzilazarou, A., Gortzi, O., Lalas, S., Zoidis, E. & Tsaknis, J. (2006). Physicochemical changes of olive oil and selected vegetable oils during frying. *Journal of Food Lipids*, 13, 27–35.

[18]   Sebedio, J. L., Catte, M., Boudier, M. A., Prevost, J. & Grandgirard, A. (1996). Formation of fatty acid geometrical isomers and of cyclic fatty acid monomers during the finish frying of frozen prefried potatoes. *Food Research International*, 29, 109-116.

[19] Márquez-Ruíz, G. & Dobarganes, C. (1996). Short-chain fatty acid formation during thermoxidation and frying. *Journal of the Science of Food and Agriculture, 70*, 120–126.

[20] Fullana, A., Carbonell-Barrachina, A. A. & Sidhu, S. (2004). Comparison of volatile aldehydes present in the cooking fumes of extra virgin olive, olive, and canola oils. *Journal of Agricultural and Food Chemistry, 52*(16), 5207-5214.

[21] Sánchez-Gimeno, A. C., Negueruela, A. I., Benito, M., Vercet, A. & Oria, R. (2008). Some physical changes in Bajo Aragón extra virgin olive oil during the frying process. *Food Chemistry, 110*, 654–658.

[22] Benedito, J., Mulet, A., Velasco, J. & Dobarganes, M. C. (2002). Ultrasonic assessment of oil quality during frying. *Journal of Agricultural and Food Chemistry, 50*(16), 4531-4536.

[23] Firestone, D. (1996). Regulation of frying fat and oil. In *Deep Frying Chemistry. Nutrition and Practical Applications*, edit by Erickson EG & Perkins MD, Champaign, USA: AOCS Press, 323-334.

[24] Jorge, N., Marquez-Ruiz, G., Martin-Polvillo, M., Ruiz-Mendez, M. V. & Dobarganes, M. C. (1996). Influence of dimethylpolysiloxane addition to edible oils; dependence on the main variables of the frying process. *Grasas y Aceitas, 47*, 14-19.

[25] Cuesta, C., Sanchez-Muniz, F. J., Hernandez, I. & Lopez-Varela, S. (1991). Changes in olive oil during frying of successive batches of potatoes. Correlations between individual indices and overall assessment of deterioration. *Revista de Agroquimica y Tecnologia de Alimentos, 31*, 523-531.

[26] Morales, M. T., Ríos, J. J. & Aparicio, R. (1997). Changes in the volatile composition of virgin olive oil during oxidation: flavors and off-flavors. *Journal of Agriculture and Food Chemistry, 45*, 2666-2673.

[27] Gunstone, F. (Editor). (2011). Vegetable Oils in *Food Technology: Composition, Properties and Uses*, 2nd Edition, Wiley-Blackwell, Hoboken, NJ, USA.

[28] Interesse, F. S., Ruggiero, P. & Vitagliano, M. (1971). Autoxidation of olive oil. Effects of chlorophyll pigments. *Industrie Agrarie, 9*, 318-324.

[29] Gómez-Alonso, S., Fregapane, G., Salvador, M. D. & Gordon, M. H. (2003). Changes in phenolic composition and antioxidant activity of virgin olive oil during frying. *Journal of Agriculture and Food Chemistry, 51*, 667-672.

[30] Mateos, R., Trujillo, M., Perez-Camino, M. C., Moreda, W. & Cert, A. (2005). Relationships between oxidative stability, triacylglycerol composition, and antioxidant content in olive oil matrices. *Journal of Agriculture and Food Chemistry, 53*, 5766–5771.

[31] Yuki, E., Agren, J. J. & Hanninen, O. (1993). Effects of cooking on the fatty acids of 3 fresh-water fish species. *Food Chemistry, 46*(4), 377–382.

[32] Brenes, M., García, A., Dobarganes, M. C., Velasco, J. & Romero, C. (2002). Influence of thermal treatments simulating cooking processes on the polyphenol content in virgin olive oil. *Journal of Agriculture and Food Chemistry, 50*(21), 5962-5967.

[33] Ishikawa, Y. (1976). Tocopherol contents of nine vegetable frying oils, and their changes under simulated deep-fat frying conditions. *Journal of the American Oil Chemists' Society, 53*, 673-676.

[34] Kalogeropoulos, N., Chiou, A., Mylona, A., Ioannou, M. S. & Andrikopoulos, N. K. (2007). Recovery and distribution of natural antioxidants (α-tocopherol, polyphenols

and terpenic acids) after pan-frying of Mediterranean finfish in virgin olive oil. *Food Chemistry, 100,* 509–517.

[35] Nieto, L. M., Hodaifa, G. & Lozano Peña, J. L. (2010). Changes in phenolic compounds and Rancimat stability of olive oils from varieties of olives at different stages of ripeness. *Journal of the Science of Food and Agriculture, 90*(14), 2393-2398.

[36] Van der Schouw, Y. T., De Kleijn, M. J., Peeters, P. H. & Grobbee, D. E. (2000). Phyto-oestrogens and cardiovascular disease risk. *Nutrition, Metabolism, and Cardiovascular Diseases, 10*(3), 154-167.

[37] Sioen, I., Haak, L., Raes, K., Hermans, C., De Henauw, S., De Smet, S. & Van Camp, J. (2006). Effects of pan-frying in margarine and olive oil on the fatty acid composition of cod and salmon. *Food Chemistry, 98,* 609–617.

[38] Candela, M., Astiasaran, I. & Bello, J. (1998). Deep-fat frying modifies high-fat fish lipid fraction. *Journal of Agricultural and Food Chemistry, 46*(7), 2793–2796.

[39] García-Arias, M. T., Álvarez Pontes, E., García-Linares, M. C., García-Fernández, M. C. & Sánchez-Muniz, F. J. (2003). Cooking–freezing–reheating (CFR) of sardine (*Sardina pilchardus*) fillets. Effect of different cooking and reheating procedures on the proximate and fatty acid compositions. *Food Chemistry, 83,* 349-356.

[40] Sacchi, R., Paduano, A., Fiore, F., Della Medaglia, D., Ambrosino, M. L. & Medina, I. (2002). Partition behavior of virgin olive oil phenolic compounds in oil-brine mixtures during thermal processing for fish canning. *Journal of Agricultural and Food Chemistry, 50*(10), 2830-2835.

[41] Perssona, E., Graziani, G., Ferracane, R., Fogliano, V. & Skoga, K. (2003). Influence of antioxidants in virgin olive oil on the formation of heterocyclic amines in fried beefburgers. *Food and Chemical Toxicology, 41,* 1587-1597.

[42] Monti, S. M., Ritieni, A., Sacchi, R., Skog, K., Borgen, E. & Fogliano, V. (2001). Characterization of phenolic compounds in virgin olive oil and their effect on the formation of carcinogenic/mutagenic heterocyclic amines in a model system. *Journal of Agricultural and Food Chemistry, 49*(8), 3969-3975.

[43] Podmore, J. (1994). Fats in bakery and kitchen products. In *Fats in Food Products.* Ed. by Moran DPJ, & Rajah KK, Springer US, 213-253.

[44] Wiedermann, L. H. (1978). Margarine and margarine oil, formulation and control. *Journal of the American Oil Chemists' Society, 55*(11), 823-829.

[45] Chiavaro, E. (2013). Crystal polymorph structure determined for extra virgin olive oil. *European Journal of Lipid Science and Technology, 115*(3), 267-269.

[46] Barba, L., Arrighetti, G. & Calligaris, S. (2013). Crystallization and melting properties of extra virgin olive oil studied by synchrotron XRD and DSC. *European Journal of Lipid Science and Technology, 115* (3), 322-329.

[47] Ascherio, A., Katan, M. B., Zock, P. L., Stampfer, M. J. & Willett, W. C. (1999). *Trans* fatty acids and coronary heart disease. *New England Journal of Medicine, 340,* 1994-1998.

[48] Duns, M. L. (1985). Palm oil in margarines and shortenings. *Journal of the American Oil Chemists' Society, 62*(2), 408-410.

[49] Ravnskov, U. (1998). The questionable role of saturated and polyunsaturated fatty acids in cardiovascular disease. *Journal of Clinical Epidemiology, 51,* 443–460.

[50] Delcuratolo, D., Gomes, T., Paradiso, V. M. & Nasti, R. (2008). Changes in the oxidative state of extra virgin olive oil used in baked Italian focaccia topped with different ingredients. *Food Chemistry*, *106*, 222–226.

[51] Caponio, F., Giarnetti, M., Paradiso, V. M., Summo, C. & Gomes, T. (2013). Potential use of extra virgin olive oil in bakery products rich in fats: a comparative study with refined oils. *International Journal of Food Science and Technology*, *48*, 82–88.

[52] Matsakidou, A., Blekas, G. & Paraskevopoulou, A. (2010). Aroma and physical characteristics of cakes prepared by replacing margarine with extra virgin olive oil. *LWT - Food Science and Technology*, *43*, 949–957.

[53] Zamora, R. & Hidalgo, F. J. (2005). Coordinate contribution of lipid oxidation and Maillard reaction to the nonenzymatic food browning. *Critical Reviews in Food Science and Nutrition*, *45*(1), 49-59.

[54] Yasuhara, A., Tanaka, Y., Hengel, M. & Shibamoto, T. (2003). Gas chromatographic investigation of acrylamide formation in browning model systems. *Journal of Agricultural and Food Chemistry*, *51*(14), 3999-4003.

[55] Friedman, M. (2003). Chemistry, biochemistry, and safety of acrylamide. A review. *Journal of Agricultural and Food Chemistry*, *51*(16), 4504-4526.

[56] Arribas-Lorenzo, G., Fogliano, V. & Morales, F. J. (2009). Acrylamide formation in a cookie system as influenced by the oil phenol profile and degree of oxidation. *European Food Research and Technology*, *229*, 63-72.

[57] Zamora, R. & Hidalgo, F. J. (2008). Contribution of lipid oxidation products to acrylamide formation in model systems. *Journal of Agricultural and Food Chemistry*, *56*(15), 6075-6080.

[58] Albi, T., Lanzón, A., Guinda, A., Pérez-Camino, M. C. & León, M. (1997). Microwave and conventional heating effects on some physical and chemical parameters of edible fats. *Journal of Agricultural and Food Chemistry*, *45*, 3000-3003.

[59] Caponio, F., Pasqualone, A. & Gomes, T. (2003). Changes in the fatty acid composition of vegetable oils in model doughs submitted to conventional or microwave heating. *International Journal of Food Science and Technology*, *38*(4), 481-486.

[60] Cerretani, L., Bendini, A., Rodriguez-Estrada, M. T., Vittadini, E. & Chiavaro, E. (2009). Microwave heating of different commercial categories of olive oil: Part I. Effect on chemical oxidative stability indices and phenolic compounds. *Food Chemistry*, *115*(4), 1381-1388.

[61] De Leonardis, A., Macciola, V. & De Felice, M. (2001). Changes in olive oils used as covering in preserves of eggplants (*Solanum Melongena*) in relation to the time and the condition of storage. *Grasas y Aceites*, *52*(2), 104-109.

[62] Baiano, A., Gomes, T. & Caponio, F. (2005). Hydrolysis and oxidation of covering oil in canned dried tomatoes as affected by pasteurization. *Grasas y Aceites*, *56*(3), 177-181.

[63] Lanza, B., Di Serio, M. G., Giansante, L., Di Loreto, G., Russi, F. & Di Giacinto, L. (2013). Effects of pasteurisation and storage on quality characteristics of table olives preserved in olive oil. *International Journal of Food Science & Technology*, *48*(12), 2630-2637.

[64] Paganuzzi, V., De Iorgi, F. & Malerba, A. (1995). Sull'olio d'oliva vergine extra impiegato in conserve alimentari confezionate in contenitori di vetro. Nota 1.

Variazione di alcuni parametri chimico-fisici al mutare delle condizioni di produzione e nel corso dell'invecchiamento. *Rivista Italiana delle Sostanze Grasse, 72*, 529-537.

[65] Paganuzzi, V., Malerba, A. & De Iorgi, F. (1996). Sull'olio d'oliva vergine extra impiegato in conserve alimentari confezionate in contenitori di vetro. Nota 2. Variazione di alcuni componenti minori al mutare delle condizioni di produzione e nel corso dell'invecchiamento. *Rivista Italiana delle Sostanze Grasse, 73*, 409-415.

[66] Bizzozero, N. & Carnelli, L. (1996). Composizione acidica e trans insaturazione dell'olio di copertura di sgombri e tonni conservati in scatola. *Industrie Alimentari, 35*, 680-683.

[67] Cucurachi, A. (1996). L'olio di oliva di copertura al tonno. *Rivista Italiana delle Sostanze Grasse, 43*, 335-342.

[68] Caponio, F., Gomes, T. & Summo, C. (2003). Quality assessment of edible vegetable oils used as liquid medium in canned tuna. *European Food Research and Technology, 216*(2), 104-108.

[69] Mayer, A. M. (1986). Polyphenol oxidases in plants. Recent progress. *Phytochemistry, 26*, 11–20.

[70] Ortega-García, F., Blanco, S., Peinado, M. Á. & Peragón, J. (2008). Polyphenol oxidase and its relationship with oleuropein concentration in fruits and leaves of olive (*Olea europaea*). cv.'Picual'trees during fruit ripening. *Tree physiology, 28*(1), 45-54.

[71] Charoenprasert, S. & Mitchell, A. (2012). Factors influencing phenolic compounds in table olives (*Olea europaea*). *Journal of Agricultural and Food Chemistry*, 60(29), 7081-7095.

[72] Georgalaki, M. D., Sotiroudis, T. G. & Xenakis, A. (1998). The presence of oxidizing enzyme activities in virgin olive oil. *Journal of the American Oil Chemists' Society, 75*, 155–159.

[73] Nicoli, M. C., Anese, M. & Manzocco, L. (1999). Oil stability and antioxidant properties of an oil-tomato food system as affected by processing. *Advanced Food Science, 21*, 10–14.

[74] Silva, L., Garcia, B. & Paiva-Martins, F. (2010). Oxidative stability of olive and its polyphenolic compounds after boiling vegetable process. *LWT—Food Science and Technology, 43*, 1336–1344.

[75] De Leonardis, A. & Macciola, V. (2011). Polyphenol oxidase from eggplant reduces the content of phenols and oxidative stability of olive oil. *European Journal of Lipid Science and Technology, 113*(9), 1124-1131.

[76] De Leonardis, A. & Macciola, V. (2013). Degradazione enzimatica dei polifenoli e stabilità ossidativa dell'olio extra vergine di oliva. *Industrie Alimentari, 52*(534), 7-14.

[77] De Leonardis, A., Angelico, R., Macciola, V. & Ceglie, A. (2013). Effects of polyphenol enzymatic-oxidation on the oxidative stability of virgin olive oil. *Food Research International, 54*(2), 2001-2007.

[78] Papadimitriou, V., Sotiroudis, T. G. & Xenakis, A. (2005). Olive oil microemulsions as a biomimetic medium for enzymatic studies: oxidation of oleuropein. *Journal of the American Oil Chemists' Society, 82*, 335–340.

[79] Tzika, E. D., Papadimitriou, V., Sotiroudis, T. G. & Xenakis, A. (2008). Oxidation of oleuropein studied by EPR and spectrophotometry. *European Journal of Lipid Science and Technology, 110*, 149–157.

[80] García-Rodríguez, R., Romero-Segura, C., Sanz, C., Sánchez-Ortiz, A. & Pérez, A. G. (2011). Role of polyphenol oxidase and peroxidase in shaping the phenolic profile of virgin olive oil. *Food Research International, 44*(2), 629-635.

[81] Weisburger, J. H. (1998). Evaluation of the evidence on the role of tomato products in disease prevention. In *Proceedings of the Society for Experimental Biology and Medicine.* Ed. by Society for Experimental Biology and Medicine (New York, NY), Vol. *218* (2), 140-143.

[82] Stahl, W. & Sies, H. (1996). Lycopene: a biologically important carotenoid for humans?. *Archives of Biochemistry and Biophysics, 336*(1), 1-9.

[83] Fielding, J. M., Rowley, K. G., Cooper, P. & O'Dea, K. (2005). Increases in plasma lycopene concentration after consumption of tomatoes cooked with olive oil. *Asia Pacific Journal of Clinical Nutrition, 14*(2), 131–136.

[84] Lesser, S., Cermak, R. & Wolffram, S. (2004). Bioavailability of quercetin in pigs is influenced by the dietary fat content. *Journal of Nutrition, 134*, 1508–1511.

[85] Tulipania, S., Martinez Huelamo, M., Rotches Ribalta, M., Estruch, R., Ferrerd, E. E., Andres-Lacuevaa, C., Illana, M. & Lamuela-Raventós, R. M. (2012). Oil matrix effects on plasma exposure and urinary excretion of phenolic compounds from tomato sauces: Evidence from a human pilot study. *Food Chemistry, 130*(3), 581-590.

[86] Graziani, G., Pernice, R., Lanzuise, S., Vitaglione, P., Anese, M. & Fogliano, V. (2003). Effect of peeling and heating on carotenoid content and antioxidant activity of tomato and tomato-virgin olive oil systems. *European Food Research and Technology, 216*(2), 116-121.

[87] Paiva-Martins, F. & Gordon, M. H. (2005). Interactions of ferric ions with olive oil phenolic compounds. *Journal of Agricultural and Food Chemistry, 53*(7), 2704-2709.

[88] Di Mattia, C. D., Sacchetti, G., Mastrocola, D. & Pittia, P. (2009). Effect of phenolic antioxidants on the dispersion state and chemical stability of olive oil O/W emulsions. *Food Research International, 42*(8), 1163-1170.

[89] Mosca, M., Cuomo, F., Lopez, F. & Ceglie, A. (2013). Role of emulsifier layer, antioxidants and radical initiators in the oxidation of Olive oil-in-Water emulsions. *Food Research International, 50*, 377-383.

[90] Mosca, M., Diantom, A., Lopez, F., Ambrosone, L. & Ceglie, A. (2013). Impact of antioxidants dispersions on the stability and oxidation of water-in-olive oil emulsions. *European Food Research and Technology 236*, 319-328.

[91] Mandala, I. G., Savvas, T. P. & Kostaropoulos, A. E. (2004). Xanthan and locust bean gum influence on the rheology and structure of a white model-sauce. *Journal of Food Engineering, 64*(3), 335-342.

[92] Salvadeo, P., Boggia, R., Evangelisti, F. & Zunin, P. (2007). Analysis of the volatile fraction of "Pesto Genovese" by headspace sorptive extraction (HSSE). *Food Chemistry, 105*(3), 1228-1235.

[93] Masino, F., Ulrici, A. & Antonelli, A. (2008). Extraction and quantification of main pigments in pesto sauces. *European Food Research and Technology, 226*(3), 569-575.

In: Virgin Olive Oil
Editor: Antonella De Leonardis

ISBN: 978-1-63117-656-2
© 2014 Nova Science Publishers, Inc.

Chapter 15

NEW BENEFITS OF THE FIBRE IN GREEN TABLE OLIVES

Sergio Lopez[1,*], Sara Jaramillo[2], Beatriz Bermudez[1], Rocio Abia[1] and Francisco J.G. Muriana[1]

[1]Laboratory of Cellular and Molecular Nutrition,
Instituto de la Grasa, Seville, Spain
[2]Phytochemicals and Food Quality Group,
Instituto de la Grasa, Seville, Spain

ABSTRACT

Green table olives are an essential component of the Mediterranean diet and are used as an ingredient in many dishes. The high nutritional value of olive fruit reflects its balanced fat content, which consists primarily of monounsaturated oleic acid and other essential polyunsaturated fatty acids, minor constituents (e.g., polyphenols), and its dietary fibre content with approximately 3 g of fibre per 100 g edible portion. The main components of the soluble fibre are pectic polysaccharides, such as homogalacturonans, rhamnogalacturonans and arabinans. Pectins have many health benefits: they are hypocholesterolemic, hypoglycemic and prebiotic, and anticancer properties have recently been attributed to them. Pectic polysaccharides can modulate the expression of the galectin-3 protein in various types of cancer. This protein is involved in numerous biological processes, such as proliferation, differentiation, cell adhesion, apoptosis and angiogenesis, and is increasingly regarded as a tumour marker.

Keywords: Green table olive, fibre, bioactivities, pectin, health

---

[*] Corresponding author email: serglom@ig.csic.es.

ABBREVIATIONS

DRA   daily recommended allowance
Gal-3   galectin-3
HDL   high-density lipoprotein
IDF   insoluble dietary fibre
IL   interleukin
LDL   low density lipoprotein
SDF   soluble dietary fibre
TDT   total dietary fibre
TNF   tumour necrosis factor

INTRODUCTION

The Mediterranean diet was recognised by UNESCO as a Common Heritage of Mankind in 2010, reflecting its incorporation of customs, knowledge of the development, production and selection of food, and its emphasis on cooking methods and balanced consumption. The nutritional pyramid of the Mediterranean diet (Figure 1) diagrams its food components according to their consumption levels and frequency of use, and the pyramid serves as a nutritional guide for the populations of the Mediterranean Basin (Spain, Italy, Greece and France). The base of the pyramid contains foods that are eaten on a daily basis, with fruits, vegetables, cereals and olive oil representing the largest fraction for meeting caloric requirements. To a lesser extent, in terms of quantity but not frequency, the Mediterranean diet includes dry fruits, olives, spices and dairy products. The pyramid progresses upward with the weekly consumption of the following, ordered in terms of frequency: fish, legumes, eggs, white meats, potatoes and red meats. Sweets are located at the top of the pyramid. One of the fundamental food elements in the pyramid is the juice derived from olives (virgin olive oil). Despite being a high-calorie food, the type of fatty acid (oleic acid) in olive oil and the presence of antioxidants (mainly polyphenols) make virgin olive oil a critical component of the diet. Because it helps prevent cardiovascular diseases [1], various types of cancer [2-4] and neurodegenerative diseases [5], the Mediterranean diet is generally associated with longevity [6] and a high quality of life [7]. The oleic acid and polyphenols contained in olive oil have positive effects on blood pressure, the thrombosis/fibrinolysis system, inflammation and various clinical parameters of lipid metabolism, such as LDL and HDL concentrations [3]. Pentacyclic triterpenes and flavonoids are present in many vegetables and are of special interest because of their chemopreventive, chemotherapeutical and adjuvant potential in cancer treatment [5-8]. Group-B vitamins, which are present in vegetables, dry fruits, dairy products, legumes and cereals, and omega-3 fatty acids, which are primarily found in dry fruits and certain fish species, have a protective effect against Alzheimer's disease [9]. In this regard, olives offer the same beneficial effects as virgin olive oil, due to their composition. The olive fruit also has the added benefit of a greater polyphenol content and high levels of dietary fibre, both of which have positive effects on human health. This chapter therefore focuses on the nutritional aspects and health effects of table olive consumption.

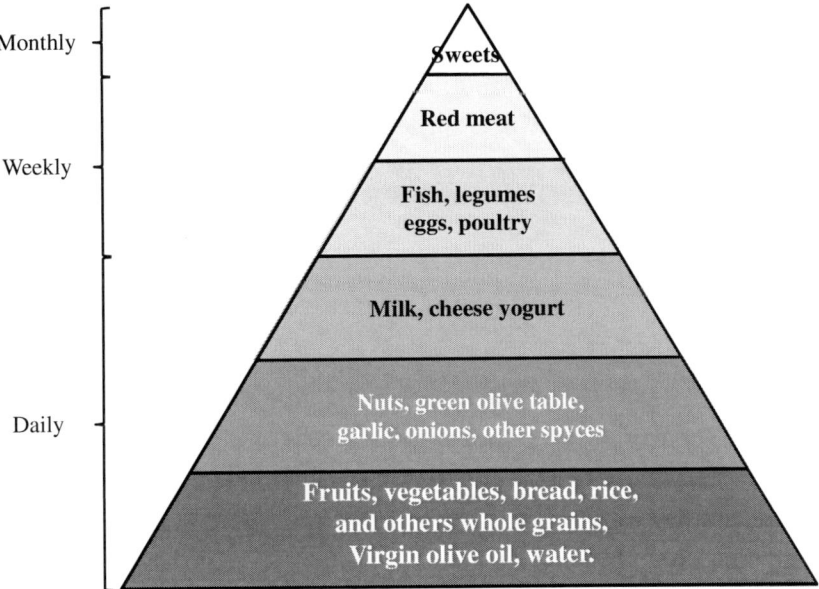

Figure 1. Mediterranean diet pyramid.

COMPOSITION OF TABLE OLIVES

Olive cultivation is extremely important for the Spanish economy, particularly in the region of Andalusia, in terms of job creation and the production of virgin olive oil and table olives. Table olives are among the most popular foods in Spain, which is currently the top producer of table olives worldwide. The 597,070 Tm produced annually accounts for 25% of global production. Spain is also the top exporter: its annual export of 280,000 Tm corresponds to 30% of the global trade of this product [10]. Although many studies have been devoted to the beneficial health effects of olive oil [11-13], relatively few research efforts have focused on the health benefits of table olives [1].

The consumption of table olives can be traced to ancient times. The first known description of the preparation of olives dates to 42 BC when Columela described various methods of olive preparation that depend on variety and ripeness [15]. In general, olives are consumed sporadically, and they are mainly served as an appetiser, for special occasions or in small quantities as part of a dish. For this reason, olives' significance in the diet is limited. In addition to enhancing dishes from a gastronomic perspective, however, olives have great nutritional value, which makes them a high-quality appetiser.

Among the various commercial presentations of table olives, the Spanish style represents more than 50% of total production. Internationally, the Manzanilla cultivar is most common, due to its high productivity and fruit quality. In the production of Spanish green olive, lye treatment is the fundamental operation and is critical for eliminating the bitter glycoside oleuropein. The process occurs under alkaline conditions, and oleuropein is hydrolysed into hydroxytyrosol and elenolic acid glycosides. The treatment of olives with sodium hypochlorite exerts a notably complete action on the fruits, primarily because it favours subsequent lactic fermentation [16]. Over the past few years, the industry has adopted the use of alkaline solutions at low temperatures and thereby less energetic treatments for obtaining a high-quality end-product, the changes were also aimed at rationalising the lye process [17].

The composition of olives depends on a number of factors, such as variety, agronomic techniques, processing and conditioning operations (e.g., pitting and stuffing). In all cases, however, the main component of olives is water (74%), and the caloric content is generally approximately 190 calories per 100 g of olives. The proportion of fat is approximately 20% of the total weight. Although olive fat exhibits a high nutritional density, the predominance of unsaturated fatty acids over saturated fatty acids contributes to the health benefits of olives and negates any negative associations with the presence of saturated fat [11, 13]. As in virgin olive oil, the most abundant fatty acid is oleic acid (~75%) followed by palmitic acid (~15%) and linoleic acid (~10%) [18]. The nutritional density of the polyunsaturated fatty acids is also significant, insofar as the consumption of table olives could, at least in part, provide the required levels of polyunsaturated fatty acids [19]. Additionally, the unsaponifiable fraction present in table olives is higher than that in virgin olive oil, resulting in a higher concentration of phytosterols [20]. Thus, the benefits associated with virgin olive oil can also be applied to olive fruits [21]. Olives also contain between 1 and 2% protein, while this content is limited, it is nevertheless interesting from a nutritional point of view. Moreover, olives contain essential amino acids, with 32% of the total amino-acid content corresponding to aspartic acid, glutamic acid and leucine [22]. In terms of vitamin content, the content of α-tocopherol, or vitamin E, in green table olives (~3.5 mg per 100 g edible portion) accounts for approximately 25% of the daily-recommended allowance. Among the various commercial types of table olives, Spanish-style green olives have a relatively high β-carotene, or pro-vitamin A, content (~300 µg per 100 g edible portion), whereas the B6 content does not exceed ~70 µg per 100 g edible portion [23]. Other nutrients present in table olives at high levels include pantothenic acid, or vitamin B5, and nicotinamide, which offer interesting benefits [24-25]. Other important components present in smaller quantities are vitamin B1 and cryptoxanthin. The organic acid content of fresh olives is reduced to notably low levels through various preparation processes. The only organic acids that can be detected are succinic acid (~9 mg per 100 g edible portion) and lactic and acetic acid (~250 and ~120 mg per 100 g edible portion, respectively), which are either produced during the fermentation process or added to control the pH [26]. The fibre content of table olives is approximately 3 g per 100 g edible portion and 1.8 g per 100 kcal. In terms of mineral content, the sodium component is notable because it is the basic ingredient of brine. The sodium content is approximately 2.3 g per 100 g edible portion. Olives also contain other minerals in smaller proportions, including calcium, potassium, magnesium, iron, phosphorus and iodine [27]. Despite all of these benefits, consumers often have an extremely negative view of table olives because of their high sodium content. However, solving this problem is easy because sodium is an added component. Moreover, there are already new presentations available aimed at satisfying consumers with a high sensitivity to the adverse effects of excess sodium consumption. The harvesting and processing of Spanish-style green olives at a very early stage of ripeness makes them a significant source of both oleuropein and hydroxytyrosol. In fact, Spanish-style table olives have a polyphenol content corresponding to approximately 2% of the pulp. Tyrosol and hydroxytyrosol constitute 70% of the total phenolic content [28-29], which is higher than the percentage in virgin olive oil.

Table 1. Nutritional composition of green table olives per 100 g edible portion

Energy (Kcal)	187
Carbohydrates (g)	1
Total lipids (g)	20
Saturated fatty acids	2.8
Monounsaturated fatty acids	14
Polyunsaturated fatty acids	2.3
• ω-3	0.134
• ω-6	2.1
Proteins (g)	1.5
Fibre (g)	2-4
Water (g)	74
Calcium (mg)	63
Iron (mg)	1.5
Iodine (mg)	1
Magnesium (mg)	12
Sodium (mg)	2250
Potassium (mg)	91
Phosphorus (mg)	17
Selenium (mg)	0.9

Finally, because the lactic acid bacteria of table olives are the most abundant microorganisms observed during fermentation, several studies are being conducted to assess the use of specific probiotic families to obtain a more functional product [30]. In fact, various strains of *Lactobacillus pentosus* and *L. plantarum* possess probiotic properties that are similar to, or even superior to, those described in *L. casei Shirota* and *L. rhamnosus GG* [31], which increase the already significant nutritional value of table olives. Table 1 shows the distribution of the major olive components.

TABLE OLIVES AS A SOURCE OF POLYPHENOLS AND FIBRE

The recognition of olive fruits as a healthy product has led to the revaluation and increased consumption of olives over the past several years from 960,000 Tm to 1,420,000 Tm (48% increase). Figures have even tripled in several "new consumer" countries, such as Canada [32]. Thus, various programmes are aimed at developing new varieties of olives that may improve the quality of table olive production [33-34]. Phenolic components are a major component of table olive processing, they give olives their bitter taste and colour and allow the fermentation process to begin [35-36]. Several studies of phenolic compounds have demonstrated that both oleuropein and its main metabolite, hydroxytyrosol, offer numerous health benefits, including antioxidant [37-38], anti-inflammatory, antiatherogenic, anti-cancer, antimicrobial [39-40], antiviral, hypolipidemic and hypoglycemic activities [41-42], along with protection against osteoporosis [43-44]. Polyphenols also provide indirect protection by increasing the levels of endogenous antioxidants that are part of the defence system [45]. A recent study also characterised the dual-action mechanism by which hydroxytyrosol prevents damage to liver cells caused by oxidative stress [46]. Thus, hydroxytyrosol represents a potential chemo-protector. As with the phenolic components of olives, triterpenic acids, particularly oleanolic acid and maslinic acid, have aroused great interest due to their health benefits [47-48]. In addition to helping prevent the proliferation of various tumour lines [49-53], triterpenic acids have also been found to have anti-inflammatory [54-55], antioxidant [51, 56-57], hypoglycemic [58] and antimicrobial [59] properties.

In addition to these components of table olives and the ways in which they promote health, one beneficial feature of table olives that is distinct from those of virgin olive oil is their high dietary fibre content. The fibre content of table olives allows them to be regarded as a "source of fibre" according to Regulation No 1924/2006 of the European Parliament and the Council of 20 December 2006 [60]. Table olives help meet the daily-recommended allowance (DRA) of fibre (30 g). The beneficial effects of fibre and certain of its components on health are well established [61-64]. Dietary fibre consists primarily of "cell wall structures", although in some cases, initially soluble components in food become associated with fibre components as the food is processed [65-66]. The cell wall is a thick membrane that surrounds all plant cells and contributes significantly to their physiology by virtue of dictating their structural and mechanical characteristics. The main polymers found in the cell wall are cellulose, hemicellulose, pectins and lignins. The proportions of these polymers vary depending on the type of plant, the developmental stage and the tissue source. There are also minor components, primarily phenolic in nature, that contribute significantly to the functional and bioactive properties of fibre [67-68]. In table olives, dietary fibre accounts for 15 to 22% of the fresh weight, with one-third of the content consisting of pectic polysaccharides with a high degree of esterification. The pectic polysaccharides are rich in arabinans, homogalacturonans and rhamnogalacturonans [69] and are located in the medium layer. The fibre also contains lignin, cellulose and hemicelluloses (rich in xylans, xyloglucans, glucuronoxylans and mannans), which are bound by the pectic polysaccharides of the primary walls of two adjacent cells (Figure 2).

The pectic polysaccharides of table olives can be divided into two groups: a water-soluble group, which consists of rhamnogalacturonans, homogalacturonans (above 50%) and arabinans, and a neutral group, which mainly consists of arabinans [70].

Regarding the hemicelluloses present in fibre, it is important to note that the presence of xylans in edible fruits is notably limited. These polysaccharides are associated with secondary cell wall growth. In table olives, the polysaccharides mainly include acid arabinoxylans. Because olive fruits are gathered at an early stage of ripeness, they also have a notable xyloglucan content.

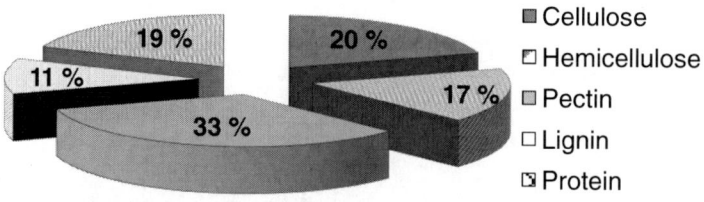

Figure 2. Composition of table olive fibre.

Homo- and/or rhamnogalacturonans are also known to be present, indicating the close relationship between xyloglucans, arabinoxylans, rhamnogalacturonans and cellulose [71-72].

During the processing of Spanish-style olives, sodium hydroxide, which is necessary for hydrolysing oleuropein and for inducing lactic fermentation, alters the integrity of the cell wall through the de-esterification and solubilisation of pectic polysaccharides and protein-pectin complexes [70]. These changes promote the β-elimination of these components, which leads to a depolymerisation of the rhamnogalacturonan chains. This depolymerisation involves the loss of arabinose short chains, arabinogalactans and free galactans [73], a decrease in the molecular weight of the xyloglucans, and the solubilisation of cellulose, arabinans and rhamnogalacturonans [74].

The lignin composition, which is primarily acetoxypinoresinol and pinoresinol, and the content of other phenolic components, such as hydroxycinnamic acids [75], in table olive fibre have also been associated with health benefits [76-77].

BIOACTIVITY OF TABLE OLIVE FIBRE

The term "dietary fibre" was used for the first time in 1953 when Hipsley defined it as the non-digestible components of the plant cell wall, which seemed to counter toxaemia related to pregnancy [78]. Many definitions of dietary fibre have subsequently emerged [79-81]. The classic definition refers to plant polysaccharides and lignins, which are resistant to the hydrolysis induced by digestive enzymes in the human body and are also beneficial for human health.

Fibre fermentation that occurs in the colon is a fundamental process. Fibre permits the maintenance and development of the bacterial microbiota and also stimulates the proliferation of epithelial cells [82]. In the colon, two main types of fermentation occur: saccharolytic fermentation and proteolytic fermentation. In terms of human health, saccharolytic fermentation is more beneficial and involves the generation of short-chain fatty acids, acetic acids, propionic acids and butyric acids. These fatty acids activate the FFA2/GPR43 and FFA3/GPR41 receptors, which are involved in metabolic diseases, such as type-2 diabetes and appetite/satiety sensation [83].

The classification of dietary fibre, total dietary fibre (TDF) that includes insoluble dietary fibre (IDF) and soluble dietary fibre (SDF), is based on solubility and/or fermentability in the intestinal lumen. Accordingly, soluble fibre is easily fermented by the bacteria in the colon, whereas insoluble fibre ferments very slowly. Soluble fibres include pectins, and insoluble fibres include cellulose, lignin and hemicellulose. As previously discussed, the definition of "fibre" has evolved, and related controversies have emerged. Several definitions recognise components that are analogous to fibre but do not derive from plants, as one example, hydrocolloids exhibit similar properties but derive from animals. With the goal of reaching a consensus, the Codex Alimentarius Commission has defined dietary fibre as a carbohydrate polymer with three or more monomeric units, which are neither digested nor absorbed in the small intestine and which belong to the following categories: polymers of edible carbohydrates naturally present in food, polymers of carbohydrates obtained from food through physical, enzymatic or chemical processes, and polymers of synthetic carbohydrates with positive effects on health that are supported by scientific evidence and generally accepted by the relevant authorities [84]. Regarding lignin and other carbohydrates, the following has been established: "Polymers of plant-derived carbohydrates may be closely related to plant lignin or other components that are distinct from carbohydrates, such as phenolic components, waxes, saponins, phytates, cutin and phytosterols. These substances, when intimately associated with polymers of plant-derived carbohydrates and when extracted with carbohydrate polymers to analyse the dietary fibre, can be considered dietary fibre. However, when they have been separated from the carbohydrate polymers and added to a food, these substances cannot be defined as dietary fibre".

Insoluble dietary fibre, particularly cellulose, has an important effect against constipation. The laxative effect of these components reflects their ability to increase the faecal mass, as the increase in volume stimulates the transit through the colon and reduces the time available for water reabsorption [85]. Additionally, insoluble fibre helps control blood cholesterol levels by allowing cholesterol to bond to bile acids and by promoting the excretion of these acids [86]. The positive effects of soluble fibre following carbohydrate metabolism are well-documented. Although few studies have provided evidence of the benefits of arabinoxylans, one has shown that the consumption of arabinoxylan-rich fibre improves the glucose and insulin postprandial response [87].

Several studies have suggested that the intestinal microbiota should be taken into consideration when evaluating risk factors for obesity and related conditions, such as dyslipidemia, inflammation, resistance to insulin and diabetes [88]. Regarding the prebiotic potential of arabinoxylans, recent findings indicate that they induce specific changes in the composition of the intestinal microbiota, which typically include an increase in bifidobacteria and a decrease in lactobacillus [89]. In the case of obesity, these effects are associated with an improvement in inflammation and integrity of the intestine barrier. According to recent descriptions, xyloglucans that are closely bonded to cellulose have an immunomodulating effect, as they increase the cytotoxic activity of macrophages against tumour cells and microorganisms, they also activate phagocytic activity and increase the levels of reactive oxygen species and nitric oxide production. Finally, xyloglucans can improve the secretion of cytokines and chemokines, including tumour necrosis factor-α (TNF-α), interleukin (IL)-1β, IL-6, IL-8, IL-12, IFN-γ and IFN-β 2 [90-91]. These results demonstrate the value of continued investigation of the effects of insoluble fibre components on tumour cells and other cell types.

Although pectic polysaccharides are not absorbed in the intestine, their known properties suggest that they may be beneficial for treating certain pathologies. For instance, pectic polysaccharides are able to form gels and absorb water, which slows gastric emptying and indicates antidiarrheal properties [92]. Over the past few years, several pectin fragments have been analysed, including those of arabinans, arabinogalactans and rhamnogalacturonan I types, these fragments exhibit an antiviral effect on diarrhoeas provoked by rotavirus by binding to it in such a way that inhibits proliferation of the virus and subsequent infections [93]. Another property of pectins is their ability to reduce intestinal monosaccharide absorption by diminishing the pace of sugar transmission from the stomach to the duodenum in diabetic individuals, effectively modulating the blood glucose level [94-95]. Pectin in the diet also affects blood cholesterol levels by virtue of its hypocholesterolemic effects [96-97]. Pectin's efficiency in this regard, as is the case for other soluble polysaccharides, reflects its ability to increase the viscosity of the intestinal content. Moreover, the characteristics of pectins in table olives in terms of physiological conditions are unaffected by depolymerisation. Thus, olives are an excellent source of soluble fibre. The prebiotic character of pectin also allows it to modify the intestinal flora, which results in a decrease in the number of intestinal bacteria whose metabolic activities produce carcinogenic components. Pectins have thus been recognised as a chemopreventive agent against colorectal cancer [98-100]. In fact, one recent study related the rhamnogalacturonan I domain to the anti-cancer properties of pectins through reduced regulation of cyclin B1 and the expression of cyclin-dependent kinase 1 [101].

Certain polysaccharides with a high content of galacturonic acid, type II arabinogalactans [102], arabinoxylans [103] and rhamnogalacturonans [104] have been found to have a gastro-protecting effect against potential ulcers. This is due to their ability to bond to the mucosal surface and act as a protecting layer, to diminish chlorhydric acid secretion and pepsin activity and to protect the mucus by increasing its secretion and/or synthesis of antioxidant species.

Significant progress has recently been made in terms of the function-structure relationship of pectin at the molecular level [105] with the possibility of designing pectins with specific properties or functions. For example, some pectins are able to inhibit the galectin-3 (Gal-3) protein, which binds carbohydrates and has a high affinity for beta-galactosidase. Gal-3 is encoded by the LGALS3 gene (14q21-q22) and is involved in many biological processes, including cell proliferation, differentiation and adhesion, angiogenesis, tumour progression, apoptosis, and metastasis. High Gal-3 expression has been observed in various neoplasias, such as thyroid, colorectal, breast and stomach cancer, and it is recognised as a tumour progression marker [106-108]. Thus, the pectin composition of table olives, which is rich in arabinogalactans and galactans, may dramatically alter the sensitivity of cells to cytotoxic medicines by inhibiting the antiapoptotic effect of Gal-3 [109]. Moreover, the ability of pectins to bind galectins and modify intercellular adhesion (cell-cell) and/or cell-matrix adhesion coincides with their inhibition of tumour extravasation and metastasis [110].

CONCLUSION

Despite the difficulty of assessing how a specific diet influences health or helps counteract certain diseases, due to variability in a large number of factors, the health benefits associated with the Mediterranean diet and virgin olive oil are generally accepted. The monounsaturated oleic acid and the antioxidant substances contained in virgin olive oil are also present in the edible fruit (table olives) from which the oil derives. Regular consumption of table olives provides an excellent source of fibre with high nutritional quality. At the epidemiological level, prior research has generally associated low dietary fibre consumption with a high incidence of digestive system chronic diseases. More recent studies have focussed on the prebiotic role of fibre in the fermentation in the colon and on its total benefit on human health (Figure 3). Due to certain aspects of the traditional processing of table olives, certain pectic polysaccharides present in table olives are able to inhibit angiogenesis in vitro by preventing Gal-3 from binding to its receptors. These data highlight the importance of fibre from table olives as a preventive agent against cancer.

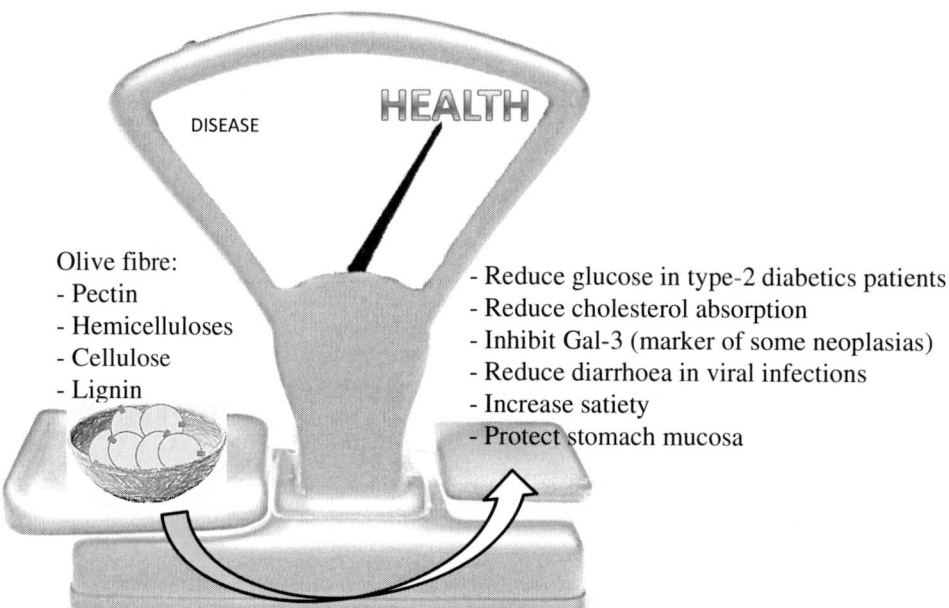

Figure 3. Summary of the beneficial effects of olive fibre.

The complex nature of carbohydrates in Spanish-style table olives should lead to the development of studies addressing the following questions: how their pectin fragments are absorbed, how they specifically bind to galectins and prevent them from promoting the adhesion of cancer cells to the walls of blood vessels and how they recognise angiogenic factors and glycoconjugate receptors involved in various types of cancer.

ACKNOWLEDGMENTS

We thank to the Spanish Ministry of Science and Innovation (MICINN) and the European Commission (EC) for financial support (AGL2011-29008 and Marie Curie PERG07-GA-2010-268413). S.L., S.J. and B.B. acknowledge financial support from the Spanish National Research Council (CSIC)/JAE-Doc Program, a contract cofounded by the European Social Fund (ESF).

REFERENCES

[1]   Lopez S, Bermudez B, Varela LM, Ortega A, Jaramillo S, Abia R, & Muriana FJG (2012) Olives and olive oil: diet and health impacts.*CAB Reviews: Perspectives in Agriculture, Veterinary Science, Nutrition and Natural Resources*, 7, 1-10.
[2]   Bosetti C, Gallus S, Trichopoulou A, Talamini R, Franceschi S, Negri E, & La Vecchia C (2003) Influence of the Mediterranean diet on the risk of cancers of the upper aerodigestive tract. *Cancer Epidemiology Biomarkers & Prevention*, 12, 1091-1094.

[3] Buckland G, Travier N, Cottet V, González CA, Luján-Barroso L, Agudo A, Trichopoulou A, Lagiou P, Trichopoulos D, Peeters PH, May A, Bueno-de-Mesquita HB, Bvan Duijnhoven FJ, Key TJ, Allen N, Khaw KT, Wareham N, Romieu I, McCormack V, Boutron-Ruault M, Clavel-Chapelon F, Panico S, Agnoli C, Palli D, Tumino R, Vineis P, Amiano P, Barricarte A, Rodríguez L, Sanchez MJ, Chirlaque MD, Kaaks R, Teucher B, Boeing H, Bergmann MM, Overvad K, Dahm CC, Tjønneland A, Olsen A, Manjer J, Wirfält E, Hallmans G, Johansson I, Lund E, Hjartåker A, Skeie G, Vergnaud AC, Norat T, Romaguera D, & Riboli E (2012) Adherence to the mediterranean diet and risk of breast cancer in the European prospective investigation into cancer and nutrition cohort study. *Inernational Journal Cancer*, 132, 2918-2927.

[4] Russo P, Del Bufalo A, & Cesario A (2012) Flavonoids acting on DNA topoisomerases: recent advances and future perspectives in cancer therapy. *Current Medicinal Chemistry*, 19, 5287-5293.

[5] Allouche Y, Warleta F, Campos M, Sánchez-Quesada C, Uceda M, Beltrán G, & Gaforio JJ (2011) Antioxidant, antiproliferative, and pro-apoptotic capacities of pentacyclic triterpenes found in the skin of olives on MCF-7 human breast cancer cells and their effects on DNA damage. *Journal Agricultural Food Chemistry*, 59, 121-130.

[6] http://www.sciencedaily.com/releases/2009/06/090624093353.htm

[7] http://www.sciencedaily.com/releases/2012/05/120529102252.htm

[8] Jaramillo S, Lopez S, Varela LM, Rodriguez-Arcos R, Jimenez A, Abia R, Guillen R, & Muriana FJG (2010) The flavonol isorhamnetin exhibits cytotoxic effects on human colon cancer cells. *Journal of Agricultural and Food Chemistry*, 58, 10869-10875.

[9] Vassallo N, & Scerri C (2013) Mediterranean diet and dementia of the Alzheimer type. *Current Aging Science*, 6, 150-162.

[10] Agencia para el Aceite de Oliva (AOO), 2011. http://aplicaciones.mapa.es/pwAgenciaAO/InfMercadosAceituna_BalanceAceituna.aao. Ministerio de Medio Ambiente, Medio Rural y Marino.

[11] Lopez S, Bermudez B, Ortega A, Varela LM, Pacheco YM, Villar J, Abia R, & Muriana FJG (2011) Effects of meals rich in either monounsaturated or saturated fat on lipid concentrations and on insulin secretion and action in subjects with high fasting triglyceride concentrations. *American Journal Clinical Nutritrion*, 93, 494-499.

[12] Lopez S, Bermudez B, Abia R, & Muriana FJG (2010) The influence of major dietary fatty acids on insulin secretion and action. *Current Opinion Lipdololgy*, 21, 15-20.

[13] Lopez S, Bermudez B, Pacheco YM, Villar J, Abia R, & Muriana FJG (2008) Distinctive postprandial modulation of beta cell function and insulin sensitivity by dietary fats: monounsaturated compared with saturated fatty acids. *American Journal Clinical Nutritrion*, 88, 638-644.

[14] Trefiletti G, Togna AR, Latina V, Marra C, Guiso M, & Togna GI (2011) 1-Phenyl-6,7-dihydroxy-isochroman suppresses lipopolysaccharide-induced pro-inflammatory mediator production in human monocytes. *British Journal Nutrition*, 106, 33-36.

[15] Rejano L, Montaño A, Casado FJ, Sanchez AH, & de Castro A (2010) Table Olives: Varieties and Variations Olives and Olive Oil en Health and Disease Prevention. Elsevier Science & Technology, USA, 5-15.

[16] Sanchez AH, Garcia P, & Rejano L (2006) Elaboration of table olives. *Grasas y Aceites*, 57, 86-94.

[17] Jaramillo S, de Castro A, & Rejano L (2011) Proceso tradicional de aderezo de aceitunas verdes de mesa: racionalización del cocido. *Grasas y Aceites*, 64, 375-382.

[18] Lopez-Lopez A, Montaño A, Garcia P, & Garrido A (2006) Fatty Acid Profile of Table Olives and Its Multivariate Characterization Using Unsupervised (PCA) and Supervised (DA) Chemometrics. *Journal Agricultural Food Chemistry*, 54, 6747-6753.

[19] Ortega A, Varela LM, Bermudez B, Lopez S, Abia R, & Muriana FJG (2012) Dietary fatty acids linking postprandial metabolic response and chronic diseases. *Food & Function*, 3, 22-27.

[20] Lopez-Lopez A, Montaño A, Ruiz-Mendez MV, & Garrido A (2008) Sterols, Fatty Alcohols, and Triterpenic Alcohols in Commercial Table Olives. *Journal American Oil Chemists' Society*. 85, 253-262.

[21] Bermudez B, Lopez S, Ortega A, Varela LM, Pacheco YM, Abia R, & Muriana FJG (2011) Oleic acid in olive oil: from a metabolic framework toward a clinical perspective. *Current Pharmaceutical Design*. 17, 831-843.

[22] Montaño A, Sanchez AH, & Castro A (2000) Changes in the amino acid composition of green olive brine due to fermentation by pure culture of bacteria. *Journal Food Science*, 65, 1022-1027.

[23] Lopez-Lopez A, Montaño A, Cortes A, & Garrido A (2008) Survey of vitamin B(6) content in commercial presentations of table olives. *Plant Foods Human Nutrition*, 63, 87-91.

[24] Gregori S, & Kelly ND (2011) Pantothenic Acid. *Alternative Medicine Review*, 16, 263-274.

[25] Damian DL (2010) Photoprotective effects of nicotinamide. *Photochemical Photobiological Sciences*, 9, 578-585.

[26] Lopez-Lopez A, Jimenez A, Garcia P, & Garrido A (2007) Multivariate analysis for the evaluation of fiber, sugars, and organic acids in commercial presentations of table olives. *Journal Agricultural Food Chemistry*, 55, 10803-10811.

[27] Lopez-Lopez A, Lopez R, Madrid F, & Garrido A (2008) Heavy metals and mineral elements not included on the nutritional labels in table olives. *Journal Agricultural Food Chemsitry*, 56, 9475-9483.

[28] Romero C, Brenes M, Yousfi K, Garcia P, Garcia A, & Garrido A (2004) Effect of cultivar and processing method on the contents of polyphenols in table olives. *Journal Agricultural Food Chemistry*, 52, 479-484.

[29] Brenes M, Rejano L, Garcia P, Sanchez AH, & Garrido A (1995) Biochemical changes in phenolic compounds during Spanish style green olive processing. *Journal Agricultural Food Chemistry*, 43, 2702-2706.

[30] Lavermicocca P, Valerio F, Lonigro SL, De Angelis M, Morelli L, Callegari ML, Rizzello CG, & Visconti A (2005) Study of adhesion and survival of lactobacilli and bifidobacteria on table olives with the aim of formulating a new probiotic food. *Applied Environmental Microbiology*, 71, 4233-4240.

[31] Argyri AA, Zoumpopoulou G, Karatzas KA, Tsakalidou E, Nychas GJ, Panagou EZ, & Tassou CC (2013) Selection of potential probiotic lactic acid bacteria from fermented olives by in vitro tests. *Food Microbiology*, 33, 282-291.

[32] Study on the promotion of consumption of olive oil and olives in the USA and Canada. A custom solution prepared by Datamonitor for the International Olive Council (2010)

– Pro.6/CO/9/09. Study on the promotion of consumption of olive oil and olives in the USA and Canada © Datamonitor Ltd.

[33] Tsantili E, Kafkaletou M, Roussos PA, & Christopoulos MV (2012) Phenolic compounds, maturation and quality in fresh green olives for table use during exposure at 20 degrees C after preharvest ReTain treatment. *Scientia Horticulturae*, 140, 26-32.

[34] Rjiba I, Dabbou S, Gazzah N, & Hammami M (2010) Effect of crossbreeding on the chemical composition and biological characteristics of Tunisian new olive progenies. *Chemistry Biodiversity*, 7, 649-655.

[35] Medina E, Garcia A, Romero C, De Castro A, & Brenes M (2009) Study of the anti lactic acid bacteria compounds in table olives. *International Journal Food Scence Technology*, 44, 1286-1291.

[36] Ciafardini G, Marsilio V, Lanza B, & Pozzi N (1994) Hydrolysis of Oleuropein by Lactobacillus plantarum Strains Associated with Olive Fermentation. *Applied Environmental Microbiology*, 60, 4142-4147.

[37] Pereira JA, Pereira APG, Ferreira ICFR, Valentao P, Andrade PB, Seabra R, Estevinho L, & Bento A (2006) Table Olives from Portugal: Phenolic Compounds, Antioxidant Potential, and Antimicrobial Activity. *Journal Agricultural Food Chemistry*, 54, 8425-8431.

[38] Boskou G, Salta FN, Chrysostomou S, Mylona A, Chiou A, & Andrikopoulos NK (2006) Antioxidant capacity and phenolic profile of table olives from the Greek market. *Food Chemistry*, 94, 558-564.

[39] Romero C, Medina E, Vargas J, Brenes M, & De Castro A (2007) In vitro activity of olive oil polyphenols against Helicobacter pylori. *Journal Agricultural Food Chemistry*, 55, 680-686.

[40] Bisignano G, Tomaino A, Lo Cascio R, Crisafi G, Uccella N, & Saija A (1999) On the in-vitro antimicrobial activity of oleuropein and hydroxytyrosol. *Journal Pharmacy Pharmacology*, 51, 971-974.

[41] Omar SH (2010) Oleuropein in Olive and its Pharmacological Effects. *Scientia Pharmaceutica*, 78, 133-154.

[42] Rafehi H, Smith AJ, Balcerczyk A, Ziemann M, Ooi J, Loveridge SJ, Baker EK, El-Osta A, & Karagiannis TC (2012) Investigation into the biological properties of the olive polyphenol, hydroxytyrosol: mechanistic insights by genome-wide mRNA-Seq analysis. *Genes Nutritrion*, 7, 343-355.

[43] Hagiwara K, Goto T, Araki M, Miyazaki H, & Hagiwara H (2011) Olive polyphenol hydroxytyrosol prevents bone loss. *European Journal Pharmacology*, 662, 78-84.

[44] Santiago-Mora R, Casado-Diaz A, De Castro MD, & Quesada-Gomez JM (2011) Oleuropein enhances osteoblastogenesis and inhibits adipogenesis: the effect on differentiation in stem cells derived from bone marrow. *Osteoporosis International*, 22, 675-684.

[45] Masella R, Di Benedetto R, Vari R, Filesi C, & Giovannini C (2005) Novel mechanisms of natural antioxidant compounds in biological systems: involvement of glutathione and glutathione-related enzymes. *Journal Nutritional Biochemistry*, 16, 577-586.

[46] Martin MA, Ramos S, Granado-Serrano AB, Rodriguez-Ramiro I, Trujillo M, Bravo L, & Goya L (2010) Hydroxytyrosol induces antioxidant/detoxificant enzymes and Nrf2 translocation via extracellular regulated kinases and phosphatidylinositol-3-

kinase/protein kinase B pathways in HepG2 cells. *Molecual Nutrition Food Research*, 54, 956-966.

[47] Romero C, Garcia A, Medina E, Ruiz-Mendez MV, De Castro A, & Brenes M (2010) Triterpenic acids in table olives. *Food Chemistry*, 118, 670-674.

[48] Guinda A, Rada M, Delgado T, Gutierrez-Adanez P, & Castellano JM (2010) Pentacyclic triterpenoids from olive fruit and leaf. *Journal Agricultural Food Chemistry*, 58, 9685-9691.

[49] Park SY, Nho CW, Kwon DY, Kang YH, Lee KW, & Park JH (2012) Maslinic acid inhibits the metastatic capacity of DU145 human prostate cancer cells: possible mediation via hypoxia-inducible factor-1α signalling. *British Journal Nutrition*, 13, 1-13.

[50] Reyes-Zurita FJ, Pachon-Peña G, Lizarraga D, Rufino-Palomares EE, Cascante M, & Lupiañez JA (2011) The natural triterpene maslinic acid induces apoptosis in HT29 colon cancer cells by a JNK-p53-dependent mechanism. *BMC Cancer*, 11, 154-166.

[51] Allouche Y, Beltran G, Gaforio JJ, Uceda M, & Mesa MD (2010) Antioxidant and antiatherogenic activities of pentacyclic triterpenic diols and acids. *Food Chemical Toxicology*, 48, 2885-2890.

[52] Petronelli A, Pannitteri G, & Testa U (2009) Triterpenoids as new promising anticancer drugs. *Anti-Cancer Drug*, 20, 880-892.

[53] Juan ME, Planas JM, Ruiz-Gutierrez V, Daniel H, & Wenzel U (2008) Antiproliferative and apoptosis-inducing effects of maslinic and oleanolic acids, two pentacyclic triterpenes from olives, on HT-29 colon cancer cells. *British Journal Nutrition*, 100, 36-43.

[54] Huang L, Guan T, Qian Y, Huang M, Tang X, Li Y, & Sun H (2011) Anti-inflammatory effects of maslinic acid, a natural triterpene, in cultured cortical astrocytes via suppression of nuclear factor-kappa B. *European Journal Pharmacology*, 672, 169-174.

[55] Dharmappa KK, Kumar RV, Nataraju A, Mohamed R, Shivaprasad HV, & Vishwanath BS (2009) Anti-inflammatory activity of Oleanolic acid by inhibition of secretory phospholipase A(2). *Planta Medica*, 75, 211-215.

[56] Tsai SJ, & Yin MC (2008) Antioxidative and anti-inflammatory protection of oleanolic acid and ursolic acid in PC12 cells. *Journal Food Science*, 73, H174-H178.

[57] Montilla MP, Agil A, Navarro MC, Jimenez MI, Garcia-Granados A, Parra A, & Cabo MM (2003) Antioxidant activity of maslinic acid, a triterpene derivative obtained from Olea europaea. *Planta Medica*, 69, 472-474.

[58] Sato H, Genet C, Strehle A, Thomas C, Lobstein A, Wagner A, Mioskowski C, Auwerx J, & Saladin R (2007) Anti-hyperglycemic activity of a TGR5 agonist isolated from Olea europaea. *Biochemical Biophysical Research Communications*, 362, 793-798.

[59] Horiuchi K, Shiota S, Hatano T, Yoshida T, Kuroda T, & Tsuchiya T (2007) Antimicrobial activity of oleanolic acid from Salvia officinalis and related compounds on vancomycin-resistant Enterococci (VRE). *Biologial Pharmaceutical Bulletin*, 30, 1147-1149.

[60] Regulation (CE) n.º 1924/2006 of the European Council.

[61] Peñalvo JL, Moreno-Franco B, Ribas-Barba L, & Serra-Majem L (2012) Determinants of dietary lignan intake in a representative sample of young Spaniards: association with

lower obesity prevalence among boys but not girls. *European Journal Clinical Nutrition*, 66, 795-798.

[62] Breneman CB, & Tucker L (2012) Dietary fibre consumption and insulin resistance - the role of body fat and physical activity. *British Journal Nutrition*, 7, 1-9.

[63] Tabung F, Steck SE, Su LJ, Mohler JL, Fontham ET, Bensen JT, Hebert JR, Zhang H, & Arab L (2012) Intake of grains and dietary fiber and prostate cancer aggressiveness by race. *Prostate Cancer*, 2012, 323296.

[64] Brownlee IA, Moore C, Chatfield M, Richardson DP, Ashby P, Kuznesof SA, Jebb SA, & Seal CJ (2010) Markers of cardiovascular risk are not changed by increased whole-grain intake: the WHOLEheart study, a randomised, controlled dietary intervention. *British Journal Nutrition*, 104, 125-34.

[65] Martinez R, Torres P, Meneses MA, Figueroa JG, Perez-Alvarez JA, & Viuda-Martos M (2012) Chemical, technological and in vitro antioxidant properties of mango, guava, pineapple and passion fruit dietary fibre concentrate. *Food Chemistry*, 135, 1520-1526.

[66] Fuentes-Alventosa JM, Rodriguez-Gutierrez G, Jaramillo S, Rodriguez-Arcos R, Fernandez-Bolaños J, Guillen R, & Jimenez A (2009) Effect of extraction method on chemical composition and functional characteristics of high dietary fibre powders obtained from asparagus by-products. *Food Chemistry*, 113, 665-671.

[67] Sapirstein HD, Wang M, & Beta T (2013) Effects of debranning on the distribution of pentosans and relationships to phenolic content and antioxidant activity of wheat pearling fractions. *LWT - Food Science Technology*, 50, 336-342.

[68] Andersson AA, Andersson R, Piironen V, Lampi AM, Nystrom L, Boros D, Fras A, Gebruers K, Courtin CM, Delcour JA, Rakszegi M, Bedo Z, Ward JL, Shewry PR, & Man P (2013) Contents of dietary fibre components and their relation to associated bioactive components in whole grain wheat samples from the HEALTHGRAIN diversity screen. *Food Chemistry*, 136, 1243-1248.

[69] Coimbra MA, Waldron KW, & Delgadillo I (1996) Selvendran RR. Effect of Processing on Cell Wall Polysaccharides of Green Table Olives. *Journal Agricultural Food Chemistry*, 44, 2394-2401.

[70] Jimenez A, Labavitch JM, & Heredia A (1994) Changes in the Cell Wall of Olive Fruit during Processing. *Journal Agricultural Food Chemistry*, 42, 1194-1199.

[71] Jimenez A, Rodriguez R, Fernandez-Caro I, Guillen R, Fernandez-Bolaños J, & Heredia A (2001) Olive Fruit Cell Wall: Degradation of Cellulosic and Hemicellulosic Polysaccharides during Ripening. *Journal Agricultural Food Chemistry*, 49, 2008-2013.

[72] Mafra I, Lanza B, Reis A, Marsilio V, Campestre C, De Angelis M, & Coimbra MA (2001) Effect of ripening on texture, microstructure and cell wall polysaccharide composition of olive fruit (Olea europaea). *Physiologia Plantarum*, 111, 439-447.

[73] Gunning AP, Bongaerts RJ, & Morris VJ (2009) Recognition of galactan components of pectin by galectin-3. *FASEB Journal*, 23, 415-424.

[74] Jimenez A, Rodriguez R, Felizon B, Fernandez-Caro I, Guillen R, Fernandez-Bolaños J, & Heredia A (2000) Cell Wall Polysaccahrides implied in green olive behaviour during the pitting process. *European Food Research Technology*, 211, 181-184.

[75] Morello JR, Vuorela S, Romero MP, Motilva MJ, & Heinonen M (2005) Antioxidant activity of olive pulp and olive oil phenolic compounds of the arbequina cultivar. *Journal Agricultural Food Chemistry*, 53, 2002-2008.

[76] Fini L, Hotchkiss E, Fogliano V, Graziani G, Romano M, De Vol EB, Qin H, Selgrad M, Boland CR, & Ricciardiello L (2008) Chemopreventive properties of pinoresinol-rich olive oil involve a selective activation of the ATM-p53 cascade in colon cancer cell lines. *Carcinogenesis*, 29, 139-146.

[77] Peterson J, Dwyer J, Adlercreutz H, Scalbert A, Jacques P, & McCullough ML (2010) Dietary lignans: physiology and potential for cardiovascular disease risk reduction. *Nutrition Reviews*, 68, 571-603.

[78] Hipsley EB (1953) Dietary "Fibre" and Pregnancy Toxaemia. *British Medical Journal*, 2, 420-422.

[79] Burkitt D, Walter ARP, & Painter NS (1972) Effect of dietary fibre on stools and transit time and its role in the causation of disease. *Lancet*, 2, 1408-1411.

[80] Trowell HC, Southgate DAT, Wolever TMS, Leeds AR, Gassull MA, & Jenkis DJA (1976) Dietary fiber redefined. *Lancet*, 1, 967.

[81] Roberfroid, M., (1993). Dietary fibre, inulin and oligofructose: a review comparing their physiological effects. *Critical Reviews Food Science Nutrition*, 33, 103-148.

[82] Flint HJ (2012) The impact of nutrition on the human microbiome. *Nutrition Reviews*, 70, 10-13.

[83] Ulven T (2012) Short-chain free fatty acid receptors FFA2/GPR43 and FFA3/GPR41 as new potential therapeutic targets. *Frontiers Endocrinology*, 3, 111.

[84] Codex Alimentarius Commission. Report of the 30th Session of the Codex Committee on Nutrition and Foods for Special Dietary Uses (CNFSDU and WHO/FAO) (2008) ALINORM 09/32/26. 27-54. p. 49 and appendix II.

[85] Topping D (2007) Cereal complex carbohydrates and their contribution to human health. *Journal Cereal Science*, 46, 220-229.

[86] Hu G, & Yu W (2013) Binding of cholesterol and bile acid to hemicelluloses from rice bran. *International Journal Food Science Nutrition*, 64, 461-466.

[87] Lu ZX, Walker KZ, Muir JG, Mascara T, & O'Dea K (2000) Arabinoxylan fiber, a byproduct of wheat flour processing, reduces the postprandial glucose response in normoglycemic subjects. *American Journal Clinical Nutrition*, 71, 1123.1128.

[88] Rothe M, & Blaut M (2012) Evolution of the gut microbiota and the influence of diet. *Beneficial Microbes*, 20, 1-7.

[89] Neyrinck AM, Van Hee VF, Piront N, De Backer F, Toussaint O, Cani PD, & Delzenne NM (2012) Wheat-derived arabinoxylan oligosaccharides with prebiotic effect increase satietogenic gut peptides and reduce metabolic endotoxemia in diet-induced obese mice. *Nutrition & Diabetes*, 2, e28.

[90] do Rosario MM, Kangussu-Marcolino MM, do Amaral AE, Noleto GR, Petkowicz CL (2011) Storage xyloglucans: potent macrophages activators. *Chemico-Biological Interactions*, 189, 127-133.

[91] Schepetkin IA, & Quinn MT (2006) Botanical polysaccharides: macrophage immunomodulation and therapeutic potential. *International Immunopharmacology*, 6, 317-333.

[92] Sanaka M, Yamamoto T, Anjiki H, Nagasawa K, & Kuyama Y (2007) Effects of agar and pectin on gastric emptying and post-prandial glycaemic profiles in healthy human volunteers. *Clinical & Expermiental Pharmacology & Physiology*, 34, 1151-1155.

[93] Baek SH, Lee JG, Park SY, Bae ON, Kim DH, & Park JH (2010) Pectic polysaccharides from Panax ginseng as the antirotavirus principals in ginseng. *Biomacromolecules*, 11, 2044-2052.

[94] Jenkins DJ, Kendall CW, Axelsen M, Augustin LS, & Vuksan V (2000) Viscous and nonviscous fibres, nonabsorbable and low glycaemic index carbohydrates, blood lipids and coronary heart disease. *Current Opinion Lipidology*, 11, 49-56.

[95] Schwartz SE, Levine RA, Weinstock RS, Petokas S, Mills CA, & Thomas FD (1988) Sustained pectin ingestion: effect on gastric emptying and glucose tolerance in non-insulin-dependent diabetic patients. *American Journal Clinical Nutrition*, 48, 1413-1417.

[96] Brouns F, Theuwissen E, Adam A, Bell M, Berger A, & Mensink RP (2012) Cholesterol-lowering properties of different pectin types in mildly hyper-cholesterolemic men and women. *European Journal Clinical Nutrition*, 66, 591-599.

[97] Theuwissen E, & Mensink RP (2008) Water-soluble dietary fibers and cardiovascular disease. *Physiology & Behavior*, 94, 285-292.

[98] Shinohara K, Ohashi Y, Kawasumi K, Terada A, & Fujisawa T (2010) Effect of apple intake on fecal microbiota and metabolites in humans. *Anaerobe*, 16, 510-515.

[99] Chen HL, Lin YM, Wang YC (2010) Comparative effects of cellulose and soluble fibers (pectin, konjac glucomannan, inulin) on fecal water toxicity toward Caco-2 cells, fecal bacteria enzymes, bile acid, and short-chain fatty acids. *Journal Agricultural Food Chemistry*, 58, 10277-10281.

[100] Waldecker M, Kautenburger T, Daumann H, Veeriah S, Will F, Dietrich H, Pool-Zobel BL, & Schrenk D (2008) Histone-deacetylase inhibition and butyrate formation: Fecal slurry incubations with apple pectin and apple juice extracts. *Nutrition*, 24, 366-374.

[101] Cheng H, Zhang Z, Leng J, Liu D, Hao M, Gao X, Tai G, & Zhou Y (2013) The inhibitory effects and mechanisms of rhamnogalacturonan I pectin from potato on HT-29 colon cancer cell proliferation and cell cycle progression. *International Journal Food Science Nutrition*, 64, 36-43.

[102] Cipriani TR, Mellinger CG, Bertolini MLC, Baggio CH, Freitas CS, Marques MCA, Gorin PAJ, Sassaki GL, & Iacomini M (2009) Gastroprotective effect of a type I arabinogalactan from soybean meal. *Food Chemistry*, 115, 687-690.

[103] Mellinger-Silva C, Simas-Tosin FF, Schiavini DN, Werner MFP, Baggio CH, Pereira IT, da Silva LM, Gorin PA, & Iacomini M (2011) Isolation of a gastroprotective arabinoxylan from sugarcane bagasse. *Bioresource Technology*, 102, 10524-10528.

[104] Nascimento AM, de Souza LM, Baggio CH, Werner MF, Maria-Ferreira D, da Silva LM, Sassaki GL, Gorin PA, Iacomini M, & Cipriani TR (2013) Gastroprotective effect and structure of a rhamnogalacturonan from Acmella oleracea. *Phytochemistry*, 85, 137-142.

[105] Morris VJ, Gromer A, Kirby AR, Bongaerts RJM, Gunning AP (2011) Using AFM and force spectroscopy to determine pectin structure and (bio) functionality. *Food Hydrocolloids*, 25, 230-237.

[106] Merlin J, Stechly L, de Beauce S, Monte D, Leteurtre E, van Seuningen I, Huet G, & Pigny P (2011) Galectin-3 regulates MUC1 and EGFR cellular distribution and EGFR downstream pathways in pancreatic cancer cells. *Oncogene*, 30, 2514-2525.

[107] Johnson KD, Glinskii OV, Mossine VV, Turk JR, Mawhinney TP, Anthony DC, Henry CJ, Huxley VH, Glinsky GV, Pienta KJ, Raz A, & Glinsky VV (2007) Galectin-3 as a

Potential Therapeutic Target in Tumors Arising from Malignant Endothelial. *Neoplasia*, 9, 662-670.

[108] Takenaka Y, Fukumori T, & Raz A (2004) Galectin-3 and metastasis. *Glycoconjugate Journal*, 19, 543-549.

[109] Glinsky VV, & Raz A (2009) Modified citrus pectin anti-metastatic properties: one bullet, multiple targets. *Carbohydrate Research*, 344, 1788-1791.

[110] Sathisha UV, Jayaram S, Harish Nayaka MA, & Dharmesh SM (2007) Inhibition of galectin-3 mediated cellular interactions by pectic polysaccharides from dietary sources. *Glycoconjugate Journal*, 24, 497-507.

In: Virgin Olive Oil
Editor: Antonella De Leonardis

ISBN: 978-1-63117-656-2
© 2014 Nova Science Publishers, Inc.

Chapter 16

FROM WASTES TO ADDED VALUE BY-PRODUCTS: AN OVERVIEW ON CHEMICAL COMPOSITION AND HEALTHY PROPERTIES OF BIOACTIVE COMPOUNDS OF OLIVE OIL CHAIN BY-PRODUCTS

Ana María Gómez-Caravaca[1,2,*], Vito Verardo[3], Alessandra Bendini[3,4] and Tullia Gallina Toschi[3,4]

[1]Department of Analytical Chemistry, University of Granada,
C/Fuentenueva s/n, Granada, Spain
[2]Research and Development of Functional Food Centre (CIDAF),
PTS Granada, Avda. Del Conocimiento s/n., Edificio Bioregión,
Granada, Spain
[3]Interdepartmental Centre of Agri-food Industrial Research
(CIRI Agroalimentare), Alma Mater Studiorum, University of Bologna,
Cesena (FC), Italy
[4]Department of Agricultural and Food Sciences, Alma Mater Studiorum,
University of Bologna, Cesena (FC), Italy

ABSTRACT

The production of olive oil originates a huge amount of residues like aqueous waste (50%) called olive mill wastewater and olive leaves that represent up to 10% of the weight of olive oil extraction and olive pomace. Moreover, other wastes were generated during the filtration processes and during the storage time when suspended solids tend to migrate at the bottom of the tanks originating sediments. However, it has been demonstrated that these wastes are rich in bioactive compounds and, thus, they can be considered as useful by-products for the recovering of these compounds to further uses.

The olive oil mill wastewater is a problematic and polluting effluent which may degrade the soil and water quality, with critical negative impacts on ecosystems functions

---

[*] E-mail: anagomez@ugr.es.

and services provided. Nevertheless, literature highlights its high concentration of antioxidants, especially phenolic compounds. However, the qualitative phenolic profiles of olive oil mill wastewater differ depending on the technological process of production of olive oil and the time of storage. Because of that, it is very important to evaluate the technological processes and the shelf life of olive oil mill wastewater to obtain a good source of bioactive compounds. In fact, the biological activities of phenolic compounds from olive mill wastewater have been studied, and they show a spectrum of highly interesting bioactivities.

Olive leaves also represent a significant by-product of the cultivation and harvesting of olive. Historically, olive leaf has also been used as a folk remedy for combating fevers and other diseases, such as malaria. Several pharmacological reports have shown that olive leaf extracts have the property to lower blood pressure, increase blood flow in the coronary arteries, and exhibit a wide antiviral activity including anti-HIV activity, or have been cited for their antitumor activity, particularly against different types of breast cancer.

Olive pomace can reach up to 30% of olive oil manufacturing, depending on the milling process which, after oil extraction, is generally distributed by means of controlled spreading on agricultural soil. However, a large quantity of olive mill solid residue remains without actual application because only small amounts are used as natural fertilizers, combustible biomass and additives in animal feeding and activated carbon. But theses residues are a valuable starting material for the production of phenol extracts that could be used in the industries. Because of that, research for valorization of olive pomaces has so far been mainly focused on panels or target known compounds considered of interest.

INTRODUCTION

The olive is one of oldest known cultivated trees in the world; in fact, some studies indicate that the earliest widespread use of olives was in the Early Bronze Age [1]. Its origins are in the region corresponding to ancient Persia and Mesopotamia, and the expansion of the Mediterranean civilizations introduced olive tree cultivation in the countries of the Mediterranean basin [2, 3].

Today, more than 98% of the world's olive oil production comes from the Mediterranean countries (especially Spain, Italy and Greece) and is around 2.5 metric tons/year. However, olive oil production has increased worldwide around 40% in the last decade. [4] This high increment can be explained by the success of the Mediterranean diet, which is linked to lower incidences of atherosclerosis, certain cancers, and cardiovascular and neurodegenerative diseases [5, 6, 7]. Nevertheless, the olive oil industry also produces huge amounts and varieties of wastes that suppose an enormous problem (Figure 1). The classic production of olive oil by the three-phase mill uses large volumes of water to aid the separation of olive oil and generates two by-products: the olive mill wastewater, vegetation water, or "alpechin", and a solid waste called pomace or "orujo". The modern two-phase processing technique, in which no water is added, generates a by-product that is a combination of liquid and solid waste called pomace or "alperujo". Other olive oil by-products generated by storage and filtration of virgin olive oil (VOO) are composed of solid and liquid storage wastes and cakes used for its filtration [8]. Besides, olive leaves are found as by-products of the farming of the olive grove, they accumulate during the pruning of olive trees and can be found in high amounts in the olive oil industries (10% of the total weight of the olives) [9].

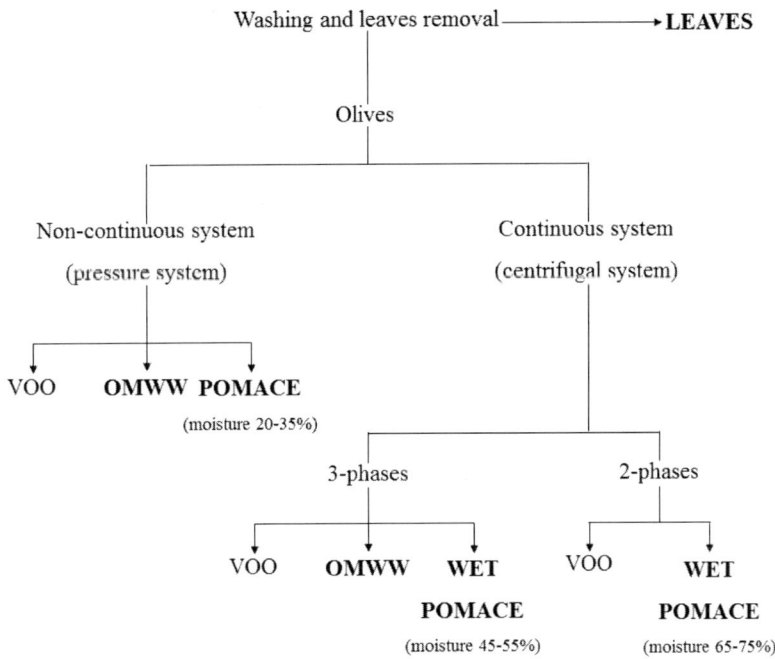

Figure 1. Olive oil mill by-products.

These by-products, specially the olive mill waste water, are detrimental to the environment due to their negative effects on soil microbial populations [10], aquatic ecosystems [11] and air through phenol and sulphur dioxide emissions [12]. The principal contributors to this pollution are the phenolic compounds contained in these by-products due to their toxicity and antimicrobial activity. Because of that, it is of great interest to find a way to process these residues from the olive oil industry

However, the phenolic compounds have numerous beneficial characteristics; they present strong antioxidant activity and different studies have demonstrated their antihypertensive, anticarcinogenic, anti-inflammatory, hypoglycemic, antimicrobial, and hypocholesterolemic effects, and all these positive effects appearing to be at least partly related to an anti-oxidative action, related primarily to low-molecular-weight polyphenols [13, 14]. All the previously stated makes phenolics a source of interest for reutilizing by food, pharmaceutical and cosmetic industry and, at the same time, it will help to diminish the volume of olive oil industry by-products.

Because of that, in this chapter it will be presented an overview of the chemical composition and healthy properties of the bioactive compounds of olive by-products.

OLIVE LEAVES

Olive leaves are an important agricultural residue obtained after the pruning step of olive trees and in the olive industry because they arrive to the olive mills mixed with the olive fruits (about 10% of the total weight of the olives corresponds to olive leaves) [15]. Traditionally, olive leaves have been widely used in popular medicine and phytotherapy as a remedy for the treatment of fever and other diseases in European and Mediterranean countries [16]. They have been well-known for their beneficial effects on metabolism when used as traditional herbal drug. Many of these health benefits are attributed to the antimicrobial properties of olive leaves [17], and because of that they have been useful to fight fever and overcome infections. All these properties have been endorsed to an important family of compounds named phenolic compounds, which have shown antioxidant, free-radical antagonism, antimicrobial activity and effects against several diseases [18].

Nowadays, the reutilization of olive leaves is still low. However, there is a great interest in olive leaves in both pharmaceutical and food industries as food additives and functional food.

Phenolic Composition

Olive leaves are rich in phenolic compounds, high-added value chemical compounds which make increment the interest in knowing the polyphenols composition of olive leaves. Besides, this fact increases the need to further investigate the potential uses of these leaves as source of antioxidants.

Many different extraction systems have been checked in order to isolate the phenolic fraction of olive leaves. The most used one is the conventional extraction, a solid-liquid extraction by maceration of the olive leaves in a solvent (usually mixtures of methanol/water or ethanol/water) [19, 20, 21].

However, other advanced extraction techniques have been used to improve the extraction recovery and to diminish the extraction time such as ultrasound-assisted extraction [22, 23, 24] superheated liquid extraction [25], microwave-assisted extraction [26, 27, 28], supercritical fluid extraction [29, 30, 31] and pressurized liquid extraction [18, 31].

Regarding the identification and quantification of phenolics contained in olive leaves different analytical approaches have been used to this end and it has been seen that olive leaves are especially rich in oleuropein, verbascoside, ligstroside, flavonoids including rutin, apigenin-7-glucoside and luteolin-7-glucoside. But also other minor phenolics can be present such as flavonoids aglycones and phenolic acids [32].

The analysis of phenolic compounds in olive leaves was first done by spectrophotometric techniques; the most used method is Folin-Ciocalteu for the determination of total phenolic compounds. However, the identification of single phenolic compounds present in olive leaves is only possible performing a previous separation of the compounds present in the samples. Because of that, different separation techniques have been applied to determine phenolics in olive leaf samples. The most used one has been the high performance liquid chromatography (HPLC) coupled to UV-Vis and mass spectrometry detectors. Nevertheless, some applications have been done by gas chromatography (GC) coupled to flame ionization and mass spectrometry detectors and capillary electrophoresis (CE) coupled to mass spectrometry. Also nuclear magnetic resonance (NMR) has been applied to identify new phenolic compounds structures.

Table 1. Determination of phenolic compounds in olive leaves

Description	Methodology	Phenolic compounds described	References
Antioxidative activities of *Olea europaea* related phenolic compounds	HPLC-DAD	Oleuropein, hydroxytyrosol and luteolin 7-glucoside	[33]
Antioxidant activity of phenolics extracted from *Olea europaea* L. leaves	HPLC-DAD	Hydroxytyrosol, tyrosol, catechin, caffeic acid, vanillic acid, vanillin, rutin, luteolin-7-glucoside, verbascoside, apigenin-7-glucoside, diosmetin-7-glucoside, oleuropein, luteolin, diosmetin	[34]
Analysis and quantification of flavanoidic compounds from Portuguese olive (*Olea europaea* L.) leaf cultivars.	HPLC-DAD and HPLC-DAD-MS(IT)	Luteolin 7,4'-O-diglucoside, luteolin 7-O-glucoside, rutin, apigenin 7-O-rutinoside, luteolin 4'-O-glucoside, luteolin, apigenin and diosmetin	[35]
Isolation and evaluation of antioxidants from leaves of a Tunisian cultivar olive tree	HPLC-DAD and HPLC-MS(Q-IT)	Oleuropein, hydroxytyrosol, luteolin 7-O-glucoside, luteolin 7-O-rutinoside, apigenin 7-O-glucoside, luteolin, apigenin, rutin	[36]
Phenolic compounds and antioxidant activity of *Olea europaea* L. fruits and leaves	Folin CiocalteuHPLC-DAD and HPLC-MS(IT)	Total phenolic compounds. Verbascoside, rutin, luteolin-7-glucoside, oleuropein and hydroxytyrosol	[37]
Characterization of new phenolic compounds from leaves of *Olea europaea* L. by high-resolution tandem mass spectrometry	HPLC-DAD and HPLC-MS(Q-TOF)	Methylated hydroxytyrosol glucoside, hydroxytyrosol glucoside, (2-methoxy)hydroxytyrosol hexose, saturated oleuropein, dimethyloleuropein	[38]
Characterization of olive-leaf phenolics by ESI-MS and evaluation of their antioxidant capacities by the CAT assay	HPLC-MS(Q-IT)	Oleuropein, oleuroside, ligstroside, verbascoside, luteolin-7-O-glucoside, quercetin, diosmetin aglycone	[39]
Composition and antioxidant activity of olive leaf extracts from Greek olive cultivars	Total phenolic compounds by Folin Ciocalteu HPLC-MS(IT)	Total phenolic compounds. Secologanoside, demethyloleuropein, oleuropein diglucoside, luteolin-7-O-glucoside, rutin, oleuropein, oleuroside, quercetin, ligstroside, verbascoside	[40]

Table 1. (Continued)

Description	Methodology	Phenolic compounds described	References
Phytochemical analysis and gastroprotective activity of an olive leaf extract	Total phenolic compounds by Folin Ciocalteu Total flavonoids HPLC-DAD	Oleuropein, caffeic acid, luteolin, luteolin-7-O-glucoside, apigenin-7-O-glucoside, quercetin, and chryseriol	[41]
Qualitative screening of phenolic compounds in olive leaf extracts by hyphenated liquid chromatography and preliminary evaluation of cytotoxic activity against human breast cancer cells	HPLC-MS(TOF) and HPLC-MS(IT)	2-(2-ethyl-3-hydroxy-6-propionylcyclohexyl)acetic acid glucoside, demethyloleuropein, 10-hydroxy-oleuropein, nüzhenide, ligstroside aglycon, ligstroside, oleuropein diglucoside, oleuropein, oleuropein aglycon, oleoside, secologanoside, elenolic acid glucoside, deacetoxy 10-hydroxyoleuropein aglycon, rutin, (+)-taxifolin, dihydroquercetin, apigenin-7-glucoside, chrysoeriol-7-O-glucoside, luteolin, quercetin, apigenin, luteolin glucoside, syringaresinol, hydroxytyrosol, hydroxytyrosol-glucoside, acteoside, isoacteoside	[42]
New possibilities for the valorization of olive oil by-products	Total phenolic compounds by Folin Ciocalteu UPLC-DAD-MS(QQQ)	Hydroxytyrosol glucoside, hydroxytyrosol, oleoside, tyrosol, coumaroyl derivative, elenolic acid glucoside, luteolin diglucoside, rutin, luteolin-rutinoside, 10-hydroxy-oleuropein, verbascoside, luteolin-7-glucoside, oleuropein-diglucoside, apigenin-rutinoside, hydroxytyrosol acetate, luteolin-glucoside, oleuropein-diglucoside, oleuropein, oleuropein isomer, oleuroside, oleuropein derivative, ligstroside, luteolin, apigenin, diosmetin	[18]
Oxidative stability of oils containing olive leaf extracts obtained by pressure, supercritical and solvent-extraction	Total phenolic compounds by Folin Ciocalteu HPLC-DAD-MS(IT)	Hydroxytyrosol, vanillic acid, hydroxytyrosol glycoside, vanillic hexoside acid, caffeic hexoside acid, vanillin, oleoside, chlorogenic acid, oleuropein aglycon, pinoresinol, caffeic acid, elenolic acid, p-coumaric acid, ferulic acid, verbacoside, ligstroside aglycon decarboxymethyl, luteolin-O-rutinoside,	

Description	Methodology	Phenolic compounds described	References
		acetoxypinoresinol, luteolin-7-glucoside, hesperitin-3-rutinoside, quercetin-3-O-galactoside, apigenin-7-rutinoside, oleuropein, oleuroside acid-10-carboxilic, apigenin-7-glucoside, oleuroside, ligstroside, luteolin-3'-7-diglucoside, luteolin-7-rutinoside, oleuropein diglucoside, oleuropein aglycon aldehyde, quercetin	[21]
Phenolic composition and in vitro antioxidant capacity of four commercial phytochemical products: Olive leaf extract (*Olea europaea* L.), lutein, sesamol and ellagic acid	Total phenolic compounds by Folin Ciocal-eu HPLC-DAD	Oleuropein, verbascoside, luteolin-7-O-glucoside, apigenin-7-O-glucoside, hydroxytyrosol and tyrosol	[43]
Kinetic and compositional study of phenolic extraction from olive leaves (var. Serrana) by using power ultrasound	Total phenolic compounds by Folin Ciocal-eu HPLC-DAD-MS(IT)	Caffeoyl quinic acid, apigenin-6,8-diglucoside, verbascoside, luteolin-7-O-rutinoside, luteolin-7-O-glucoside, luteolin-7-O-glucoside isomer, oleuropein glucoside, apigenin rutinoside, apigenin-7-O-glucoside, oleuropein, luteolin	[24]
Optimization of microwave-assisted extraction for the characterization of olive leaf phenolic compounds by using HPLC-ESI-TOF-MS/IT-MS2	HPLC-MS(TOF) and HPLC-MS(IT)	Quinic acid, secologanoside, vanillin, hydroxytyrosol, elenolic acid glucoside isomer, oleuropein aglycom derivative, luteolin diglucoside isomer, elenolic acid glucoside isomer, luteolin diglucoside isomer, 2-(2-ethyl-3-hydroxy-6-propionylcyclohexyl)Ac Ac glucoside, rutin, luteolin rutinoside isomer, 10-hydroxy-oleuropein, luteolin glucoside isomer, oleuropein glucoside, apigenin rutinoside, syringaresinol, diosmin isomer, luteolin rutinoside isomer, diosmin isomer, taxifolin, luteolin glucoside isomer, apigenin-7- glucoside, luteolin glucoside isomer, chryseriol-7-O-glucoside, 2''-methoxyoleuropein isomer, luteolin glucoside isomer, 2''-methoxyoleuropein isomer 2, oleuropein, oleuropein isomer, luteolin, quercetin, pinoresinol, acetoxypinoresinol, apigenin, diosmetin	[28]

Table 1. (Continued)

Description	Methodology	Phenolic compounds described	References
Use of advanced techniques for the extraction of phenolic compounds from Tunisian olive leaves: Phenolic composition and cytotoxicity against human breast cancer cells	HPLC–MS(TOF) and HPLC–MS(IT)	Quinic acid, hydroxytyrosol glucoside, secologanoside, vanillin, hydroxytyrosol, elenolic acid glucoside isomer 1, oleuropein aglycone derivative, luteolin diglucoside isomer 1, elenolic acid glucoside isomer 2, luteolin diglucoside isomer 2, 2-(2-Et-Hy-6-propionylcyclohexyl)acetic acid glucoside, rutin, luteolin-rutinoside isomer 1, 10-hydroxy-oleuropein, luteolin glucoside isomer 1, oleuropein diglucoside, apigenin rutinoside, syringaresinol, diosmin isomer 1, luteolin rutinoside isomer 2, diosmin isomer 2, taxifolin, luteolin glucoside isomer 2, apigenin-7-O-glucoside, luteolin glucoside isomer 3, chryseriol-7-O-glucoside, 2''-methoxyoleuropein isomer 1, luteolin glucoside isomer 4, 2''-methoxyoleuropein isomer 2, oleuropein, oleuropein isomer, luteolin, quercetin, pinoresinol, acetoxypinoresinol, apigenin, diosmetin	[31]
HPLC–ESI–QTOF–MS as a powerful analytical tool for characterising phenolic compounds in olive-leaf extracts	HPLC–MS(Q-TOF)	Quinic acid, oleoside/secologanoside, p-hydroxybenzoic acid, elenolic acid diglucoside, p-coumaric acid, vanillin, oleoside methyl ester, 7-epiloganin, elenolic acid glucoside, luteolin-7,4-O-diglucoside, hydroxyoleuropein, luteolin-7-O-rutinoside, rutin, verbascoside, hydroxytyrosol acetate, luteolin-7-O-glucoside, oleuropein diglucoside, oleuropein diglucoside (isomer 2), apigenin-7-O-rutinoside, apigenin-7-O-glucoside, luteolin-4-O-glucoside/luteolin-3-O-glucoside, oleuropein diglucoside (isomer 3), oleuropein, oleuropein isomer, oleuroside, 6'-O-[2,6-dimethyl-8-hydroxy-2-octenoyloxy]secologanoside, ligstroside, luteolin	[44]

Description	Methodology	Phenolic compounds described	References
Quantitative seasonal changes in the leaf phenolic content related to the alternate-bearing patterns of olive (*Olea europaea* L. cv. Gemlik)	HPLC-DAD	Chlorogenic acid, caffeic acid, 3-hydroxycinnamic acid, p-coumaric acid, scopolin, oleuropein	[45]
Variations in phenolic compounds and antiradical scavenging activity of *Olea europaea* leaves and fruits extracts collected in two different seasons	Total phenols and o-diphenols Total flavonoids Condensed tannins HPLC-DAD	Gallic acid, hydroxytyrosol, chlorogenic acid, protocatechuic acid, hydroxyphenylacetic acid, 4-hydroxybenzoic acid, catechin, oleuropein, p-coumaric acid, ferulic acid, rosmarinic acid, vanillic acid, m-coumaric acid, o-coumaric acid, phenylacetic acid, cinnamic acid, luteolin, apigenin	[46]
Olea europaea L. leaf extract and derivatives: antioxidant properties	GC-FID	Tyrosol, hydroxytyrosol, syringic acid, gallic acid, ferulic acid, oleuropein, oleuropein aglycon	[47]
Antibacterial activity and chemical constitutions of *Olea europaea* L. Leaf extracts	HPLC-MS(IT) and GC-MS	Oleuropein, tyrosol, 4-hydroxy benzoic acid, 3,4-dimethoxy benzoic acid, vanillic acid (3-methoxy,4-hydroxy benzoic acid), 3,4-dihydroxy benzoic acid, 4-hydroxy,3,5 dimethoxy benzoic acid (syringic acid), p-coumaric acid, ferulic acid, caffeic acid	[48]
Identification of phenolic compounds in olive leaves using CE-ESI-TOF-MS	CE-ESI-TOF-MS	Tyrosol, hydroxytyrosol acetate, hydroxytyrosol, ligstroside aglycone, oleuropein aglycon, oleuropein, apigenin 7-glucoside, demethyloleuropein, vanillic acid, o-coumaric acid, caffeic acid, rutin, verbascoside, oleoside, diosmetin, apigenin, luteolin	[49]
Isolation and characterization of a new hydroxytyrosol derivative from olive (*Olea europaea*) leaves	NMR	3,4-dihydroxyphenylethyl [(2,6-dimethoxy-3-ethylidene)-tetrahydropyran-4-yl]acetate (3,4-DHPEA-DETA)	[50]

The most representative methodologies and the phenolic compounds determined by each one have been summarized in the Table 1.

Healthy Properties Linked to Phenolic Compounds of Olive Leaves

Traditionally, olive leaves have been used as an herbal tea or as a powder because of the benefits reported in folk medicine. The bioactivity of the phenolic extracts of olive leaves and some individual phenolics such as oleuropein and hydroxytyrosol have been tested *in vitro* and *in vivo*. Different studies have showed that phenolic compounds of olive leaves can act against some of the main diseases that affect humans. Among these beneficial effects, olive leaf extracts are protective against cardiovascular diseases, certain types of cancer, osteoporosis and neurological diseases; they also have hypolipidemic, hypoglycemic and hepatoprotective effects.

Table 2. Health effects attributed to olive leaf phenolics

Bioactive effect	Disease	Responsible compounds/Extracts	Reference
Acute doxorubicin cardiotoxicity side effects	Cardiovascular	Oleuropein	[52]
Inhibition of platelet aggregation	Atherosclerosis	50% water-extracted olive leaf extract and 50% vegetable glycerine	[53]
Reduction of blood pressure	Hypertension	Olive leaf extract EFLA®943	[54]
Inhibition of platelet aggregation	Atherosclerosis	Olive leaf extract rich in hydroxytyrosol (22%)	[55]
Oxidative injury in vascular endothelial cells	Cardiovascular	Hydroxytyrosol	[56]
Induction of vascular smooth muscle cells apoptosis	Cardiovascular	Hydroxytyrosol	[57]
Reduction of intracellular reactive oxygen species levels in vascular endothelial cells	Cardiovascular	Hydroxytyrosol	[58]
Restoring of biological functions of endothelial progenitor cells	Cardiovascular	Oleuropein and oleacein	[59]
Hypoglycemic activity	Diabetes	Oleuropeoside	[60]
Inhibition of hyperglycemia and oxidative stress induced by diabetes	Diabetes	Oleuropein	[61]
Inhibitory effect on postprandial blood glucose	Diabetes	Oleanolic acid, luteolin-7-O-β glucoside and luteolin-4'-O- β glucoside	[62]

Bioactive effect	Disease	Responsible compounds/Extracts	Reference
Inhibition of advanced glycation end products	Diabetes	Aqueous and methanolic olive leaf extracts	[63]
Induction of growth inhibition and differentiation of human leukemia HL-60 cells	Cancer	70% ethanolic olive leaf extracts	[64]
Inhibitory effects on the metabolic status (cell viability) of three breast cancer models in vitro	Cancer	Methanolic olive leaf extracts	[42]
Cytotoxic activity against breast-cancer-cell line SKBR3	Cancer	Oleuropein and the flavones luteolin and apigenin	[66]
Antioxidant/prooxidant effects on the Cu+2-induced oxidation of human low-density lipoprotein (LDL)	Cholesterol	Oleuropein and hydroxytyrosol	[67]
Hypocholesterolemic and antioxidant activities	Cholesterol	Oleuropein, oleuropein aglycone and hydroxytyrosol-rich extracts	[68]
Extension of the cellular lifespan and delays the appearance of the senescence morphology of human embryonic fibroblasts	Anti-aging	Oleuropein	[69]
Inhibitory effects on xanthine oxidase (XO),	Gout	80% ethanolic *Olea europaea* leaf dry extract and single compounds: oleuropein, luteolin-7-O-β -d-glucoside, caffeic acid, luteolin, apigenin, tyrosol	[70]
Hepatoprotective effects against ethanol-induced damages	Liver disease	Oleuropein	[71]
Effect on antioxidant status in postmenopausal women compared to premenopausal women	Osteoporosis	Oleuropein	[72]
Inhibition of 6-hydroxydopamine-induced PC12 cell apoptosis	Parkinson	Oleuropein	[73]

As reported in literature, one of these potentially bioactive compounds is the secoiridoid oleuropein, which can constitute up to 6–9% of dry matter in the leaves. Another important phenolic is hydroxytyrosol; however other bioactive components found in olive leaves include related secoiridoids and flavonoids [17].

The interest in a further knowing of the effects on health of olive leaf extracts is very important, and many authors have put their efforts to do research about these topics. Table 2 shows some of the most interesting effects of olive leaf extracts in the different areas of medicine.

OLIVE MILL WASTE WATER

Olive oil mill wastewater (OMWW) has been the most pollutant and troublesome waste produced by olive mills in all Mediterranean countries where they are generated in huge quantities in short periods of time [74].

It is very difficult to calculate the volume of OMWW produced worldwide, as this depends on many parameters, such as the olive variety, olive seed maturity, cultivation techniques, and geological-climatic conditions.

However, it is strongly dependent on the processing system: in pressure systems the volume produced varies from 40 to 60 L per 100 kg of olives, while in two-phase systems it is approximately 10 L per 100 kg of olives and in three-phase centrifugation systems it varies from 80 to 120 L per 100 kg of olives [75].

They are constituted by vegetable water of the fruit and the water used in different stages of oil extraction. They hold olive pulp, mucilage, pectin, oil, etc., suspended in a relatively stable emulsion [74]. The main characteristic of OMWW is the presence of organic compounds such as organic acids, lipids, alcohols and polyphenols. Due to the phenolic content, OMWW is an exploitable source of natural bioactive compounds, which could be used in energy recovery, and construction industry, agriculture, food, cosmetic and pharmaceutical industry, as antioxidants, preservatives or even prophylactic agents against certain human diseases [75].

Table 3. Determination of phenolic compounds in OMWW

Description	Methodology	Phenolic compounds described	Reference
A comparison of OMWW from three different processes in Morocco	Spectrophotometric determination by Folin-Ciocalteu reagent	Total phenolic content	[77]
A new process for the management of OMWW and recovery of natural antioxidants	HPLC-UV	Hydroxytyrosol, tyrosol	[78]
Antioxidant activity of OMWW obtained from different thermal treatments	HPLC-UV	Hydroxytyrosol, tyrosol, vanillic acid, ferulic acid, p-cumaric acid, 3,4-DHPEA-EA, HPEA-EDA	[79]

Description	Methodology	Phenolic compounds described	Reference
Antioxidant and other biological activities of OMWW	HPLC-DAD	Hydroxytyrosol, tyrosol, elenolic acid, oleuropein derivatives, luteolin 7-glucoside, quercetin, cinnamic acid derivatives, hydroxytyrosol derivatives.	[80]
Qualitative phenolic profile (HPLC-DAD-MS) from OMWW at different states of storage and evaluation of hydrolysis process as a pretreatment to recover their antioxidants	HPLC-DAD-MS(TOF)	Vanillin, vanillic acid, hydroxy DEA, OxDEA, caffeic acid, DEA, syringaresinol, EA, oleuropein, luteolin, ac-pin, apigenin	[81]
Characterization and fractionation of phenolic compounds extracted from OMWW	HPLC-DAD-MS	Hydroxytyrosol, tyrosol, caffeic acid, vanillic acid, verbascoside, luteolin-7-glucoside, elenolic acid, dialdehydic form of decarboxymethyl oleuropein aglycon, ligstroside, luteolin	[82]
Determination of polyphenolic compounds in OMWW by gas chromatography–mass spectrometry	GC-MS	p-vanillin, tyrosol, 3-hydroxyphenylacetic acid, protocatetic aldehyde, p-hydroxybenzoic acid, vanillol, p-hydroxyphenylacetic acid, syringic aldehyde, di-hydroxyphenylethanol, p-hydroxyphenylpropionic acid, vanillic acid, protocatequic acid, 3,4,5-trimetoxibenzoic acid, syringic acid, p-coumaric acid, gallic acid, ferulic acid, esculetine, caffeic acid, sinapic acid, epicatechin	[83]
Extraction of antioxidants from OMWW and electro-coagulation of exhausted fraction to reduce its toxicity on anaerobic digestion	GC-MS	Tyrosol, hydroxytyrosol, protocatechuic acid, 3,4-dihydroxymandelic acid, p-coumaric acid, homovanillic acid, ferulic acid, caffeic acid	[84]
Fractionation of OMWW by membrane separation techniques	HPLC-UV	Hydroxyl-tyrosol, protocatechuic acid, catechol, tyrosol, caffeic acid, p-cumaric acid	[85]
Partition of phenolic compounds during the virgin olive oil industrial extraction process	HPLC-DAD	3,4-DHPEA derivative, 3,4-DHPEA, p-HPEA, vanillic acid, homovanillic acid, vanillin	[86]

Table 3. (Continued)

Description	Methodology	Phenolic compounds described	Reference
Phenolic profile and antioxidant activities of OMWW	HPLC-UV	Gallic acid, hydroxytyrosol-4-β-glucoside, hydroxytyrosol, tyrosol, caffeic acid, *p*-coumaric acid, oleuropein aglycone	[87]
Qualitative and quantitative evolution of polyphenolic compounds during composting of an olive-mill waste–wheat straw mixture	HPLC-DAD	Gallic acid, hydroxytyrosol, 4-hydroxyphenylacetic acid, caffeic acid, syringic acid, 4-methylcatechol, *p*-coumaric acid, oleuropein	[88]
Recovery and concentration of polyphenols from OMWW by integrated membrane system	HPLC-UV	Hydroxytyrosol, protocatechin acid, tyrosol, caffeic acid, *p*-coumaric acid, oleuropein	[89]
Stability and antioxidant potential of purified OMWW extracts	HPLC-DAD-MS	Gallic acid, hydroxytyrosol glucoside, hydroxytyrosol, 3,4-dihydroxyphenyl acetic acid, DEDA, oleoside, tyrosol, 5-caffeoylquinic acid, caffeic acid, cinnamic acid derivatives, verbascoside, *p*-coumaric acid, rutin, isoquercitrin, luteolin 7-O-glucoside, elenolic acid, caffeic acid derivative, cinnamic acid derivative, Hy-DEDA, oleosidic derivative, luteolin	[90]
The fate of olive fruit phenols during commercial olive oil processing: Traditional press versus continuous two- and three-phase centrifuge	HPLC-DAD-FLD-MS	Hydroxytyrosol glucoside, hydroxytyrosol, tyrosol, verbascoside, verbascoside derivative, luteolin-7-O-glucoside, rutin, *p*-HPEA-DEDA, comselogoside	[91]
Ultrasonic extraction of phenols from OMWW: comparison with conventional methods	HPLC-DAD-FLD-MS	Hydroxytyrosol, tyrosol, chlorogenic acid, vanillin, demethylligstroside, β-hydroxy verbascoside, *p*-HPEA-DEDA, comselogoside	[92]
Verbascoside, isoverbascoside, and their derivatives recovered from OMWW as possible food antioxidants	HPLC-DAD-MS	Verbascoside residue, isoverbascoside residue, oxidized dimeric caffeic acid, oxidized verbascoside, β-hydroxyverbascoside diastereoisomer, β-hydroxyisoverbascoside	[93]

Description	Methodology	Phenolic compounds described	Reference
		diastereoisomer, verbascoside, oxidized isoverbascoside, isoverbascoside	

Phenolic Composition

As reported for olive leaves, also OMWW are rich in phenolic compounds. OMWW typically contains 98% of the total phenols in the olive fruit. Recovery of phenols from OMWW is a difficult analytical task for several reasons: phenols are reactive chemical species, vulnerable to oxidation, conjugation, hydrolysis, polymerization, and complexation [76]. Table 3 showed the phenolic compounds identified in OMWW.

Just as varietal, seasonal, geographical, and agricultural factors affect the phenol profile in olive fruit, so they will affect the profile in OMWW too. However, benzoic acid derivatives (4-hydroxybenzoic, protocatechuic, vanillic acids), hydroxycinnamic acid derivatives (ferulic, caffeic acids) and secoiridoid derivatives (particularly tyrosol and hydroxytyrosol) were dominant in OMWW. Particularly, during the olive oil extraction, hydroxytyrosol is produced as a result of hydrolysis of this compound by esterase action and it is present in high amounts in OMWW.

Due to the high phenolic content in OMWW, they can be isolated to produce phenolic extracts. To this aim, recent research studies employing physical, chemical, biological and combined technologies. About physical technologies, membrane technology offers several advantages (low energy consumption, no additive requirements, no phase change) over traditional techniques to recover phenolic compounds [85, 94].

The proposed system included some well-known membrane operations such as microfiltration (MF) and nanofiltration (NF), as well as others not yet investigated for this specific application, such as osmotic distillation (OD) and vacuum membrane distillation (VMD) [89]. Depending on the initial wastewater and its polyphenol content the concentration in the obtained solution ranges from 0.5-30.0 g L^{-1} polyphenols. [95].

A combined physical/chemical technology was also used by several authors that employed different sorbents (such as clayey diatomite and zeolitic volcanic tuffs, Azolla and granular activated carbon) or other resins (Amberlite XAD7, XAD16, IRA96 and Isolute ENV+) to isolate the phenolic compounds [96, 97, 98].

Mazzei and co-workers [99] demonstrate the catalytic performance of a biocatalytic membrane reactor using β-glucosidase immobilized in polysulfone capillary membranes to hydrolyze the oleuropein into aglycon. Finally, a hydroxytyrosol triacetyl derivative and new alkyl hydroxytyrosyl ethers were obtained by chemical treatments of OMWW [100, 101].

Healthy Properties Linked to Phenolic Compounds of OMWW

The biological activities of phenolic compounds from olive mill wastewater have been studied and shown interesting bioactivities.

Table 4. Healthy properties linked to phenolic compounds of OMWW

Bioactive effect	Disease	Responsible compounds/Extracts	Reference
Anti-apoptotic activity of hydroxytyrosol and hydroxytyrosyl laurate	Cancer	Hydroxytyrosol and hydroxytyrosyl laurate	[105]
Assessment of verbascoside absorption in human colonic tissues using the chamber model	Oxidative stress	Verbascoside	[106]
Biological activity of high molecular weight phenolics from OMWW	Oxidative stress	OMWW extracts	[107]
Chemical and cellular antioxidant activity of phytochemicals purified from OMWW	Oxidative stress	Hydroxytyrosol, tyrosol and dialdehydic form of decarboxymethyl elenolic acid	[108]
Hydroxytyrosol-rich OMWW extract protects brain cells in vitro and ex vivo	Oxidative stress	Hydroxytyrosol	[109]
Major phenolic compounds in olive oil modulate bone loss in an ovariectomy/inflammation experimental model	Osteopenia	Tyrosol and hydroxytyrosol	[110]
Modulation of the expression of the proinflammatory IL-8 gene in cystic fibrosis cells by extracts deriving from OMWW	Cystic fibrosis	OMWW extract	[111]
Olive phenol hydroxytyrosol prevents passive smoking–induced oxidative stress	Oxidative stress	Hydroxytyrosol	[112]
Olive phenolics increase glutathione levels in healthy volunteers	Oxidative stress	Hydroxytyrosol and oleuropein	[113]
The postprandial inflammatory response after ingestion of heated oils in obese persons is reduced by the presence of phenol compounds	Oxidative stress	Hydroxytyrosol, tyrosol, oleuropein aglycone, oleuropein aglycone dialdehyde form, luteolin, apigenin, p-coumaric acid, vanillic acid, ferulic acid	[114]
Verbascosides from OMWW: assessment of their bioaccessibility and intestinal uptake using an in vitro digestion/Caco-2 model system	Oxidative stress	Isoverbascoside, verbascoside	[115]
A thromboxane effect of a hydroxytyrosol-rich OMWW extract in patients with uncomplicated type I diabetes	Thrombotic and microthrombotic processes	Hydroxytyrosol	[116]
Decreased superoxide anion production in cultured human promonocyte cells (THP-1) due to polyphenol mixtures from OMWW processing	Oxidative stress	OMWW extract	[117]

The major phenolic compounds in OMWW are oleuropein, hydroxytyrosol and tyrosol. These phenolic compounds possess a high spectrum of biological activities, including antioxidant, anti-inflammatory, antibacterial, and antiviral functions [102].

Researchers [103, 104] have demonstrated *in vitro* and *in vivo* antimicrobial activity of OMWW against several phytopathogenic bacteria and fungi. Most of these works have associated the antimicrobial activity of OMWW with its content in phenolic compounds and in some cases with specific phenolic substances such as hydroxytyrosol, methylcatechol, and others.

Visioli and co-workers [80] noticed that OMWW extracts are able to inhibit human LDL oxidation (a process involved in the pathogenesis of atherosclerosis) and to scavenge superoxide anions and hypochlorous acid and, also, inhibited the production of leukotrienes by human neutrophils.

Table 4 reports the principal studies about the bioactive effects of OMWW against different diseases.

POMACE

Olive pomace is the solid by-product obtained in the oil mill when olive oil is mechanically extracted from olive fruits. In the triphasic and pressure process the resultant pomace is quite dry and is usually used for the extraction of olive pomace oil with hexane, which has a high added value [118]. However, the same use for the pomace obtained by the 2-phases decanter, is very difficult because of the high content of water and low content of oil [119]. The quantities of pomace depend by the olive oil processing; the amount expressed as kg/t of olives was 250-330, 450-550 and 800-850 for pressure system, 3-phases and 2-phases decanter, respectively [119].

In Spain, olive oil is commonly extracted using a two-phase centrifugation system, a process that generates a semisolid waste that is called two-phase pomace or "alperujo". It has been estimated that the generation of alperujo as olive processing by-product in Spain is approximately 4–6 million tonnes every year. However, alperujo presents a serious environmental problem for Mediterranean countries due to its highly polluting organic load that includes lipids, pectins, polyalcohols, sugars, tannins and phenolic compounds, which limit its biodegradation because of their high toxicity [120].

The exploitation of "alperujo" from an environmental point of view may be approached in a number of ways, such as composting, gasification, steam explosion treatment for obtaining hydroxytyrosol or the extraction of oils as presented above. The by-products generated are the stone, the fat-free solid (or exhausted olive cake) and the residual wastewater [121].

Phenolic Composition

The quantity of secoiridoids can be diminished by the malaxation process of the olives during olive oil production [118].

Olive pomace resulting from the two-phase olive oil extraction method is a cheap source of natural antioxidants, in concentrations up to 100 times higher than in olive oil, which results from the polar nature of both this residue and olive phenols, but also from the low-polar nature of oil [122]. The polyphenol extract from the waste water of olive oil's pomace significantly inhibited lipid oxidation in pre-cooked ground beef and pork. The antioxidant effect increased with the dose and was higher in beef than in pork. In comparison with commercial antioxidants made from tea or wine, the ranking of efficacy was: tea>olive > wine [123]. The phenolic extract from pomace was used to enrich refined oils and was noticed that the resistance to oxidation during the heating process is superior for sunflower oil enriched with natural phenolic antioxidants than for extravirgin olive oil [124].

Table 5 shows the phenolic compounds identified in olive pomace.

Table 5. Phenolic compounds identified in olive pomace

Description	Methodology	Phenolic compounds described	Reference
The concentration of oleocanthal in olive oil waste	HPLC-MS	Oleocanthal	[125]
A study of the precursors of the natural antioxidant phenol 3,4-dihydroxyphenylglyc ol in olive oil waste	HPLC-MS	3,4-Dihydroxyphenylglycol	[120]
Characterisation of phenolic extracts from olive pulp and olive pomace by electrospray mass spectrometry	HPLC-MS	Verbascoside, rutin, caffeoyl-quinic acid, luteolin-4-glucoside, 11-methyl-oleoside, hydroxytyrosol-1-β-glucoside, luteolin-7-rutinoside and oleoside, 6-β-glucopyranosyloleoside, 6-β-rhamnopyranosyl-oleoside, 10-hydroxy-oleuropein, oleuropein glucoside	[118]
Chemical characterization and properties of a polymeric phenolic fraction obtained from olive oil waste	HPLC-DAD	Hydroxytyrosol, tyrosol	[126]
Hydroxytyrosol 4-β-D-Glucoside, an important phenolic compound in olive fruits and derived products	HPLC-MS	Tyrosol, verbascoside, luteolin 7-O-glucoside, rutin, oleuropein, hydroxytyrosol 4-β-D-Glucoside, hydroxytyrosol	[127]

Description	Methodology	Phenolic compounds described	Reference
Identification of oleuropein oligomers in olive pulp and pomace	HPLC-MS	Oleuropein dimers, oleuropein trimers, oleuropein tetramers, oleuropein pentamers	[128]
Isolation and identification of phenolic glucosides from thermally treated olive oil byproducts	HPLC-MS, NMR	Verbascoside, oleuropeinic acid, tyrosol derivative, hydroxytyrosol derivative	[129]
Methods for preparing phenolic extracts from olive cake for potential application as food antioxidants	HPLC DAD UPLC-MS	Tyrosol, hydroxytyrosol, vanillin, p-coumaric acid, vanillic acid, caffeic acid, 3,4-DHPEA-AC, 3,4-DHPEA-EDA, 3,4-DHPEA-EA, methyl 3,4-DHPEA-EA, oleuropein derivative, oleuropein, elenolic acid, verbascoside, p-HPEA-EA, p-HPEA-EDA, ligstroside derivative, pinoresinol, acetoxypinoresinol, rutin, apigenin, luteolin, apigenin-7-glucoside, luteolin-7-glucoside	[130]
New phenolic compounds hydrothermally extracted from the olive oil byproduct alperujo and their antioxidative activities	HPLC-MS	3,4-dihydroxyphenylglycol, hydroxytyrosol, protocatechuic acid, elenoic acid derivative, vanillic acid, caffeic acid, elenoic acid derivative, hydroxytyrosol acetate.	[131]
Partition of phenolic compounds during the virgin olive oil industrial extraction process	HPLC-DAD	3,4-DHPEA derivate, 3,4-DHPEA, p-HPEA, demethyloleuropein, demethyl-ligstroside, 3,4-DHPEA-EDA, verbascoside, vanillic acid, homovanillic acid, vanillin, luteolin-7-O-G, rutin, apigenin-7-O-G, luteolin, apigenin	[86]
Phenolic extract obtained from steam-treated olive oil waste: characterization and antioxidant activity	HPLC-MS	3,4-Dihydroxyphenylglycol, hydroxytyrosol, tyrosol, hydroxytyrosol acetate, protocatechuic acid, caffeic acid, 4-hydroxybenzoic acid, p-coumaric acid, vanillic acid, oleuropein, oleuropein aglycone hemiacetal, oleuropein aglycone, ligstroside, comsegoloside, luteolin-7-O-glucoside, luteolin-7-	[132]

Table 5. (Continued)

Description	Methodology	Phenolic compounds described	Reference
		O-rutinoside, apigenin-7-O-rutinoside, verbascoside.	
Recovery of hydroxytyrosol rich extract from two-phase Chemlali olive pomace by chemical treatment	HPLC-UV	Hydroxytyrosol, 3,4-DHPA, tyrosol, p-hydroxybenzoic acid, ferulic acid, oleuropein aglycon, oleuropein	[133]
Tentative identification of phenolic compounds in olive pomace extracts using liquid chromatography–tandem mass spectrometry with a quadrupole–quadrupole-time-of-flight mass detector	HPLC-MS	Hydroxytyrosol, hydroxytyrosol glucoside, hydroxytyrosol diglucoside, hydroxytyrosol rhamnoside, tyrosol, tyrosol glucoside, loganin, loganin glucoside, loganin acid, loganin acid glucoside, 7-deoxyloganic acid, secologanic acid, secologanoside, secologanin, eleanolic acid, oleoside, oleoside glucoside, oleoside diglucoside, oleoside riboside, oleoside-11-methylester, oleoside dimethylester, oleuropein, 10-hydroxy-oleuropein, oleuropein aglycone, verbascoside, 3,4-DHPEA-EDA, p-HPEA-EDA, 3,4-DHPEA-EA, p-HPEA-EA, oleuropein derivatives, rutin, apigenin, luteolin, apigenin glucoside, luteolin glucoside, taxifolin, diosmetin, quercetin, pinoresinol, hydroxypinoresinol, acetoxypinoresinol, cinnamic acid, p-coumaric acid, caffeic acid, protocatechuic acid, vanillic acid, ferulic acid, gallic acid	[134]
Valorization of olive oil solid waste using high pressure–high temperature reactor	Spectrophotometric methods and HPLC-UV	Hydroxytyrosol, protocatechuic acid, tyrosol, vanillic acid, caffeic acid, syringic acid, vanillin, p-coumaric acid, oleuropein, apigenin	[135]

Healthy Properties Linked to Phenolic Compounds of Pomace

As previously reported, a variety of substances with proven antioxidant and radical scavenging activity, such as hydroxytyrosol (3,4-DHPEA), tyrosol (p-HPEA) and their secoiridoid derivatives (dialdheydic form of decarboxymethyl elenolic acid, 3,4-DHPEA-EDA or p-HPEA-EDA) as well as verbascoside, is also contained in olive pomace.

Moreover, the antifungal and antibacterial activity of phenolic compounds extracted from olive pomace against *Alternaria solani*, *Botrytis cinerea* and *Fusarium culmorum* was investigated [136, 137].

The data showed by Dal Bosco and co-workers [138, 144] noticed that the nutritional quality of rabbit meat can be improved by adding a 5% concentration of olive pomace to the standard diet. To achieve this goal, only olive pomaces of high quality in terms of pro-oxidant/antioxidant content should be used. Specifically, olive pomaces with a peroxide value less than 10 meq kg^{-1} and an ortho-diphenol concentration higher than 20 g kg^{-1} can guarantee a satisfactory meat oxidative stability.

Table 6 reports the principal studies about the bioactive effects of pomace phenolics against different diseases.

BY-PRODUCTS GENERATED DURING OLIVE OIL STORAGE AND ITS FILTRATION

Recently, Lozano-Sanchez et al. [145] evaluated also the waste generated during the filtration process of virgin olive oils, and different classes of hydrophilic phenolic compounds retained in filter aids have been identified, including phenolic acids, phenolic alcohols, secoiridoids, lignans, and flavones. In the Mediterranean area, olive oil is usually produced from September to February and stored by the mill companies in large tanks until the oil is filtered, bottled and then marketed.

During this storage time, suspended solids tend to migrate at the bottom of the tanks originating sediments. At this purpose, Lozano-Sánchez and co-authors [146] characterized by RRLC-ESI-TOF/IT-MS the phenol composition of the mix of solid and aqueous wastes generated during nine months of virgin olive oil (Hojiblanca variety olives) storage in a tank of an industrial plant.

Results evidenced that elenolic acid derivatives and phenolic alcohols, ranged respectively from 514 to 601 mg kg^{-1} and from 159 to 194 mg kg^{-1} being the dialdehydic form of decarboxymethyl elenolic acid and hydroxytyrosol the main phenolic components in solid waste extracts. Among the secoiridoids, oleuropein aglycone and its hydroxylated and decarboxymethyl derivatives were the most abundant compounds, but also significant amounts of (+)-acetoxypinoresinol, luteolin, and apigenin were detected in lignans and flavones fractions, respectively.

The concentrations of these three compounds in solid waste were higher than previously described in VOO from the same variety of olives [147].

Phenolic and other polar compounds present in aqueous waste extract were also determined and, among the polar molecules, quinic acid was the most abundant, whereas in the phenolic fraction hydroxytyrosol and tyrosol were the major compounds.

Table 6. Healthy effects attributed to pomace phenolics

Bioactive effect	Disease	Responsible compounds/Extracts	Reference
Development of a phenol-enriched olive oil with phenolic compounds from olive cake	Oxidative stress	Tyrosol, hydroxytyrosol, vanillin, p-coumaric acid, vanillic acid, caffeic acid, oleuropein, 3,4-DHPEA-AC, elenolic acid, p-HPEA-EDA, 3,4-DHPEA-EDA, ligstroside derivative, p-HPEA-EA, oleuropein derivative, 3,4-DHPEA-EA, ME 3,4-DHPEA-EA, pinoresinol, acetoxypinoresinol, apigenin, luteolin, apigenin-7-glucoside, luteolin-7-glucoside, rutin, verbascoside.	[138]
Antioxidant activity and biological evaluation of olive pomace extract	Oxidative stress	Olive pomace extract	[139]
Effects of olive polyphenols administration on nerve growth factor and brain-derived neurotrophic factor in the mouse brain	Brain cell development	Phenolea ® Active Complex	[140]
Effects of polyphenol extract from olive pomace on anoxia-induced endothelial dysfunction	Endothelial dysfunction	Oleuropein and tyrosol	[141]
Phenolic compounds in olive oil and olive pomace from Cilento (Campania, Italy) and their antioxidant activity	Oxidative stress	Gallic acid, hydroxytyrosol, tyrosol, caffeic acid, syringic acid, oleuropein, ligstroside aglycone, oleuropein aglycone, ferulic acid	[142]
Valorization of olive mill residues: Antioxidant and breast cancer antiproliferative activities of hydroxytyrosol-rich extracts derived from olive oil by-products	Cancer	Hydroxytyrosol-1-glucoside, hydroxytyrosol, tyrosol, oleuropein aglycone isomer, Verbascoside, oleuropein, de(carboxymethyl)oleuropein aglycone isomer in aldehyde form	[143]

At this regard, hydroxytyrosol in aqueous waste was significantly higher than in the solid waste as well as the amount of tyrosol was six times higher than in solid waste. On the other hand, the contents of luteolin and apigenin in aqueous waste were significantly lower than in solid waste and, among secoiridoid derivatives, only the decarboxymethylated form of oleuropein aglycone was detected in significant amount but its presence was less abundant than in the solid extract. Wastes generated during the storage of VOO contain mainly polar molecules of different molecular masses, which are related to the phenolic composition of oil that change during storage process. In fact, complex phenolic compounds undergo modifications due to hydrolysis, increasing in decarboxymethylated secoiridoids and phenylethylalcohols and, as suggested by some authors, also oxidation products of phenols appear [148].

The combination of hydrolysis and oxidation could justify the presence of elenolic acid derivatives in olive oil wastes after storage. Oxidation of secoiridoid aglycones involves the acidic portion and not the aromatic alcoholic moiety, which is characterized by the conversion of the aldehydic group of elenolic acid to the carboxylic group making the molecules more polar thus more water soluble.

In a very recent work [149], authors studied the changes in the phenolic patterns of a VOO and in its relative by-products during storage for ten months in tanks. Given that storage by-products become enriched in polar phenols over time, authors proposed that the relationship between the status of molecules in the oil and wastes could be useful for understanding the changes of these bioactive compounds over time. Thanks to the analysis by high-performance liquid chromatography (HPLC) coupled to electrospray time-of-flight mass spectrometry (TOF-MS) of the phenolic fraction once a month, they suggested degradation pathways for the complex phenols and their oxidized and hydrolyzed derivatives. In fact, when the correlation for the pair oleuropein aglycone or decarboxymethyl oleuropein aglycon and their hydrolyzed derivatives was evaluated in VOO, the determination coefficients proved higher than 0.950.

On the other hand, by-products at all stages of the storage exhibited higher contents in hydroxytyrosol than olive oil because its appearance rate constant (Ka) was higher in the by-products than in the olive oil. The obtained results can help olive-oil producers to establish the optimal time for collecting by-product as an alternative source of polar phenols with bioactive properties.

CONCLUSION

The bioactivities exerted by olive polyphenols highlighted once again that the olive tree by-products, including both olive leaves and OMWW, can be considered as a natural source of antioxidants able to perform advantageous effects to prevent oxidative stress related disorders. Conventional processes for olive oil production originate large amounts of wastes (leaves, oil pomace, olive mill waste water, sediments) and, at now, very few industrial plants invest in purification and utilization of those by-products, preventing also environmental pollution.

Dealing with by-products of olive oil processing is a very important issue for the modern agriculture thus more attention in potential "green technologies" (bioconversions, molecular distillation, use of membranes for osmosis and ultrafiltration, application of separation, concentration and purification steps) in olive oil industry is needed. Food industries are continually involved in research and developing of new products that can be marketed as "functional" thanks to the presence of specific bioactive phytochemical compounds. At this regard, processes aiming the recovery and purification of antioxidant polyphenols from olive by-products are a promising alternative for the valorization of such agro-industrial wastes. In this way, several technologies/methodologies have been carried out to isolate the phenolic bioactive compounds. However, further studies will be necessary to optimize the phenolic recovery in a sustainable context.

REFERENCES

[1]   Liphschitz N., Gophna R., Hartmana M., and Biger G. (1991) The beginning of olive (*Olea europaea*) cultivation in the old world: A reassessment. *Journal of Archaeological Science*, 18, 441-453.

[2]   Kiritsakis A. K. (1998) Olive oil. From the Tree to the Table, 2nd edn. Trumbull, Conn.: Food and Nutrition Press.

[3]   Di Giovacchino L. (2000) In: Hardwood J., Aparicio R. (eds) Technological aspects. Handbook of olive oil. Analysis and properties. *Aspen Publications*, pp 17–59.

[4]   Dermeche S., Nadour M., Larroche C., Moulti-Mati F., and Michaud P. (2013) Olive mill wastes: Biochemical characterizations and valorization strategies. *Process Biochemistry*, 48, 1532–1552.

[5]   Owen R. W., Giacosa A., Hull W. E., Haubner R., Spiegelhalder B., and Bartsch H. (2000) The antioxidant/anticancer potential of phenolic compounds isolated from olive oil. *European Journal of Cancer*, 36, 1235-1247.

[6]   López S., Pacheco Y. M., Bermudez B., Abia R., and Muriana F. J. G. (2004) Olive oil and cancer. *Grasas y Aceites*, 55, 33–41.

[7]   Perona J. S., Cabello-Moruno R., and Ruiz-Gutierrez V. (2006) The role of virgin olive oil components in the modulation of endothelial function. *The Journal of Nutritional Biochemistry*, 17, 429-445.

[8]   Fernández-Bolaños J., Rodríguez G., Rodríguez R., Guillén R., and Jiménez A. (2006) A potential use of olive by-products. Extraction of interesting organic compounds from olive oil waste. *Grasas y Aceites*, 57(1), 95-106.

[9]   Tabera J., Guinda A., Ruiz-Rodriguez A., Señorans J. F., Ibañez E., Albi T., and Reglero G. (2004) Counter-current supercritical fluid extraction and fractionation of high-added-value compounds from a hexane extract of olive leaves. *Journal of Agricultural and Food Chemistry*, 52, 4774-4779.

[10]   Paredes M. J., Moreno E., Ramos-Cormenzana A., and Martinez J. (1987) Characteristics of soil after pollution with wastewaters from olive oil extraction plants. *Chemosphere*, 16(7), 1557-1564.

[11]   Della Greca M., Monaco P., Pinto G., Pollio A., Previtera L., and Temussi F. (2001) Phytotoxicity of low-molecular-weight phenols from olive mill waste waters. *Bulletin of Environmental Contamination and Toxicology*, 67, 352-359.

[12] Rana G., Rinaldi M., and Introna M. (2003) Volatilisation of substances after spreading olive oil waste water on the soil in a Mediterranean environment. *Agriculture, Ecosystems and Environment*, 96, 49-58.

[13] Martín-Peláez S., Covas M. I., Fitó M., Kušar A., and Pravst I. (2013) Health effects of olive oil polyphenols: Recent advances and possibilities for the use of health claims. *Molecular Nutrition and Food Research*, 57, 760–771.

[14] Raederstorff D. (2009) Antioxidant activity of olive polyphenols in humans: a review. *International Journal for Vitamin and Nutrition Research*, 79, 152–165.

[15] Taamalli A., Arráez-Román D., Zarrouk M., Valverde J., Segura-Carretero A., and Fernández-Gutiérrez A. (2012) The occurrence and bioactivity of polyphenols in tunisian olive products and by-products: A review. *Journal of Food Science*, 77, R83-R92.

[16] Ghanbari R., Anwar F., Alkharfy K. M., Gilani A-H., and Saari N (2012) Valuable nutrients and functional bioactives in different parts of olive (*Olea europaea* L.)-A review. *International Journal of Molecular Sciences*, 13, 3291-3340.

[17] El S. N., and Karakaya S. (2009) Olive tree (*Olea europaea*) leaves: potential beneficial effects on human health. *Nutrition Reviews,* 67(11), 632–638.

[18] Herrero M., Temirzoda T. N., Segura-Carretero A., Quirantes R., Plaza M, and Ibañez E. (2011) New possibilities for the valorization of olive oil by-products. *Journal of Chromatography A*, 1218, 7511-7520.

[19] De Nino A., Lombardo N., Perri E., Procopio A., Raffaelli A., and Sindona G. (1997) Direct identification of phenolic glucosides from olive leaf extracts by atmospheric pressure ionization tandem mass spectrometry. *Journal of Mass Spectrometry,* 32, 533-541.

[20] Lalas S., Athanasiadis V., Gortzi O., Bounitsi M., Giovanoudis I., Tsaknis J., and Bogiatzis F. (2011) Enrichment of table olives with polyphenols extracted from olive leaves. *Food Chemistry,* 127(4), 1521-1525.

[21] Jiménez P., Masson L., Barriga A., Chávez J., and Robert P. (2011) Oxidative stability of oils containing olive leaf extracts obtained by pressure, supercritical and solvent-extraction. *European Journal of Lipid Science and Technology*, 113, 497–505.

[22] Japón-Luján R., Luque-Rodríguez J. M., and Luque de Castro M. D. (2006) Dynamic ultrasound-assisted extraction of oleuropein and related biophenols from olive leaves. *Journal of Chromatography A*, 1108, 76–82.

[23] Şahin S., and Şamlı R. (2013) Optimization of olive leaf extract obtained by ultrasound-assisted extraction with response surface methodology. *Ultrasonics Sonochemistry,* 20, 595–602.

[24] Ahmad-Qasem M. H., Cánovas J., Barrajón-Catalán E., Micol V., Cárcel J. A., and García-Pérez J. V. (2013) Kinetic and compositional study of phenolic extraction from olive leaves (var. Serrana) by using power ultrasound. *Innovative Food Science and Emerging Technologies*, 17, 120–129.

[25] Japón-Luján R., and Luque de Castro M. D. (2006) Superheated liquid extraction of oleuropein and related biophenols from olive leaves. *Journal of Chromatography A*, 1136, 185–191.

[26] Japón-Luján R., Luque-Rodríguez J. M., and Luque de Castro M. D. (2006) Multivariate optimisation of the microwave-assisted extraction of oleuropein and

related biophenols from olive leaves. *Analytical and Bioanalytical Chemistry*, 385, 753-759.

[27] Rafiee Z., Jafari S. M., Alami M., and Khomeiri M. (2011) Microwave-assisted extraction of phenolic compounds from olive leaves; a comparison with maceration. *The Journal of Animal and Plant Sciences*, 21(4), 738-745.

[28] Taamalli A., Arráez-Román D., Ibañez E., Zarrouk M., Segura-Carretero A., and Fernández-Gutiérrez (2012) A optimization of microwave-assisted extraction for the characterization of olive leaf phenolic compounds by using HPLC-ESI-TOF-MS/IT-MS^2. *Journal of Agricultural and Food Chemistry*, 60, 791−798.

[29] Xynos N., Papaefstathiou G., Psychis M., Argyropoulou A., Aligiannis N., and Skaltsounis A-L. (2012) Development of a green extraction procedure with super/subcritical fluids to produce extracts enriched in oleuropein from olive leaves. *Journal of Supercritical Fluids*, 67, 89-93.

[30] Şahin S., and Bilgin M. (2012) Study on oleuropein extraction from olive tree (*Olea europaea*) leaves by means of SFE: comparison of water and ethanol as co-solvent. *Separation Science and Technology*, 47, 2391-2398.

[31] Taamalli A., Arráez-Román D., Barrajón-Catalán E., Ruiz-Torres V., Pérez-Sánchez A., Herrero M., Ibañez E., Micol V., Zarrouk M., Segura-Carretero A., and Fernández-Gutiérrez A. (2012) Use of advanced techniques for the extraction of phenolic compounds from Tunisian olive leaves: Phenolic composition and cytotoxicity against human breast cancer cells. *Food and Chemical Toxicology*, 50, 1817−1825.

[32] Tsimidou M. Z., and Papoti V. T. (2010) Bioactive ingredients in olive leaves. In: Olives and olive oil in health and disease prevention. Preedy V. R., and Watson R. R. *Academic Press*, pp 349-356.

[33] Le Tutour B., and Guedon D. (1992) Antioxidative activities of *Olea europaea* related phenolic compounds. *Phytochemistry*, 31(4), 1173-1178.

[34] Benavente-García O., Castillo J., Lorente J., Ortuño A., and Del Rio J. A. (2000) Antioxidant activity of phenolics extracted from *Olea europaea* L. leaves. *Food Chemistry*, 68, 457-462.

[35] Meirinhos J., Silva B. M., Valentão P., Seabra R. M., Pereira J. A., Dias A., Andrade P. B., and Ferreres F. (2005) Analysis and quantification of flavonoidic compounds from Portuguese olive (*Olea europaea* L.) leaf cultivars. *Natural Product Research*, 19, 189-195.

[36] Bouaziz M., and Sayadi S. (2005) Isolation and evaluation of antioxidants from leaves of a Tunisian cultivar olive tree. *European Journal Lipid Science Technology*, 107, 497−504.

[37] Silva S., Gomes L., Leitão F., Coelho A. V., and Vilas Boas L. (2006) Phenolic compounds and antioxidant activity of *Olea europaea* L. fruits and leaves. *Food Science and Technology International*, 12, 385-395.

[38] Di Donna L., Mazzotti F., Salerno R., Tagarelli A., Taverna D., and Sindona G. (2007) Characterization of new phenolic compounds from leaves of *Olea europaea* L. by high-resolution tandem mass spectrometry. *Rapid Communications in Mass Spectrometry*, 21, 3653-3657.

[39] Laguerre M., López Giraldo L. J., Piombo G., Figueroa-Espinoza M. C., Pina M., Benaissa M., Combe A., Rossignol Castera A., Lecomte J., and Villeneuve P. (2009) Characterization of olive-leaf phenolics by ESI-MS and evaluation of their antioxidant

capacities by the CAT assay. *Journal of the American Oil Chemists' Society*, 86, 1215-1225.

[40] Kiritsakis K., Kontominas M. G., Kontogiorgis C., Hadjipavlou-Litina D. Moustakas A., and Kiritsakis A. (2010) Composition and antioxidant activity of olive leaf extracts from greek olive cultivars. *Journal of the American Oil Chemists' Society*, 87, 369-376.

[41] Dekanski D., Janiçijevic-Hudomal S., Tadic V., Markovic G., Arsic I., and Mitrovic D. M. (2009) Phytochemical analysis and gastroprotective activity of an olive leaf extract. *Journal of the Serbian Chemical Society*, 74 (4), 367-377.

[42] Fu S., Arráez-Roman D., Segura-Carretero A., Menéndez J. A., Menéndez-Gutiérrez M. P., Micol V., and Fernández-Gutiérrez A. (2010) Qualitative screening of phenolic compounds in olive leaf extracts by hyphenated liquid chromatography and preliminary evaluation of cytotoxic activity against human breast cancer cells. *Analytical and Bioanalytical Chemistry*, 397, 643-654.

[43] Hayes J. E., Allen P., Brunton N., O'Grady M. N., and Kerry J. P. (2011) Phenolic composition and in vitro antioxidant capacity of four commercial phytochemical products: Olive leaf extract (*Olea europaea* L.), lutein, sesamol and ellagic acid. *Food Chemistry*, 126, 948-955.

[44] Quirantes-Piné R., Lozano-Sánchez J., Herrero M., Ibáñez E., Segura-Carretero A., and Fernández-Gutiérrez A. (2013) HPLC–ESI–QTOF–MS as a powerful analytical tool for characterising phenolic compounds in olive-leaf extracts. *Phytochemical Analysis*, 24, 213-223.

[45] Mert C., Barut E., and Ipek A. (2013) Quantitative seasonal changes in the leaf phenolic content related to the alternate-bearing patterns of olive (*Olea europaea* L. cv. Gemlik). *Journal of Agricultural Science and Technology*, 15, 995-1006.

[46] Brahmi F., Mechri B., Dhibi M., and Hammami M. (2013) Variations in phenolic compounds and antiradical scavenging activityof *Olea europaea* leaves and fruits extracts collected in two different seasons. *Industrial Crops and Products*, 49, 256-264.

[47] Briante R., Patumi M., Terenziani S., Bismuto E., Febbraio F., and Nucci R. (2002) *Olea europaea* L. leaf extract and derivatives: antioxidant properties. *Journal of Agricultural and Food Chemistry*, 50, 4934-4940.

[48] Korukluoglu M., Sahan Y., Yigit A., Ozer E. T., and Gücer S. (2010) Antibacterial activity and chemical constitutions of *Olea europaea* L. leaf extracts. *Journal of Food Processing and Preservation*, 34, 383-396.

[49] Arráez-Román D., Sawalha S., Segura-Carretero A., Menendez J., and Fernández-Gutiérrez A. (2008) Identification of phenolic compounds in olive leaves using CE-ESI-TOF-MS. *Agro Food Industry Hi-Tech*, 19(6), 18-22.

[50] Paiva-Martins F., and Pinto M. (2008) Isolation and characterization of a new hydroxytyrosol derivative from olive (*Olea europaea*) leaves. *Journal of Agricultural and Food Chemistry*, 56, 5582-5588.

[51] El S. N., Karakaya S. (2009) Olive tree (*Olea europaea*) leaves: potential beneficial effects on human health. *Nutrition Reviews*, 67(11), 632-638.

[52] Andreadou I., Sigala F., Iliodromitis E. K., Papaefthimiou M., Sigalas C., Aligiannis N., Savvari P., Gorgoulis V., Papalabros E., and Kremastinos D. T. (2007) Acute doxorubicin cardiotoxicity is successfully treated with the phytochemical oleuropein through suppression of oxidative and nitrosative stress. *Journal of Molecular and Cellular Cardiology*, 42, 549-558.

[53] Singh I., Mok M., Christensen A. M., Turner A. H., and Hawley J. A. (2008) The effects of polyphenols in olive leaves on platelet function. *Nutrition, Metabolism and Cardiovascular Diseases*, 18, 127-132.

[54] Perrinjaquet-Moccetti T., Busjahn A., Schmidlin C., Schmidt A., Bradl B., and Aydogan C. (2008) Food supplementation with an olive (*Olea europaea* L.) leaf extract reduces blood pressure in borderline hypertensive monozygotic twins. *Phytotherapy Research*, 22, 1239-1242.

[55] Wang L., Geng C., Jiang L., Gong D., Liu D., Yoshimura H., and ZhongThe L. (2008) Anti-atherosclerotic effect of olive leaf extract is related to suppressed inflammatory response in rabbits with experimental atherosclerosis. *European Journal of Nutrition*, 47, 235-243.

[56] Zrelli H., Matsuoka M., Kitazaki S., Araki M., Kusunoki M., Zarrouk M., and Miyazaki H. (2011) Hydroxytyrosol induces proliferation and cytoprotection against oxidative injury in vascular endothelial cells: role of Nrf2 activation and HO-1 induction. *Journal of Agricultural and Food Chemistry*, 59, 4473-4482.

[57] Zrelli H., Matsuka M., Araki M., Zarrouk M., and Miyazaki H. (2011) Hydroxytyrosol induces vascular smooth muscle cells apoptosis through NO production and PP2A activation with subsequent inactivation of akt. *Planta Medica*, 77, 1680-1686.

[58] Zrelli H., Matsuka M., Kitazaki S., Zarrouk M., and Miyazaki H. (2011) Hydroxytyrosol reduces intracellular reactive oxygen species levels in vascular endothelial cells by upregulating catalase expression through the AMPK–FOXO3a pathway. *European Journal of Pharmacology*, 660, 275-282.

[59] Parzonko A., Czerwinska M. E., Kiss A. K., and Naruszewicz M. (2013) Oleuropein and oleacein may restore biological functions of endothelial progenitor cells impaired by angiotensin II via activation of Nrf2/hemeoxygenase-1 pathway. *Phytomedicine*, 20, 1088-1094.

[60] Gonzalez M., Zarzuelo A., Gamez M. J., Utrilla M. P., Jimenez J., and Osuna I. (1992) Hypoglycemic activity of olive leaf. *Planta Medica*, 58(6), 513-515.

[61] Al-Azzawie H. F., Alhamdani M. S. S. (2006) Hypoglycemic and antioxidant effect of oleuropein in alloxan-diabetic rabbits. *Life Sciences*, 78, 1371-1377.

[62] Komaki E., Yamaguchi S., Maru I., Kinoshita M., Kakehi K., Ohta Y., and Tsukada Y. (2003) Identification of anti-α-amylase components from olive leaf extracts. *Food Science and Technology Research*, 9 (1), 35-39.

[63] Kontogianni V. G., Charisiadis P., Margianni E., Lamari F. N., Gerothanassis I. P., and Tzakos A. G. (2013) Olive leaf extracts are a natural source of advanced glycation end product inhibitors. *Journal of Medicinal Food*, 16(9), 817-822.

[64] Abaza L., Talorete T. P. N., Yamada P., Kurita Y., Zarrouk M., and Isoda H. (2007) Induction of growth inhibition and differentiation of human leukemia HL-60 cells by a tunisian Gerboui olive leaf extract. *Bioscience, Biotechnology, and Biochemistry*, 71, 1306-1312.

[65] Quirantes-Piné R., Zurek G. Barrajón-Catalán E., Bäßmann C., Micol V., Segura-Carretero A., and Fernández-Gutiérrez A. (2013) A metabolite-profiling approach to assess the uptake and metabolism of phenolic compounds from olive leaves in SKBR3 cells by HPLC–ESI-QTOF-MS. *Journal of Pharmaceutical and Biomedical Analysis*, 72, 121-126.

[66]   Briante R., Febbraio F., and Nucci R. (2004) Antioxidant/prooxidant effects of dietary non-flavonoid phenols on the Cu^{+2}-induced oxidation of human low-density lipoprotein (LDL). *Chemistry and Biodiversity*, 1, 1716-1729.

[67]   Jemai H., Bouaziz M., Fki I., El Feki A., and Sayadi S. (2008) Hypolipidimic and antioxidant activities of oleuropein and its hydrolysis derivative-rich extracts from Chemlali olive leaves. *Chemico-Biological Interactions*, 176, 88-98.

[68]   Chondrogianni N., Chinou I., and Gonos E. S. (2010) Anti-aging properties of the olive constituent oleuropein in human cells. In: Olives and olive oil in health and disease prevention. Prccdy VR, Watson RR. *Academic press,* 1335-1343.

[69]   Flemmig J., Kuchta K., Arnhold J., and Rauwald H. W. (2011) *Olea europaea* leaf (Ph.Eur.) extract as well as several of its isolated phenolics inhibit the gout-related enzyme xanthine oxidase. *Phytomedicine*, 18, 561-566.

[70]   Alirezaei M., Dezfoulian O., Kheradmand A., Neamati S., Khonsari A., and Pirzadeh A. (2012) Hepatoprotective effects of purified oleuropein from olive leaf extract against ethanol-induced damages in the rat. *Iranian Journal of Veterinary Research*, 13, Ser. No. 40.

[71]   García-Villalba R., Larrosa M., Possemiers S., Tomás-Barberán F. A., and Espín J. C. (in press) Bioavailability of phenolics from an oleuropein-rich olive (*Olea europaea*) leaf extract and its acute effect on plasma antioxidant status: comparison between pre- and postmenopausal women. *European Journal of Nutrition*, DOI 10.1007/s00394-013-0604-9.

[72]   Pasban-Aliabadi H., Esmaeili-Mahanim S., Sheibani V., Abbasnejad M., Mehdizadeh A., and Mehdi Yaghoobi M. (2013) Inhibition of 6-hydroxydopamine-induced PC12 cell apoptosis by olive (*Olea europaea* L.) leaf extract is performed by its main component oleuropein. *Rejuvenation Research*, 16, 134-142.

[73]   Roig A., Cayuela M. L., and Sanchez-Monedero M. A. (2006) An overview on olive mill wastes and their valorisation methods. *Waste Management*, 26, 960–969.

[74]   Kapellakis I. E., Tsagarakis K. P., and Crowther J. C. (2008) Olive oil history, production and by-product management. *Reviews in Environmental Science and Bio/Technology*, 7, 1–26.

[75]   Obied H. K., Allen M. S., Bedgood D. R., Prenzler P. D., Robards K., and Stockmann R. (2005) Bioactivity and analysis of biophenols recovered from olive mill waste. *Journal of Agricultural and Food Chemistry*, 53, 823-837.

[76]   Ben Sassi A., Boularbah A., Jaouad A., Walker G., and Boussaid A. (2006) A comparison of olive oil mill wastewaters (OMW) from three different processes in Morocco. *Process Biochemistry*, 41, 74–78.

[77]   Agalias A., Magiatis P., Skaltsounis A. L., Mikros E., Tsarbopoulos A., Gikas E., Spanos I., and Manios T. (2007) A new process for the management of olive oil mill waste water and recovery of natural antioxidants. *Journal of Agricultural and Food Chemistry*, 55, 2671-2676.

[78]   Giuffrè A. M., Sicari V., Piscopo A., and Louadj L. (2012) Antioxidant activity of olive oil mill wastewater obtained from different thermal treatments. *Grasas y Aceites*, 63, 209-213.

[79]   Visioli F., Romani A., Mulinacci N., Zarini S., Conte D., Vincieri F. F., and Galli C. (1999). Antioxidant and other biological activities of olive mill waste waters. *Journal of Agricultural and Food Chemistry*, 47, 3397-3401.

[80] Gómez-Caravaca A. M., Cerretani L., Segura-Carretero A., Fernández-Gutiérrez A., Lercker G., and Gallina Toschi T. (2011) Qualitative phenolic profile (HPLC-DAD-MS) from olive oil mill waste waters at different states of storage and evaluation of hydrolysis process as a pretreatment to recover their antioxidants. *Progress in Nutrition,* 13, 22-30.

[81] De Marco E., Savarese M., Paduano A., and Sacchi R. (2007) Characterization and fractionation of phenolic compounds extracted from olive oil mill wastewaters. *Food Chemistry,* 104, 858-867.

[82] Zafra A., Juarez M. J. B., Blanc R., Navalon A., Gonzalez J., and Vılchez J. L. (2006) Determination of polyphenolic compounds in wastewater olive oil by gas chromatography–mass spectrometry. *Talanta,* 70, 213–218.

[83] Khoufi S., Aloui F., and Sayadi S. (2008) Extraction of antioxidants from olive mill wastewater and electro-coagulation of exhausted fraction to reduce its toxicity on anaerobic digestion. *Journal of Hazardous Materials,* 151, 531-539.

[84] Cassano A., Conidi C., Giorno L., and Drioli E. (2013) Fractionation of olive mill wastewaters by membrane separation techniques. *Journal of Hazardous Materials*, 248-249, 185-193.

[85] Artajo L. S., Romero M. P., Suarez M., and Motilva M. J. (2007) Partition of phenolic compounds during the virgin olive oil industrial extraction process. *European Food Research and Technology*, 225, 617-625.

[86] El-Abbassi A., Kiai H., and Hafidi A. (2012) Phenolic profile and antioxidant activities of olive mill wastewater. *Food Chemistry*, 132, 406-412.

[87] Ait Baddi G., Cegarra J., Merlina G., Revel J. C., and Hafidia M. (2009) Qualitative and quantitative evolution of polyphenolic compounds during composting of an olive-mill waste–wheat straw mixture. *Journal of Hazardous Materials*, 165, 1119-1123.

[88] Garcia-Castello E., Cassano A., Criscuoli A., Conidi C., and Drioli E. (2010) Recovery and concentration of polyphenols from olive mill wastewaters by integrated membrane system. *Water Research*, 44, 3883-3892.

[89] He J., Alister-Briggs M., de Lyster T., and Jones G. P. (2012) Stability and antioxidant potential of purified olive mill wastewater extracts. *Food Chemistry*, 131, 1312-1321.

[90] Klen T. J., and Vodopivec B. M. (2012) The fate of olive fruit phenols during commercial olive oil processing: traditional press versus continuous two- and three-phase centrifuge. *LWT - Food Science and Technology*, 49, 267-274.

[91] Klen T. J., and Vodopivec B. M. (2011) Ultrasonic extraction of phenols from olive mill wastewater: comparison with conventional methods. *Journal of Agricultural and Food Chemistry*, 59, 12725-12731.

[92] Cardinali A., Pati S., Minervini F., D'Antuono I., Linsalata V., and Lattanzio V. (2012) Verbascoside, isoverbascoside, and their derivatives recovered from olive mill wastewater as possible food antioxidants. *Journal of Agricultural and Food Chemistry*, 60, 1822-1829.

[93] Cassano A., Conidi C., and Drioli E. (2011) Comparison of the performance of UF membranes in olive mill wastewaters treatment. *Water research*, 45, 3197-3204.

[94] Mudimu O. A., Peters M., Brauner F., and Braun G. (2012) Overview of membrane processes for the recovery of polyphenols from olive mill wastewater. *American Journal of Environmental Sciences*, 8, 195-201.

[95] Stamatakis G., Tsantila N., Samiotaki M., Panayotou G. N., Dimopoulos A. C., Halvadakis C. P., and Demopoulos C. A. (2009) Detection and isolation of antiatherogenic and antioxidant substances present in olive mill wastes by a novel filtration system. *Journal of Agricultural and Food Chemistry*, 57, 10554-10564.

[96] Bertin L., Ferri F., Scoma A., Marchetti L., and Fava F. (2011) Recovery of high added value natural polyphenols from actual olive mill wastewater through solid phase extraction. *Chemical Engineering Journal*, 171, 1287-1293.

[97] Ena A., Pintucci C., and Carlozzi P. (2012) The recovery of polyphenols from olive mill waste using two adsorbing vegetable matrices. *Journal of Biotechnology*, 157, 573-577.

[98] Mazzei R., Giorno L., Piacentini E., Mazzuca S., and Drioli E. (2009) Kinetic study of a biocatalytic membrane reactor containing immobilized β-glucosidase for the hydrolysis of oleuropein. *Journal of Membrane Science*, 339, 215-223.

[99] Capasso R., Sannino F., De Martino A., and Manna C. (2006) Production of triacetylhydroxytyrosol from olive mill waste waters for use as stabilized bioantioxidant. *Journal of Agricultural and Food Chemistry*, 54, 9063-9070.

[100] Madrona A., Pereira-Caro G., Mateos R., Rodríguez G., Trujillo M., Fernández-Bolaños J., and Espartero J. L. (2009) Synthesis of hydroxytyrosyl alkyl ethers from olive oil waste waters. *Molecules*, 14, 1762-1772.

[101] Zbakh H., and E. L. Abbassi A. (2012) Potential use of olive mill wastewater in the preparation of functional beverages: A review. *Journal of Functional Foods*, 4, 53-65.

[102] Tafesh A., Najami N., Jadoun J., Halahlih F., Riepl H., and Azaizeh H. (2011) Synergistic antibacterial effects of polyphenolic compounds from olive mill waste water. Evidence-Based Complementary and Alternative Medicine, Article ID 431021, 9 pages.

[103] Medina E., Romero C., de los Santos B., de Castro A., García A., Romero F., and Brenes M. (2011) Antimicrobial activity of olive solutions from stored alpeorujo against plant pathogenic microorganisms. *Journal of Agricultural and Food Chemistry*, 59, 6927-6932.

[104] Burattini S., Salucci S., Baldassarri V., Accorsi A., Piatti E., Madrona A., Espartero J. L., Candiracci M., Zappia G., and Falcieri E. (2013) Anti-apoptotic activity of hydroxytyrosol and hydroxytyrosyl laurate. *Food and Chemical Toxicology*, 55, 248-256.

[105] Cardinali A., Rotondo F., Minervini F., Linsalata V., D'Antuono I., Debellis L., and Ferruzzi M. G. (2013) Assessment of verbascoside absorption in human colonic tissues using the Using chamber model. *Food Research International*, 54, 132-138.

[106] Cardinali A., Cicco N., Linsalata V., Minervini F., Pati S., Pieralice M., Tursi N., and Lattanzio V. (2010) Biological activity of high molecular weight phenolics from olive mill wastewater. *Journal of Agricultural and Food Chemistry*, 58, 8585-8590.

[107] Angelino D., Gennari L., Blasa M., Selvaggini R., Urbani S., Esposto S., Servili M., and Ninfali P. (2011) Chemical and cellular antioxidant activity of phytochemicals purified from olive mill waste waters. *Journal of Agricultural and Food Chemistry*, 59, 2011-2018.

[108] Schaffer S., Podstawa M., Visioli F., Bogani P., Muller W. E., and Eckert G. P. Hydroxytyrosol-rich olive mill wastewater extract protects brain cells in vitro and ex vivo. *Journal of Agricultural and Food Chemistry*, 55, 5043-5049.

[109] Puel C., Mardon J., Agalias A., Davicco M. J., Lebecque P., Mazur A., Horcajada M. N., Skaltsounis A. L., and Coxam V. (2008) Major phenolic compounds in olive oil modulate bone loss in an ovariectomy/inflammation experimental model. *Journal of Agricultural and Food Chemistry*, 56, 9417-9422.

[110] Lampronti I., Borgatti M., Vertuani S., Manfredini S., and Gambari R. (2013) Modulation of the expression of the proinflammatory il-8 gene in cystic fibrosis cells by extracts deriving from olive mill waste water. Evidence-Based Complementary and Alternative Medicine, Article ID 960603, 11 pages.

[111] Visioli F., Galli C., Plasmati E., Viappiani S., Hernandez A., Colombo C., and Sala A. (2000) Olive phenol hydroxytyrosol prevents passive smoking–induced oxidative stress. *Circulation*, 102, 2169-2171.

[112] Visioli F., Wolfram R., Richard D., Abdullah M. I. C. B., and Crea R. (2009) Olive phenolics increase glutathione levels in healthy volunteers. *Journal of Agricultural and Food Chemistry*, 57, 1793-1796.

[113] Perez-Herrera A., Delgado-Lista J., Torres-Sanchez L. A., Rangel-Zuniga O. A., Camargo A., Moreno-Navarrete J. M., Garcia-Olid B., Quintana-Navarro G. M., Alcala-Diaz J. F., Munoz-Lopez C., Lopez-Segura F., Fernandez-Real J. M., Luque de Castro M. D., Lopez-Miranda J., and Perez-Jimenez F. (2012) The postprandial inflammatory response after ingestion of heated oils in obese persons is reduced by the presence of phenol compounds. *Molecular Nutrition and Food Research*, 56, 510-514.

[114] Cardinali A., Linsalata V., Lattanzio V., and Ferruzzi M. G. (2011) Verbascosides from olive mill waste water: assessment of their bioaccessibility and intestinal uptake using an in vitro digestion/Caco-2 model system. *Journal of Food Science*, 76, H48-H54.

[115] Leger C. L., Carbonneau M. A., Michel F., Mas E., Monnier L., Cristol J. P., and Descomps B. (2005) A thromboxane effect of a hydroxytyrosol-rich olive oil wastewater extract in patients with uncomplicated type I diabetes. *European Journal of Clinical Nutrition*, 59, 727-730.

[116] Leger C. L., Kadiri-Hassani N., and Descomps B. (2000) Decreased superoxide anion production in cultured human promonocyte cells (THP-1) due to polyphenol mixtures from olive oil processing wastewaters. *Journal of Agricultural and Food Chemistry*, 48, 5061-5067.

[117] Cardoso S. M., Guyot S., Marnet N., Lopes-da-Silva J. A., Renard C. M. G. C., and Coimbra M. A. (2005) Characterisation of phenolic extracts from olive pulp and olive pomace by electrospray mass spectrometry. *Journal of the Science of Food and Agriculture*, 85, 21-32.

[118] Di Serio M. G., Lanza B., Iannucci E., Russi F., and Di Giovacchino L. (2001) Valorization of wet olive pomace produced by 2 and 3-phases centrifugal decanter. *La Rivista Italiana delle Sostanze Grasse*, 88, 111-117.

[119] Lama-Muñoz A., Rodríguez-Gutiérrez G., Rubio-Senent F., Palacios-Díaz R., and Fernández-Bolaños J. (2013) A study of the precursors of the natural antioxidant phenol 3,4-dihydroxyphenylglycol in olive oil waste. *Food Chemistry*, 140, 154-160.

[120] Sánchez-Moral P., and Ruiz-Méndez M. V. (2006) Production of pomace olive oil. *Grasas y Aceites*, 57, 47-55.

[121] Sánchez de Medina V., Priego-Capote F., and Luque de Castro M. D. (2012) Characterization of refined edible oils enriched with phenolic extracts from olive leaves and pomace. *Journal of Agricultural and Food Chemistry*, 60, 5866-5873.

[122] DeJong S., and Lanari M. C. (2009) Extracts of olive polyphenols improve lipid stability in cooked beef and pork: contribution of individual phenolics to the antioxidant activity of the extract. *Food Chemistry,* 116, 892-897.

[123] Orozco-Solano M. I., Priego-Capote F., and Luque de Castro M. D. (2011) Influence of simulated deep frying on the antioxidant fraction of vegetable oils after enrichment with extracts from olive oil pomace. *Journal of Agricultural and Food Chemistry*, 59, 9806-9814.

[124] Cicerale S., Conlan X. A., Barnett N. W., and Keast R. S. J. (2011) The concentration of oleocanthal in olive oil waste. *Natural Product Research*, 25, 542-548.

[125] Rubio-Senent F., Rodríguez-Gutiérrez G., Lama-Muñoz A., and Fernández-Bolaños J. (2013) Chemical characterization and properties of a polymeric phenolic fraction obtained from olive oil waste. *Food Research International*, 54, 2122-2129.

[126] Romero C., Brenes M., Garcia P., and Garrido A. (2002) Hydroxytyrosol 4-β-D-glucoside, an important phenolic compound in olive fruits and derived products. *Journal of Agricultural and Food Chemistry*, 50, 3835-3839.

[127] Cardoso S. M., Guyot S., Marnet N., Lopes-da-Silva J. A., Silva A. M. S., Renard C. M. G. C., and Coimbra M. A. (2006) Identification of oleuropein oligomers in olive pulp and pomace. *Journal of the Science of Food and Agriculture,* 86, 1495-1502.

[128] Rubio-Senent F., Lama-Muñoz A., Rodríguez-Gutiérrez G., and Fernández-Bolaños J. (2013) Isolation and identification of phenolic glucosides from thermally treated olive oil byproducts. *Journal of Agricultural and Food Chemistry*, 61, 1235-1248.

[129] Suarez M., Romero M. P., Ramo T., Macia A., and Motilva M. J. (2009) Methods for preparing phenolic extracts from olive cake for potential application as food antioxidants. *Journal of Agricultural and Food Chemistry*, 57, 1463-1472.

[130] Rubio-Senent F., Rodríguez-Gutíerrez G., Lama-Muñoz A., and Fernández-Bolaños J. (2012) New phenolic compounds hydrothermally extracted from the olive oil byproduct alperujo and their antioxidative activities. *Journal of Agricultural and Food Chemistry*, 60, 1175-1186.

[131] Rubio-Senent F., Rodríguez-Gutiérrez G., Lama-Muñoz A., and Fernández-Bolaños J. (2013) Phenolic extract obtained from steam-treated olive oil waste: Characterization and antioxidant activity. *LWT - Food Science and Technology*, 54, 114-124.

[132] Rigane G., Bouaziz M., Baccar N., Abidi S., Sayadi S., and Ben Salem R. (2012) Recovery of hydroxytyrosol rich extract from two-phase Chemlali olive pomace by chemical treatment. *Journal of Food Science*, 77, C1077-C1083.

[133] Peralbo-Molina A., Priego-Capote F., and Luque de Castro M. D. (2012) Tentative identification of phenolic compounds in olive pomace extracts using liquid chromatography−tandem mass spectrometry with a quadrupole−quadrupole-time-of-flight mass detector. *Journal of Agricultural and Food Chemistry*, 60, 11542-11550.

[134] Aliakbarian B., Casazza A. A., and Perego P. (2011) Valorization of olive oil solid waste using high pressure–high temperature reactor. *Food Chemistry,* 128, 704-710.

[135] Medina E., Romero C., de los Santos B., de Castro A., García A., Romero F., and Brenes M (2011) Antimicrobial activity of olive solutions from stored alpeorujo against plant pathogenic microorganisms. *Journal of Agricultural and Food Chemistry*, 59, 6927-6932.

[136] Winkelhausen E., Pospiech R., and Laufenberg G. (2005) Antifungal activity of phenolic compounds extracted from dried olive pomace. *Bulletin of the Chemists and Technologists of Macedonia*, 24, 41-46.

[137] Dal Bosco A., Mourvaki E., Cardinali R., Servili M., Sebastiani B., Ruggeri S., Mattioli S., Taticchi A., Esposto S., and Castellini C. (2012) Effect of dietary supplementation with olive pomaces on the performance and meat quality of growing rabbits. *Meat Science*, 92, 783-788.

[138] Suarez M., Romero M. P., and Motilva M. J. (2010) Development of a phenol-enriched olive oil with phenolic compounds from olive cake. *Journal of Agricultural and Food Chemistry*, 58, 10396-10403.

[139] Aliakbarian B., Palmieri D., Casazza A. A., Palombo D., and Perego P. (2012) Antioxidant activity and biological evaluation of olive pomace extract. *Natural Product Research*, 26, 2280-2290.

[140] De Nicolò S., Tarani L., Ceccanti M., Maldini M., Natella F., Vania A., Chaldakov G. N., and Fiore M. (2013) Effects of olive polyphenols administration on nerve growth factor and brain-derived neurotrophic factor in the mouse brain. *Nutrition*, 29, 681-687.

[141] Palmieri D., Aliakbarian B., Casazza A. A., Ferrari N., Spinella G., Pane B., Cafueri G., Perego P., and Palombo D. (2012) Effects of polyphenol extract from olive pomace on anoxia-induced endothelial dysfunction. *Microvascular Research*, 83, 281-289.

[142] Cioffi G., Pesca M. S., De Caprariis P., Braca A., Severino L., and De Tommasi N. (2010) Phenolic compounds in olive oil and olive pomace from Cilento (Campania, Italy) and their antioxidant activity. *Food Chemistry*, 121, 105-111.

[143] Ramos P., Santos S. A. O., Guerra Â. R., Guerreiro O., Felício L., Jerónimo E., Silvestre A. J. D., Neto C. P., and Duarte M. (2013) Valorization of olive mill residues: antioxidant and breast cancer antiproliferative activities of hydroxytyrosol-rich extracts derived from olive oil by-products. *Industrial Crops and Products*, 46, 359-368.

[144] Lozano-Sanchez J., Segura-Carretero A., and Fernandez-Gutierrez A. (2011) Characterisation of the phenolic compounds retained in different organic and inorganic filter aids used for filtration of extra virgin olive oil. *Food Chemistry*, 124, 1146-1150.

[145] Lozano-Sánchez J., Giambanelli E., Quirantes-Piné R., Cerretani L., Bendini A., Segura-Carretero A., and Fernández-Gutiérrez A. (2011) Wastes generated during the storage of extra virgin olive oil as a natural source of phenolic compounds. *Journal of Agricultural and Food Chemistry*, 59, 11491-11500.

[146] Lozano-Sanchez J., Segura-Carretero A., Menendez J. A., Oliveras-Ferraros C., Cerretani L., and Fernandez-Gutierrez A. (2010) Prediction of extra virgin olive oil varieties through their phenolic profile. Potential cytotoxic activity against human breast cancer cells. *Journal of Agricultural and Food Chemistry*, 58, 9942-9955.

[147] Lerma-Garcia M. J., Simo-Alfonso E. F., Chiavaro E., Bendini A., Lercker G., and Cerretani L. (2009) Study of chemical changes produced in virgin olive oils with different phenolic contents during an accelerated storage treatment. *Journal of Agricultural and Food Chemistry*, 57, 7834-7840.

[148] Lozano-Sánchez J., Bendini A., Quirantes-Piné R., Cerretani L., Segura-Carretero A., and Fernández-Gutiérrez A. (2013) Monitoring the bioactive compounds status of extra-virgin olive oil and storage by-products over the shelf life. *Food Control*, 30, 606-615.

In: Virgin Olive Oil
Editor: Antonella De Leonardis

ISBN: 978-1-63117-656-2
© 2014 Nova Science Publishers, Inc.

Chapter 17

RECOVERY OF POLYPHENOLS FROM OLIVE OIL MILL WASTES: A SELECTIVE APPROACH ANCHORED TO MOLECULARLY IMPRINTING TECHNOLOGY

Raquel Garcia, Nuno Martins and Maria João Cabrita*

ICAAM — Instituto de Ciências Agrárias e Ambientais Mediterrânicas,
Universidade de Évora, Évora, Portugal

ABSTRACT

Olive oil is the main fat ingredient in the traditional Mediterranean diet mainly due to their antioxidant properties, the benefits associated with the regulation of cholesterol levels and prevention of cardiovascular diseases [1].

During the production of Virgin Olive Oil (VOO) large amounts of wastes, namely olive oil mill wastewaters (OMWW) are produced constituting a severe environmental problem in Mediterranean countries [2]. The main organic composition of OMWW comprises sugars, nitrogenous compounds, volatile acids, polyalcohols, pectins, fats and polyphenols [3]. The increasing interest on natural antioxidants combined with the recognized bioactivity of olive polyphenols makes OMWW very attractive as a potent source of natural antioxidants [1,4]. The valuable healthy benefits associated to polyphenols demonstrated by several studies that support their biological properties, in particular on the reduction of oxidative stress, have stimulated the research on technological strategies to extract and isolate those bioactive compounds.

The development of analytical methodologies that enable the recovery of phenolic compounds from OMWW represents an analytical challenge mainly due to the inherent complexity of the matrix. In recent years, aiming to introduce reliable methodologies to achieve that purposes, several attempts have been performed mainly focused on extraction procedures, membrane separation, centrifugation and chromatography, being these techniques used individually or in a combined form [5]. Most of the analytical procedures developed have been mainly focused on the recovery of some polyphenols that possess a more powerful antioxidant, free radical scavenging, antimicrobial and anticarcinogenic properties, such as hydroxytyrosol, tyrosol, oleuropein and phenolic acids. Nowadays, the research on analytical methodologies based on molecularly

---

*Corresponding author: raquelg@uevora.pt; rmartagarcia@yahoo.com.

imprinted polymers (MIPs) for the selective recovery of phenolic compounds from OMWW has emerged as an innovative technology enabling the selective extraction and isolation of target molecules from the OMWW complex matrix which represents a valuable process in terms of further enhancing the waste valorization [6-8].

The scope of this review is to provide a general overview on the most used techniques for the recovery of polyphenols from olive oil mill waste with special emphasis on the application of MIP technology and also its potential impact on the field of olive oil waste valorization.

Keywords: Wastewaters, antioxidants, polyphenols, molecularly imprinted polymers

INTRODUCTION

Virgin olive oil (VOO) is one of the major constituents of the Mediterranean diet constituting the main source of fat [9]. In the last years, several works have explored the beneficial properties of olive oil being speculated that the consumption of olive oil contributes to a lower incidence of coronary heart disease and some cancers. Nowadays, the well known health properties of VOO seems to be linked to a high level of monounsaturated fatty acids (MUFA) which play a significant role on the cardioprotective action and also to the presence of a great variety of minor components with relevant biological properties, making VOO a functional food [10,11]. Thus, the consumption of VOO has increased greatly in the last years. The minor component levels are strongly dependent of the cultivar, climate, ripeness of the olive at harvesting as well as the processing system employed [9-12].

The manufacturing process to produce olive oil has undergone significant changes during the last years since the traditional discontinuous pressing process has been replaced by continuous centrifugation leading to the production of different kinds of wastes. The introduction of this processing technique enables the separation of the oil, the vegetation water and solid phase of the olive by means of a continuous process. During the process, a new by-product which is a combination of liquid and solid waste called as "alperujo" or "olive mill waste" is generated. This high-humidity residue which has a thick sludge consistency contains approximately 80% of the olive fruit, including skin, seed, pulp and pieces of stones [9]. The majority of the bioactive compounds remain in the wastes or alperujo, subsequently the development of suitable strategies which contributes for the recovery of these added value products circumventing the waste disposal problem, could be considered as an emerging issue.

1. OLIVE OIL MILL WASTEWATER

During the olive processing wastewaters are produced, like Olive Oil Mill Wastewaters (OMWW) which is considered as one of the strongest industrial effluents due to its high organic content as well as the presence of phytotoxic components [13]. The amounts and characteristics of the OMWW produced are largely dependent of the technology employed on the extraction of Virgin Olive Oil (VOO), being approximately 50% and in the range 80-110% of the initial olive weight for the traditional and the continuous processes, respectively

[14]. Other factors, such as type and maturity of olives, region of origin, climatic conditions and cultivation/ processing methods have also a great influence on the final characteristics of OMWW [15]. These wastewaters are considered unavoidable by-products constituting an ecological and economical problem for the producers since they are a source of pollution due to its relatively high organic load, having influence on the biochemical oxygen demand (BOD) and chemical oxygen demand (COD) which show values up to 100 gL^{-1} and 200 gL^{-1}, respectively [16]. Thus, in olive oil-producing countries, namely Spain, Italy and Greece, OMWW disposal represents a significant environmental concern [17]. Since OMWW are a powerful source of organic matter and nutrients a common practice was the application of such wastewater in the soils, particularly in arid and semi-arid regions [15].

In the last years, several efforts have been developed for the pretreatment of OMWW aiming to improve the quality of the wastewater allowing the removal of some of its toxicity. Some of these approaches relies with physical, chemical and biological treatments as well as combinations of thereof [15].

Recently, the potential use of such wastewaters as a promising source of bioactive compounds with high value has attracted the attention of the scientific community which has done efforts on the research of their chemical composition and on the investigation of the related health benefits. Thus, these research studies have contributed to the environmental protection and also for the isolation of added-values chemicals which are now recognized as potential targets for the food, cosmetic and pharmaceutical industries [18,19].

1.1. Chemical Composition

OMWW are emulsions of oils, mucilage, pectins, sugars, nitrogen containing compounds and polyphenols varying the composition with the olives cultivar, ripening stage and also the technological extraction systems. Due to the high concentrations of recalcitrant compounds, namely lignins and tannins OMWW presents a characteristic dark color. The phenolic fraction contained in OMWW, which concentration varies significantly from 0.5 to 24gL^{-1} comprises simple phenols and flavonoids, or polyphenols which result from the polymerization of the simple phenols.

2. BIOACTIVE COMPONENTS OF OLIVE MILL WASTEWATER

The olive fruit is very rich in phenolic compounds hence from the total phenolic content of the fruit only 2% passes into the oil phase, getting approximately 53% of the remaining amount lost in the OMWW [20]. Thus, OMWW display a very complex chemical composition being characterized by an extremely rich composition on phenolic compounds for which are assigned antioxidant activity, cardioprotective, and cancer preventive activities in humans [21-23]. Then, in this sense the recovery of the bioactive components contained in OMWW has aroused a growing interest, being currently recognized as potential targets for the food and pharmaceutical industries.

Generally, the amount of specific biophenols in OMWW depends on the cultivar and maturity of the olive fruit, their climatic conditions, storage time, processing technique and deeply on the extraction conditions of VOO [22].

The hydroxytyrosol (figure 1) is the most exhaustively studied among bioactive compounds due to its presence in all OMWW extracts [21-23] and its higher and various biological activities, namely antioxidant, cardioprotective, antiatherogenic, chemopreventive, anti-inflammatory, antimicrobial and skin-bleaching [22-24].

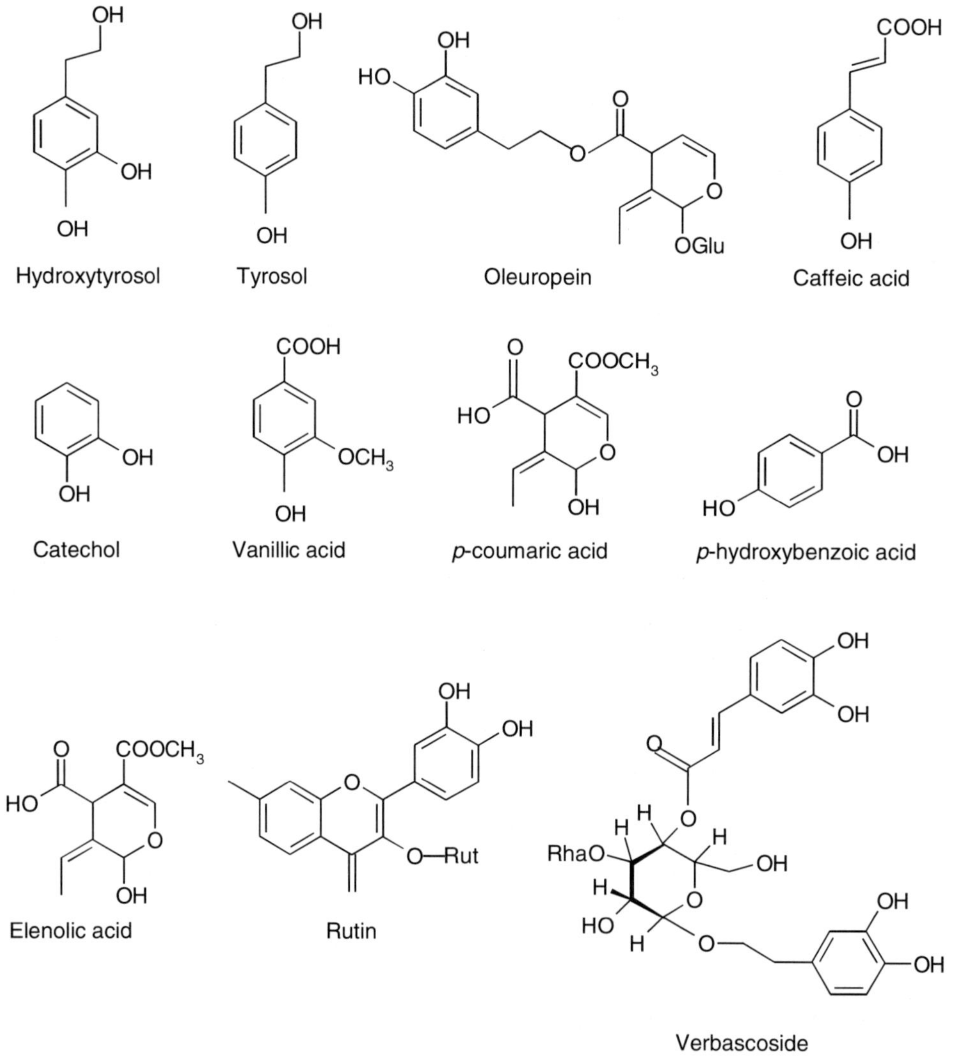

Figure 1. Chemical structures of bioactive components present in OMWW.

The chemical composition of OMWW also comprises tyrosol (Figure 1) [22], which possess several bioactive actions, namely antioxidant, anti-inflammatory [23], antiatherogenic and cardioprotective [25]. Another relevant OMWW component is the oleuropein (figure 1) [20,26] to which are attributed some biological properties, such as antioxidant, hypotensive, antiviral, cytostatic, anti-inflammatory, antihypertensive, enzyme modulation [22], cardioprotective, antiatherogenic, antimicrobial, endocrinal [27], hypoglycemic, hypolipidemic, hypocholesterolemic [28], anti-aging and anti-wrinkle [29]. The biological activities of caffeic acid (figure 1), often present in OMWW, are broadly described in the literature. Some of those properties are related with antioxidant, antiatherogenic, chemopreventive, anti-inflammatory [23], antimicrobial, antifungal [30], anti-aging and anti-wrinkle [29] benefits. Other phenolic acids such as vanillic [26], p-coumaric, elenolic [23] and p-hydroxybenzoic [6] acids (figure 1) were also found in OMWW, being also assigned into vanillic acid some relevant bioactive properties, namely antioxidant [26] and antimicrobial activities [30]. In addition, to p-coumaric acid, some authors claimed antimicrobial, antioxidant and chemopreventive [22] activities. For elenolic acid is attributed antioxidant [31], antiviral [23] and antimicrobial [32] activities while for p-hydroxybenzoic acid is assigned antioxidant activity [6]. Verbascoside [33], rutin [26], and catechol [34] (figure 1) are other OMWW bioactive constituents being the former characterized by chemoprevention, antioxidant [26], cardioprotective, antihypertensive and anti-inflammatory [33] activities. Rutin possesses antiatherogenic, antioxidant, chemoprevention and anti-inflammatory [34] activities. Lastly, catechol is characterized by antimicrobial, carcinogenic, antioxidant and anticancer activities [35].

Besides the great variety of bioactive compounds represented in Figure 1, other components such as tannins, anthocyanins and flavonoids with potential biological interest due to their antioxidant activities [23] are also identified on OMWW, which are beyond the scope of this review.

3. RECOVERY OF POLYPHENOLS FROM OLIVE MILL WASTEWATER

The powerful pollutant effect of OMWW forbidden its direct discharged into water or land. Hence, several procedures based on physical, chemical, biological or combined technologies have been investigated aiming to reduce the undesirable properties of OMWW prior to disposal. However, the discovery that OMWW is a rich source of phenolic compounds which possess remarkable biological properties namely antioxidant, free radical scavenging, antimicrobial and anticarcinogenic activities [21], as recently reviewed by Obied and co-workers, has encouraged the scientific community to develop effective processes enabling the recovery of these compounds from such wastes. Thus, some methodologies have been developed mainly focused on extraction, membrane separation, centrifugation and chromatographic techniques.

3.1. Recovery of Phenolic Compounds from OMWW by Traditional Methodologies

3.1.1. Porous Materials (Inorganic, Activated Carbons and Resins)

The removal of phenolic compounds from OMWW has been performed by means of inorganic materials [36-38] and activated carbons [6,39,40]. Furthermore, Niebelschuetz and Liedeck have evaluated the adsorption of those bioactive compounds into ion-exchange resins after a biological treatment and mechanical separation [41]. Alternatively, the use of resins during OMWW processing enables simultaneously the water reuse and the recovery of antioxidants [42]. Recently, Agalias and coworkers have developed an experimental procedure for the treatment of OMWW based on three main successive steps involving successive filtration stages to remove suspended solids followed by the application of adsorbent resins (XAD-16 and XAD-7HP) to achieve deodoring and decolorization of the wastewater and the removal/recovery of the polyphenol and lactone contents. Then, an thermal evaporation has been applied leading to the recovery of the organic solvents mixture used in the resin regeneration process pursued by the separation of the polyphenols and other organic compounds using fast centrifuge partition chromatography, which allows the recovery of pure hydroxytyrosol [43].

3.1.2. Solvent Extraction

In spite of the high cost owing to the large consumption of organic solvents, Solvent Extraction is the most widely used technique for the recovery of phenolic compounds from OMWW. However, the disadvantageous use of organic solvents could be overtaken by the employment of supercritical fluids (SCFs), which implies the use of expensive high pressure equipment. In the recent years, several patented methods have been implemented allowing the isolation of oleuropein aglycon [44], triacetylhydroxytyrosol [45], pectins and valuable polyphenols [46] from OMWW.

Recently, the potential of ultrasonic-assisted extraction has been evaluated on the recovery of phenols from OMWW [47]. Despite the low cost and the easy mode of operation while compared to other methodologies, namely membrane separations and ultrasonic extracting efficiency in aqueous OMWW is limited owing to the water nonmiscible solvent use. However, the application of ultrasonic-assisted extraction into freeze-dried solid OMWW is not so limited and has proven to be a suitable alternative to conventional solvent extraction techniques gathering several advantages, which includes increase extraction yields, faster kinetics and simplicity. Although, it high solvent consumption and some limited capacity, if not operated in a continuous mode, represent some of their disadvantages [47].

3.1.3. Solid Phase Extraction (SPE)

Recently, a Solid Phase Extraction (SPE)-based procedure for the recovery of OMWW phenolic compounds has been developed for water solutions and further applied and assessed on OMWW [48,49]. Aiming to achieve that purpose, the suitability of resins with different physical properties, namely Amberlite XAD-4, XAD-7, XAD-16, IRA-96 and Isolute ENV+ has been investigated by means of molecular adsorption studies. As expected, the polarity of the target analyte has a great influence on the adsorption ratios of the adsorbents. In these studies, the highest phenol adsorption (76%) was achieved with IRA-96 polar resin. By contrast, the application of Isolute ENV+ as adsorbent on SPE procedure allows the recovery of almost 60% of the overall phenols [48]. Particularly, Isolute ENV+ retains almost completely hydroxytyrosol, which is one of the most abundant and valuable phenolic compound of OMWW, when non-acidified ethanol was used as the desorbing phase [49]. Thus, these works have contributed for the development of an effective SPE procedure for the recovery of natural phenols from OMWW based on SPE approach.

3.1.4. Membrane Separations

Since the chemical composition of OMWW comprises phenolic compounds of a wide range of molecular masses from low molecular weight phenolics, such as benzoic acids and derivatives into high molecular weight phenolics, namely secoiridoid aglycons and lignans [35], their recovery with high purity constitute a demanding task, which makes membrane technology a powerful tool on the recovery of biophenols from OMWW. Therefore, the application of this methodology on the recovery of antioxidants from OMWW has emerged as a very advantageous technique mainly due to the low energy consumption, no additive requirements and no phase change. Particularly, the treatment of OMWW by means of membrane techniques leads to the separation of suspended solids, which are retained on membrane surface allowing its separation from the wastewater.

Application of conventional filtration membranes on the treatment of OMWW is still applied allowing to separate the different components according to its particle sizes (microfiltration, ultrafiltration and nanofiltration), the molecular weight of the components that affect the permeability (Reverse Osmosis) or its aggregation (Membrane Distillation and Osmotic Distillation) [50]. These methods are commonly applied in a sequential form enabling the recovery, purification and concentration of antioxidants.

Several methodologies based on membrane technologies for the recovery of natural antioxidants from OMWW has been recently investigated [51-54] and patented [55].

In the future, recovery of antioxidants from OMWW seems to be toward the implementation of methodologies involving membranes in a sequential design. However, the application of selective methodologies that enable a selective recovery of bioactive compounds could be a feasible and versatile alternative on the achievement of extracts with high purity.

3.2. Selective Methodologies

The huge complexity of OMWW, in particular the presence of high fat contents makes the recovery of polyphenols from OMWW a very demanding task due to the eventual co-extraction of lipids with the analytes, hampering the development of suitable methodologies. Thus, the implementation of analytical approaches that enable a high recovery of polyphenol fraction and, additionally, a high degree of purity is highly desirable. In this context, the application of molecularly imprinting technology which is based on molecular recognition mechanism [56] could open new avenues on the development of promising alternative methodologies with higher selectivity and furthermore overcoming some of the limitations of the common methodologies.

The application of molecularly imprinting technology comprises the development of molecularly imprinted polymers (MIPs) considered as artificially-engineering receptors, which are specifically designed to possess a predefined selectivity for the target analyte. Figure 2 shows a schematic representation of the molecular imprinting process.

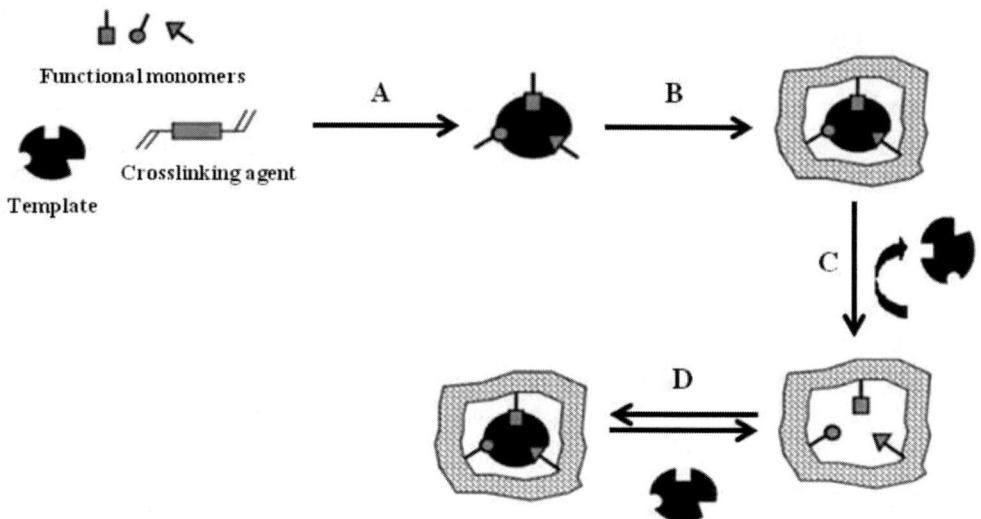

Figure 2. General scheme of the principle of molecularly imprinted polymer technique (Adapted from [57]). (A) Functional monomers, crosslinking agent, radical initiator and template are mixed in a porogenic solvent to allow complexation between the functional monomers and the imprint molecule; (B) polymerization; (C) removal of the template, which leaves available on the polymer structure recognition sites complementary to the template; (D) analyte binding on the specific imprinted sites.

Generally, MIP synthesis is carried out *in situ* by a co-polymerization process using a functional monomer, a template molecule in the presence of a crosslinking agent and an initiator in a porogenic solvent [56]. However, it is necessary to be aware that many variables of the imprinting process influence the molecular recognition of MIP and subsequently their selectivity. Complementary interactions between the template, the functional monomer and crosslinker are mandatory on the process of molecular organization of the artificial receptor binding sites being also significant others factors, such as the stoichiometry and concentration of the template and monomers since it influences both polymer morphology and selectivity for the target molecule. The porogen used in the polymerization process plays a remarkable role on the mediation of the interactions between the functional groups and the template molecule, which is overriding for the polymer morphology, porosity and accessibility of the binding sites of the tridimensional polymer network [58]. Firstly, pre-polymerization step enables the generation of porous on the imprinting material that will allow the diffusion of the analyte into the specific generated binding sites [59]. The polymerization process occurs usually in solution and is initiated by using a free radical initiator. According to the chemical interactions between the functional monomers and crosslinker, several polymerization methods have been explored, namely non-covalent, covalent or semi-covalent interactions [58]. After polymerization, the template molecules are removed from the tridimensional polymer structure leaving available selective recognition sites with shape, size and functionalities complementary to the template [60]. During this ingenious process of synthesis is introduced a molecular memory into the MIP allowing the selective rebinding of the target molecule.

However, the production of MIPs with suitable properties involves several steps, namely the selection and development of an appropriate synthetic method as well as the choice of the chemical entities involved on the synthesis (functional monomer, crosslink, porogen) which constitute some of the critical points on the overall process requiring most often optimization procedures [61]. The achievement of the "ideal" MIP is not limited to the formation on the polymeric network of cavities with a complementary size, shape and chemical functionality of the template but also that enable an efficient extraction through an increase in the accessibility of the porous structure with a narrow interaction with interferent compounds, showing also other properties, such as high chemical resistance and thermal stability [61].

Recently, different methods have been explored for the development of MIPs [62-64], each of these mechanisms involves the control of different parameters during the synthesis leading to the formation of MIPs with different properties, namely morphology.

The peculiar characteristics of these biomimetic materials, in particular their stability, low cost and ease of preparation have contributed to the widespread use of the molecularly imprinting technology in several applications, such as environmental, biomedical and food analysis [56,62]. Despite the widespread and successfully use of MIPs in various applications, this approach remains almost unexplored on the development of selective MIPs for the isolation/ pre-concentration of bioactive polyphenols of OMWW being these studies limited to a restrict number of compounds.

Aiming to introduce selective methodologies for the recognition of caffeic and *p*-hydroxybenzoic acids from OMWW, the pioneering studies of Michailof and co-workers [6] have explored several synthetic procedures for the preparation of selective MIPs for the molecular recognition of those target molecules. Thus, for the preparation of MIP for the molecular recognition of caffeic acid has been tested as functional monomers 4-vinylpyridine, allylurea, allylaniline and methacrylic acid, ethylene glycol dimethylacrylate (EDMA), pentaerythritol trimethylacrylate (PETRA) and divinylbenzene 80 (DVB80) as crosslinkers and tetrahydrofuran as porogen; whereas, for the synthesis of MIPs selective for *p*-hydroxybenzoic acid has been evaluated 4 vinylpyridine, allylurea and allylaniline as functional monomers, EDMA and PETRA as crosslinkers and acetonitrile as porogen. In order to evaluate the molecular recognition of the synthesized imprinted material some chromatographic studies have been performed by means of imprinting factor determinations and adsorption assays. Within these studies it was possible to conclude that the more selective MIPs for the molecular recognition of caffeic and *p*-hydroxybenzoic acids were prepared with the following chemical entities: 4-vinylpyridine as monomer (M), PETRA as crosslinker (C), ratio (T:M:C) of 1:4:12 and 4-vinylpyridine as monomer (M), EDMA as crosslinker (C), ratio (T:M:C) of 1:4:20, respectively.

The suitability of the selected MIPs for the isolation/pre-concentration of caffeic and *p*-hydroxybenzoic acids was evaluated using those imprinting materials as adsorbents in solid phase extraction (MISPE). After optimization of the MISPE procedure, the selected MIPs have proved ability to the selective recognition of caffeic and *p*-hydroxybenzoic acids being promising for the quick and efficient isolation of those bioactive compounds from OMWW.

More recently, Puoci and co-workers [8] have been devoted to the development of an innovative biocompatible approach for the selective recovery of gallic acid (GA) from OMWW based on molecularly imprinting technology. To achieve this goal, a selective MIP for GA has been prepared by non-covalent imprinting approach using GA as template (T), methacrylic acid (MAA) as functional monomer (M), ethylene glycol dimethacrylate (EGDMA) as crosslinker, 2,2'-azoisobutyronitrile (AIBN) as radicalar initiator and tetrahydrofuran (THF) as porogen and using a ratio of 1:16 (T:M). The molecular recognition of the MIPs has been assessed by rebinding experiments using different solvents (ethanol, water and ethanol/water mixtures) and further, Langmuir and Freundlich models have been employed to study the adsorption process. The studies were also extended to the evaluation of MIPs selectivity using structurally analogues of the template molecule. The results obtained on these studies have shown that the proposed synthetic conditions are suitable for the formation of imprinted cavities in the MIP structure allowing a selective recognition for the template molecule. Finally, taking into account the main goal of this study, the selective extraction of GA from an actual site OMWW has been performed by a modified SPE process using the synthesized MIP as adsorbent, which allows the selective recovery of 87% of pure GA. The results have shown the suitability of the proposed procedure for the selective recovery of the target molecule from OMWW.

CONCLUSION

The recovery of high added value compounds from OMWW is an attractive and versatile way of reuse these wastes which is particularly relevant for an economical and environmental point of view. In particular, the extraction of bioactive compounds has a double impact - the detoxification of wastes and their potential use as functional compounds in food or cosmetic industries and on pharmacological applications.

Particularly, OMWW can be seen as a cheap source of phenolic compounds with strong antioxidant and radical scavenger properties. The recovery of polyphenols from OMWW by conventional methods has been investigated during the last years. However, the development of selective approaches for the recovery of polyphenols based on molecularly imprinting technology (MIT) is limited to a few works. Nevertheless, within the application of this technology an improvement on selectivity is achieved, due to the pre-defined selectivity inherent to the MIT process, allowing the recovery of the target bioactive compounds with high purity and efficiency. The implementation of this methodology represents also other advantages, such as their scale up feasibility and eco-friendly character. Thus, the promising features attributed to MIT will lead to a growing interest on the implementation of this innovative and versatile approach for the isolation /pre-concentration of other bioactive compounds from OMWW. Briefly, the use of molecularly imprinted polymers on the development of suitable methodologies for the recovery of bioactive compounds from OMWW seems to be very attractive, since it combines high selectivity and affinity for the adsorbents complemented with some other interesting properties, namely low cost, physical resistance and reusing.

ACKNOWLEDGMENTS

This work has been supported by FEDER and National Funds, through the Programa Operacional Regional do Alentejo (InAlentejo) Operation ALENT-07-0262-FEDER-001871/ Laboratório de Biotecnologia Aplicada e Tecnologias Agro-Ambientais and FEDER Funds through the Operational Programme for Competitiveness Factors - COMPETE and National

Funds through FCT - Foundation for Science and Technology under the Strategic Project PEst-C/AGR/UI0115/2011 and Project PTDC/AGR-ALI/117544/2010.

REFERENCES

[1]   Goméz-Romero M, García-Villalba R, Carrasco-Pancorbo A, & Fernández-Gutiérrez A (2012) Metabolism and Bioavailability of olive oil polyphenols. In *Olive Oil-Constituents, Quality, Health Properties and Bioconversions*. Edited by Dimitrios Boskou, InTech.

[2]   Frankel E, Bakhouche A, Lozano-Sánchez J, Segura-Carretero A, & Fernández-Gutiérrez A (2013) Literature Review on Production Process To Obtain Extra Virgin Olive Oil Enriched in Bioactive Compounds. Potential Use of Byproducts as Alternative Sources of Polyphenols. *Journal of Agricultural and Food Chemistry*, 61, 5179-5188.

[3]   Lafka TI, Lazou AE, Sinanoglou VJ, & Lazos ES (2011) Phenolic and antioxidant potential of olive oil mil wastes. *Food Chemistry*, 125, 92-98.

[4]   DellaGreca M, Previtera L, Temussi F, & Zarrelli (2004) Low-molecular-weight Components of Olive oil Mill Waste-waters. *Phytochemical Analysis*, 15, 184-188.

[5]   Takaç S, & Karakaya A (2009) Recovery of Phenolic Antioxidants from Olive Mill Wastewater. *Recent Patents on Chemical Engineering*, 2, 230-237.

[6]   Michailof C, Manesiotis P, & Panayiotou C (2008) Synthesis of caffeic acid and p-hydroxybenzoic acid molecularly imprinted polymers and their application for the selective extraction of polyphenols from olive mill waste waters. *Journal of Chromatography A*, 1182, 25-33.

[7]   Sergeyeva TA, Gorbach LA, Slinchenko OA, Goncharova LA, Piletska OV, Brovko OO, Sergeeva LM, & Elska GV (2010) Towards development of colorimetric test-systems for phenols detection based on computationally-designed molecularly imprinted polymer membranes. *Materials Science and Engineering C*, 30, 431-436.

[8]   Puoci F, Scoma A, Cirillo G, Bertin L, Fava F, & Picci N (2012) Selective extraction and purification of gallic acid from actual site olive mil wastewaters by means of molecularly imprinted microparticles. *Chemical Engineering Journal*, 198, 529-535.

[9]   Rodríguez-Gutiérrez G, Lama-Munoz A, Ruiz-Méndez MV, Rubio-Senent F, Fernández-Bolanos J (2012) New Olive-Pomace Oil Improved by Hydrothermal Pre-Teatments. In Olive Oil-Constituents, Quality, Health properties and Bioconversions. Edited by Dimitrius Boskou, InTech.

[10] Covas MI, Ruiz-Gutiérrez V, Torre R, Kafatos A, Lamuela-Raventós RM, Osada J, Owen RW, & Visioli F (2006) Minor Components of Olive Oil: Evidence to Date of Health Benefits in Humans. *Nutrition Reviews*, 64, 20-30.

[11] Pérez-Jiménez F, Ruano J, Pérez-Martínez P, López-Segura F, & López-Miranda J (2007) The influence of olive oil on human health: not a question of fat alone. *Molecular Nutrition and Food Research*, 51, 199-1208.

[12] Lesage-Meessen L, Navarro D, Maunier S, Sigoillot JC, Lorquin J, Delattre M, Simon JL, Asther M, & Labat M (2001) Simple phenolic content in olive oil residues as a function of extraction systems. *Food Chemistry*, 75, 501-507.

[13] Rodríguez G, Rodríguez R, Guillén R, Jiménez A, & Fernández-Bolanos J (2007) Effect of steam treatment of alperujo on the enzymatic saccharification and in vitro digestibility. *Journal of Agriculture and Food Chemistry*, 55, 136-142.

[14] Mulinacci N, Romani A, Galardi C, Pinelli P, Giaccherini C, & Vincieri F (2001) Polyphenolic content in olive oil waste waters and related olive samples. *Journal of Agricultural and Food Chemistry*, 49, 3509-3514.

[15] Paraskeva P, & Diamadopoulos E (2006) Technologies for olive mill wastewater (OMW) treatment: a review. *Journal of Chemical Technology and Biotechnology*, 81, 1475-1485.

[16] Dhaouadi H, & Marrot B (2008) Olive mill wastewater treatment in a membrane bioreactor: process feasibility and performances. *Chemical Engineering Journal*, 45, 225-231.

[17] Zunin P, Fusella GC, Leardi R, Boggia R, Bottino A, & Capannelli G (2011) Effect of the Addition of Membrane Processed Olive Mill Waste Water (OMWW) to Extra Virgin Olive Oil. *Journal of the American Oil Chemists Society*, 88, 1821-1829.

[18] El-Abbassi A, Kiai H, & Hafidi A (2012) Phenolic profile and antioxidant activities of olive mil wastewater. *Food Chemistry*, 132, 406-412.

[19] Zbakh H, & El Abbassi A (2012) Potential use of olive mill wastewater in the preparation of functional beverages: A review. *Journal of Functional Foods*, 4, 53-65.

[20] Rodis PS, Karathanos VT, & Mantzavinou A (2002) Partitioning of olive oil antioxidants between oil and water phases. *Journal of Agricultural and Food Chemistry*, 50, 596-601, and references therein.

[21] Obied HK, Allen MS, Bedgood DR, Prenzler PD, Robards, K., & Stockmann, R. (2005). Bioactivity and analysis of biophenols recovered from olive mill waste. *Journal of Agricultural and Food Chemistry*, 53, 823-837.

[22] Visioli F, & Galli C (1998) Olive Oil Phenols and Their Potential Effects on Human Health. *Journal of Agricultural and Food Chemistry*, 46, 4292-4296.

[23] Visioli F, Romani A, Mulinacci N, Zarini S, Conte D, Vincieri FF, & Galli C (1999) Antioxidant and other biological activities of olive oil mill waste water. *Journal of Agricultural and Food Chemistry*, 47, 3397-3401.

[24] Chikamatsu Y, Ando H, Yamamoto A, Kyo S, Yamashita K, & Dojo K (1996) Hydroxytyrosol as melanin formation inhibitor and lipid peroxide formation inhibitor and its application to topical preparations and bath preparations. *Jpn. Patent* 8119825, 1-10.

[25] Marrugat J, Covas MI, Fito M, Schroder H, Miro-Casas E, Gimeno E, Lopez-Sabater MC, De La Torre R, & Farre M (2004) Effects of differing phenolic content in dietary olive oils on lipids and LDL oxidation - a randomized controlled trial. *European Journal of Nutrition*, 43, 140-147.

[26] Servili M, Baldioli M, Selvaggini R, Miniati E, Macchioni A, & Montedoro G (1999) High-performance liquid chromatography evaluation of phenols in olive fruit, virgin olive oil, vegetation waters, and pomace and 1D-and 2D-nuclear magnetic resonance characterization. *Journal of the American Oil Chemists' Society*, 76, 873-882.

[27] Visioli F, & Galli C (1998) The effect of minor constituintes of olive oil on cardiovascular disease: new findings. *Nutrition Reviews*, 56, 142-147.

[28] Driss F, Duranthon V, & Viard V (1996) Effects biologique des compose´s polyphenoliques de l'olivier. *Corps Gras*, 3, 448-451.

[29] Kim BJ, & Kim JH (1997) Biological screening of 100 plant extracts for cosmetic use (II): anti-oxidative activity and free radical scavenging activity. *International Journal of Cosmetic Science,* 19, 299-307.

[30] Aziz NH, Farag SE, Mousa, & Abo Zaid MA (1998) Comparative antibacterial and antifungal effects of some phenolic compounds. *Microbios*, 93, 43-54.

[31] Fleming HP, Walter WM, & Etchells JL (1973) Antimicrobial properties of oleuropein and products of its hydrolysis from green olives. *Journal of* Applied Microbiology, 26, 777-782.

[32] Cardinali A, Pati S, Minervini F, D'Antuono I, Linsalata V, & Lattanzio V (2012) Verbascoside, Isoverbascoside, and Their Derivatives Recovered from Olive Mill Wastewater as Possible Food Antioxidants. *Journal of Agricultural and Food Chemistry,* 60, 1822-1829, and references therein.

[33] Fiorentino A, Gentili A, Isidori M, Monaco P, Nardelli A, Parrella A, & Temussi F (2003) Environmental effects caused by olive mill wastewaters: toxicity comparison of low-molecular weight phenol components. *Journal of Agricultural and Food Chemistry, 51,* 1005-1009.

[34] Lee OH, & Lee BY (2010). Antioxidant and antimicrobial activities of individual and combined phenolics in Olea europaea leaf extract. *Bioresource Technology*, 101, 3751-3754, and references therein.

[35] Bendini A, Cerretani L, Carrasco-Pancorbo A, Gómez-Caravaca AM, Segura-Carretero A, Fernández-Gutiérrez A, & Lercker G (2007) Phenolic Molecules in Virgin Olive Oils: a Survey of Their Sensory Properties, Health Effects, Antioxidant Activity and Analytical Methods. An Overview of the Last Decade, Review. *Molecules, 12,* 1679-1719 and references therein.

[36] Al-Malah K, Azzam KOJ, & Abu-Lail NI (2000) Olive mills effluent (OME) wastewater post-treatment using activated clay. *Separation and Purification Technology,* 20, 225-234.

[37] Ugurlu M, & Hazirbulan S (2007) Removal of some organic compounds from pre-treated olive mil wastewater by sepiolite. *Fresenius Environmental Bulletin,* 16, 887-895.

[38] Santi CA, Cortes S, D'Acqui LP, Sparvoli E, & Pushparaj B (2008) Reduction of organic pollutants in olive mil wastewater by using different mineral substrates as adsorbents. *Bioresource Technology*, 99, 1945-1951.

[39] Galiatsatou P, Metaxas M, Arapoglou D, & Kasselouri-Rigopoulo V (2002) Treatment of olive mil waste water with activated carbons from agricultural by-products. *Waste Manage,* 22, 803-812.

[40] Achak M, Hafidi A, Ouazzani N, Sayadi S, & Mandi L (2009) Low cost biosorbent "banana peel" for the removal of phenolic compounds from olive mil wastewater: kinetic and equilibrium studies. *Journal of Hazardous Materials,* 166, 117-125.

[41] Niebelschuetz H, & Liedecke H (2000) A method of purifying waste water containing phenolics. EP1041044.

[42] Kastanas I, Andrikopoulos N, & Vercauteren J (2005) A method for the clearance of olive mill waste waters. GR2003100295.

[43] Agalias A, Magiatis P, Skaltsounis AL, Mikros E, Tsarbopoulos A, Gikas E, Spanos I, & Manios T (2007) A New Process for the Management of Olive Oil Mill Waste Water

and Recovery of Natural Antioxidants. *Journal of Agriculture and Food Chemistry*, 55, 2671-2676.

[44] Emmons W, & Guttersen C (2005) Isolation of oleuropein aglycon from olive vegetation water. US20050103711.

[45] De Martino A, Sannino F, Manna C, Gianfreda L, & Capasso R (2007) Process for producing triacytylhydroxytyrosol from olive oil mill waste waters for use as stabilized antioxidant. WO2007074490.

[46] Tornberg E, & Galanakis C (2008) Olive waste recovery. WO2008082343.

[47] Klen TJ, & Vodopivec BM (2011) Ultrasonic Extraction of Phenols from Olive Mill WasteWater: Comparison with Conventional Methods. *Journal of Agricultural and Food Chemistry,* 59, 12725-12731.

[48] Ferri F, Bertin L, Scoma A, Marchetti L, & Fava F (2011) Recovery of low molecular weight phenols through solid-phase extraction. *Chemical Engineering Journal,* 166, 994-1001.

[49] Bertin L, Ferri F, Scoma A, Marchetti L, & Fava F (2011) Recovery of high added value natural polyphenols from actual olive mill wastewater through solid phase extraction. *Chemical Engineering Journal,* 171, 1287-1293.

[50] Mudimu OA, Peters M, Brauner F, & Braun G (2012) Overview of Membrane Processes for the Recovery of Polyphenols from Olive Mill Wastewater. *American Journal of Environmental Sciences,* 8, 195-201.

[51] Russo C (2007) A new membrane process for the selective fractionation and total recovery of polyphenols, water and organic substances from vegetation waters. *Journal of Membrance Science,* 288, 239-246.

[52] Turano E, Curcio S, De Paola MG, Calabrò V, & Iorio G (2002) An integrated centrifugation-ultrafiltration system in the treatment of olive mil wastewater. *Journal of Membrance Science,* 209, 519-531.

[53] Paraskeva CA, Papadakis VG, Tsarouki E, Kanneloupoulou DG, & Koutsoukos PG (2007) Membrane processing for olive mill wastewater fractionation. *Desalination* 213, 218-229.

[54] Ochando-Pulido JM, & Martinez-Ferez A (2012) A focus on pressure-driven membrane technology in olive mil wastewater reclamation: state of the art. *Water Science & Technology*, 66, 2505-2516.

[55] Anselmi C, Bottino A, Capannelli G, Centini C, & Maffei FR (2006) Procedimento per il trattamento delle acque reflue da lavorazione degli oli di oliva e simili per l'ottenimento, per via naturale, di antiossidanti per impieghi farmaceutico, cosmetico ed alimentare, IT Patent FI2006A000318.

[56] Vasapollo G, Del Sole R, Mergola L, Lazzoi MR, Scardino A, Scorrano S, & Mele G (2011) Molecularly Imprinted Polymers: Present and Future Prospective. *International Journal of Molecular* Sciences, 12, 5908-5945.

[57] Garcia R, Cabrita MJ, & Freitas AMC (2011) Application of Molecularly Imprinted Polymers for the Analysis of Pesticide Residues in Food—A Highly Selective and Innovative Approach. *American Journal of Analytical Chemistry*, 2, 16-25.

[58] Batra D, & Shea KJ (2003) Combinatorial methods in molecular imprinting. *Current Opinion in Chemical Biology*, 7, 434–442.

[59] Moreno-Bondi MC, Navarro-Villoslada F, Benito-Peña E, Urraca JL & (2008) Molecularly Imprinted Polymers as Selective Recognition Elements in Optical Sensing. *Current Analytical Chemistry*, 4, 316-340.

[60] Qi P, Wang J, Jin J, Su F, & Chen J (2010) 2,4-Dimethylphenol imprinted polymers as a solid-phase extraction sorbent for class-selective extraction of phenolic compounds from environmental water. *Talanta*, 81, 1630-1635.

[61] Kloskowski A, Pilarczyk M, Przyjazny A, & Namiesnik J (2009) Progress in Development of Molecularly Imprinted Polymers as Sorbents for Sample Preparation. *Critical Reviews in Analytical Chemistry*, 39, 43-58.

[62] Chen L, Xu S, Li J (2011) Recent advances in molecular imprinting technology: current status, challenges and highlighted applications. *Chemical Society Reviews*, 40, 2922-2942.

[63] Pérez-Moral N, & Mayes AG (2004) Comparative study of imprinted polymer particles prepared by different polymerisation methods. *Analytica Chimica Acta*, 504, 15-21.

[64] Hosoya K, Kageyama Y, Kimata K, Araki T, Tanaka N, & Frechet JMJ (1996) preparations and properties of uniform size macroporous polymer beads prepared by two–step swelling and polymerization method utilizing divinyl succinate or divinyl adipate as a crosslinking agent. *Journal of Polymer Science A*, 34, 2767–2774.

In: Virgin Olive Oil
Editor: Antonella De Leonardis

ISBN: 978-1-63117-656-2
© 2014 Nova Science Publishers, Inc.

Chapter 18

DISPOSAL AND TREATMENT OF OLIVE OIL INDUSTRY WASTES (OLIVE MILL WASTEWATERS, SOLID WASTES) FOR THE REMEDIATION OF POLLUTED CULTIVATED SOILS AND INDUSTRIAL WASTEWATERS

Zacharias Ioannou[1,2] and Victor Kavvadias[2]*

[1] Department of Food Science & Nutrition, School of the Environment,
University of the Aegean, Myrina, Lemnos, Greece
[2] Soil Science Institute of Athens, National Agricultural Research Foundation,
ELGO Demeter, Lycovrissi, Attiki, Greece

ABSTRACT

Olive oil industry grows constantly, especially in countries with a Mediterranean climate such as Spain, Italy and Greece, which are leading producers of olives. Olive oil production results in an annual generation of more than 30 million m^3 of olive mill wastewaters (OMWs) and 4 million tons of olive stone by-products (OLS), respectively.

Solid wastes of oil industry such as olive stones (OLS) or dried olive pomace (OP) are low-cost lignocellulosic materials. They are used alone or in combination with OMWs as an efficient fuel with low amounts of N and S. Their heating power can contribute to power generation in the electricity sector and space heating in residential and commercial buildings. Researchers have also been focused on the gasification treatment of olive tree cuttings and OLS, which leads to syngas and H_2-rich gas production. Moreover, OLS and OP are promising materials alone or in combination with polymeric materials as precursors for the production of activated carbons through pyrolysis and activation (chemical or physical) processes. Activated carbons are promoted for the removal of contaminants (e.g. phenolic compounds, dyes) or radioactive

---

* Corresponding Author address: 2 Mitropolite Ioakeim St., 81400, Myrina, Lemnos, Greece; Email: zioan@teemail.gr.

materials due to their microporous structure having extensive applications as filters for the purification of potable water and industrial wastewaters. Moreover, OLS and OP can also act as metal biosorbents, i.e. heavy metals, from aqueous effluents. Furthermore, other studies have indicated the production of humic fertilizers by oxiammoniation of hydrolyzed OLS and soil amendments by olive husk increasing heavy metal solubility in soils, enhancing metal absorption by plants and removing them from soils through phytoextraction. Other applications of OLS are as plastic fillers, abrasives, in cosmetic industry and as a dietary animal supplementation.

The disposal and treatment of olive mill wastewaters are a major problem of the olive oil industry. Studies related with agronomic application of raw OMWs often produce diverging or even contradictory results mainly due to the large variability in OMWs composition. OMWs have favorable chemical properties (organic carbon, potassium and phosphorus content) and disposal on soils may be considered as an appropriate option to solve management problems, restore soil fertility and promote productivity, provided that land spreading is controlled and soil is not associated with sensitive aquifers. However, uncontrolled disposal of OMW can cause several ecological problems due to seasonal production, strong and unpleasant odor, and potential toxicity to soil microflora, surface- and groundwater. OMWs are often discharged directly into sewer systems and water streams or disposed of in lagoons despite the fact that such management options are not allowed in most Mediterranean countries. Although several techniques have been developed for OMW management, detoxification and valorization their application is often too expensive for most olive-oil mills. It is underlined that no specific European Commission legislation exists today for OMW management and each country issues different guidelines. Continuous and reliable monitoring of soil and water quality can provide useful data regarding environmental quality in the wider OMW disposal areas and assist scientists to implement feasible preventive and remedial actions when required.

The assessment of existing knowledge on practices and policies for the recycling of OMWs and solid wastes will promote the long term sustainability of OMW and solid wastes management.

Keywords: Olive mill wastewaters, olive stones, oil industry wastes, by-products

INTRODUCTION

Olive oil is produced in those regions of the world where climatic conditions are as favorable as those prevailing in the Mediterranean countries. The Mediterranean region accounts for not less than 98% of the world olive oil production [1]. The main olive-producing countries are Spain, Italy and Greece. Global olive oil production is according to the International Olive Council estimation, at least 3.1 million tons in 2013/2014 (November 2013 IOC estimation).

The European Union (E.U.) is expected to provide 2.3 million tons of global output, with Spain alone to deliver 1.5 million, Italy 450,000, and Greece 230,000 tons. Moreover, the global table olive production is estimated around 2.6 million tons in 2013/2014 with the highest production of 700,000 tons to come from E.U. Spain produced around 513,000 tons of table olives while Greece and Italy produced around 94,000 and 74,000 tons respectively (November 2013 IOC estimation) [2].

Producer prices for extra virgin olive oil were estimated to 2.36 €/kg in Spain and Greece and 3.04 €/kg in Italy in the last week of October 2013. The difference between the price of refined olive oil and extra virgin olive oil currently lies at around 0.20 €/kg in Spain and 0.68 €/kg in Italy [3].

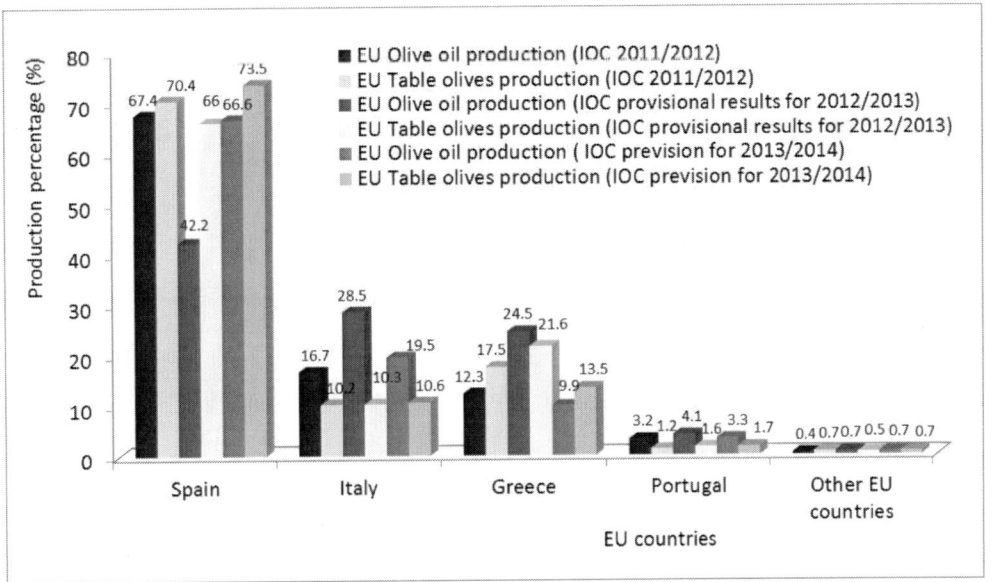

Figure 1. Percentage of European Union production of olive oil and table olives from the leading countries for 2011/2012, provisional results for 2012/2013 and a prevision for 2013/2014 according to IOC (November 2013 Newsletter).

The production of high-quality oil from olives consists of the following stages: leaf removal and washing, milling or crushing, mixing of the olive paste (malaxation), olive oil extraction processes, oil filtration, storage and bottling [4]. After their collection from the orchards and leaf removal, olives are washed to remove impurities that can contaminate the final product. Then, olives are crushed either to stone mills or hammer mills so as to break the cells and facilitate the release of oil. Malaxation prepares the paste for separation of oil, increasing the percentage of available oil. The main constituents of olive paste are: olive oil, small pieces of kernel (pit), water and cellular debris of the crushed olives. The oil can be extracted from the paste by pressing, centrifugal decanters (two phase or three phase systems), selective filtration (Sinolea process) and vertical centrifuge or through the combination of methods. After oil extraction, oil is stored in bulk up to three months and then is bottled and sold. Some oils are filtered before bottling to remove any suspended solids and residual fruit-water [5].

During the above process of oil production from olives, by-products are generated such as the olive pomace, the twigs and leaves and the olive mill wastewaters (OMWs). Olive pomace (OP) contains the olive fruit skin, pulp and olive stone (OLS) fragments as a result of oil extraction from olive paste [6]. The composition of OMWs varies depending on olive variety, the ripeness of fruit and the extraction process, i.e. centrifuge or press. The typical composition of OMWs is water, organic compounds and mineral salts in a proportion of 83-94, 4-16 and 0.4-2.5 % w./w respectively. It has a dark red to black color with high conductivity and phenolic compounds which are responsible for its toxicity towards animals and plants. The high content of organic matters and polyphenols in combination with the large quantities produced and the seasonal character of the industry has led to considerable pollution [7].

Worldwide the interest in the recovery, recycling and upgrading of residues from plant food processing has increased drastically. These food industries produce large volume of wastes both solid and liquid, which represent a potential environmental pollution problem. These wastes are also promising sources of compounds that can be recovered and used as valuable substances by developing new processes.

SOLID OLIVE OIL INDUSTRY WASTES

Partially Pyrolyzed Solid Wastes for Wastewaters Applications

Solid wastes, which were derived from olive industry, consist primarily of dried olive pomace, twigs and leaves. Olive pomace (OP) sorbents were tested for preconcentration of Cd(II), Zn(II) and Cu(II) from waters. They were prepared by heat pretreatment under inert atmosphere (partial pyrolysis) at various temperatures (100-300°C). Partial pyrolysis is expected to change the chemical and textural characteristics of the sorbent without forming activated carbon increasing also the permeability of olive pomace [8]. Other studies have shown that the speciation of Cr(III)/Cr(VI) from industrial and natural waters using partially pyrolyzed olive pomace was improved. Cr(III) was selectively adsorbed at pH 3 on pyrolyzed OP at 150°C while total Cr was determined with flame atomic absorption spectrometry. The concentration of Cr(VI) was calculated from the difference between total Cr and Cr(III) [6]. Consequently, partially pyrolyzed OP can be used as a novel sorbent for heavy metal adsorption from industrial wastewaters.

Activated Carbons Derived from Solid Wastes for Wastewaters and Gases Applications

One of the most popular and extensively researched areas is the production of activated carbons from solid wastes derived from olive oil industry. Typical activated carbons consist of particles with porous structure having a network of interconnected macropores, mesopores and micropores resulting to high surface area and good capacity. Their surface chemistry and the chemical characteristics of the adsorbate, such as polarity, ionic nature and functional groups determine the nature of bonding mechanisms and the extent of adsorption [9-17]. Researchers have shown the production of activated carbon from olive husk by chemical activation with KOH. The adsorption ability of activated carbon was tested to phenols [18] and to a mixture of polyphenols consisting of caffeic acid, vanillin vanillic acid, 2-hydroxybenzoic acid and gallic acid at two temperatures [19]. Novel activated carbons presented higher adsorption ability compared to a commercial carbon called Acticarbon CX.

Similar studies reported the adsorption of heavy metals from activated carbons, i.e. cadmium and zinc ions from activated olive stone by heating in CO_2 at 1123 K [20] or by $ZnCL_2$ activation [21], copper ions from activated olive cake by chemical activation with phosphoric acid [22], lead ions removal from potable water using carbonized olive stones by heating in N_2 at 700-800°C and activated by $ZnCL_2$ and KOH [23]. A novel research [24] proposed the production of activated carbon from olive stone using a microwave for the heating process. These carbons were used for cadmium removal from aqueous solutions in a percentage of 95.32%. Heating process through microwaves requires significantly lesser holding time compared to conventional heating methods. Moreover, the removal of radionuclide and toxic metals such as uranium and thorium from waste solutions on activated carbons is a significant subject for environmental control. Activated carbons were prepared by chemical activation of olive stones using $ZnCL_2$ as chemical agent and then heated to temperatures from 500 to 700°C. The maximum adsorption occurred at pH 6 and 4, respectively [25]. Furthermore, the use of activated carbons has expanded to the management of dye pollutants in wastewaters especially in the textile industry, i.e. methylene blue, remazol red B removal from wastewaters through activated carbons derived from olive stones either pyrolyzed in N_2 atmosphere up to 1000°C and activated with water vapor at 930°C [26,27] or chemically activated by $ZnCL_2$ [28]. Olive stone by-products can be combined with organic compounds, e.g. phenol-formaldehyde resin following appropriate pyrolysis and activation of the composite material. The originality of the produced material lies in the modification of its pore structure in comparison with activated carbons, which derived from typical raw materials, or carbon molecular sieves, that come from carbonization of polymer materials. These activated carbons, which derived from the raw composite material, were used for the adsorption of dyes and phenols from aqueous solutions [18,29,30,31]. Other toxic compounds such as cyanides are discharged by various industries particularly the chemical synthesis plants (nylon, fibers, resins and herbicides), metallurgical processes (extraction of gold and silver), plating and surface finishing [32]. Free cyanide can be removed from aqueous solutions by oxidation with hydrogen peroxide, H_2O_2 in the presence of activated carbon prepared from olive stones. Microporous carbons prepared from lignocellulosic materials are effectively used as adsorbents for different gases. Activated carbon pellets without binder have been prepared by chemical activation of olive stones with H_3PO_4 for methane storage. The activation temperature varied from 350 to 1000°C [33]. Moreover, activated carbons were also used for flue gas desulphurization because of their ability to adsorb SO_2 [34]. SO_2 is a widespread pollutant and its adsorption on activated carbons can be done in two different forms, one weakly bound on the surface of the carbonaceous matrix (physisorption process) and the other more strongly bound (chemisorption process). Carbonized and activated olive stones with CO_2 or a mixture of CO_2/H_2O up to 1100 K are also ideal adsorbents for nitrogen and carbon dioxide [35].

Lignocellulosic Wastes for Wastewaters Applications through Biosorption Process

Lignocellulosic materials, especially olive stones, were investigated as biosorbents to heavy metals and dyes. Typical SEM micrograph of olive stone biomass, which was ground and sieved to yield grains with a diameter less than 300μm, is presented in Figure 2(A). Impurities in olive stones, which derived from soils with high content of calcium carbonate or calcium fertilizers, are presented in Figure 2(B). The data of the chemical characteristics (elemental analysis) of olive stone are determined according to Energy Dispersive X-ray analysis.

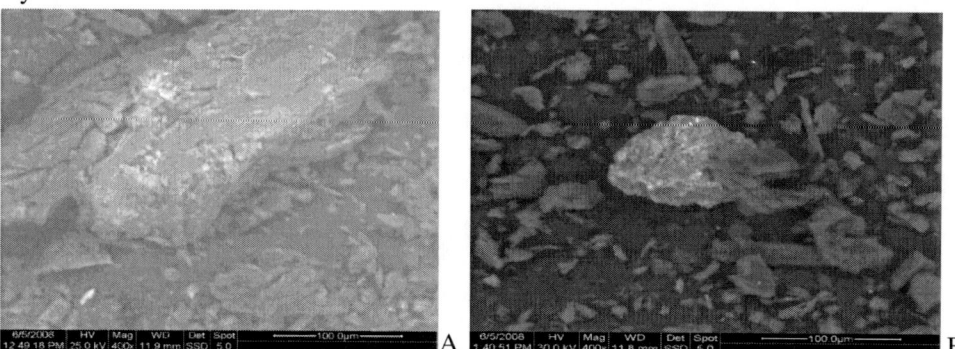

Figure 2. SEM micrographs of olive stone (A) x400 and olive stone impurities, (B) x400.

EDAX analysis of olive stone, Figure 2(A), led to the following elemental proportion: 68% C, 29% O and 3% w./w. other elements such as K, Ca, Cu while olive stone impurities, Figure 2(B), presented the following proportion: 24.5% C, 42% O, 32% Ca in the form of CaO or $CaCO_3$ and 1.5% w./w. other elements. The proportion of C/O is equal to 2.35 indicating the existence of surface functional groups, e.g. carboxyl, carbonyl groups, which help to the biosorption of heavy metals and dyes. Biosorption is an alternative technique for the removal of heavy metals from wastewaters based on the adsorption of pollutants through mechanisms such as physical adsorption, complexing and ion exchange [36].

Major advantages of biosorption compared to conventional treatment methods include: low cost, high efficiency of metal removal from wastewaters, minimization of chemical or biological sludge, no additional nutrient requirements, regeneration of biosorbent and possibility of metal recovery [37]. In a biosorption process, the main treatment of adsorptive material, e.g. olive stone, includes rinsing with boiling and cold water, drying and sieving. Olive stone waste separated from pulp were used as sorbent material for the removal of Pb(II), Ni(II), Cu(II) and Cd(II) from single and binary aqueous solution. Olive wastes were rinsed with boiling and cold water and then dried in an oven at 105 °C until constant weight. Once dried, they were ground and sieved. Metal sorption is pH-dependent and maximum sorption was found to occur at initial pH around 5.5–6.0 [38]. Moreover, the biosorption of trivalent and hexavalent chromium on olive stone materials has been studied. The percentage of adsorbed Cr(III) reached values equal to 90% at pH between 4 and 6 while the percentage of adsorbed Cr(VI) reached values equal to 80% at pH lower than 2 [39]. Another study [40] has shown the biosorption of Fe(III) ions from olive stones, which were washed with water, were boiled for two hours in order to remove the remaining organic matter, were dried and finally were ground in the mill, in a batch system. Iron adsorption was investigated with respect to initial metal ion concentration, adsorbent dose and adsorbent size. According to the results, the equilibrium adsorption capacity increased when the adsorbent particles' size decreased and the percentage of iron adsorption increased when the initial concentration of biomass increased. Other studies have shown the biosorption of cadmium [36], lead [41] and copper [37] ions from olive stone and two-phase olive mill solid wastes. Except for the biosorption of metals, olive solid wastes and especially olive pomace can be used for the biosorption of dyes such as RR198, a reactive textile dye [42]. The biosorption experiments were conducted in various operating conditions such as pH, biosorbent dosage, time, ionic strength and temperature. The results indicated that the highest dye biosorption capacity was found at pH 2 and the needed time to reach equilibrium was 40 min.

Solid Wastes for the Remediation of Polluted Soils

A promising field for solid olive oil industry wastes is the production of soil amendments and fertilizers. Researchers [43] have indicated the production of humic fertilizers by oxiammoniation of hydrolyzed olive pits residues. The raw material, which was acid-hydrolyzed olive pits, was oxidized with nitric acid and then ammoniation reaction occurred in a heated batch reactor. Functional groups such as –COOH, -OH and –OCH$_3$ were created with oxidization and then was substituted by nitrogen compounds through ammoniation reaction. During the process two fractions were produced: a soluble fraction that can be used as nitrogen enriched humic fertilizer and a solid product with less than 4% w. of nitrogen that can be applied as soil conditioner. Another study [44] has investigated the effect of olive husk and cow manure as soil amendments on heavy metal availability such as Zn(II) and Cu(II) in a contaminated calcareous soil. Such soil amendments (olive husk) increased heavy metal solubility in soils, enhanced metal absorption by plants and removed them from soils through phytoextraction.

Other Applications of Solid Wastes

Apart from the remediation of wastewaters and polluted soils, olive oil industry by-products are ideal raw materials for energy purposes, liquid and gas new products. A technical and financial analysis of thermal disposal plant solutions with energy recovery has been carried out with the combined treatment of olive mill wastewaters and olive husk [45]. Other researchers have studied the energy potential of the industrial olive by-products in order to produce electricity [46].

Due to the high heating power of OLS (heat of combustion equal to 4.1kcal/kg), they find application in the majority of thermal processes for space calefaction in residential and commercial buildings [47].

Researchers have also focused on the gasification treatment of olive tree cuttings and OLS, which led to syngas and H$_2$-rich gas production [48,49]. Moreover, olive residues can be pyrolyzed in a fixed bed reactor under specific conditions for the production of bio-oil, a fuel with similarities with petroleum with fewer harmful emissions such as SO$_x$, NO$_x$ gases and no CO$_2$ emissions to the atmosphere which accelerate the greenhouse effect [50]. Furthermore, the production of furfural has been studied by acid hydrolysis of olive stones in a tubing-bomb reactor system [51]. Furfural has mainly industrial uses as solvent. Other products derived from olive stones, are biopolyols [52] and xylo-oligosaccharides [53], which are used in food and pharmaceutical industries.

Olive by-products can successfully be used as a dietary animal supplementation. Olive leaves and stones are fibrous with a low digestibility, especially of crude protein. When they supplemented, they may be successfully used in animal diets. The rich oil content of olive leaves increases the efficiency of microbial protein synthesis in the rumen and improves milk fat quality produced from lactating animals [1,54]. Other applications of olive stones are as abrasives because they resist to rupture and deformation and do not produce contamination through their residues and in cosmetic industry due to their exfoliation qualities [55]. Market products include olive stones as a component to aid in skin exfoliation.

OLIVE MILL WASTEWATERS

Most olive oil (98%) is produced in the Mediterranean region, mainly between October and February [56]. Spain is the world's biggest producer of olive oil producing 20% of the wastes of the Mediterranean basin ($2x10^6$-$3x10^6$ m^3/year) before the implementation of the two-phase extraction process in most of the Spanish olive mills. Italy is the second European producer with 1.16 million ha of olive grove while Greece ranks third in the world in terms of olive oil production with 0.81 million ha [57].

The effect of the olive oil production is due to the great relevance of the quantity and difficulty of treatment of the related wastes (LIFE OLEICO+):

- The liquid residues (Olive Mill Waste Waters-OMWW): a mixture of vegetation water directly extracted from olives and oil production water, coming from washing and processing. OMWW can be reused for different purposes, like irrigation, compost, etc., only after suitable treatments to reduce salinity, organic content and acidity;
- pomaces: composed by skins, pits, pulp and a variable percentage of wastewater, depending from the type of extraction (2-3 phases). The pomaces typically have acidic pH and contain high concentrations of polyphenols and organic fatty matter and if properly treated, the pomaces become suitable for the production of amenders, compost and fuel;
- Two Phase Olive Mill Waste-TPOMW (alperujo): residue resulting from two-phase extraction, composed by both OMW and solid waste;
- pulp: a residual paste, which is produced if the whole olive stones are removed from the paste prior to processing.
- pits: due to high energy/power, the less expensive raw olive stones are commonly used instead of pellets for combustion in dedicated burners;
- pruning: branches, leaves and woods from the olive orchard are the only by-products which could be easily reclaimed as biomass.

Quantity and quality of residues are strongly influenced by the oil extraction method, but also from type and maturity of olives, region of origin, climatic conditions and associated cultivation and processing methods [58]. The volume of generated OMWW depends on the extraction process and varies between 40 L and 100 L per 100 kg of milled olives [59]. OMWW is an aqueous, easily fermentable, foul-smelling and turbid liquid. Its quantitative and qualitative characteristics differ broadly and depend mainly on climate, soil type, variety and ripeness of olives, physiological stage, harvesting time, as well as on the extraction method used [60,61]. It has a high biological pollutant load (BOD$_5$ in the range 20–120 g/L) which is 25–80 times greater than the pollution level of common urban wastewater and a high chemical pollutant load (COD in the range 40–240 g/L) [56]. It is characterized by relatively acidic pH (4.5-5.5), high electrical conductivity (3.5-12.5 dS/m), high organic content (40-165 g/L) [62]. In addition, it has high free polyphenol concentrations (between 3.0 and 24 g/L) due to olive pulp esters and glycoside hydrolysis, produced during oil extraction [58,62]. Its inorganic constituents may potentially serve as sources of plant nutrients, improve soil fertility [63] and render this effluent suitable for re-use as soil amendment [64].

Landspreading of Wastes

The key tenet in support of land spreading of agricultural wastes is that it recycles nutrients and organic matter to the land, which would otherwise be lost in disposal to landfill or thermal destruction. Provided that benefit to agriculture can be demonstrated, land spreading of wastes is considered preferable to thermal destruction or landfilling in the ranking of options in the Waste Framework Directive issued by the EU. Land spreading is recognized as waste utilization and not disposal, and thus, farmers that recognize and adopt this precondition should be eager to improve the management of wastes by investment as appropriate in storage at the point of production, dewatering and other treatment, monitoring and analysis, and field trials to quantify the agricultural benefit of their wastes. Two of the benefits of land spreading of wastes are that it is often an economic route for the waste producer compared with the other options available, and for the farmer it usually represents a free or competitively-priced source of nutrients and/or soil conditioner. Without doubt, landspreading is an environmentally and economically beneficial option of wastes management.

Uncontrolled disposal of olive mill wastes (OMW) causes several ecological problems due to, strong and unpleasant odor, potential toxicity to soil microflora, surface- and groundwater [65,66]. The toxicity of OMW is mainly related to its high content of phenolic compounds [60,67]. OMW is often discharged directly into sewer systems and water streams or disposed of in lagoons despite the fact that such management options are not allowed in most Med countries [65]. It is underlined that no specific European Commission legislation exists today for OMW management and each country issues different guidelines.

Although the disposal of mills wastes in the environment is not permitted, it is estimated that up to 1.5 million tons of olive oil mills' wastes are disposed untreated each year. Due to the chemical-physical characteristics olive wastes can cause effects on the environment (LIFE OLEICO+):

- into the water: the discharge of olive waste into water bodies may worryingly increase the phosphorous content, cause color alteration and bad smell;
- into the air: the discharge of olive waste in open spaces may stimulate microbial fermentation, with the production of methane and a wide array of harmful or simply bad smelling gases.
- and into the soil.

In this case and for Mediterranean countries disposal of OMW constitute a major environmental problem. In some Mediterranean countries, untreated OMW are often discharged directly into sewer systems and water streams or disposed in evaporation ponds/lagoons and in soils despite the fact that such management options are not allowed in most Mediterranean countries [65]. Moreover, no specific European Commission legislation exists today for OMW management and each country issues different guidelines [67,68].

Application of rich in organic matter OMW to soils closes the residue-resource gap and may play a fundamental role in preservation of olive tree ecosystems. However, application of insufficiently stable organic matter may induce a number of negative effects such as increase in mineralization rate of native soil organic C, induction of anaerobic conditions and release of phytotoxic compounds that may adversely affect plant growth [69,70].

Higher levels of total nitrogen and NO_3-N were measured in soils treated with OMW [71] while levels of NH_4-N appeared lower probably due to low nitrogen mineralization as a result of high C/N ratios present in OMW. The positive effect of OMW application on the levels of total organic nitrogen was related to addition of a high C source and immobilization of inorganic or mineralized N by N-fixing microflora [72,73].

Controversial effects have been reported concerning OMW disposal on soil pH [65]. Researchers have mentioned [74] that soil pH decreased after amendment with OMW. However, the application of agricultural wastes has a temporary acidifying effect shortly after application [75]. On the other hand, the disposal of agricultural wastes on soils that have high clay content or are rich in carbonates has no significant effect on pH [62]. A slight increase in soil pH values after application of non-buffered OMW has been reported [76]. Sierra et al. [77] suggested that this increase in pH was due to generation of Na_2CO_3 from Na-rich wastewaters while pH values came back to normal two months after OMW application.

Increased leaching of sodium and high salinity values were observed in soils after OMW application [65,74,78]. It seems that the increase in EC values, mainly in the upper soil layers, when excessive load of OMW is applied, is irreversible [79,80]. Contrary to the above, Chartzoulakis et al. [81] reported that after 3 years soil salinity remains almost unaffected in soil fertilized with raw OMW.

López-Piñeiro et al. [82] noted that high levels of soluble organic carbon present in OMW facilitate transport and accelerate ion exchange in soils, while contact of OMW with clayey soils promotes dissolution of carbonates and increases permeability. In addition, incorporation of OMW organic matter in soil activates soil microbial activity and causes degradation of organic matter, however, application at high rates inhibits microbial growth and slows down degradation processes [62]. The content of exchangeable soil K increases in general after application of OMW [59,79,83]. However, overloading of soil with K^+ and Na^+ impairs the composition of the exchange complex and causes leaching of calcium and magnesium [84] resulting thus in increased osmotic pressure, alteration of soil structure and higher salinity levels [74,85]. The content of soil available P also increases after OMW application [59,83]. This increase seems to be a serious issue in areas of uncontrolled OMW disposal [62]. On the contrary, extractable phosphorus levels in soils decrease after OMW application presumably due to the formation of insoluble Ca-P compounds or the inhibition of microbial activity that facilitates phosphate release [86].

The presence of recalcitrant and phytotoxic metals in soils endangers the quality of the ecosystems. Different processes that occurred in soil systems affect metals mobility by positive or negative way. Thus, although OMW is not rich in heavy metals [87], application of OMW was found to increase metal availability in soils and mobilize heavy metals due to its acidity. OMW is slightly acidic and so can reduce soil pH when used as an organic soil amendment [88]. This, in turn, may reduce metals' immobilization in soil [87]. Moreover, the type of organic matter is a significant factor which affects metals' mobility, since the presence of high water soluble organic matter fraction enhances metal solubility and thus leaching [89]. Bejarano and Madrid [90] found that the released amounts of metals due to pH and OMW effect, were comparable to the exchangeable fractions. On the other hand, metals are fixed by carbonates, while, the degradation process of organic matter of OMW leads to the formation of inorganic compounds (such as phosphates) that can precipitate metals [91]. In addition, the CO_2 produced can enhance the dissolution of $CaCO_3$ in calcareous soils, which enters the carbonate/bicarbonate equilibrium in the soil solution, thus affecting metal precipitation. Application of OMW may also increase metal availability if enhanced microbial activity is seen in soils [92]. De la Fuente et al. [93] reported increased availability of Zn and Pb shortly after application of OMW. Similarly, increased concentrations of available Zn and Fe in soil were recorded after OMW application by Tisdale et al. [94] and Aqeel et al. [95], which were attributed to enhanced soil microbial activity producing natural chelates that enhance micronutrient availability. Although organic matter mineralization can release the associated metals in the soil solution, the presence of $CaCO_3$ in calcareous soils can lead to metal precipitation as CO_3 salts [96].

Treatment of OMW

OMW are considered as valuable resources because they are very rich in organic matter and contain nutrients that can be easily recycled. Besides, OMW may be considered as a significant water source [62,63]. These factors underline the important recycling potential of OMW as inexpensive fertilizer that can be used to maintain and also restore the quality of degraded agro-ecosystems [65]. Remediation of OMW can be assisted by soil biotic and abiotic processes. OMW acidity can be buffered by the alkalinity provided by soil carbonates [66,67]. Soil also acts as natural catalyst and promotes oxidation and polymerization of phenolic compounds into less toxic products [81]. Soil microflora, mainly fungi, is also able to detoxify oxidized OMW [82]. Thus, OMW can be efficiently recycled in Med countries by taking into account soil properties and climatic conditions to maximize anticipated benefits.

Although several techniques have been developed for OMW management, detoxification and valorization their application is often too expensive for most olive-oil mills [97]. Treatment of OMW is complex due to its high COD and content of phenolic compounds which inhibit biodegradation [98]. The presence of polysaccharides, lipids, proteins may inhibit growth of certain specific anaerobic microorganisms [99]. Various treatment methods have been proposed to detoxify OMW prior to discharge in evaporation ponds or speading on land [100,101]. These include physicochemical, biological, thermal and combined methods [102,103,104]. Advanced oxidation of OMW has been also studied in detail [105,106].

Soil biotic and abiotic components play very important role during OMW remediation. Hanifi and El Hadrami [107] suggested valorization of soil degradation potency as an efficient detoxification strategy. Mediterranean soils offer ideal conditions for efficient recycling of OMW since its acidity can be buffered by the alkalinity provided by soil carbonates [108]. Storage in evaporation ponds enables rather fast OMW degradation and is considered as an efficient and cost effective pre-treatment option [109]. Moreover indigenous micro-organisms present in OMW exhibit noticeable potential for degradation of phenolic compounds.

Biotreatment of OMW can be carried out both aerobically [69] and anaerobically [110]. Among the different anaerobic treatment alternatives studied so far, the use of upflow anaerobic sludge blanket reactor (UASB) offers significant advantages [111]. Biogas, mainly methane, may be also produced [66]. Several microorganisms are able to remove phenolic compounds from OMW [112]. Composting is recognized as an environment friendly and economically viable technology for utilizing organic wastes such as OMW [63]. Experiments have indicated the need for lignin-cellulosic bulking agents and nitrogen sources for the development of a process-controlled composting protocol, so that phytotoxicity is eliminated and a final product with stabilized and humified organic matter is obtained [113]. Another treatment option involves mixing of destoned OMW with appropriate hygroscopic natural organic wastes, packaging in sacks and outdoor storage for a period of three months to stimulate natural aerobic microbial activity [114].

In addition, traditional methods used for the treatment of OMW in the Med region include: a) disposal in shallow evaporation ponds, b) disposal on soil, c) incineration and d) production of fermentation products, fat and oils preservatives (i.e. phenolic compounds) [62]. The first two are the most common low cost management options considering the fact that olive oil production units are widely scattered in most Med countries.

The main limiting factors for the development of rather cheap and simple methods for remediation and safe reuse of OMW are heterogeneity in composition, potential toxicity and high concentration of recalcitrant hazardous compounds [115].

CONCLUSION

Olive oil industries produce large volume of wastes both solid and liquid, which represent an environmental pollution problem. These wastes are also promising sources of compounds that can be recovered and used as valuable substances. Solid olive oil industry wastes, especially dried olive pomace and olive stone, are used as filters for the removal of heavy metals and dyes from waters and industrial wastewaters. They are formed either to partially pyrolyzed solids or activated carbons through carbonization and activation processes. Except for pyrolyzed solids and carbons, olive stones through a soft process, i.e. covering rinsing, drying and sieving, can be used as biosorbent materials for the purification of wastewaters from heavy metals and dyes. Microporous activated carbons derived especially from olive stones, are applied as filters for the storage of gases e.g. CO_2, SO_2. A developing field is the production of new soil amendments and fertilizers through lignocellulosic by-products derived from olive oil industry, leading either to the remediation of polluted soils from heavy metals or to the enrichment of soil with crucial nutrients, e.g. nitrogen. Olive oil industry by-products are ideal raw materials for energy purposes, liquid and gas new products, e.g. bio-oil, furfural, or as plastic fillers, abrasives and in cosmetic industry. Furthermore, leaves and olive stones contribute also to dietary animal supplementation.

There is no specific European Commission legislation exists today for OMW management, so each country issues different guidelines that may differ between regions. OMW is often discharged directly into sewer systems and water streams or disposed of in lagoons despite the fact that such management options are not allowed in most Mediterranean countries. OMW have favorable chemical properties and disposal on soils may be considered as an appropriate option to solve management problems, restore soil fertility and promote productivity. OMW may be also used for restoration of degraded soils in arid and semi-arid climates as well as for maintenance of groundwater levels. Complete treatment and detoxification of OMW is rather complex and expensive especially at mill scale in most Med countries. Composting seems to be a feasible low cost approach that allows recycling of OMW and production of an agronomically acceptable end-product, free of phytotoxic substances and heavy metals. OMW can be efficiently recycled in the Med countries since soil properties and climatic conditions are favorable.

REFERENCES

[1]   Alcaide E M, & Nefzaoui A (1996) Recycling of olive oil by-products: Possibilities of utilization in animal nutrition. *International Biodeterioration & Biodegradation*, 38(3–4), 227-235.

[2]   International Olive Oil Council, World Olive Oil Figures, http://www. internationaloliveoil. org/estaticos/view/131-world-olive-oil-figures, Acces-sed 20/12/ 2013

[3]   http://www.oliveoiltimes.com/olive-oil-business/world-olive-oil-production-tops-3-million-tons/36985, Accessed 20/12/2013

[4]   Vossen P (2007) Olive Oil: History, Production and Characteristics of the World's Classic Oils. *Horticultural Science*, 42(5), 1093-1100.

[5]   Kapellakis I E, Tsagarakis K P, & Crowther J C (2008) Olive oil history, production and by-product management. *Reviews in Environmental Science & Biotechnology*, 7, 1–26.

[6]   El-Sheikh A H, Abu Hilal M M, & Sweileh J A (2011) Bio-separation, speciation and determination of chromium in water using partially pyrolyzed olive pomace sorbent. *Bioresource Technology*, 102, 5749-5756.

[7]   Davies L C, Vilhena A M, Novais J M, & Martins-Dias S (2004) Olive mill wastewater characteristics: modelling and statistical analysis. *Grasas y Aceites*, 55(3), 233-241.

[8]   El-Sheikh A H, Sweileh J A, & Saleh M I (2009) Partially pyrolyzed olive pomace sorbent of high permeability for preconcentration of metals from environmental waters. *Journal of Hazardous Materials*, 169, 58-64.

[9]   Li W, Yang K, Pehg J, Zhang L, Guo S, & Xia H (2008) Effects of carbonization temperatures on characteristics of porosity in coconut shell chars and activated carbons derived from carbonized coconut shell chars. *Industrial Crops and Products*, 28, 190-198.

[10]   Aksu Z, & Yener J (2001) A comparative adsorption/biosorption study of mono-chlorinated phenols onto various sorbents. *Waste Management*, 21, 695-702.

[11] Vijayalakshmi P R, Raksh V J, & Rodriguez J (1998) Adsorption of phenol, cresol isomers and benzyl alcohol from aqueous solution on activated carbon at 278, 298 and 323 K. *Journal of Chemical Technology and Biotechnology*, 71, 173-179.

[12] Calleja G, Serna J, & Thirumaleswara S G B (1993) Kinetics of adsorption of phenolic compounds from wastewater onto activated carbon. *Carbon*, 31, 691-697.

[13] Streat M, Patrick J W, & Camporro-Perez M J (1995) Sorption of phenol and para-chlorophenol from water using conventional and novel activated carbons, *Water Research*, 29, 467-472.

[14] Edgehill R U, & Lu G Q (1998) Adsorption characteristics of carbonized bark for phenol and pentachlophenol. *Journal of Chemical Technology and Biotechnology*, 71, 27-34.

[15] Brasquet C, Roussy J, Subrenat E, & Le Cloirec P (1996) Adsorption and selectivity of activated carbon fibers application to organics. *Environmental Technology*, 17, 1245-1252.

[16] Srivastava S K, Tyagi R, Pal N, & Mohan D (1997) Process development for removal of substituted phenol by carbonaceous adsorbent obtained from fertilizer waste. *Journal of Environmental Engineering*, 123, 842-851.

[17] Ferro-Garcia M A, Rivera-Ultrilla J, Bautista-Toledo I, & Moreno-Castilla C M (1996) Chemical and thermal regeneration of an activated carbon saturated with chlorophenols. *Journal of Chemical Technology and Biotechnology*, 67, 183-189.

[18] Ioannou Z, & Simitzis J (2009) Adsorption kinetics of phenol and 3-nitrophenol from aqueous solutions on conventional and novel carbons *Journal of Hazardous Materials*, 171, 954–964.

[19] Michailof C, Stavropoulos G C, & Panayiotou C (2008) Enhanced adsorption of phenolic compounds, commonly encountered in olive mill wastewaters, on olive husk derived activated carbons. *Bioresource Technology*, 99, 6400-6408.

[20] Ferro-Garcia M A, Rivera-Utrilla J, Rodríguez-Gordillo J, & Bautista-Toledo I, (1988) Adsorption of zinc, cadmium and copper on activated carbons obtained from agricultural by-products. *Carbon*, 26(3), 363-373.

[21] Kula I, Uğurlu M, Karaoğlu H, & Çelik A (2008) Adsorption of Cd(II) ions from aqueous solutions using activated carbon prepared from olive stone by ZnCL$_2$ activation, *Bioresource Technology*, 99, 492-501.

[22] Baccar R, Bouzid J, Feki M, & Montiel A (2009) Preparation of activated carbon from Tunisian olive-waste cakes and its application for adsorption of heavy metal ions. *Journal of Hazardous Materials*, 161, 1522-1529.

[23] Spahis N, Addoun A, Mahmoudi H, & Ghaffour N (2008) Purification of water by activated carbon prepared from olive stones. *Desalination*, 222, 519-527.

[24] Alslaibi T M, Abustan I, Ahmad M A, & Foul A A (2013) Cadmium removal from aqueous solution using microwaved olive stone activated carbon. *Journal of Environmental Chemical Engineering*, 1(3), 589-599.

[25] Kütahyali C, & Eral M (2010) Sorption studies of uranium and thorium on activated carbon prepared from olive stones: Kinetic and thermodynamic aspects. *Journal of Nuclear Materials*, 396, 251-256.

[26] Berrios M, Ángeles Martín M, & Martín A (2012) Treatment of pollutants in wastewater: Adsorption of methylene blue onto olive-based activated carbon. *Journal of Industrial and Engineering Chemistry*, 18, 780-784.

[27] Simitzis J, & Sfyrakis J (1993) Pyrolysis of lignin biomass-novolac resin for the production of polymeric carbon adsorbents. *Journal of Analytical and Applied Pyrolysis*, 26, 37-52.

[28] Uğurlu M, Gürses A, & Acukyildiz M (2008) Comparison of textile dyeing effluent adsorption on commercial activated carbon and activated carbon prepared from olive stone by $ZnCL_2$ activation. *Microporous and Mesoporous Materials*, 111, 228-235.

[29] Simitzis J, & Ioannou Z (2011) Activated carbonaceous materials based on thermosetting binder precursors in: Activated Carbon: Classifications, Properties and Applications, Chap. 12, J. F. Kwiatkowski (editor), Nova Science Publishers, Inc. p.p. 377-392.

[30] Ioannou Z, & Simitzis J (2013) Production of carbonaceous adsorbents from agricultural by-products and novolac resin under a continuous countercurrent flow type pyrolysis operation. *Bioresource Technology*, 129, 191–199.

[31] Ioannou Z, & Simitzis J (2013) Adsorption of methylene blue dye onto activated carbons based on agricultural by-products: equilibrium and kinetic studies. *Water Science & Technology*, 67(8), 1688-1694.

[32] Yeddou R, Nadjemi B, Halet F, Ould-Dris A, & Capart R (2010) Removal of cyanide in aqueous solution by oxidation with hydrogen peroxide in presence of activated carbon prepared from olive stones. *Minerals Engineering*, 23, 32-39.

[33] Djeridi W, Ouederni A, Wiersum A D, Llewellyn P L, & El Mir L (2013) High pressure methane adsorption on microporous carbon monoliths prepared by olives stones. *Materials Letters*, 99, 184-187.

[34] Carrasco-Marin F, Utrena-Hidalgo E, Rivera-Utrilla J, & Moreno-Castilla C (1992) Adsorption of SO_2 in flowing air into activated carbons from olive stones. *Fuel*, 71, 575-580.

[35] Iley M, Marsh H, & Rodriguez-Reinoso F (1973) The adsorptive properties of carbonized olive stones. *Carbon*, 11, 633-638.

[36] Blázquez G, Hernáinz F, Calero M, & Ruiz-Núñez L F (2005) Removal of cadmium ions with olive stones: the effect of some parameters. *Process Biochemistry*, 40, 2649-2654.

[37] Blázquez G, Martín-Lara M A, Dionisio-Ruiz E, Tenorio G, & Calero M (2011) Evaluation and comparison of the biosorption process of copper ions onto olive stone and pine bark. *Journal of industrial and Engineering Chemistry*, 17, 824-833.

[38] Fiol N, Villaescusa I, Martínez M, Miralles N, Poch J, & Serarols J (2006) Sorption of Pb(II), Ni(II), Cu(II) and Cd(II) from aqueous solution by olive stone waste. *Separation and Purification Technology*, 50, 132-140.

[39] Blázquez G, Hernáinz F, Calero M, Martín-Lara M A, & Tenorio G (2009) The effect of pH on the biosorption of Cr(III) and Cr(VI) with olive stone. *Chemical Engineering Journal*, 148, 473-479.

[40] Nieto L M, Alami S B D, Hodaifa G, Faur C, Rodríguez S, Giménez J A, & Ochando J (2010) Adsorption of iron on crude olive stones. *Industrial Crops and Products*, 32, 467-471.

[41] Martín-Lara M A, Hernáinz F, Calero M, Blázquez G, & Tenorio G (2009) Surface chemistry evaluation of some solid wastes from olive-oil industry used for lead removal from aqueous solutions. *Biochemical Engineering Journal*, 44, 151-159.

[42] Akar T, Tosun I, Kaynak Z, Ozkara E, Yeni O, Sahin E N, & Akar S T (2009) An attractive agro-industrial by-product in environmental cleanup: Dye sorption potential of untreated olive pomace. *Journal of Hazardous Materials*, 166, 1217-1225.

[43] Riera F A, Alvarez R, & Coca J (1991) Humic fertilizers by oxiammoniation of hydrolyzed olive pits residues. *Fertilizer Research*, 28, 341-348.

[44] Clemente R, Paredes C, & Bernal M P (2007) A field experiment investigating the effects of olive husk and cow manure on heavy metal availability in a contaminated calcareous soil from Murcia (Spain). *Agriculture, Ecosystems and Environment*, 118, 319-326.

[45] Caputo A C, Scacchia F, & Pelagagge P M (2003) Disposal of by-products in olive oil industry: waste-to-energy solutions. *Applied Thermal Engineering*, 23, 197-214.

[46] Celma A R, Rojas S, & López-Rodriguez F (2007) Waste-to-energy possibilities for industrial olive and grape by-products in Extremadura. *Biomass & Bioenergy*, 31, 522-534.

[47] Duran C Y (1985) Propriedades termoquimicas del arujo de aceitona, Poder calorifico. *Grasas y Aceites*, 36, 45–47.

[48] Skoulou V, Zabaniotou A, Stavropoulos G, & Sakelaropoulos G (2008) Syngas production from olive tree cuttings and olive kernels in a downdraft fixed-bed gasifier. *International Journal of Hydrogen Energy*, 33, 1185-1194.

[49] Skoulou V, Koufodimos G, Samaras Z, & Zabaniotou A (2008) Low temperature gasification of olive kernels in a 5-kW fluidized bed reactor for H_2-rich producer gas. *International Journal of Hydrogen Energy*, 33, 6515-6524.

[50] Pütün A E, Uzum B B, Apaydin E, & Pütün E (2005) Bio-oil from olive oil industry wastes: Pyrolysis of olive residue under different conditions. *Fuel Processing Technology*, 87, 25-32.

[51] Montané D, Salvadó J, Torras C, & Farriol X (2002) High-temperature dilute-acid hydrolysis of olive stones for furfural production. *Biomass & Bioenergy*, 22, 295-304.

[52] Matos M, Filomena-Barreiro M, & Gandini A (2010) Olive stone as a renewable source of biopolyols. *Industrial Crops and Products*, 32 7-12.

[53] Nabarlatz D, Ebringerová A, & Montané D (2007) Autohydrolysis of agricultural by-products for the production of xylo-oligosaccharides. *Carbohydrate Polymers*, 69, 20-28.

[54] Molina-Alcaide E, & Yáñez-Ruiz D R (2008) Potential use of olive by-products in ruminant feeding: A review. *Animal Feed Science and Technology*, 147, 247-264.

[55] Rodríguez G, Lama A, Rodríguez R, Jiménez A, Guillén R, & Fernández-Bolaños J (2008) Olive stone an attractive source of bioactive and valuable compounds. *Bioresource Technology*, 99, 5261-5269.

[56] Alburquerque J A, Gonzalvez J, Garcia D, & Cegarra J (2004) Agrochemical characterization "alperujo", a solid by-product of the two-phase centrifugation method of olive oil extraction. *Bioresource Technology*, 91, 195-200.

[57] EUROSTAT (2012). Available from: http://epp.eurostat.ec.europa.eu/. Accessed 20/12/2013

[58] Paraskeva P, & Diamadopoulos E (2006) Technologies for olive mill wastewater (OMW) treatment: a review. *Journal of Chemical Technology and Biotechnology*, 81, 1475-1485.

[59] Ferri D, Convertini G, Montemurro F, Rinaldi M, & Rana G (2002) Olive Wastes Spreading in Southern Italy: Effects on Crops and Soil. In: Proceedings of 12th International Soil Conservation Conference, (ISCO): Sustainable Utilization of Global Soil and Water Resources. Beijing 2002. pp. 593-600.

[60] Kotsou M, Mari I, Lasaridi K, Chatzipavlidis I, Balis C, & Kyriacou A (2004) The effect of olive oil mill wastewater (OMW) on soil microbial communities and suppressiveness against Rhizoctonia solani. *Applied Soil Ecology*, 26, 113-121.

[61] Altieri R, & Esposito A (2008) Olive orchard amended with two experimental olive mill wastes mixtures: Effects on soil organic carbon, plant growth and yield. *Bioresource Technology*, 99, 8390-8393.

[62] Kavvadias V, Doula M K, Komnitsas K, & Liakopoulou N (2010) Disposal of olive oil mill wastes in evaporation ponds: Effects on soil properties. *Journal of Hazardous Materials,* 182, 144-155.

[63] Paredes C, Cegarra J, Bernal M P, & Roig A (2005) Influence of olive mill wastewater in composting and impact of the compost on Swiss chard crop and soil properties. *Environment International*, 31, 305-312.

[64] Rinaldi M, Rana G, & Introna M (2003) Olive-mill wastewater spreading in southern Italy: effects on a durum wheat crop. *Field Crops Research*, 84, 319–326.

[65] Hanifi S, & El Hadrami I (2009) Olive Mill Wastewaters: Diversity of the Fatal Product in Olive Oil Industry and its Valorisation as Agronomical Amendment of Poor Soils. A Review. *Journal of Agronomy*, 8, 1-13.

[66] Khatib A, Aqra F, Yaghi N, Subuh Y, Hayeek B, Musa M, Basheer S, & Sabbah I (2009) Reducing the environmental impact of olive mill wastewater. *American Journal of Environmental Sciences*, 5, 1-6.

[67] Komnitsas K, Zaharaki D, Doula M, & Kavvadias V (2011) Origin of Recalcitrant Heavy Metals Present in Olive Mill Wastewater Evaporation Ponds and Nearby Agricultural Soils. *Environmental Forensics*, 12, 319–326.

[68] McNamara C J, Anastasiou C, O'Flaherty V, & Mitchell R (2008) Bioremediation of Olive Mill Wastewater. *International Biodeterioration & Biodegradation*, 61, 127-134.

[69] Komilis E, Karatzas C P, & Halvadakis D P (2005) The effect of olive mill wastewater on seed germination after various pretreatment techniques. *Journal of Environmental Management*, 74, 339-348.

[70] Martinez-Garcia G, Williams C J, Burgoyne A, & Edyvean R G J (2003) Aerobic treatment of alpechin by Candida tropicalis. http://www.expoliva.com/expoliva2003/ simposium/comunicaciones/ Tec-18-Texto.pdf.

[71] Nikolaidis N, Kalogerakis N, Psyllakis E, Tzorakis O, Moraitis D, Stamati F, Valta K, Peroulaki E, Vozinakis I, & Papadoulakis V (2008) Agricultural Product Waste Management in Evrotas River Basin. EnviFriendly Technology Report #3 May 2008, LIFE05ENV/GR/00045, pp. 21.

[72] Bengtson P, & Bengtsson G (2005) Bacterial immobilization and re-mineralization of N at different growth rates and N concentrations. *Microbial Ecology*, 54, 13-19.

[73] Aguilar M J (2009) Olive oil mill wastewater for soil nitrogen and carbon conservation. *Journal of Environmental Management*, 90, 2845-2848.

[74] Zenjari A, & Nejmeddine A (2001) Impact of spreading olive mill wastewater on soil characteristics: laboratory experiments. *Agronomie*, 21, 749–755.

[75] Levi-Minzi R, Saviozzi A, Riffaldi R, & Falzo L (1992) Land application of vegetable water: effects on soil properties. *Olivae*, 40, 20-25.

[76] Sierra J, Marti E, Montserrat G, Cruanas R & Garau M A (2000) Aprovechamiento del alpechin a traves de suelo. Estimacion del posible impacto sobre la saguas de infiltracion. *Edafologia*, 72, 91-102.

[77] Sierra J, Marti E, Garau M A, & Cruanas R (2007) Effects of the agronomic use of olive oil mill wastewater: Field experiment. *Science of the Total Environment*, 378, 90-94.

[78] Sierra J, Marti E, Montserrat G, Cruanas R, & Garnu M A (2001) Characterisation and evolution of a soil affected by olive oil mill wastewater disposal. *Science of the Total Environment*, 279, 207-214.

[79] Mechri B, Mariem F B, Baham M, Elhadj S B, & Hammami M (2008) Change in soil properties and the soil microbial community following land spreading of olive mill wastewater affects olive trees key physiological parameters and the abundance of arbuscular mycorrhizal fungi. *Soil Biology & Biochemistry*, 40, 152-161.

[80] Di Serio M G, Lanza B, Mucciarella M R, Russi F, Iannucci E, Marfisi P, & Madeo A (2008) Effects of olive mill wastewater spreading on the physico-chemical and microbiological characteristics of soil. *International Biodeterioration & Biodegradation*, 62, 403–407.

[81] Chartzoulakis K, Psarras G, Moutsopoulou M, & Stefanoudaki E (2010) Application of olive mill wastewater to a Cretan olive orchard: Effects on soil properties, plant performance and the environment. *Agriculture Ecosystems & Environment*, 138, 293-298.

[82] López-Piñeiro A, Murillo S, Barreto C, Munoz A, Rato J M, Albarran A, & Garcia A (2007) Changes in organic matter and residual effect of amendment with two-phase olive-mill waste on degraded agricultural soils. *Science of the Total Environment*, 378, 84-89.

[83] López-Piñeiro A, Albarrán A, Nunes J M R, & Barreto C (2008) Short and medium-term effects of two-phase olive mill waste application on olive grove production and soil properties under semiarid Mediterranean conditions. *Bioresource Technology*, 99, 7982-7987.

[84] Cabrera F, López R, Martinez-Bordiú A, Dupuy de Lome E, & Murillo J M (1996) Land treatment of olive oil wastewater. *International Biodeterioration & Biodegradation*, 38, 215-225.

[85] Mekki A, Dhouib A, & Sayadi S (2006) Changes in microbial and soil properties following amendment with treated and untreated olive mill wastewater. *Microbiological Research*, 161, 93-101.

[86] Mechri B, Attia F, Braham M, Ben Elhadj S, & Hammami M (2007) Agronomic application of olive mill wastewaters with phosphate rock in a semi-arid Mediterranean soil modifies the soil properties and decreases the extractable soil phosphorus. *Journal of Environmental Management*, 85, 1088-1093.

[87] Martinez-Garcia G, Bachmann R T, Williams C J, Burgoynr A, & Edyvean R G J (2006) Olive oil waste as a biosorbent for heavy metals. *International Biodeterioration and Biodegradation,* 58, 231–238.

[88] Clemente R, Paredes C, Bernal M P (2007) A field experiment investigating the effects of olive husk and cow manure on heavy metal availability in a contaminated calcareous soil. *Agriculture Ecosystems & Environment*, 118, 319–326.

[89] Bernal M P, Clemente R., Walker D J (2009) Interactions of heavy metals with soil organic matter in relation to phytoremediation, in: Navarro-Avi, J.P., (Ed.), Phytoremediation: The Green Salvation of the World, Research Signpost, Kerala, pp. 109–129.

[90] Bejarano M, Madrid L (1996) Solubilization of heavy metals from a river sediment by an olive oil mill effluent at different pH values. Environmental Technology, 17, 427–432.

[91] Walker D J, Clemente R, Roig A, Bernal M P (2003) The effects of soil amendments on heavy metal bioavailability in two contaminated Mediterranean soils. Environmental Pollution, 122, 303–312.

[92] Kavvadias V, Komnitsas K, Doula M (2011) Long term effects of olive mill wastes disposal on soil fertility and productivity. In: Satinder Kaur Brar (Ed.) Hazardous Materials: Types, Risks and Control, Nova Science Publishers Inc., pp.433-471.

[93] De la Fuente C, Clemente R, Bernal M P (2008) Changes in metal speciation and pH in olive processing waste and sulphur-treated contaminated soil. Ecotoxicology *and* Environmental Safety, 70, 207-215.

[94] Tisdale S L, Nelson W L, Beaton J B, Havlin J L (2003) Soil fertility and fertilizers. 5th ed., Pearson Education, Inc.

[95] Aqeel A M, Hameed K M, Alaudatt M (2007) Effect of Olive Mill By-products on Mineral Status, Growth and Productivity of Faba Bean. *Journal of Agronomy,* 6, 403-408.

[96] Usman A R A, Kuzyakov Y, Stahr K (2004) Dynamics of organic C mineralization and the mobile fraction of heavy metals in a calcareous soil incubated with organic wastes. *Water Air & Soil Pollution,* 158, 401–418.

[97] Ouzounidou G, Zervakis G I, & Gaitis F (2010) Raw and microbiologically detoxified olive mill waste and their impact on plant growth. *Terrestrial and Aquatic Environmental Toxicology*, 4, 21-38.

[98] Ramos-Comenzana A, Monteolica-Sanchez M, & Lopez M J (1995) Bioremediation of alpechin. *Biosorption and Bioremediation*, 35, 249-268.

[99] Echaliotis C, Papadopoulou K, Kotsou M, Mari I, & Balis C (1999) Adaptation and population dynamics of Aztobacter vinelandii during aerobic biological treatment of olive mill wastewater. *FEMS Microbiology Ecology*, 30, 301-311.

[100] El Hadrami A, Belaqziz M, El Hassni M, Hanifi S, Abbad A, Capasso R, Gianfreda L, & El Hadrami I (2004) Physico-chemical characterization and effects of olive oil mill wastewaters fertirrigation on the growth of some Mediterranean crops. *Journal of Agronomy* 3, 247-254.

[101] Bonanomi G, Giorgi V, Giovanni D, Neri Scala D, & Banaomi F (2006) Olive mill residues affect saprophytic growth and disease incidence of foliar and soilborne plant fungal pathogens. *Agriculture, Ecosystems & Environment*, 115, 194-200.

[102] Israelides C, Vlyssides A, Galiatsatou P, Iconomou D, Arapoglou D, Christopoulou N, & Bocari M (2006) Methods of Integrated Management of Olive Oil Mill Wastewater (OMW). EU Environmental Quality Standards (Eqs), International Conference

'Protection and Restoration of the Environment VIII, Chania, Crete, Greece, July 3-7, 2006.

[103] Roig A, Cayuela M L, & Sánchez-Monedero M A (2006) An overview on olive mill wastes and their valorisation methods. *Waste Management*, 26, 960-969.

[104] Arvanitoyannis I S, Kassaveti A, & Stefanatos S (2007) Olive oil waste treatment: A comparative and critical presentation of methods, advantages and disadvantages. *Critical Reviews in Food Science and Nutrition*, 47, 187-229.

[105] El Hajjouji H, Barje F, Pinelli E, Bailly J R, Richard C, Winterton P, Revel J C, & Hafidi M (2008) Photochemical UV/TiO$_2$ treatment of olive mill wastewater (OMW). *Bioresource Technology*, 99, 7264-7269.

[106] Kallel M, Belaid C, Boussahel R, Ksibi M, Montiel A, & Elleuch B (2009). Olive mill wastewater degradation by Fenton oxidation with zero-valent iron and hydrogen peroxide. *Journal of Hazardous Materials*, 163, 550–554.

[107] Hanifi S, & El Hadrami I (2008) Olive Mill Wastewaters fractioned soil-application for safe agronomic reuse in date palm (Phoenix dactylifera L.) fertilization. *Journal of Agronomy*, 7, 63-69.

[108] Senesi N, Plaza C, Brunetti G, & Polo A (2007) A comparative survey of recent results on humic-like fractions in organic amendments and effects on native soil humic substances. *Soil Biology & Biochemistry*, 39, 1244-1262.

[109] Saadi I, Laor Y, Raviv M, & Medina S (2007) Land spreading of olive mill wastewater: effects on soil microbial activity and potential phytotoxicity. *Chemosphere*, 66, 75-83.

[110] Borja R, Garrido E S, Martinez L, Cormenzana R A, & Martin A (1993) Kinetic study of anaerobic digestion of olive mill wastewater previously fermented with Aspergillus terreus. *Process Biochemistry*, 28, 397-404.

[111] Ergüder T H, Güven E, & Demirer G N (2000) Anaerobic treatment of olive mill wastes in batchreactors. *Process Biochemistry*, 36, 243-248.

[112] Ramos-Cormenzana A, Juárez-Jiménez B, & Garcia-Pareja M P (1996) Antimicro-bial activity of olive mill wastewaters (alpechin) and biotransformed olive oil millwastewater. *International Biodeterioration & Biodegradation*, 38, 283-290.

[113] Vlyssides G A, Bouranis L D, Loizidou M, & Karvouni G (1996) Study of a demonstration plant for the co-composting of olive-oil-processing wastewater and solid residue. *Bioresource Technology*, 56, 187-193.

[114] Altieri R, Pepi M, Esposito A, & Fontanazza G (2005) Chemical and microbiological characterization of olive-mill waste-based substrata produced by the O.Mi.By.P. technology and their grounds amendment. In: J. Benitez, (Ed.), Integrated Soil and Water Management for Orchard Development. Role and Importance, FAO Land and Water Bulletin, No. 10, Roma, pp. 91-101.

[115] Kamaya Y, Tsuboi S, Takada T, & Suzuki K (2006) Growth stimulation and inhibition effects of 4-hydroxybenzoic acid and some related compounds on the freshwater green alga Pseudokirchnerielta subcapitata. *Archives of Environmental Contamination and Toxicology*, 51, 537-541.

INDEX

B

C

D

F

G

H

N

O

Q

R

U

V

W

X

Y

Z